35 Springer Series in Chemical Physics
Edited by Robert Gomer

W0079697

Springer Series in Chemical Physics

Editors: V. I. Goldanskii   R. Gomer   F. P. Schäfer   J. P. Toennies

Chemistry and Physics of Solid Surfaces V

Editors: R. Vanselow and R. Howe

With 303 Figures

Springer-Verlag Berlin Heidelberg GmbH 1984

Professor Ralf Vanselow

Department of Chemistry and Laboratory for Surface Studies,
The University of Wisconsin-Milwaukee,
Milwaukee, WI 53201, USA

Professor Russell Howe

Department of Chemistry, University of Auckland,
Private Bag, Auckland, New Zealand

Series Editors

Professor Vitalii I. Goldanskii

Institute of Chemical Physics
Academy of Sciences
Vorobyevskoye Chaussee 2-b
Moscow V-334, USSR

Professor Dr. Fritz Peter Schäfer

Max-Planck-Institut für
Biophysikalische Chemie
D-3400 Göttingen-Nikolausberg
Fed. Rep. of Germany

Professor Robert Gomer

The James Franck Institute
The University of Chicago
5640 Ellis Avenue
Chicago, IL 60637, USA

Professor Dr. J. Peter Toennies

Max-Planck-Institut für Strömungsforschung
Böttingerstraße 6-8
D-3400 Göttingen
Fed. Rep. of Germany

ISBN 978-3-642-82255-1   ISBN 978-3-642-82253-7 (eBook)
DOI 10.1007/978-3-642-82253-7

Library of Congress Cataloging in Publication Data. (Revised for volume V) Main entry under title: Chemistry and physics of solid surfaces. (Springer series in chemical physics ; v. 20) At head of title, v. 1: CRC. Vol. 3- edited by: Ralf Vanselow, Walter England; Vol. 5 edited by R. Vanselow and R. Howe. Vol. 3 published in: Boca Raton, Fla. Vol. 4- published: Berlin ; New York : Springer-Verlag. Includes bibliographical references and indexes. 1. Surface chemistry. 2. Solid state chemistry. I. Vanselow, Ralf. II. Tong, S. Y. III. Chemical Rubber Company. IV. Series: Springer series in chemical physics ; v. 20, etc. QD508.C48   541'.3453   77-25890

© by Springer-Verlag Berlin Heidelberg 1984

Originally published by Springer-Verlag Berlin Heidelberg New York Tokyo in 1984
Softcover reprint of the hardcover 1st edition 1984

2153/3130-543210

We dedicate this volume to the memory of our friend and co-author

Gosse A. Bootsma

who passed away on August 29, 1982

Gosse Age Bootsma was born in 1933 in the town of Hilversum, near Amsterdam. He studied chemistry at Utrecht University, where he did his thesis work on the crystal structures of mesotartaric acid under the supervision of Professor J.M. Bijvoet. At the same time he pursued one of his hobbies: Slavic languages. He acquired a good command of Russian which enabled him to help many colleagues with the translation of Russian scientific literature.

In 1964 he joined the surface chemistry group at Philips Research Laboratories in Eindhoven, where he and F. Meijer pioneered the application of ellipsometry to surface studies.

Gosse's work on low coverage physisorption, and on chemisorption on semiconductor surfaces, is well-known.

In 1969 he changed his research interests to the field of whisker growth and properties. In 1971 he left the Philips company to become a lecturer at the University of Utrecht. His main theme was the plane specificity of metal surfaces with regard to gas adsorption and its relation to catalysis. Ellipsometry was an important tool in these studies. In 1978 he became full professor in the Department of Technical Physics at the Twente University of Technology. Here he participated in work on the thermodynamic basis of the ion sensitive field effect transistor (ISFET). The last thesis under his supervision was written by L.J. Hanekamp. It was near completion when Gosse died on August 29, 1982.

Gosse was regional editor of Surface Science. He also was a contributor to *Chemistry and Physics of Solid Surfaces*.

With his death, we, and his many colleagues in the scientific community, lost a good scientist and particularly a humane and likable man.

Onno L.J. Gijzeman

Van't Hoff Laboratory
University of Utrecht
Padualaan 8

NL-3584 Utrecht

Preface

This volume contains review articles which were written by the invited speakers of the Sixth International Summer Institute in Surface Science (ISISS), held at the University of Wisconsin-Milwaukee in August 1983.

The objective of ISISS is to bring together a group of internationally recognized experts on various aspects of surface science to present tutorial review lectures over a period of one week. Each speaker is asked, in addition, to write a review paper on his lecture topic. The collected articles from previous Institutes have been published under the following titles:

Surface Science: Recent Progress and Perspectives, Crit. Rev. Solid State Sci. 4, 124-559 (1974).

Chemistry and Physics of Solid Surfaces, Vol. I (1976), Vol. II (1979), Vol. III (1982) (CRC Press, Boca Raton, FL), and Vol. IV (1982), Springer Ser. Chem. Phys., Vol. 20 (Springer-Verlag Berlin, Heidelberg, New York 1982)

No single collection of reviews (or one-week conference for that matter) can possibly cover the entire field of modern surface science, from heterogeneous catalysis through semiconductor surface physics to metallurgy. It is intended, however, that the series *Chemistry and Physics of Solid Surfaces* as a whole should provide experts and students alike with a comprehensive set of reviews and literature references on as many aspects of the subject as possible, particular emphasis being placed on the gas-solid interface. Each volume is introduced with a historical review of the development of one aspect of surface science by a distinguished participant in that development.

The historical introduction to the present volume is given by *Somorjai* who reviews the major contribution made by his research group over the past 20 years in applying the techniques of surface physics to understand heterogeneous catalysis at a molecular level. The catalytic theme is continued by *Grätzel*, who discusses the topic of photocatalysis, which has recently become important in the context of solar energy utilization, and by *Bell*, who reviews the application of infrared Fourier transform spectroscopic methods to study practical catalysts. The characterization of small catalytic particles by electron microscopy is reviewed by *Yacaman*.

The field of surface spectroscopy continues to contain a mixture of established techniques applied to new systems, and new techniques applied to long standing problems. *Tatarchuk* and *Dumesic* discuss the new development of an old technique, Mössbauer spectroscopy, as a surface sensitive probe through the detection of back scattered electrons. *Howe* describes developments in the application of magnetic resonance spectroscopy to surfaces since Lunsford's

1975 review. A new technique is described by *Cavanagh* and *King*: high-resolution laser spectroscopy applied to surface chemical reactions. *Winograd* discusses the recent extension of secondary ion mass spectrometry, angle-resolved SIMS, while *Gibson* considers ion scattering from surfaces. The pioneer of electron energy loss spectroscopy, *Ibach*, reviews recent applications in the study of surface phonons. A coutionary note is sounded to all users of particle beams to analyze surfaces by *Kelly*, who considers the question of composition changes induced by particle bombardment.

Surface crystallography was at one time the exclusive domain of low-energy electron diffraction. The alternative use of atom diffraction to determine surface structure is described by *Engel*. The techniques of SEXAFS and NEXAFS are shown by *Stöhr* to give structural information on surfaces where long-range order is not necessarily present. *Estrup* reviews the phenomenon of reconstruction of metal surfaces, and *Kleban* considers the theoretical implications of surface steps.

Crystal growth is considered from an experimental viewpoint by *Ehrlich*, and from a theoretical viewpoint by *Gilmer*. Recent developments in the theory of epitaxy are reviewed by van der Merwe, and the phenomena of commensurate-incommensurate transitions on surfaces are discussed by *Bak*.

Finally, the electronic properties of surfaces are considered by *Mönch*, who described work-function measurements, and *Dow*, who discussed surface state dispersion relations and Schottky barrier heights in semiconductors.

As in previous volumes, an extensive subject index is provided.

We should like to thank the sponsors of ISISS: The Air Force Office of Scientific Research and the Office of Naval Research (Grant No. N00014-83-G0046) as well as the College of Letters and Sciences, the Laboratory for Surface Studies, and the Graduate School at UWM for making both the conference and the publication of this volume possible. The cooperation of authors and publisher in achieving rapid publication is also acknowledged.

On 29 August 1982, Gosse A. Bootsma, a man with exceptional human qualities and a highly regarded colleague, died after a short period of illness. Professor Bootsma had contributed a review paper on "Chemisorption Investigated by Ellipsometry" to volume four of this series. The authors and editors wish to dedicate this publication to his memory.

Milwaukee, Auckland   *R. Vanselow • R. Howe*
April 1984

Contents

List of Contributors

Allen, Roland E.

Department of Physics, Texas A&M University
College Station, TX 77843, USA

Bak, Per

Physics Department, Brookhaven National Laboratory
Upton, NY 11973, USA

Bell, Alexis T.

Department of Chemical Engineering, University of California
Berkeley, CA 94720, USA

Cavanagh, Richard R.

Center for Chemical Physics, Molecular Spectroscopy Division,
National Bureau of Standards, Washington, D.C. 20234, USA

Dow, John D.

Department of Physics, University of Notre Dame
Notre Dame, IN 46556, USA

Dumesic, James A.

Department of Chemical Engineering, University of Wisconsin
Madison, WI 53706, USA

Ehrlich, Gert

Coordinated Science Laboratory and Department of Metallurgy,
University of Illinois at Urbana-Champaign, 1101 West Springfield Avenue
Urbana, IL 61801, USA

Engel, Thomas

Department of Chemistry BG-10, University of Washington
Seattle, WA 98195, USA

Estrup, Peder J.

Department of Physics and Department of Chemistry, Brown University
Providence, RI 02912, USA

Gibson, Walter M.

Department of Physics, SUNY ALBANY, Albany, NY 12222, USA

Gilmer, George H.

Bell Laboratories, Murray Hill, NJ 07974, USA

Grätzel, Michael

Institut de Chimie Physique, Ecole Polytechnique Fédérale
CH-Lausanne, Switzerland

Howe, Russell F.

Chemistry Department, University of Auckland, Private Bag
Auckland, New Zealand

Ibach, Harald

Institut für Grenzflächenforschung und Vakuumphysik, Kernforschungsanlage
Jülich, Postfach 19 13, D-5170 Jülich, Fed. Rep. of Germany

Kalyanasundaram, Kuppuswamy

Institut de Chimie Physique, Ecole Polytechnique Fédérale
CH-1015 Lausanne, Switżerland

Kelly, Roger

IBM Thomas J. Watson Research Center, P.O. Box 218
Yorktown Heights, NY 10598, USA

King, David S.

Center for Chemical Physics, Molecular Spectroscopy Division,
National Bureau of Standards, Washington, D.C. 20234, USA

Kleban, Peter H.

Laboratory for Surface Science and Technology and
Department of Physics and Astronomy, University of Maine
Orono, ME 04469, USA

Mönch, Winfried

Laboratorium für Festkörperphysik, Universität Duisburg,
Bismarckstr. 81, D-4100 Duisburg, Fed. Rep. of Germany

Rahman, Talat S.

Department of Physics, Kansas State University
Manhattan, KS 66506, USA

Sankey, Otto F.

Department of Physics, Arizona State University
Tempe, AZ 85287, USA

Somorjai, Gabor A.

Materials and Molecular Research Division, Lawrence Berkeley Laboratory, and
Department of Chemistry, University of California, Berkeley, CA 94720, USA

Stöhr, Joachim

Corporate Research Science Laboratories, Exxon Research and Engineering
Company, Clinton Township, Annandale, NJ 08801, USA

Tatarchuk, Bruce J.

Department of Chemical Engineering, Auburn University
Auburn, AL 36849, USA

van der Merwe, Jan H.

Department of Physics, University of Pretoria,
Pretoria 0002, Republic of South Africa

Winograd, Nicholas

Department of Chemistry, The Pennsylvania State University,
152 Davey Laboratory, University Park, PA 16802, USA

Yacaman, Miguel J.

Instituto de Fisica, Universidad Nacional Autonoma de Mexico,
Apartado Postal 20-364, Mexico 20, D.F.

1. The Molecular Surface Science of Heterogeneous Catalysis: History and Perspective

G. A. Somorjai

With 21 Figures

It is a rare occasion that one is invited to prepare a personal and historical review and discussion of one's research. I am grateful for this oppor-
tunity and shall attempt to describe how I became involved with modern surface science and how it was employed in my laboratory for studies of the
chemistry of surfaces and heterogeneous catalysis. Out of this research came
new approaches that have impact on catalysis science and technology.

In 1957, when I entered graduate school for chemistry at Berkeley, having
just arrived from Hungary, there were two fields of chemistry, heterogeneous
catalysis and polymer science, that caught my imagination and I wanted to
pursue. None of the faculty members at Berkeley were carrying out research
in these areas. Nevertheless, Richard Powell, an inorganic chemist and kine-
ticist, was willing to give me a research project in catalysis. As a result
my Ph.D. research explored changes of particle size and shape of dispersed
platinum catalysts supported on γ alumina when subjected to oxidizing and
reducing atmospheres. The technique I used for these studies was small angle
X-ray scattering [1.1]. During the three years it took me to finish the project, I learned a great deal about catalysis, and I also learned that important advances in our understanding in physical chemistry can be made only if
techniques are available for molecular-level scrutiny of catalysts. It was
not too difficult to see that the fields that entered the mainstream of research of modern physical chemistry were those that permitted molecular
level studies via spectroscopy or diffraction. When I finished my Ph.D. research in 1960, surface chemistry and catalysis were not among these fields.
With the exception of field emission and field ion microscopies, there were
no techniques available for atomic-scale studies of surfaces that could be
broadly employed. So I left the field and carried out solid-state chemistry
and physics research at IBM in New York for the next four years, concentrat-
ing on studies of the vaporization mechanisms of single crystals of cadmium
sulfide and other II-VI compounds.

In the meantime, low-energy electron diffraction (LEED) made its appearance as a practical tool to detect the structure of ordered surfaces and adsorbed monolayers as a result of innovations in instrumentation by Germer in the Bell Laboratories. I was instantly fascinated with the opportunities this technique presented for definitive atomic-level studies of surface structure and the ability to obtain clean surfaces of single crystals when coupled with ultrahigh vacuum and sputter etching that was developed a few years before by Farnsworth. Upon moving to Berkeley in 1964 as an Assistant Professor, I started LEED studies on platinum (Pt) single-crystal surfaces in the hope that we could learn what is so unique about them that makes this metal such an excellent catalyst for hydrocarbon conversion reactions, especially for dehydrocyclization of aliphatic molecules to aromatic species. In early 1965, our first {100} Pt single-crystal surfaces exhibited strange diffraction patterns which led us to the discovery of surface reconstruction [1.2] of metal surfaces (Fig.1.1). While semiconductor surfaces were known by that time to have different surface structures from those expected from the surface projection of the bulk unit cell, this was the first time a metal has shown similar features. Shortly thereafter, iridium [1.3] {100} and gold {100} surfaces [1.4] were reported to exhibit similar reconstructed surface structures. It took fifteen years to develop the theory of low-energy electron diffraction and surface crystallography to solve this structure [1.5], which we did in 1980 (Fig.1.2). The square unit cell of the Pt {100} surface is buckled into an hexagonal arrangement. The periodic coincidence of atoms in the hexagonal layer with atoms in the underlying square layer produces the busy diffraction pattern and complex unit cell shown in Fig.1.1.

The next seven years saw an explosive development of new techniques to study the properties of surface monolayers. Auger electron spectroscopy (AES) was developed by Harris and others to determine surface composition and X-ray photoelectron spectroscopy (XPS) was developed to identify the oxidation states of surface atoms. These three techniques, LEED, AES and XPS, contributed the most of all techniques available at that time to the unraveling of the molecular properties of the surface monolayer [1.6,7]. Simultaneously experimental studies were constantly coming up with newly discovered surprising surface properties, mostly of low Miller index flat surfaces. Among these properties, perhaps the most striking were the following. (1) The prevalence of ordering of clean surfaces and adsorbed layers of atoms and molecules [1.8]. Hundreds of ordered monolayer surface structures were discovered that undergo alterations as a function of temperature and coverage. Surface

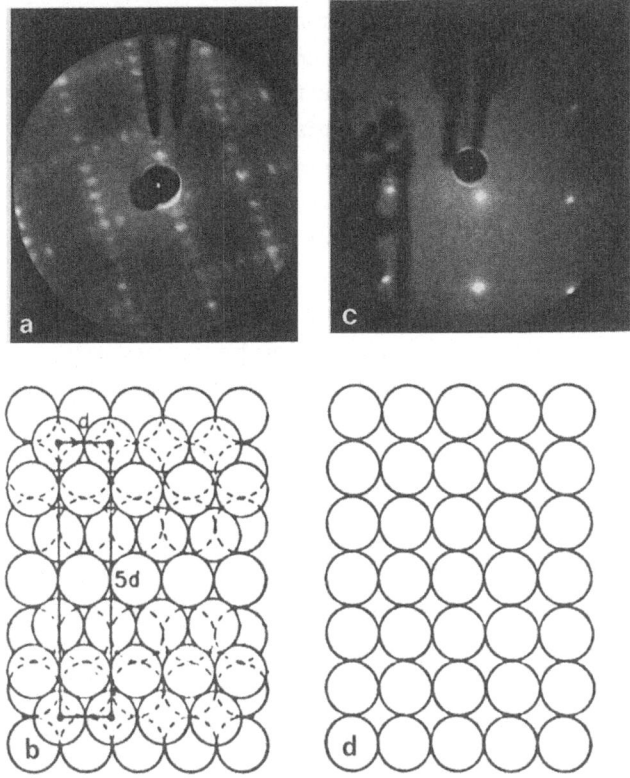

<u>Fig.1.1.</u> a) Diffraction pattern from reconstructed Pt{100} surface. b) Schematic representation of {100} surface with (5×1) surface structure. c) Diffraction pattern from unreconstructed Pt{100} surface. d) Schematic representation of unreconstructed {100} surface

fcc (100) : buckled hexagonal top layer

two-bridge   top/center

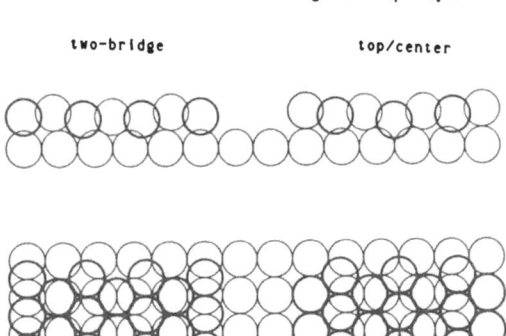

<u>Fig.1.2.</u> Structure of reconstructed Pt{100} crystal face as solved by surface crystallography

phases with order-order and order-disorder transformations were uncovered that yield complex adsorption isotherms where the relative strength of adsorbate-substrate and adsorbate-adsorbate interactions controls the structural changes. Good examples of these are the surface structures that form on adsorption of CO on RH{111} [1.9] and other transition metal surfaces, O on Rh{111} surfaces [1.10] and sulfur on the molybdenum {100} surface [1.11]. (2) Reconstruction and relaxation of clean surfaces is more of a rule than an exception. Surface-structure analysis by LEED surface crystallography indicates that the surface atoms in many solid surfaces seek new equilibrium positions as a result of the anisotropy of the surface environment [1.6]. This of course leads to surface reconstruction. An equally common observation is relaxation where the interlayer distance between the first and second atomic layers is shorter than the interlayer distances in subsequent layers below the surface. Surface relaxation can be readily rationalized if one assumes the surface atoms to be intermediate between a diatomic molecule and atoms in a bulk surrounded by many neighbors. The interatomic distances in diatomic molecules are much shorter than those in the bulk solid. Thus, such a contraction between the first and second layers of atoms at the surface is just what one expects, based on such a simple model. It is also found, in agreement with the predictions of this model, that the more open the surface the larger the contraction between the first and second layers. (3) The surface composition of multicomponent systems (alloys, oxides, etc.) is different from the bulk composition [1.12,13]. There is surface segregation of one of the constituents that lowers the surface free energy of the multicomponent system. The contamination of most surfaces with carbon, sulfur, silicondioxide and other impurities can be attributed to the surface thermodynamic driving force behind surface segregation. The thermodynamics of surface segregation was developed and good agreement between experimental and calculated surface compositions has been obtained for many binary alloy systems [1.14]. In Table 1.1, several binary metallic systems are listed for which there is surface segregation of one of the constituents. For a two-component system that behaves as an ideal or as a regular solution, the atom fractions at the surface can be related to the atom fractions in the bulk as shown in Fig.1.3. We can also predict how the changed surface composition in the surface monolayer is modified layer by layer as the bulk composition is approached. This is shown for the gold-silver alloy that forms a regular solid solution in Fig.1.4. The first surface layer is rich in silver [1.15]; the second layer is rich in gold, the third again rich in silver, and by the fourth layer the bulk alloy composition is reestablished. We can actually

Table 1.1. Surface composition of alloys: experimental results and predictions of the regular solution models

Alloy Systems	Phase Diagram	Segregating Constituent	
		Predicted	
		Regular Solution	Experimental
Ag-Pd	simple	Ag	Ag
Ag-Au	simple	Ag	Ag
Au-Pd	simple	Au	Au
Ni-Pd	simple	Pd	Pd
Fe-Cr	low-T phase	Cr	Cr
Au-Cu	low-T ordered phase	Cu	Au, none, or Cu, depending on composition
Cu-Ni	low-T miscibility gap	Cu	Cu
Au-Ni	miscibility gap	Au	Au
Au-Pt	miscibility gap	Au	Au
Pb-In	intermediate phase	Pb	Pb
Au-In	complex	In	In
Al-Cu	complex	Al	Al
Pt-Sn	complex	Sn	Sn
Fe-Sn	complex	Sn	Sn
Au-Sn	complex	Sn	Sn

For an ideal solid solution:

$$\frac{x_2^s}{x_1^s} = \frac{x_2^b}{x_1^b} \exp\left[\frac{(\sigma_1 - \sigma_2)a}{RT}\right]$$

For a regular solid solution:

$$\frac{x_2^s}{x_1^s} = \frac{x_2^b}{x_1^b} \exp\left[\frac{(\sigma_1 - \sigma_2)a}{RT}\right] \exp\left\{\frac{\Omega(\ell + m)}{RT}\left[(x_1^b)^2 - (x_2^b)^2\right]\right.$$
$$\left. \frac{\Omega \ell}{RT}\left[(x_2^s)^2 - (x_1^s)^2\right]\right\}$$

where Ω = regular solution parameter = $\dfrac{\Delta H_{mixing}}{x_1^b \cdot x_2^b}$

ℓ = fraction of nearest neighbors in surface layer.

m = fraction of nearest neighbors in adjacent layer.

Fig.1.3

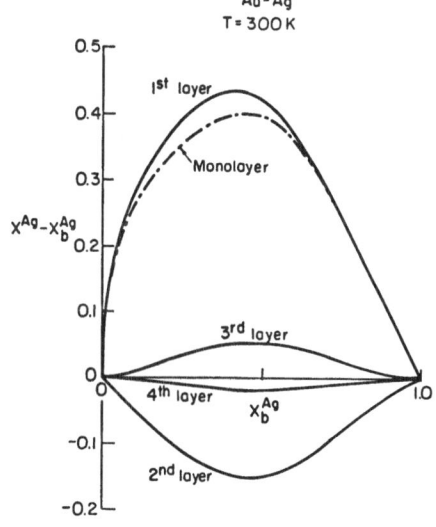

Au-Ag
T = 300 K

Fig.1.4

Fig.1.3. Ideal and regular solid-solution models for a binary system predicting surface segregation of the constituent with lower surface free energy

Fig.1.4. Surface excess of Ag as a function of bulk composition and layer by layer in Ag-Au alloys

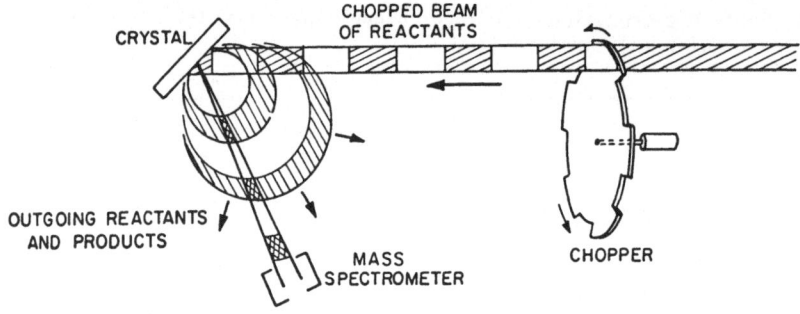

Fig.1.5. Scheme of the molecular-beam surface-scattering experiment

see the beginning of compound formation by the alternation of excess constituents for alloys with exothermic heat of mixing, layer by layer.

It is often found that the oxidation states of the surface atoms are different than for atoms in the bulk. Aluminum oxide and vanadium pentoxide
[1.16,17] are two examples where definitive experimental surface-science
studies indicated reduced oxidation states for the metal ions in the surface
layers.

The techniques of LEED, AES, XPS are static. Since there was a need to
develop dynamic measurements to study the kinetics of surface processes, we
developed molecular-beam surface scattering [1.18]. The scheme of the molecular-beam surface scattering experiment is shown in Fig.1.5. This is one of
the most rapidly developing areas of surface chemical physics today. It can
determine the energy states of molecules before and after scattering, or
after a reaction at the surface, when combined with laser spectroscopy
[1.19].

Most surface studies, including the adsorption of organic molecules of
ever increasing size, were carried out on low Miller index single-crystal
surfaces of metals or semiconductors. For example, phthalocyanine adsorbs
on the copper {111} surface and from an ordered layer [1.20] that can be
readily studied by surface-science techniques. However, these surfaces are
quite inert when viewed from the chemical and especially catalytic points
of view. They are unreactive with many reactive molecules at around 300 K
and at low pressures of 10^{-8} to 10^{-6} torr. Even molecular-beam studies using
mixed H_2 and D_2 beams failed to detect significant H_2/D_2 exchange on the
platinum {111} surface at low pressures. Anyone working in the field of catalysis would know that this reaction at atmospheric pressures occurs readily
on any dispersed transition metal surface below 300 K. The reason for the
lack of reactivity could be the surface structure or the low-pressure condi-

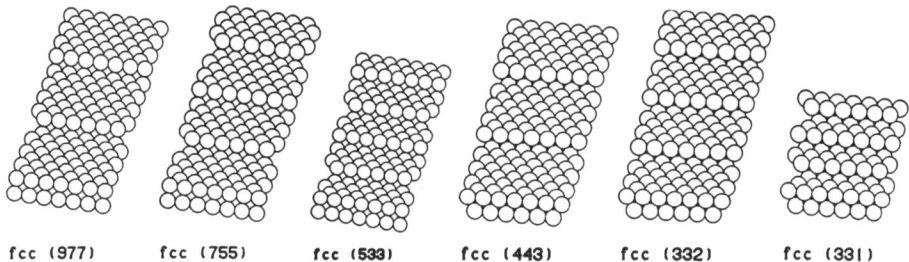

fcc (977)   fcc (755)   fcc (533)   fcc (443)   fcc (332)   fcc (331)

Fig.1.6. Structure of several high Miller index stepped surfaces with different terrace widths and step orientations

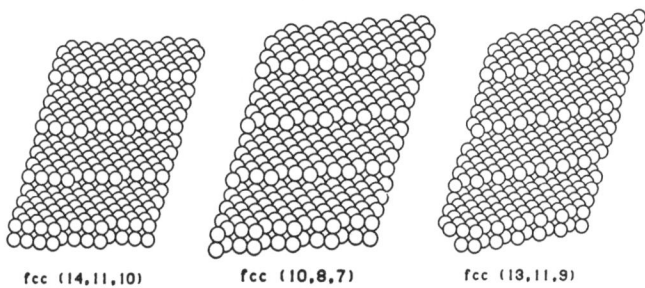

fcc (14,11,10)   fcc (10,8,7)   fcc (13,11,9)

Fig.1.7. Surface structures of several high Miller index surfaces with de-
ferring kink concentrations in the steps

tions or both. So we did two things. We have started studies of high Miller
index surfaces and developed the high-pressure low-pressure apparatus for
combined surface-science and high-pressure catalytic studies.

By cutting flat low Miller index orientation single crystals at some
angle with respect to the low-index orientation, high Miller index surfaces
were obtained [1.21]. These surfaces exhibit ordered step terrace arrangements where the terraces are of variable width, depending on the angle of
cut, and are separated by periodic steps of usually one atom in height
(Fig.1.6). The step periodicity as well the step height can be readily determined by LEED studies [1.22]. These surfaces can be cut so that the steps
also have a large concentration of kinks (Fig.1.7). Surfaces may have as
much as 40% of their atoms in step sites, and the kink concentration can
reach 10%. As compared to these high concentrations of line defects, point
defects, such as adatoms or vacancies, have very small concentrations (less
than 1%) when in equilibrium with the bulk and with other surface defects.
These steps and kinks are stable under the conditions of most catalytic surface reactions. Figures 1.6,7 show some of the stepped and kinked surfaces
that were prepared, which exhibited very different reactivities compared to

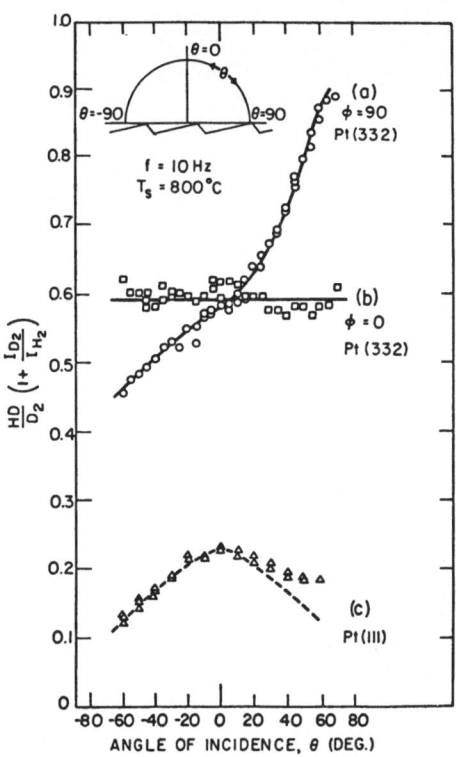

Fig.1.8. HD production during H_2/D_2 mixed molecular-beam scattering, as a function of angle of incidence. Intensity of HD molecular beam is normalized to the incident D_2 intensity. a) Pt{332} surface with stepped edges perpendicular to incident beam ($\phi = 90°$) shows the highest reactivity. b) Pt{332} surface where the beam projection on the surface is parallel to the step edges, $\phi = 0°$. c) Pt{111} surface exhibits almost an order of magnitude lower activity

the flat surfaces. Adsorption studies of hydrocarbons and carbon monoxide revealed preferential bond breaking at these defect sites: C-H and C-C bond breaking was readily detectable on stepped or kinked platinum surfaces upon adsorption of organic molecules, even at 300 K and low pressures [1.23], while under the same circumstances the {111} surface was unreactive. Molecular-beam studies of H_2/D_2 exchange on stepped surfaces showed a 7- to 10-fold higher dissociation probability [1.24] of the hydrogen molecule than on the flat {111} crystal face (Fig.1.8).

During the period between 1972-1976, the importance and research of surface defects, steps and kinks in surface chemical phenomena of all kinds has become well documented. Of course we were lucky that the stepped surfaces were also ordered and exhibited periodicities that were readily detectable by low-energy electron diffraction before adsorption and after adsorption of various atoms and molecules.

To combine ultrahigh vacuum surface-science and high-pressure catalytic studies, the following apparatus was developed in my laboratory (Fig.1.9). After suitable surface characterization in ultrahigh vacuum by LEED, AES and other surface-sensitive techniques the small area sample, often a single

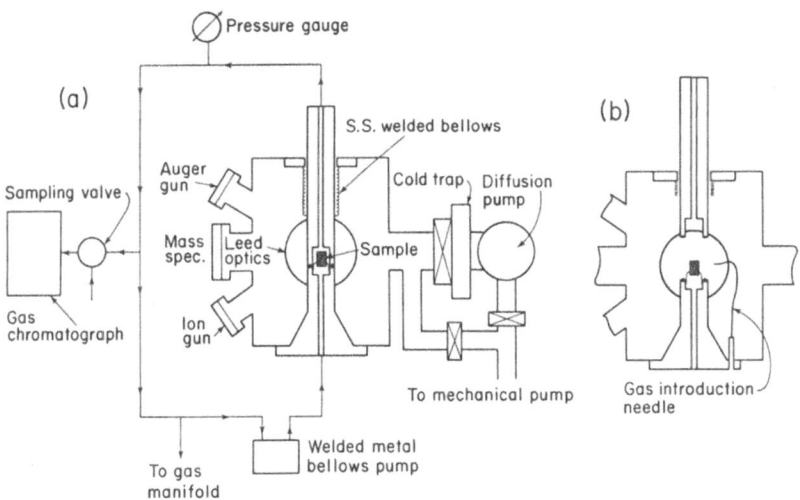

Fig.1.9a,b. Experimental apparatus for the catalytic reaction rate studies
on single-crystal or polycrystalline surfaces of low surface area at low and
high pressures in the 10^{-7} to 10^{+4} torr range

crystal or polycrystalline foil, is enclosed in an isolation cell that can
be pressurized with the reactants [1.25]. The sample is then heated to the
reaction temperature and the products that form are analyzed by a gas chromatograph connected to the high-pressure loop. The high-pressure reactor can
be used in batch or in flow modes [1.26]. The detection sensitivity of the
gas chromatograph is sufficiently high for a 1 cm^2 surface area to be adequate to monitor the product distribution as long as the turnover rates over
the catalysts are greater than 10^{-4} molecules per site per second. Using
this high-pressure apparatus, we can carry out catalytic reactions under
conditions that are virtually identical to those used in chemical technology.
We can then evacuate the high-pressure cell, open it and analyze the surface
properties of the working catalyst in ultrahigh vacuum using the various
techniques of surface science. Then, the isolation cell may be closed again
and the high-pressure reaction may be continued and again interrupted for
surface analysis in vacuum. Using this apparatus we showed not only that
catalytic reactions can readily be investigated using small area single-
crystal surfaces, but that these surfaces can be used as model heterogeneous
catalysts.

    This approach was being used in my laboratory to study hydrocarbon con-
version reactions on platinum, the hydrogenation of carbon monoxide on rhodium and iron, and for the ammonia synthesis from nitrogen and hydrogen on
iron.

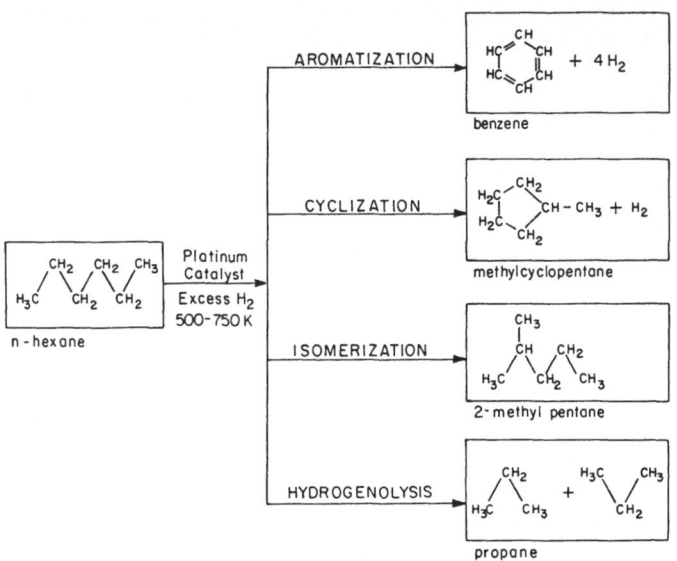

Fig.1.10. Skeletal rearrangement reactions of hydrocarbons are catalyzed by Pt with high activity and unique selectivity. Depicted here are the several reaction pathways which occur simultaneously during the catalyzed conversion of n-hexane (C_6H_{14}). The isomerization, cyclization and aromatization reactions that produce branched or cyclic products are important in the production of high octane gasoline from petroleum naphta. The hydrogenolysis reaction involving C-C bond breaking yields undesirable gaseous products

We discovered rapidly that the selectivity is structure sensitive during hydrocarbon conversion reactions. In these studies, the use of flat, stepped and kinked single-crystal surfaces with variable surface structure was very useful indeed. Figure 1.10 shows how n-hexane may undergo several different reactions over platinum. For the important aromatization reaction of n-hexane to benzene and n-heptane to toluene we discovered that the hexagonal platinum surface where each surface atom is surrounded by six nearest neighbors is three to seven times more active than the platinum surface with square unit cell [1.27], Fig.1.11. Optimum aromatization activity is achieved on stepped surfaces with terraces about five atoms wide with hexagonal orientation as indicated by reaction studies over more than ten different crystal surfaces with varied terrace orientation and step and kink concentrations. The reactivity pattern displayed by platinum crystal surfaces for alkane isomerization reactions is completely different from that for aromatization. Our studies revealed that the maximum rate and selectivity (rate of desired reaction divided by total rate) for butane isomerization reactions are obtained on the flat crystal face with square unit cell [1.28]. Isomerization rates for this surface are four to seven times higher than those for the hexagonal

STRUCTURE SENSITIVITY OF ALKANE AROMATIZATION

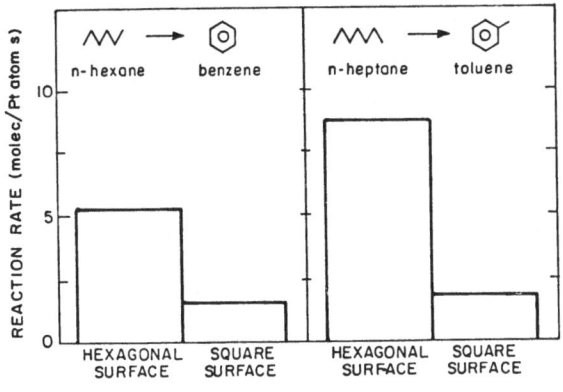

Fig.1.11. Dehydrocycliz-ation of alkanes to aromatic hydrocarbons is one of the most important petroleum reforming reactions. Comparison of reaction rates for n-hexane at atmospheric pressure over the two flat platinum single-crystal faces with different atomic structure. The platinum surface with hexagonal atomic arrangement is several times more active than the surface with a square unit cell over a wide range of reaction conditions

STRUCTURE SENSITIVITY OF LIGHT ALKANE SKELETAL REARRANGEMENT

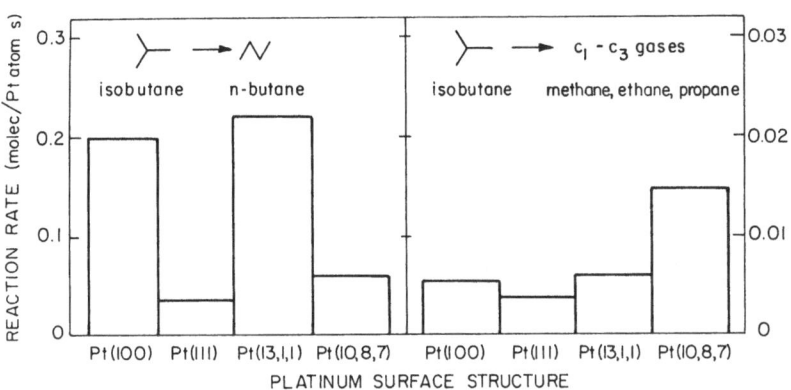

Fig.1.12. Reaction rates as a function of surface structure for isobutane isomerization and hydrogenolysis catalyzed at 570 K and atmospheric pressure over the four Pt surfaces. The rates for both reaction pathways are very sensitive to structural features of the model single-crystal surfaces. Isomerization is favored on Pt surfaces with a square (100) atomic arrangement. Hydrogenolysis rates are maximized when kink sites are present in high concentration as in the Pt{10 8 7} crystal surface

surface and are increased only slightly by surface irregularities on the platinum surfaces as shown in Fig.1.12.

For the undesirable hydrogenolysis reactions that require C-C bond scission, we found that the two flat surfaces with highest atomic density exhibit very similar low reaction rates. However, the distribution of hydrogen-

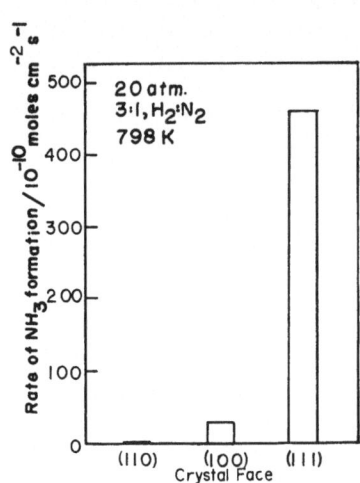

olysis products varies sharply over these surfaces. The hydrogenolysis rates
increase markedly three- to fivefold when kinks are present in high concentrations on the Pt surfaces [1.29] (Fig.1.12).

Figure 1.13 compares the rates of ammonia synthesis over the {111}, {100}
and {110} crystal faces of iron [1.30]. The {111} crystal face is 420 times
more active and the {100} face is 32 times more active than the {110} iron
surface. This reaction was studied by several outstanding researchers that
include Emmett, Boudart, and Ertl, who worked for a long time on the mechanism of ammonia synthesis. Although the extreme structure sensitivity of
this reaction was predicted by many, these data are the first clear experimental demonstration of this structure sensitivity.

We found that the catalytically active surface is covered by a tenacious
carbonaceous deposit that stays on the metal surface during many turnovers
of the catalytic reactions. To determine the surface residence time of this
carbonaceous deposit, the platinum surface was dosed with ^{14}C-labelled organic molecules under reaction conditions [1.31]. Carbon 14 is a β particle
emitter, so using a particle detector we monitored its surface concentration
as a function of time during the catalytic reaction. Using another surface-
science technique, thermal desorption, we could determine the hydrogen content of the adsorbed organic layer by detecting the amount of desorbing hydrogen with a mass spectrometer. From these investigations we found that the
residence time of the observed carbonaceous layer depends on its hydrogen
content, which in turn depends on the reaction temperature as shown in Fig.
1.14. While the amount of deposit does not change much with temperature, its
composition does. It becomes much poorer in hydrogen as the reaction tem-

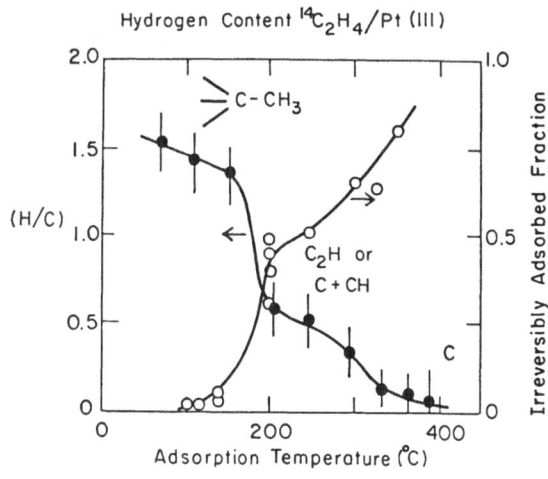

Hydrogen Content $^{14}C_2H_4$/Pt (III)

C–CH$_3$

C$_2$H or C+CH

C

(H/C)

Adsorption Temperature (°C)

Irreversibly Adsorbed Fraction

Fig.1.14. ^{14}C-labelled ethylene C$_2$H$_4$ chemisorbed as a function of temperature on a flat Pt surface with hexagonal orientation. The hydrogen H/C composition of the adsorbed species was determined from hydrogen thermal desorption studies. The amount of preadsorbed ethylene that could not be removed by subsequent treatment in one atmosphere of hydrogen represents the "irreversibly adsorbed fraction." The adsorption reversibility decreases markedly with increasing adsorption temperature as the surface species become more hydrogen deficient. The irreversibly adsorbed species have very long surface residence times of the order of days in the reactant mixture

perature is increased. The adsorption reversibility decreases markedly with increasing temperature, as the carbonaceous deposit becomes more hydrogen deficient. As long as the composition is about $C_nH_{1.5n}$ and the temperature below 450 K, the organic deposit can be readily removed in hydrogen and the residence time is in the range of the turnover time of the catalytic reaction. As this deposit loses hydrogen with increasing reaction temperatures above 450 K, it converts to an irreversibly adsorbed deposit with composition of $C_{2n}H_n$ that can no longer be removed readily (hydrogenated) in the presence of excess hydrogen. Nevertheless, the catalytic reaction proceeds readily in the presence of this active carbonaceous deposit. Around 750 K, this active carbon layer is converted to a graphitic layer that deactivates the metal surface and all chemical activity for any hydrocarbon reaction ceases. Hydrogen-deuterium exchange studies indicate rapid exchange between the hydrogen atoms in the adsorbed reactive molecules and the hydrogen in the active but irreversibly adsorbed deposit. Only the carbon atoms in this layer do not exchange. Thus one important property of the carbonaceous deposit is its ability to store and exchange hydrogen. The structure of adsorbed hydrocarbons, the fragments that form, and changes of their stoichiometry and structure can readily be studied by combined LEED and high-resolution electron energy loss (HREELS) techniques [1.32]. The surface-structure determination of organic monolayers is

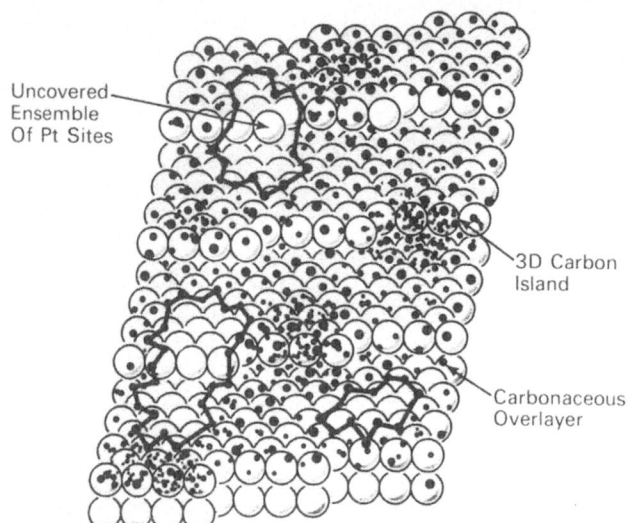

Uncovered Ensemble Of Pt Sites

3D Carbon Island

Carbonaceous Overlayer

Fig.1.15. Model of the working Pt catalyst developed from our combination of surface studies using single-crystal surfaces and hydrocarbon reaction rate studies on the same surfaces

one of the major directions of research that I pursue in my laboratory and it will remain so for many years to come.

How is it possible that the hydrocarbon conversion reactions exhibit great sensitivity to surface structure of platinum, while under reaction conditions the metal surface is covered with a near monolayer carbonaceous deposit [1.2]? In fact, often more than a monolayer amount of carbon-containing deposit is present as indicated by surface-science measurements. To determine how much of the platinum surface is exposed and remains uncovered we utilize the adsorption and subsequent thermal desorption of carbon monoxide, which, while readily adsorbed on the metal surface at 300 K and at low pressures, does not adsorb on the carbonaceous deposit. Our results indicated up to 10%-15% of the surface remains uncovered while the rest of the metal surface is covered by the organic deposit [1.31]. The fraction of uncovered metal sites decreases slowly with increasing reaction temperature. The structure of these uncovered metal islands is not very different from the structure of the initially clean metal surface during some of the organic reactions, while thermal desorption studies indicate that the steps and ledges become preferentially covered in others.

As a result of our catalyzed hydrocarbon conversion reaction studies on platinum crystal surfaces, a model has been developed for the working platinum reforming catalyst, is Fig.1.15. Between 80% and 90% of the catalyst

14

surfaces is covered with an irreversibly adsorbed carbonaceous deposit that stays on the surface for times much longer than the reaction turnover time [1.31]. The structure of this carbonaceous deposit varies continuously from two-dimensional to three-dimensional with increasing reaction temperature, and there are platinum patches that are not covered by the deposit. These metal sites can accept the reactive molecules and are responsible for the observed structure sensitivity and turnover rates. While there is evidence that the carbonaceous deposit participates in some of the reactions by hydrogen transfer by providing sites for rearrangement and desorption while remaining inactive in other reactions, its chemical role requires further exploration.

The oxidation state of surface atoms is also very important in controlling both the activity and selectivity of catalytic reactions. For example, rhodium was reported to yield predominantly acetaldehyde and acetic acid, when prepared under appropriate experimental conditions during the reaction of carbon monoxide and hydrogen [1.33]. Our studies using unsupported polycrystalline rhodium foils detected mostly methane [1.34], along with small amounts of ethylene and propylene under very similar experimental conditions. It appears that most of the organic molecules form following the dissociation of carbon monoxide by the rehydrogenation of CH_x units similarly to alkane and alkene production from CO/H_2 mixtures over other more active transition metal catalysts, e.g., iron, ruthenium and nickel. However, when rhodium oxide Rh_2O_3 was utilized as a catalyst, large concentrations of oxygenated C_2 or C_3 hydrocarbons were produced, including ethanol, acetaldehyde and propionaldehyde [1.35]. Furthermore, the addition of C_2H_4 to the CO/H_2 mixture yielded propionaldehyde, indicating the carbonylation ability of Rh_2O_3. Under similar experimental conditions over rhodium metal, C_2H_4 was quantitatively hydrogenated to ethane and the carbonylation activity was totally absent. Clearly, higher oxidation state rhodium ions are necessary to produce the oxygenated organic molecules. Unfortunately Rh_2O_3 reduced rapidly in the CO/H_2 mixture to a metallic state with drastic alteration of the product distribution from oxygenated hydrocarbons to methane. To stabilize the rhodium ion, lanthanum rodate ($LaRhO_3$) was prepared by incorporating Rh_2O_3 into La_2O_3 at high temperatures [1.36]. Over this stable catalyst the formation of oxygenated products from CO and H_2 predominated. The reason for the change of selectivity in CO/H_2 reactions with alteration of the oxidation state of the transition metal is due largely to the change of heats of adsorption of CO and H_2 as the oxidation state of the transition metal ion is varied, Fig.1.16. The CO adsorption energy is decreased upon oxidation and

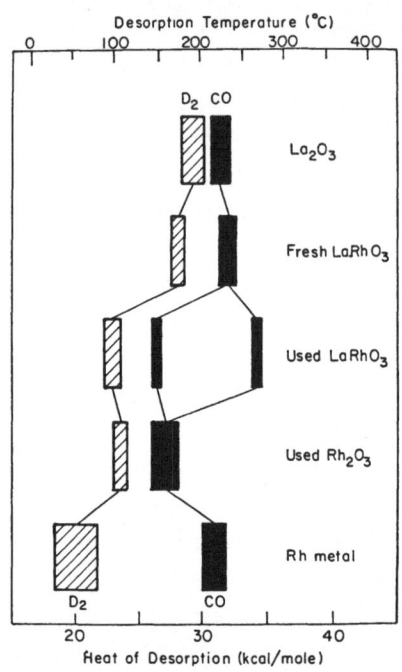

Fig.1.16. Heat of desorption of CO and D₂ from La₂O₃ fresh and used, LaRhO₃ fresh and used, Rh₂O₃ fresh and used and Rh metal. The spread of each value represents the variation with surface coverage rather than experimental uncertainty

the heat of adsorption of D_2 is increased presumably due to the formation of a hydroxide. In addition, the metal is primarily active for hydrogenation and CO dissociation while the oxide can perform carbonylation and has reduced hydrogenation activity. There are many examples of changing selectivity and activity of the catalyst as the oxidation state of the transition metal ion is varied. This is clearly one of the important ingredients for heterogeneous catalysis.

As a result of these findings and others during the period from 1976 to the present, we could identify three molecular ingredients that control catalytic properties. These are the *atomic surface structure*, an *active carbonaceous deposit* and the proper *oxidation state of surface atoms*. As a result of these studies we have also developed a model for the working platinum catalyst.

Much of the research in physical sciences and certainly everything I described as our work during the period 1965-1978 uses the *passive* approach. The purpose, in my case, is to understand the molecular ingredients of heterogeneous catalysis. In all these investigations very little if any thought is given as to how to use the understanding obtained. During the last three years I have attempted to carry out more and more *active* research in two ways. First to use the knowledge gained to build a better catalyst system.

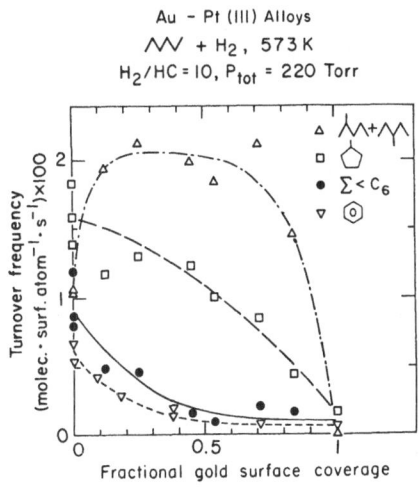

Au − Pt (III) Alloys
∧∧∧ + H₂, 573 K
H₂/HC = 10, P_tot = 220 Torr

Turnover frequency
(molec.·surf. atom⁻¹.s⁻¹)×100

2

1

0

0   0.5   1
Fractional gold surface coverage

△ ∧∧+∧∧
□ ⬠
● Σ < C₆
▽ ⬡

Fig.1.17. The rate of formation of
various products from n-hexane as a
function of fractional Au surface
coverage or Au/Pt alloys prepared by
vaporizing and diffusing Au into Pt
{111} crystal surfaces

Additives are being used to alter beneficially the surface structure to reduce the amount of carbon deposit or to slow down its conversion to the inactive graphitic form. Bimetallic or multimetallic catalyst systems have
been prepared and studied in my laboratory and elsewhere. By the addition
of one or more other transition metals (paladium, iridium, rhenium or gold)
to platinum, the alloy-catalyst system can be operated at higher reaction
temperatures to obtain higher reaction rates. They show slower rates of de-
activation (have longer lifetimes) and can also be more selective for a
given chemical reaction (dehydrocyclization or isomerization) than the one-
component catalyst. For example, the addition of small amounts of gold to
platinum increased the isomerization activity [1.37] while drastically reducing the dehydrocyclization activity of this catalyst (Fig.1.17). The addition of potassium changes the binding energy of carbon monoxide drastically by changing the charge density of the transition metal surface atoms which
in turn, are transferred to the molecular orbitals of carbon monoxide [1.38].
We have explored in great detail the effect of alkali atoms on transition metal surfaces both in accelerating or inhibiting various chemical reactions.
In Figs.1.18,19 we show examples of how alkali metal addition changes the bonding of carbon monoxide on platinum surfaces [1.39] while increasing the activation energy for C-H bond breaking on platinum surfaces [1.40]. In the former case the carbon monoxide dissociation reaction is definitely facilitated
by the presence of potassium. In the latter case potassium is a uniform poi-
son for hydrocarbon conversion reactions on platinum surfaces. We are also
attempting to formulate and build new catalysts by incorporating transition

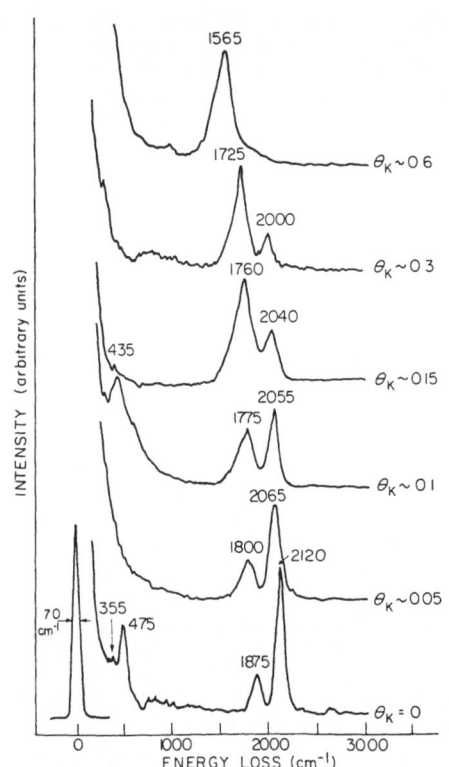

Fig.1.18. Vibrational spectra of the saturation CO coverage adsorbed on Pt{111} at 300 K as a funcion of pre-adsorbed K coverage

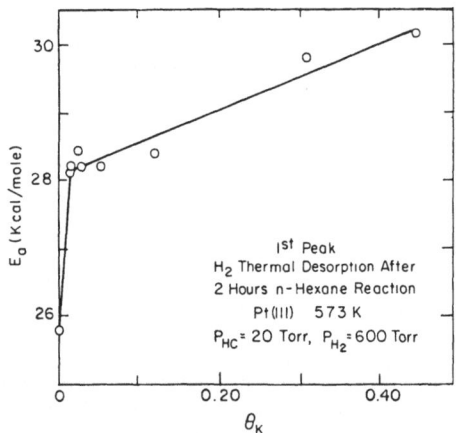

Fig.1.19. Activation energy of the hydrogen β-elimination from carbonaceous deposits after n-hexane reactions over Pt{111} surfaces as a function of K coverages

metal ions into refractory oxide lattices. In this way we stabilize the transition metal high oxidation state under reducing reaction conditions.

As combined surface-science and catalytic reaction studies develop and working models for catalysts help to build new catalysts, the field becomes high-technology catalysis science. This transition from art to catalysis science could not have come soon enough. The rapidly rising cost of petroleum necessitates the use of new fuel sources such as coal, shale and tar sand and the use of new feedstocks for chemicals including CO and H_2 and coal liquids. The new fuel and chemical technologies based on these feedstocks require the development of an entirely new generation of catalysts. Ultimately, our fuel and chemicals must be produced from the most stable and abundant molecules on our planet, among them carbon dioxide, water, nitrogen and oxygen. To build the catalytic chemistry starting from these species is a considerable challenge that I believe will be met by catalysis science in the future.

This brings me to the second new approach of active research in catalysis science which is really a very old one indeed. We are attempting to find catalysts for chemical reactions of small molecules that have not been explored.

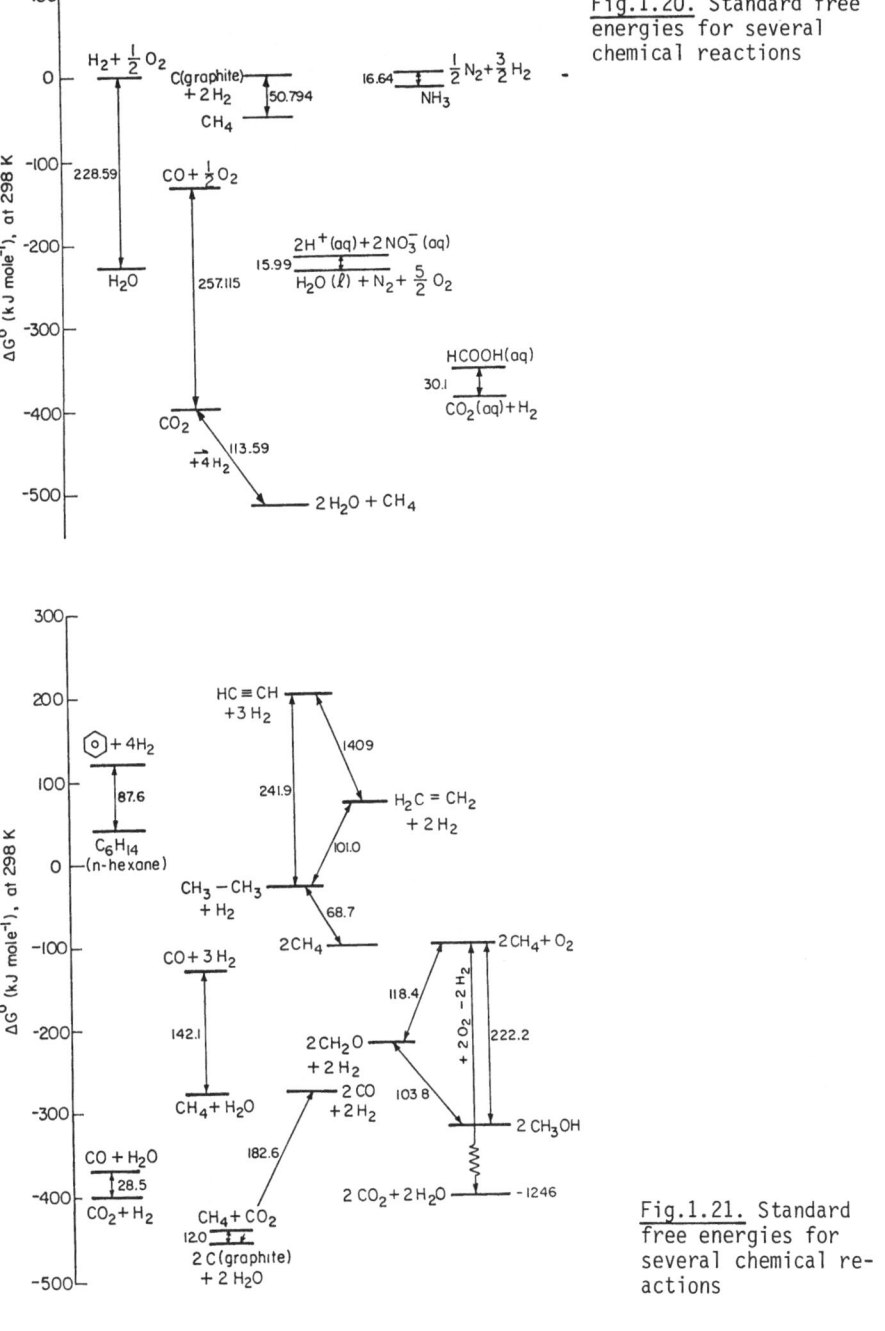

Fig.1.20. Standard free energies for several chemical reactions

Fig.1.21. Standard free energies for several chemical reactions

In Figs.1.20,21 we show several chemical reactions of small molecules, most of which are thermodynamically feasible, yet, their catalytic science has not been explored. The two reactions that require the input of extra energy, the splitting of water to hydrogen and oxygen by solar radiation or the photochemical splitting of carbon dioxide to carbon monoxide and oxygen, are important reactions and major challenges that could be met by clever catalytic chemists and surface scientists. The reduction of carbon dioxide to formic acid, the reactions of nitrogen with oxygen and water to form nitric acid and the partial oxidation of methane to formaldehyde and methanol and the reaction of water with graphite to produce methane are among those very important reactions to produce fuels and chemicals that are yet to be explored from the point of view of catalytic scientists. At the beginning of the century, much of the science was developed by searching for catalysts to carry out important reactions that were thermodynamically feasible, but this approach was abandoned in the last thirty to forty years. I believe that with our increasingly more detailed knowledge of catalysts and catalysis science, perhaps we can embark again on searching for catalysts for important reactions that have not been utilized in chemical technology as yet. The future is indeed unlimited and bright for molecular surface science and its application to catalysis science.

There are many other applied areas of surface science where molecular surface science has hardly been employed: lubrication, chemical corrosion, radiation damage and adhesion are among them. It is my hope that by combining techniques and knowledge in molecular surface science and in one of these applied fields, major advances will be made that will bring these fields to high technology where science leads the development of the technology. The solid-liquid interface is an area that requires intense exploration in the near future. Clearly, developments in electrochemistry and biology depend on our ability to scrutinize on the molecular level the solid-liquid interface. Surface-science studies of solid-solid interfaces could produce major advances in the fields of coatings and composites. The opportunities are virtually unlimited for clever surface scientists who are willing to be broadminded enough to learn about applications of their molecular level studies to important fields and technologies based on surface science.

Acknowledgements. This work was supported by the Director, Office of Energy Research, Office of Basic Energy Sciences, Chemical Sciences and Materials Sciences Divisions of the U.S. Department of Energy under Contract Number DE-AC0376SF00098.

References

1.1 R.E. Powell, P.W. Montgomery, G. Jura, G.A. Somorjai: *Small-Angle X-ray Scattering*, ed. by H. Brumberger (Gordon and Breach, London 1967) p.449
1.2 S. Hagstron, H.B. Lyon, G.A. Somorjai: Phys. Rev. Lett. **15**, 491 (1965); H.B. Lyon, G.A. Somorjai: J. Chem. Phys. **46**, 2539 (1967)
1.3 J.T. Grant: Surf. Sci. **18**, 282 (1969)
1.4 D.G. Fedak, N.A. Gjostein: Surf. Sci. **8**, 77 (1967)
1.5 M.A. Van Hove, R.J. Koestner, P.C. Stair, J.P. Biberian, L.L. Kesmodel, G.A. Somorjai: Surf. Sci. **103**, 189, 218 (1981);
1.6 G.A. Somorjai: *Chemistry in Two Dimensions: Surfaces* (Cornell University Press 1981)
1.7 G.A. Somorjai: In *50 Years of Electron Diffraction*, ed. by Brockway (Reidel, Dordrecht 1981)
1.8 G.A. Somorjai, M.A. Van Hove: *Adsorbed Monolayers on Solid Surfaces*, Structure and Bonding, Vol.38 (Springer, Berlin, Heidelberg, New York 1979)
1.9 M.A. Van Hove, R.J. Koestner, G.A. Somorjai: Phys. Rev. Lett. **50**, (12) 903 (1982)
D.G. Castner, G.A. Somorjai: Chem. Rev. **79**, 233 (1979)
1.10 D. Castner, B.A. Sexton, G.A. Somorjai: Surf. Sci. **71**, 519 (1978)
D.G. Castner, G.A. Somorjai: Surf. Sci. **83**, 60 (1979)
1.11 M. Salmeron, R. Chianelli, G.A. Somorjai: Surf. Sci. **127**(3), 526 (1983)
1.12 S.H. Overbury, G.A. Somorjai: Discussion Meeting, Faraday Soc. **60**, 279 (1975)
1.13 S.H. Overbury, P.A. Bertrand, G.A. Somorjai: Chem. Revs. **75**, 547 (1975)
1.14 A. Jablonski, S.H. Overbury, G.A. Somorjai: Surf. Sci. **65**, 578 (1977)
1.15 S.H. Overbury, G.A. Somorjai: Surf. Sci. **55**, 209 (1976)
1.16 T.M. French, G.A. Somorjai: J. Phys. Chem. **74**, 2489 (1970)
1.17 F. Szalkowski, G.A. Somorjai: J. Chem. Phys. **56**, 6097 (1972)
1.18 G.A. Somorjai: Angew. Chemie 89, 94 (1977)
1.19 M. Asscher, W.L. Guthrie, T.-H. Lin, G.A. Somorjai: J. Chem. Phys. **78** (11), 6992 (1983)
1.20 J.C. Buchholz, G.A. Somorjai: J. Chem. Phys. **66**, 573 (1977)
1.21 G.A. Somorjai: Adv. Catal. **26**, 1 (1977)
1.22 D.W. Blakely, G.A. Somorjai: Surf. Sci. **65**, 419 (1977)
1.23 D.W. Blakely, G.A. Somorjai: J. Catal. **42**, 181 (1976)
1.24 M. Salmeron, R.J. Gale, G.A. Somorjai: J. Chem. Phys. **67**, 5324 (1977)
1.25 D.W. Blakely, E. Kozak, B.A. Sexton, G.A. Somorjai: J. Vac. Sci. Technol. **13**, 1091 (1976)
1.26 A.L. Cabrera, N.D. Spencer, E. Kozak, P.W. Davies, G.A. Somorjai: Rev. Sci. Instrum. **53**(12), 1888-1893 (1982)
1.27 S. Mark Davis, Francisco Zaera, G.A. Somorjai: J. Catal. (to be published)
1.28 S.M. Davis, F. Zaera, G.A. Somorjai: J. Am. Chem. Soc. **104**, 7453-7461 (1982)
1.29 W.D. Gillespie, R.K. Herz, E.E. Petersen, G.A. Somorjai: J. Catal. **70**, 147 (1981)
1.30 N.D. Spencer, R.C. Schoonmaker, G.A. Somorjai: J. Catal. **74**, 129 (1982)
1.31 S.M. Davis, F. Zaera, G.A. Somorjai: J. Catal. **77**, 439-459 (1982)
1.32 J.E. Crowell, R.J. Koestner, L.H. Dubois, M.A. Van Hove, G.A. Somorjai: *Recent Advances in Analytical Spectroscopy*, ed. by K. Fuwa (Pergamon, London 1982) p.211
1.33 M.M. Bhasin, W.J. Barkley, P.C. Ellgen, T.P. Wilson: J. Catal. **54**, 120 (1980)
1.34 B.A. Sexton, G.A. Somorjai: J. Catal. **46**, 167 (1977)
1.35 P.R. Watson, G.A. Somorjai: J. Catal. **72**, 347 (1981)
1.36 P.R. Watson, G.A. Somorjai: J. Catal. **74**, 282-295 (1982)

1.37 J.W.A. Sachtler, G.A. Somorjai. J. Catal. **81**, 77-94 (1983)
1.38 E.L. Garfunkel, J.E. Crowell, G.A. Somorjai: J. Phys. Chem. **86**, 310
     (1982)
1.39 J.E. Crowell, E.L. Garfunkel, G.A. Somorjai: Surf. Sci. **121**, 303-320
     (1982)
1.40 F. Zaera, G.A. Somorjai: J. Catal. (to be published)

2. Fourier-Transform Infrared Spectroscopy in Heterogeneous Catalysis

A. T. Bell

With 13 Figures

2.1 Introduction

Infrared spectroscopy has proven to be an extremely valuable method for obtaining information about the structure of species present at a catalyst
surface [2.1-9]. Because infrared spectra can be acquired in the presence
of a gas phase, and over a broad range of pressures and temperatures, it is
often possible to obtain information about the working state of the catalyst.
Such in situ studies have helped to establish whether an adsorbate undergoes rearrangement or decomposition upon adsorption, and in a limited number
of instances has led to the identification of reaction intermediates.

Most of the published work on infrared spectroscopy of surface structures
has been carried out with dispersive spectrometers. More recently, though,
there has been an increasing trend towards the use of Fourier-transform (FT)
spectrometers [2.8,9]. The principal advantage of FT spectrometers over dispersive instruments is a greatly reduced time required to acquire a spectrum
with a given signal-to-noise ratio. The advent of reliable commerical FT
spectrometers during the past decade or so has helped to broaden the scope
of materials that can be characterized by infrared spectroscopy and has
greatly facilitated the acquisition of spectra under dynamic conditions.

This chapter begins with a brief review of the optical principles underlying FT spectroscopy and summarizes the advantages of FT over dispersive
spectrometers. Next, the acquisition of infrared spectra by transmission,
diffuse reflectance, and photoacoustic techniques is discussed and the techniques compared. Finally, a series of illustrations is presented to demonstrate the advantages and disadvantages of each technique. Since virtually
all of the published work has concerned powdered catalysts, the discussions
given here have been limited to materials in this form.

2.2 Optical Principles

A brief review of the optical principles governing the operation of a FT
spectrometer is presented in this section. This discussion will serve as a
basis for understanding what factors govern instrument resolution and signal-
to-noise ratio. The reader interested in more detailed information is referred to [2.10-13].

The optical arrangement of a typical FT spectrometer is shown in Fig.2.1.
An infrared beam enters the source side of a Michelson interferometer where
part of the beam is reflected to the fixed mirror by the beam splitter and
part is transmitted to the movable mirror. Following reflection, both beams
recombine at the beam splitter. Since the path length of the beam reflected
from the movable mirror is in general slightly different from that of the
beam reflected from the fixed mirror, the two beams interfere either con-
structively or destructively. Displacement of the movable mirror at a fixed
velocity v modulates the beam exiting from the interferometer. The modulation frequency for each wave number $\bar{\nu}$ is given by $2v\bar{\nu}$.

The signal observed by the detector depends on the displacement or retardation of the movable mirror s and is known as the interferogram $I(s)$. To
obtain a spectrum $I(\bar{\nu})$, it is necessary to take the Fourier transform of the
fluctuating portion of the interferogram, $\bar{I}(s) = [I(s) - 0.5I(0)]$. Thus,

$$I(\bar{\nu}) = \int_{-\infty}^{\infty} \bar{I}(s)\cos(2\pi\bar{\nu}s)ds \quad . \tag{2.1}$$

An illustration of the relationship between $\bar{I}(s)$ and $I(\bar{\nu})$ for a spectrum
consisting of three narrow lines is shown in Fig.2.2a.

In practice, the Fourier integral indicated in (2.1) cannot be executed
over s from $-\infty$ to $+\infty$, since the interferogram can be determined experimentally only over a finite range $(-s_{max} \leq s \leq + s_{mx})$. As a consequence, the integration can be performed only over a finite range. Figure 2.2b shows that
truncation of the limits of integration results in a broadening of the spectral peaks.

An additional consequence of finite retardation is the appearance of secondary extrema or "wings" on either side of the primary features. The presence of these features is disadvantageous, especially for observing a weak
absorbance in proximity to a strong one. To diminish this problem the interferogram is usually multiplied by a triangular apodization function which
forces the product to approach zero continuously for $s = \pm s_{max}$. Fourier
transformation of the apodized interferogram produces a spectrum such as
that shown in Fig.2.2c.

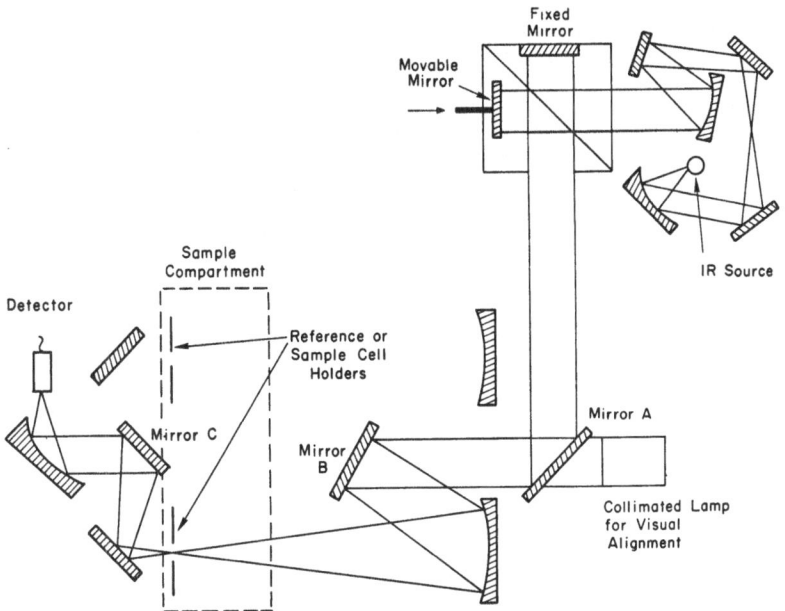

Fig.2.1. Optical arrangement of an FT spectrometer

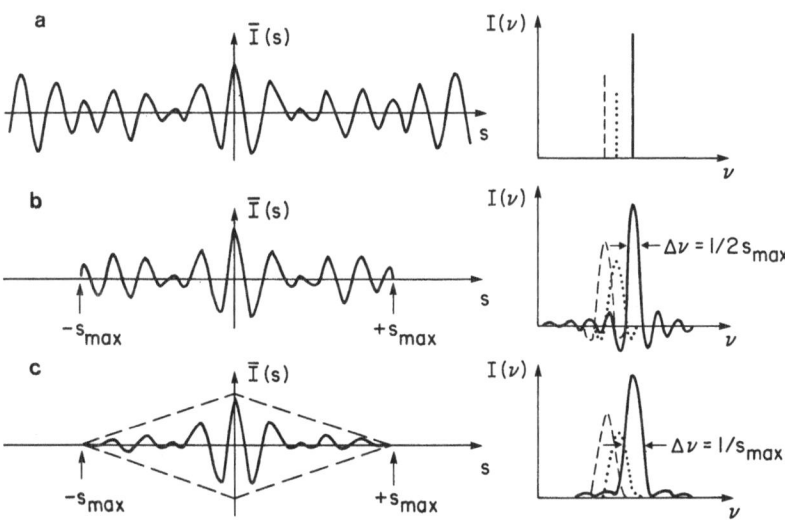

Fig.2.2a-c. Fourier transformed spectrum of: (a) an infinite interferogram; (b) a finite interferogram; and (c) a finite interferogram with triangular apodization

The resolution of an FT spectrometer is dictated by the maximum retardation of the movable mirror. Thus,

$$\Delta\bar{\nu} = 1/s_{max} \quad .$$

(2.2)

The resolving power of the spectrometer R is given by

$$R = \bar{\nu}/\Delta\bar{\nu}$$

$$= \bar{\nu}s_{max} \quad .$$

(2.3)

It follows, therefore, that the resolving power increases with increasing wave number.

The signal-to-noise ratio (SNR) for any type of spectrometer is dictated by (2.4):

$$SNR = B(T,\bar{\nu})\theta\eta\Delta\bar{\nu}t_m^{\frac{1}{2}}/NEP \quad ,$$

(2.4)

where $B(T,\bar{\nu})$ is the source brightness, which is a function of the source temperature and the frequency; θ is the optical throughput; η is the optical efficiency; $\Delta\bar{\nu}$ is the resolution; t_m is the observation time of a single spectral element of width $\Delta\nu$; and NEP is the noise equivalent power of the detector. Equation (2.4) indicates that high SNR's are favored by use of a bright source, high throughput optics, low resolution (i.e., a large value of $\Delta\bar{\nu}$), long spectral collection times, and a low-noise detector.

The advantages of FT over dispersive spectrometers can be discussed in terms of (2.4) [2.8,12-14]. For this purpose, advantage is defined either by the ratio of SNR's for a fixed data-collection time, or by the ratio of data-collection times for a fixed SNR. One of the principal advantages of FT over dispersive spectrometers, known as Fellgett's advantage, results from the fact that an FT spectrometer observes all spectral elements simultaneously, whereas a monochromer spends only a fraction of the total data-collection time observing each resolution element. Table 2.1 illustrates several examples of Fellgett's advantage. It is apparent that the multiplexing capabilities of FT spectrometers offer, in principle, very significant advantages, particularly in the speed of spectral acquisition. *Griffiths* et al. [2.14] have shown that Fellgett's advantage is never fully realized in practice and that the comparison of FT and dispersive instruments must take into consideration all factors entering into (2.4). When this is done, an FT spectrometer still outperforms dispersive spectrometers by a substantial margin. Thus, for a fixed resolution, FT instruments offer SNR advantages of 10 to 10^2 (assuming a fixed data-collection time) and data-collection-time advantages of 10^2 to 10^3 (assuming a fixed SNR). It must be recognized, of course, that these

Table 2.1. Tabulated values of Fellgett's advantage

Resolution $\Delta\bar{\nu}$ [cm^{-1}]	Range $\bar{\nu}_R$ [cm^{-1}]	t_m Advan. M	SNR Advan. M$^{\frac{1}{2}}$
8	3600	450	21
2	3600	1800	42
0.5	3600	7200	85
0.5	400	800	28

advantages are determined on the assumption that the full midinfrared range of frequencies (4000 to 400 cm^{-1}) is of interest. For situations in which information is required from a narrower spectral range, the advantages of FT spectroscopy diminish as either M or M$^{\frac{1}{2}}$.

2.3 Techniques for the Acquisition of Spectra

There are three primary techniques for obtaining infrared spectra from powdered samples: transmission, diffuse reflectance, and photoacoustic spectroscopy. Attenuated total reflection and emission spectroscopy have also been applied, but these techniques suffer from a number of disadvantages which limit their applicability [2.15]. This section outlines the theory underlying transmission, diffuse reflectance, and photoacoustic spectroscopy, and describes the types of cells needed to practice each of these techniques.

2.3.1 Transmission Spectroscopy

Transmission spectroscopy has been the most widely used technique to study the structure of adsorbed species and the structure of compound catalysts (e.g., metal oxides, metal sulfides, zeolites, etc.) [2.1-6]. The sample is usually prepared by pressing a fine powder of the solid into a self-supporting disk. To assure adequate transmission of infrared radiation through the disk, its thickness is usually limited to less than ~200 μm. The use of powder-particle diameters below a few microns helps to minimize the loss of radiation by scattering.

A large number of cells has been developed for transmission spectroscopy [2.1-9]. Some are designed to allow pretreatment of the sample in a furnace located to one side of the observation cell. While simple in construction, these cells require transfer of the sample back and forth between the furnace and the observation cell, and do not permit in situ examination of the sample. More recently, a number of designs have appeared for cells which permit observation of the sample under pretreatment or reaction conditions.

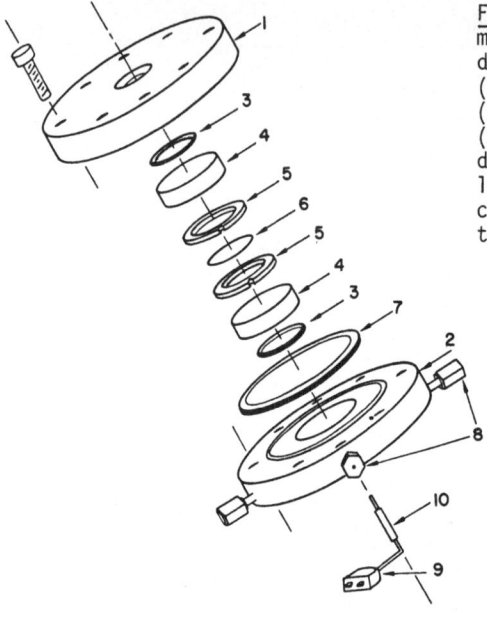

<u>Fig.2.3.</u> Exploded view of the transmission infrared spectroscopy cell developed by Hicks et al. [2.16]: (1) top flange; (2) bottom flange; (3) Kalrez O ring; (4) CaF$_2$ window; (5) sample holder; (6) catalyst disk; (7) copper gasket; (8) Swagelok fitting; (9) sheathed thermocouple; (10) sleave attached to thermocouple sheath

An example of such a cell, developed in the author's laboratory, is shown in Fig.2.3 [2.16]. This cell can operate at pressures between 1 and 20 atm and at temperatures up to 300°C. Because the dead volume of the cell is small, 0.25 cm^3, it is ideally suited for transient response experiments. The small optical path length through the gas inside the cell ($\sim$ 1 mm) makes this design ideal for high-pressure studies where interference from gas-phase absorption must be minimized.

For transmission spectroscopy, the relationship between absorbance A, and the concentration of absorbing material c, is given by

$$A = -\ln(I/I_0)$$

$$= \varepsilon dc \quad , \tag{2.5}$$

where I_0 and I are the intensities of incident and transmitted beams, ε is the integrated absorbtion coefficient, and d is the sample thickness. In the case of solid materials, ε is constant, and hence A is a linear function of c. This relationship breaks down, though, for adsorbed species. Thus, for example, several authors [2.17-19] have demonstrated that the value of ε for adsorbed CO decreases progressively with increasing coverage. This means that quantitative interpretation of infrared spectra of adsorbed species must be carried out with care.

2.3.2 Diffuse-Reflectance Spectroscopy

To take a diffuse-reflectance spectrum of a powder, the sample is placed in a shallow cup and irradiated with radiation from the interferometer. The incident radiation passes into the bulk of the sample, undergoes reflection, refraction, scattering, and absorption by varying degrees before reemerging at the sample surface. The diffusely reflected radiation from the sample is collected by a spherical or elliptical mirror and focused onto the detector.

A number of different cells for diffuse-reflectance spectroscopy have been described in the literature [2.12,15]. The arrangement shown in Fig. 2.4, recently reported by *Hamadeh* [2.20], is particularly well suited for in situ studies of catalysts. Gases can be passed directly through the powdered sample permitting intimate contact, and the sample can be heated to temperatures as high as $600^\circ C$. The range of pressures over which the cell can be used is 10^{-6} torr to 1 atm.

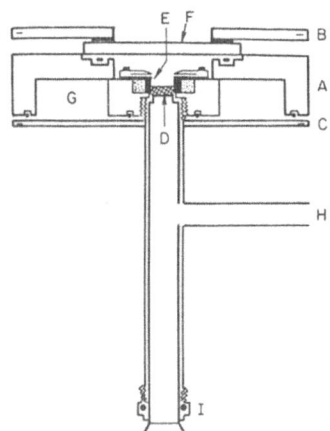

Fig.2.4. Cross-sectional view of the diffuse-reflectance cell developed by Hamadeh [2.20]: (A) main body; (B) cover plate for the sample compartment; (C) cover plate for the cooling water cavity; (D) frit to support the sample; (E) Sample heater; (F) KCl window; (G) cavity for cooling water; (H) gas outlet line; (i) nut to vacuum-tighten the sample holder to the housing tube

Unlike transmission spectra, diffuse-reflectance spectra cannot be determined directly, and are usually obtained using a nonabsorbing powder as a reference [2.15]. Common reference materials for measurements at infrared frequencies are alkali halides. To avoid problems of sampling depth, the beds of sample and reference materials must be sufficiently deep so that an increase in their depths causes no further change in the signal. The ratio of the diffuse reflectance from the sample to that of the nonabsorbing powdered reference at "infinite depth" is designated R_∞. When the amount of specularly reflected radiation is small and the scattering coefficient s is constant, R_∞ is related to the concentration of absorbers by the Kubelka-

Munk function, $f(R_\infty)$. Thus

$$f(R)_\infty = \frac{(1 - R_\infty)^2}{2R_\infty} = \frac{\varepsilon c}{s} \quad . \tag{2.6}$$

The linearity of $f(R_\infty)$ in c depends on whether or not ε is independent of c, as in the case of transmission spectroscopy.

2.3.3 Photoacoustic Spectroscopy

The application of photoacoustic spectroscopy to obtain infrared spectra of powdered samples is a relatively recent development [2.21-23]. To take a photoacoustic spectrum, the sample is placed in a sealed cell containing an inert gas and a very sensitive microphone, and exposed to radiation emanating from the interferometer of an FT spectrometer. For frequencies where the sample absorbs radiation, the radiant energy is converted to thermal energy. This causes the gas above the sample to heat up and expand, thereby setting up a pressure wave which is detected by the microphone.

As yet, there have been relatively few cell designs proposed for taking photoacoustic spectra of powdered samples [2.15,21]. An example of a cell which has been used quite successfully is shown in Fig.2.5 [2.24,25]. In this design the sample is contained in a rectangular holder which can easily be introduced into the cell. The cell is coupled acoustically to the sample compartment by a narrow channel.

In contrast to the cells used for transmission and diffuse-reflectance spectroscopy, it is not possible to acquire photoacoustic spectra in the presence of the adsorbate gas or at elevated temperatures. The first of these constraints arises from the fact that the photoacoustic effect is much stronger for gases than for solids. As a consequence, it is necessary to flush the adsorbate gas from the cell with a nonabsorbing gas such as N_2 or He, prior to taking a spectrum. Thus, only strongly adsorbed species can be studied. The constraint on temperature is due to the fact that the SNR decreases significantly as the temperature increases above ambient. It is also appropriate to note that even under ideal circumstances the SNR of a photoacoustic detector is significantly lower than that of a HgCdTe detector, used for transmission or diffuse-reflectance spectroscopy.

The signal put out by the microphone of a photoacoustic cell is proportional to the temporal changes in the gas pressure within the cell, $\Delta P(t)$. *Rosencwaig* and *Gersho* [2.26] have shown that $\Delta P(t)$ is related to the heat released to the gas, Q, by the following expression

$$\Delta P(t) = Q \exp\left[i(\omega t - \frac{\pi}{4})\right] \quad , \tag{2.7}$$

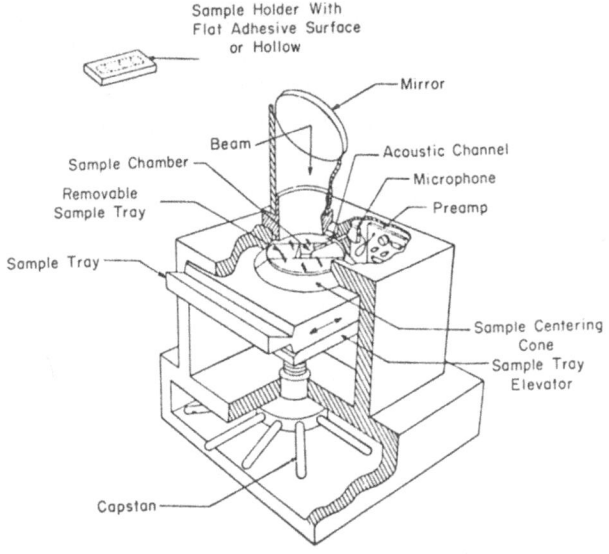

Sample Holder With
Flat Adhesive Surface
or Hollow

Mirror

Beam

Sample Chamber

Acoustic Channel

Removable
Sample Tray

Microphone

Preamp

Sample Tray

Sample Centering
Cone

Sample Tray
Elevator

Capstan

Fig.2.5. Cross-sectional view of the Gilford NIR-Vis-UV photoacoustic cell [2.24]

where ω is the modulation frequency of the infrared radiation. The quantity Q is a function of the optical and thermal properties of the sample, and of the thermal properties of the gas over the sample. For thermally and optically thick solids [2.23,26]

$$Q = - \frac{iP_0 I_0 \gamma}{4 \times 2 \, \ell_g} T_0 \left(\frac{\mu_s^2 \mu_g}{k_s} \right) \epsilon c \quad , \tag{2.8}$$

where I_0 is the intensity of the incident radiation, P_0 is the ambient gas pressure, T_0 its temperature, γ the ratio of specific heats of the gas at constant pressure and volume, ℓ_g is the height of the ambient gas layer, k_s is the specific heat of the solid, and μ_s and μ_g are the thermal diffusion lengths of the solid and gas, respectively. Equations (2.7,8) indicate that when ϵ is independent of c, the amplitude of $\Delta P(t)$ will be proportional to c.

2.4 Applications

As noted earlier, the principal advantages of FT spectroscopy over dispersive spectroscopy are a higher SNR for a fixed period of data collection and significantly a lower data-collection time to achieve a fixed SNR. Because of this, FT spectrometers are used increasingly for studies in which the amount of radiation transmitted through the sample, or reflected from it, is low, and for studies in which the sample undergoes a rapid change in composition. This section briefly reviews the applications of FT spectroscopy to study the dynamics of adsorbed species using transmission techniques and the characterization of solids and adspecies by diffuse reflectance and photoacoustic spectroscopy.

2.4.1 Transmission Spectroscopy

The application of FT spectrometers to dynamic studies holds considerable promise for obtaining insight into the kinetics of adsorption, desorption, and surface reaction processes. Several illustrations taken from the author's recent work will be described here.

Savatsky and *Bell* [2.27,28] have studied the kinetics of NO adsorption onto, and desorption from, a silica-supported Rh catalyst. Figure 2.6 illustrates the temporal development of the bands observed at 1680, 1830, and 1910 cm^{-1}. The band at 1680 cm^{-1}, assigned to N-O vibrations of $NO_a^{\delta-}$, grows to a maximum intensity and then diminishes for exposures above 100 s. During this latter period, the band at 1910 cm^{-1}, assigned to $NO_a^{\delta+}$, intensifies and, after exposure times of 800 s or more, becomes the dominant spectral feature. Since the band at 1680 cm^{-1} is characteristic of NO adsorbed on a reduced rhodium surface and the band at 1910 cm^{-1} is characteristic of NO adsorbed on an oxidized surface, the relative intensities of these bands provide an indication of the extent to which the rhodium surface is oxidized by NO. The band at 1830 cm^{-1} is due to NO adsorbed in a neutral state, and, as seen in Fig.2.1, this feature is relatively insensitive to the progressive oxidation of the catalyst. The effect of temperature on the intensities of the bands for $NO_a^{\delta-}$ and $NO_a^{\delta+}$ is shown in Fig.2.7, for a fixed exposure time. It is apparent that with increasing temperature the high-frequency band for $NO^{\delta+}$ increases monotonically as the low-frequency band for $NO^{\delta-}$ decreases. These trends indicate that the oxidation of Rh by NO is accelerated at higher temperatures.

The kinetics of NO desorption from silica-supported Rh were also studied [2.28]. Figure 2.8 shows how the integrated intensity of the band at 1680 cm^{-1} decreases with time, for several temperatures. Over the temperature range considered the intensities of the bands at 1830 and 1910 cm^{-1} did not change with time, indicating that NO_a and $NO_a^{\delta+}$ are more strongly adsorbed than $NO_a^{\delta-}$. Using an equilibrium desorption model to simulate the data in Fig.2.8, it was determined that the rate coefficient for NO desorption is $k_d = 1.2 \times 10^{15} \exp(-25,000/RT)$. Both the preexponential factor and the ac-

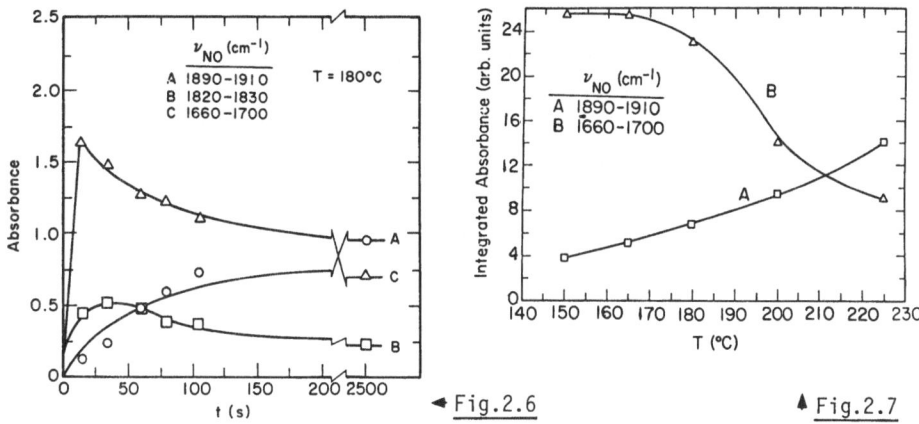

Fig.2.6   Fig.2.7

Fig.2.6. Time evolution of bands associated with NO adsorption on a 4% Rh/
SiO$_2$ catalyst: adsorption temperature, 150 C; NO partial pressure, 0.03 atm;
25 scans per spectrum taken at a rate of 1 scan/s

Fig.2.7. Temperature effects on the intensity of the bands observed for NO
adsorption on 4% Rh/SiO$_2$ following exposure to NO for 10 min

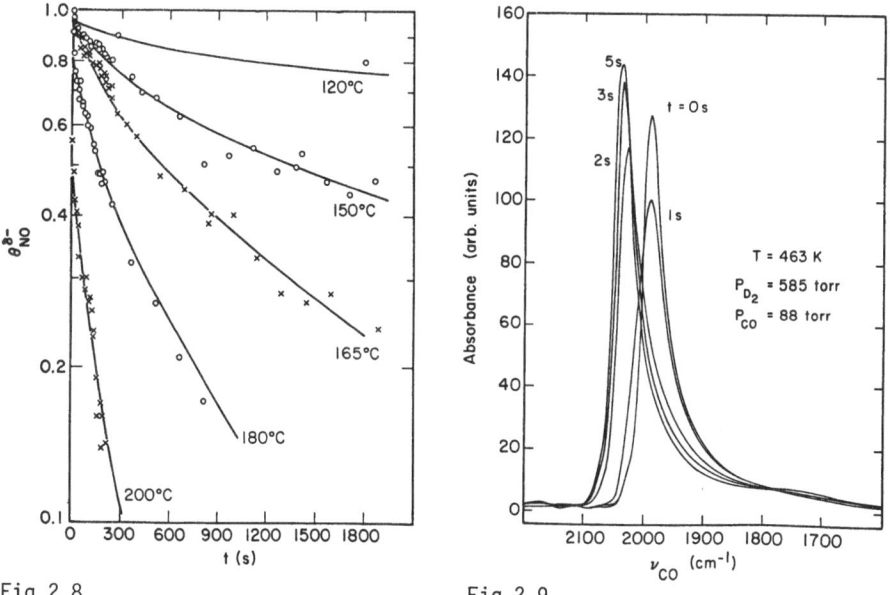

Fig.2.8   Fig.2.9

Fig.2.8. Decrease in coverage by NO$_a^{\delta-}$ with time at different temperatures

Fig.2.9. Infrared spectra of adsorbed CO illustrating the displacement of
^{13}CO during the steady-state reaction of CO and D$_2$ over a 4% Ru/SiO$_2$ catalyst

33

tivation energy for desorption are in good agreement with values determined from temperature-programmed desorption experiments performed with the same catalyst sample [2.29].

Another illustration of the application of FT transmission spectroscopy to dynamic studies is shown in Fig.2.9. These spectra are taken from a recent investigation reported by *Winslow* and *Bell* [2.19]. The rate of CO exchange between the gas phase and the surface during CO hydrogenation over a Ru/SiO_2 catalyst was observed by making a step-function change in the reactant feed from $^{13}CO/D_2$ to $^{12}CO/D_2$. The band in Fig.2.9 is due to linearly adsorbed CO. Initially, the band appears at 2000 cm^{-1}, as expected for adsorbed ^{13}CO. When the $^{12}CO/D_2$ is introduced, the band rapidly moves up scale, and within 4 s reaches a frequency of 2030 cm^{-1}, characteristic of adsorbed ^{12}CO. This sequence, taken at about one interferogram per second, demonstrates that relatively rapid surface reactions can be followed by FT spectroscopy.

2.4.2 Diffuse-Reflectance Spectroscopy

The earliest efforts to obtain diffuse-reflectance spectra from powdered samples were carried out using dispersive spectrometers. *Kortüm* and *Delfts* [2.30] reported spectra of ethylene and hydrogen cyanide on various metal oxides. The quality of the spectra was sufficient to identify the formation of C_2H_5 surface species from adsorbed C_2H_4 and hydrogen donated by the surface OH groups. Spectra of adsorbed HCN revealed evidence for HCN decomposition, polymer buildup, and dicyan formation. The spectra of pyridine adsorbed on HY zeolite were obtained by *Niwa* et al. [2.31]. While noisy, these spectra could still be used to identify a variety of surface structures.

Hamadeh [2.20] has recently used the cell shown in Fig.2.4 in conjunction with FT spectroscopy to obtain high-quality spectra of CO adsorbed on Rh/ Al_2O_3. An example of these spectra is shown in Fig.2.10. The bands at 2085 and 2027 cm^{-1} are assigned to the symmetric and antisymmetric stretches of CO in gem-dicarbonyl structures. The band 2060 cm^{-1} is attributed to linearly adsorbed CO and the broad band at 1875-1890 cm^{-1} to bridge-bonded CO. The weak feature at 1625 cm^{-1} is believed to be due to a formate structure.

To assess the sensitivity of his apparatus, *Hamadeh* [2.20] also examined the spectrum of C_2H_4 adsorbed on Pd/Al_2O_3, since the extinction coefficients for adsorbed C_2H_4 are significantly lower than those for adsorbed CO. A representative spectrum is shown in Fig.2.11. The absence of any bands below 3000 cm^{-1} suggests that the adsorbed species is ethylenic rather than paraffinic. The band at 2140 cm^{-1} and the shoulder at 2090 cm^{-1} are assigned to

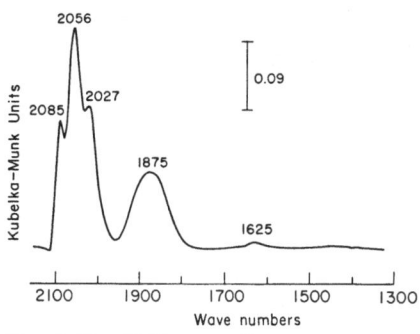

Fig.2.10. Diffuse-reflectance spectrum of CO adsorbed on 5% Rh/Al$_2$O$_3$ following catalyst reduction at 1000 torr of H$_2$ at 350°C: CO pressure = 5.84 torr [2.20]

Fig.2.11 ➙

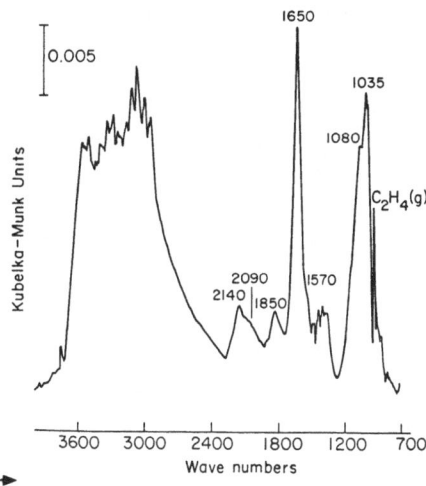

Fig.2.11. Diffuse-reflectance spectrum of C$_2$H$_4$ adsorbed on 5% Pd/Al$_2$O$_3$: C$_2$H$_4$ pressure = 6.19 torr [2.20]

two types of weakly adsorbed hydrogen atoms, and the band at 1850 to strongly adsorbed hydrogen. The band at 1650 cm^{-1} is assigned to the C=C stretch of a disubstituted vinyl group such as

> CH $\overset{\cdots}{=\!=\!=}$ CH$_2$.

The shoulder at 1570 cm^{-1} is attributed to C=C stretching vibrations of a vinyl group adsorbed at a threefold site, i.e.,

→ CH $\overset{\cdots}{=\!=\!=}$ CH$_2$.

The remaining two bands at 1035 and 1080 cm^{-1} are assigned to the out-of-plane deformations of a vinyl group. The spectrum in Fig.2.11 suggests, therefore, that upon adsorption both hydrogen atoms at one end of the C$_2$H$_4$ molecule are transferred to the metal and the residual hydrocarbon species becomes bridge bonded to either two or three Pd atoms.

2.4.3 Photoacoustic Spectroscopy

Very limited application of photoacoustic spectroscopy has been made thus far to studies of catalysts and adsorbed species, and none of the reported studies have used an FT spectrometer. *Low* and *Parodi* [2.21] demonstrated the feasibility of observing surface species on fused silica. Examples of their spectra are presented in Fig.2.12. Upon contacting the silica with HSiCl$_3$, spectra a and b show the conversion of ≡Si-OH groups to ≡Si-O-SiHCl$_2$ groups. Treatment of the silanized sample with NH$_3$ leads to the appearance of bands

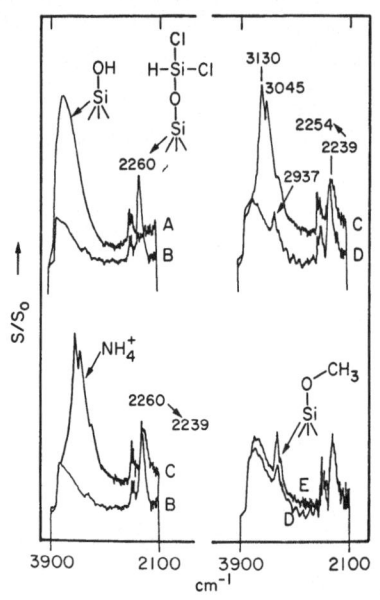

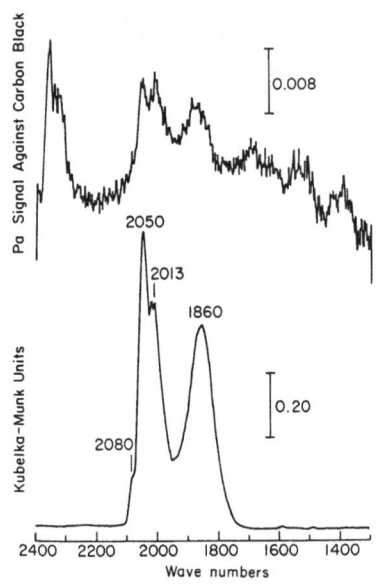

Fig.2.12   Fig.2.13

Fig.2.12. Photoacoustic spectra of: (A) silica, after degassing at 500°C; (B) after reaction with HSiCl3; (C) after exposure to NH3; (D) after degassing to remove sorbed NH3; (E) after reaction with CH3OH [2.21]

Fig.2.13. Spectra of CO adsorbed on 5% Rh/Al2O3 obtained using diffuse-reflectance (lower trace) and photoacoustic spectroscopy (upper trace) [2.20]

for NH_4^+ and a shift in the position $\equiv$Si-H band (spectrum c). The reversibility of NH_3 adsorption is illustrated by spectrum d. Finally, treatment of the sample with CH_3OH leads to the formation of $\equiv$Si-O-CH_3 groups, as evidenced by spectrum e.

Hamadeh [2.20] has recently compared the diffuse-reflectance and photoacoustic techniques for acquiring the spectrum of CO adsorbed on Rh/Al_2O_3. Figure 2.13 illustrates the spectra obtained by the two methods. It is seen that the SNR of the diffuse-reflectance spectrum is considerably better than that of the photoacoustic spectrum even though with the former only one scan was collected, whereas 1024 scans were averaged to obtain the photoacoustic spectrum.

2.5 Conclusions

Work accomplished during the past several years has amply demonstrated the utility of FT spectroscopy for characterizing catalyst surfaces and adsorbed species. The most promising methods for acquiring spectra appear to be transmission and diffuse-reflectance spectroscopy. Cells capable of operation over a broad range of temperatures and pressures have been developed and can be used for in situ studies. Since FT spectra with good SNR can be acquired in a second or so, it is now possible to study many processes under dynamic conditions. Such investigations are particularly exciting and hold promise for providing data on the kinetics of adsorption, desorption, decomposition, and reaction of various gases. The kinetics of solid-state reactions can also be studied using similar techniques.

Acknowledgement. I acknowledge support for this work by the Division of Chemical Sciences, Office of the Basic Energy Sciences, U.S. Department of Energy under Contract DE-AC03-76SF-00098 and by the Engineering Directorate of the National Science Foundation under Grant CPE-7826352.

References

2.1 R.P. Eischens, W.A. Pliskin: Adv. Catal. **10**, 1 (1958)
2.2 L.H. Little: *Infrared Spectra of Adsorbed Species* (Academic, New York 1966)
2.3 M.L. Hair: *Infrared Spectroscopy in Surface Chemistry* (Dekker, New York 1967)
2.4 G. Blyholder: In *Experimental Methods in Catalytic Research*, ed. by R.B. Anderson (Academic, New York 1968)
2.5 M.R. Basila: Appl. Spectrosc. Rev. **1**, 289 (1968)
2.6 A.V. Kiselev, V.I. Lygin: *Infrared Spectra of Surface Compounds* (Wiley, New York 1975)
2.7 W.N. Delgass, G.L. Haller, R. Kellerman, J.H. Lunsford: *Spectroscopy in Heterogeneous Catalysis* (Academic, New York 1979)
2.8 A.T. Bell, M.L. Hair (eds.): "Vibrational Spectroscopies for Adsorbed Species", ACS Symposium Series 137 (American Chemical Society, Washington, D.C. 1980)
2.9 C.L. Angell: In *Fourier Transform Infrared Spectroscopy — Techniques Using Fourier Transform Interferometry*, Vol.3, ed. by J.R. Ferraro, L.H. Basile (Academic, New York 1982)
2.10 R.J. Bell: *Introductory Fourier Transform Spectroscopy* (Academic, New York 1972)
2.11 R. Geick: In *New Theoretical Aspects*, Topics in Current Chemistry, Vol.58 (Springer, Berlin, Heidelberg, New York 1975)
2.12 P.R. Griffiths: *Chemical Infrared Fourier Transform Spectroscopy* (Wiley, New York 1975)
2.13 T. Hirschfeld: In *Fourier Transform Infrared Spectroscopy — Applications to Chemical Systems*, Vol.2 (Academic, New York 1979)
2.14 P.R. Griffiths, H.J. Sloane, R.W. Hannah: Appl. Spectrosc. **31**, 485 (1977)
2.15 P.R. Griffiths, M.P. Fuller: *Advances in Infrared and Raman Spectroscopy*, Vol.9, ed. by R.E. Hester, R.J.H. Clark (Hyden, London 1981)
2.16 R.F. Hicks, C.S. Kellner, B.J. Savatsky, W.C. Hecker, A.T. Bell: J. Catal. **71**, 216 (1981)
2.17 R.P. Eischens: Acc. Chem. Res. **5**, 74 (1972)
2.18 D.A. Seanor, C.H. Amberg: J. Chem. Phys. **42**, 2967 (1965)

2.19 P. Winslow, A.T. Bell: J. Catal. (in press)
2.20 I.M. Hamadeh: Ph.D. Dissertation, Ohio University, Athens, Ohio (1982)
2.21 M.J.D. Low, G.A. Parodi: Appl. Spectrosc. **34**, 76 (1980)
2.22 Y.-H. Pao (ed.): *Optoacoustic Spectroscopy and Detection* (Academic, New York 1977)
2.23 A. Rosencwaig: *Photoacoustics and Photoacoustic Spectroscopy* (Wiley, New York 1980)
2.24 R.E. Blank, T.W. Wakefield: Anal. Chem. **51**, 50 (197????)
2.25 D.W. Vidrine: Appl. Spectrosc. **34**, 405 (1980)
2.26 A. Rosencwaig, A. Gersho: J. Appl. Phys. **47**, 64 (1976)
2.27 B.J. Savatsky, A.T. Bell: In *Catalysis Under Transient Conditions*, ed. by A.T. Bell, L.L. Hegedus, ACS Symposium Ser. 178 (American Chemical Society, Washington, D.C. 1982)
2.28 B.J. Savatsky: Ph.D. Dissertation, University of California, Berkeley, CA (1981)
2.29 A.A. Chin, A.T. Bell: J. Phys. Chem. **87**, 3700 (1983)
2.30 G. Kortüm, H. Delfts: Spectrochim. Acta **20**, 405 (1964)
2.31 M. Niwa, T. Hattori, M. Takahashi, K. Shirai, M. Watanabe, Y. Murakami: Anal. Chem. **51**, 46 (1979)

3. Magnetic Resonance in Surface Science

R. F. Howe

With 16 Figures

3.1 Introduction

The spectroscopic techniques of nuclear magnetic resonance (NMR) and electron
paramagnetic resonance (EPR or ESR) have been known for more than 30 years.
Only in the past 15 years, however, have they been widely applied in the
field of surface science. In principle EPR spectroscopy is limited to observation of surface species containing unpaired electrons. These may include adsorbed radicals, molecules containing unpaired electrons (e.g., NO), transitional metal ions and surface defects of various types. Furthermore, EPR is
a highly sensitive technique, able to detect (in favourable cases) as few as
10^{11} spins. NMR spectroscopy, on the other hand, is much broader in scope,
since there are many different nuclei possessing non-zero nuclear spins.
As a surface technique however, NMR suffers the disadvantage of having a
relatively low sensitivity. Both EPR and NMR provide information about the
structure and environment of the species being observed, in a completely nondestructive manner.

The application of magnetic resonance techniques to surface analysis was
reviewed by *Lunsford* in [3.1]. Several other reviews since then have described
more recent applications of EPR [3.2-5] and NMR [3.6-10] to surface chemistry,
and the proceedings of two international conferences on magnetic resonance
in colloid and interface science have been published [3.11,12].

It is the purpose of this review to describe the present status of EPR
and NMR as surface-science techniques. This will be done by means of selected
examples, with particular emphasis placed on developments in the field since
Lunsford's 1976 review. No attempt is made at complete literature coverage,
for which the earlier reviews should be consulted. An outline is first given
of the important theoretical aspects of EPR and NMR spectroscopy.

3.2 Background

Specialised textbooks should be consulted for full details of the theory
and practice of EPR [3.13-16] and NMR [3.16-19]. The following is intended
as a brief summary.

3.2.1 EPR Spectroscopy

The interaction of a paramagnetic species containing a single unpaired electron with an external magnetic field H is described approximately by the following spin Hamiltonian:

$$\mathscr{H} = \beta H \cdot \underline{g} \cdot S + \Sigma\, S \cdot \underline{A} \cdot I \qquad\qquad (3.1)$$

where   is the Bohr magneton, **g** the so-called g tensor, **S** the electron spin,
A the nuclear hyperfine tensor and **I** a nuclear spin. The first term describes
the interaction of the electron with the field, and the second term contains
hyperfine interactions between the electron spin and any nuclear spins in
its vicinity. Equation (3.1) is approximate in that it neglects the direct
interaction of the nuclear spins with the field, as well as any interaction
of the electron spin with nuclear quadrupoles (for those nuclei having $I > \frac{1}{2}$).
The neglected terms are small, but not necessarily negligible. An additional
term (the zero field splitting term) must be added to (3.1) for species containing more than one unpaired electrons, although such species have rarely
been studied on surfaces.

In the conventional EPR experiment the magnetic field is varied to induce
transitions between eigenstates of (3.1) with microwave radiation of fixed
frequency (usually about 9.5 GHz). The value of g determines the field at
which absorbtion occurs, while the hyperfine interactions cause multiplet
splitting of the resonance. Since both of the interactions described in (3.1)
are anisotropic, **g** and **A** are second-rank tensors which can be characterised
by their principal components and by the orientations of their principal axes
with respect to a molecular coordinate system. Experiments with single-crystal samples in which all the paramagnetic species are aligned in the same
direction allow complete determination of the tensor components and the orientations of their principal axes from measurements of the spectrum as a function of the orientation of the crystal in the magnetic field.

The surface scientist is usually faced with a situation in which the paramagnetic species are randomly oriented in a polycrystalline powder sample,
and the observed EPR spectrum is an envelope of spectra from all orientations.
Determination of the spin Hamiltonian parameters from such a powder spectrum

is a more difficult problem. In simple cases, measurement of the positions
of turning points (maxima, minima or points of inflexion) in the first derivative absorbtion spectrum can give the magnitudes of the principal components of the **g** and **A** tensors, since the turning points correspond to alignment of the field along principal axes. In practice, many of the expected
features may overlap and be poorly resolved. The orientations of the tensor
axes cannot be obtained from a powder spectrum.

The reverse procedure, calculation of the powder spectrum expected for a
given set of spin Hamiltonian parameters, is readily achieved with computer
programs which compute the spectrum for one orientation and integrate over
all orientations [3.20]. Examples of such computed powder spectra will be
considered later. It should be noted here that successful computer simulation
of the observed spectrum is an important and necessary test of the correct-
ness of **g** and **A** tensor components estimated from inspection of the observed
spectrum.

In situations where a surface species changes its orientation on the time
scale of the EPR experiment (circa 10^{-9} s), the anisotropy in the spectrum
may be partially or completely averaged to zero. An example of the effects
of surface mobility will be considered later. The normal EPR spectrum is insensitive to motion occurring on a time scale longer than about 10^{-7} s.

Two additional parameters that characterise the interaction of a spin system with its surroundings are the spin-lattice and spin-spin relaxation times
T_1 and T_2, where T_1 is the time constant for transfer of energy from the spin
system to its surroundings, and T_2 is the corresponding time constant for
the transfer of energy from one spin system to another. These parameters are
not readily measured in the normal EPR experiment, although they may be estimated from measurements of saturation of the EPR signal at high microwave
powers. The relaxation times are in principle easily obtained from pulsed
EPR experiments, but these techniques have not yet found wide application.

3.2.2 NMR Spectroscopy

The approximate spin Hamiltonian for interaction of a nuclear spin I with
an external magnetic field, analogous to equation (3.1), is

$$\mathcal{H} = -g_n \beta_n \mathbf{I} \cdot (1 - \boldsymbol{\sigma}) \cdot \mathbf{H} + \sum_{i<k} \mathbf{I}_i \cdot \mathbf{J}_{ik} \cdot \mathbf{I}_k + \sum_{i<k} \mathcal{H}_{ik}^{dip} \quad , \tag{3.2}$$

where g_n is the nuclear g value, β_n the nuclear magneton, $\boldsymbol{\sigma}$ the chemical
shift tensor, and $\mathbf{J}_{ik}$ the indirect nuclear coupling tensor. The first term
describes the nuclear Zeeman interaction with the magnetic field. The field
at the nucleus may differ from the external field due to screening effects

from surrounding electrons, which are incorporated into the chemical shift.
Since screening effects will in general be anisotropic, the chemical shift
is expressed as a second-rank tensor which is formally analogous to the **g**
tensor in EPR spectroscopy. However, it should be noted that in NMR the effect of the electronic environment is to modify the field rather than the
value of g. The second term describes the magnetic interaction between different nuclei via bonding electrons. The coupling constant J_{ik} is often
treated as a scalar quantity, although it is formally a second-rank tensor.
The third term, which dominates in NMR of solids, describes the direct dipolar interaction between nuclear spins. The dipolar Hamiltonian has the form

$$\mathscr{H}_{ik}^{dip} = g_i g_k \beta_n^2 \left[\frac{\mathbf{I}_i \cdot \mathbf{I}_k}{r^3} - \frac{3(\mathbf{I}_i \cdot \mathbf{r})(\mathbf{I}_k \cdot \mathbf{r})}{r^5} \right] \quad , \tag{3.3}$$

where r is the radius vector connecting spins i and k. The dipolar Hamiltonian
is anisotropic, and may be expressed alternatively in terms of a dipolar
coupling tensor which has the important property of being

$$\mathscr{H}_{ik}^{dip} = \mathbf{I}_i \cdot \mathbf{D} \cdot \mathbf{I}_k \tag{3.4}$$

traceless. The second and third terms together in (3.2) are formally analogous to the hyperfine coupling term in (3.1). Equation (3.2) neglects any
contribution from nuclear quadrupole moments and from interaction of the
nuclear spin with unpaired electrons.

In solution, rapid motion of molecules on the NMR time scale (circa 10^{-6} s)
averages the dipolar coupling tensor, the chemical shift anisotropy and any
anisotropy in the indirect coupling term to zero. A solution NMR spectrum
then provides the average chemical shift [$\sigma = \frac{1}{3} (\sigma_{xx} + \sigma_{yy} + \sigma_{zz})$] and the
isotropic coupling constants J_{ik}. In solid samples, on the other hand, all
of the anisotropic interactions will in general be present, and the NMR
spectrum obtained from a polycrystalline powder will be a broad envelope of
spectra from all possible orientations in the magnetic field.

A number of successful NMR studies of molecules absorbed on high surface
area solids have been made using the so-called wide-line NMR technique, in
which the sample is continuously irradiated with single-frequency RF radiation while the magnetic field is swept through resonance (analogous to the
conventional EPR experiment) [3.21]. Pulsed NMR spectrometers are now used
more routinely [3.19]. In the usual pulse experiment, a short intense pulse
of RF radiation generates a component of nuclear spin magnetisation in a
plane perpendicular to the direction of the static magnetic field, which in
turn generates a signal in a receiver coil in this plane. When the RF pulse

is turned off, the perpendicular component of magnetisation decays to zero (the free induction decay) with a time constant $T_2 \leq T_1$, where T_1 and T_2 are the nuclear spin-lattice and spin-spin relaxation times. Fourier transformation of the free induction decay produces a conventional absorption versus frequenc spectrum.

The sensitivity of the pulsed NMR technique exceeds that of wide-line NMR by several orders of magnitude. *Pfeiffer* [3.6] estimated the limiting sensitivity of a modern pulsed instrument for protons in a field of 10000 Gauss (the number of protons needed to give a signal-to-noise ratio of 10:1) to be about 10^{19}. For other nuclei, with smaller magnetic moments, the sensitivity will be correspondingly lower, although the advent of superconducting solenoid magnets (magnetic fields up to 140000 Gauss) gives some improvement.

Earlier reviews [3.1,6-10] described NMR studies of physically adsorbed molecules, in which the rapid motion of the adsorbed molecules averages the anisotropic interactions to zero, giving a solution-like NMR spectrum. Such studies have been particularly useful in observing ^{13}C NMR spectra of weakly adsorbed organic molecules. Solution-like spectra are also obtained in situations where a strongly chemisorbed species can rapidly exchange with a physically adsorbed phase [3.22]. The earlier reviews have also covered the use of pulse NMR to measure the relaxation times T_1 and T_2, and this aspect will not be considered here.

The most important advance of the past 10 years as far as NMR of surfaces is concerned is the development of high resolution NMR of solids. The dipolar contribution to the spin Hamiltonian, (3.2), in most cases swamps the terms which provide more directly useful information about the structure and environment of the species under observation, i.e., the chemical shift and the indirect nuclear coupling constants, and often broadens the signal from polycrystalline powders beyond detection. Methods are now available for removing dipolar broadening from NMR spectra of solids, as well as broadening due to anisotropy in the chemical shift tensor. A brief outline of these methods is given here; several recent reviews on high resolution NMR of solids should be consulted for further details [3.23-25].

The dipolar interaction between dissimilar nuclei is readily removed by applying intense RF radiation at the resonance frequency of the second nucleus. This effectively decouples it from the first. The decoupling technique can also be used to enhance the sensitivity of the first nucleus where it is magnetically dilute, through cross polarisation. This has found particular use in ^{13}C NMR of organic molecules, where polarisation transfer from the more abundant ^1H nuclei significantly enhances the ^{13}C NMR signal.

The dipolar interaction between similar nuclei is more difficult to remove. If the nucleus of interest is magnetically dilute (e.g., ^{13}C, 1.1% natural abundance), then the r^{-3} dependence of the dipolar interaction results in very little homonuclear broadening. For magnetically concentrated nuclei (e.g., ^{1}H), the so-called magic angle spinning (MAS) technique can in principle be employed. This relies on the fact that dipolar interaction is, to first order, proportional to $(1-3 \cos^2 \theta)$, where θ is the angle between the vector connecting the dipoles and the external field. The dipolar interaction therefore vanishes at $\theta = 54°44'$. Rapid rotation of a powder sample about an axis which makes an angle of $54°44'$ to the magnetic field direction should then remove dipolar broadening from the spectrum, provided the rotation rate exceeds the magnitude of the broadening (in frequency units). In practice, however, the homonuclear dipolar broadening usually exceeds the highest rotation rates that can be achieved (circa 5000 s^{-1}). An interesting exception is the recently reported observation [3.26] of ^{1}H NMR spectra of protons in zeolites. In this case the protons are far enough apart so that MAS at 2500 s^{-1} could remove dipolar broadening sufficiently to allow resolution of two different types of protons in the NMR spectra.

An alternative approach to the removal of dipolar broadening is the use of multiple pulse sequences to achieve the equivalent of motional averaging of the anisotropic interactions in spin space rather than coordinate space. Two recent monographs describe this method in detail [3.27,28]. Sequences of 4, 8 or even more high intensity RF pulses are applied to the spin system between each measurement of the free induction decay signal voltage in the receiver coil. These have the effect of averaging the dipolar interactions to zero between each sampling of the free induction decay; thus the spectrum obtained by Fourier transformation of the free induction decay shows no dipolar broadening. The chemical shift anisotropy is reduced in magnitude, but not completely removed. The requirements of very short high intensity RF pulses and exact pulse timing have to date restricted the multiple pulse technique to a few specialised laboratories, but its general application appears promising.

Magic angle spinning, on the other hand, is now widely used in commercial spectrometers to suppress the chemical shift anisotropy in spectra of magnetically dilute nuclei, particularly ^{13}C, where dipolar broadening is minimal. The chemical shift anisotropy can be averaged to zero by MAS in the same way as dipolar broadening. Although potentially useful information is lost, the resulting solution-like spectra can be more readily interpreted.

3.3 Applications of EPR Spectroscopy

3.3.1 Identification of Adsorbed Radicals

Over 400 papers have been published on the EPR spectra of radicals adsorbed on solid surface since 1972, as described in earlier reviews [3.1-5]. The particular topic of oxygen containing radicals has been reviewed by *Lunsford* [3.29] and more recently by *Che* and *Tench* [3.30].

Spectra of radicals adsorbed on polycrystalline surfaces are often poorly resolved, for the reasons discussed above. Complete and unambiguous identification of an adsorbed radical can become a difficult problem. Measurement of the g-tensor components is not sufficient for identification, since the surface environment may alter the g tensor significantly from that of a known radical in the gas phase or the solid state.

The procedure that must be followed to identify an unknown surface species is illustrated by a recent study of radicals produced photochemically on silica-supported MoO_3 catalysts [3.31]. Irradiation of these catalysts with ultraviolet light in the presence of H_2 at 77 K or below causes photoreduction of Mo^{+6} to Mo^{+5}. When the irradiation was carried out in the presence of H_2 and O_2 the EPR spectrum shown in Fig.3.1a was obtained. This complex spectrum consists of 2 overlapping signals. The central components are the 3 features (turning points) associated with the principal g-tensor components of a known species, the O_2^- radical. This species can be formed independently by adsorbing O_2 on a thermally reduced MoO_2-SiO_2 catalyst. The remaining features in the spectrum are due to a second signal which had not been observed before. Figure 3.1b shows the corresponding spectrum obtained on irradiation in D_2 and O_2. The O_2^- signal is unchanged, but the changes in the second signal show clearly that it contains hyperfine splitting from a single ^1H or ^2H nucleus. The powder spectrum of a radical containing a single ^1H nucleus should show 3 sets of doublets, 1 set associated with each principal tensor component, since the proton has a nuclear spin of ½. The ^2H nucleus has a spin of 1, and a nuclear moment about 6 times smaller than that of the proton. The powder spectrum of the deuterated radical should thus show 3 sets of triplets, with about one sixth of the splitting observed with H_2. The observed spectra are entirely consistent with this analysis. Furthermore, the spin Hamiltonian parameters estimated from the observed spectra produced satisfactory computer simulations of the observed spectra, as shown in Fig.3.2.

A crucial experiment in the identification of the unknown radical in Figs.3.1,2 was determination of the number of oxygen atoms it contains. The

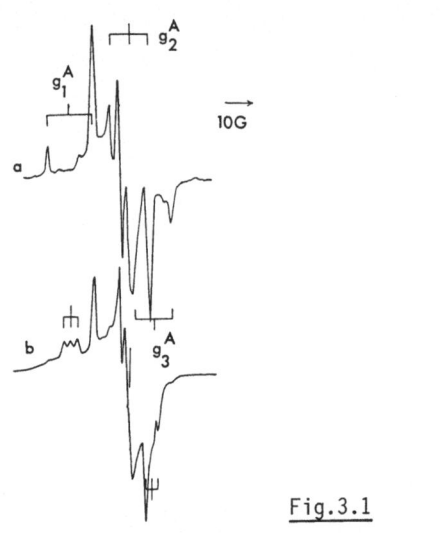

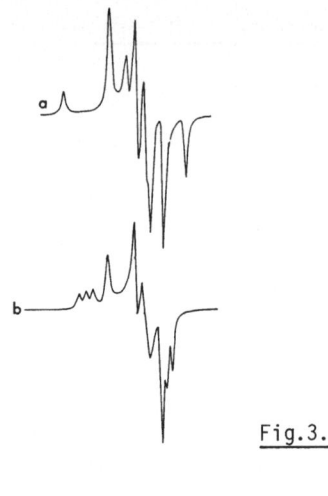

Fig.3.2

Fig.3.1

Fig.3.1a,b. EPR spectra obtained by irradiation of MoO_3-SiO_2 catalyst at 77 K with light of wavelength 340 nm in: a) $H_2 + O_2$; b) $D_2 + O_2$

Fig.3.2a,b. Computer simulations of the spectra in Fig.3.1

first assignment made (on the basis of the ^{1}H hyperfine splitting) was to the hydroxyl radical, HO [3.32]. Natural abundance $^{16}O_2$ has no nuclear spin, but can be enriched in the odd isotope ^{17}O, which has a nuclear spin of 5/2. A radical containing a single ^{17}O nucleus (such as HO·) would then give in principle 3 sets of 6 ^{17}O hyperfine lines in the powder spectrum. Radicals containing 2 or more ^{17}O nuclei would give more complex hyperfine patterns. The situation simplifies somewhat in practice, since oxygen radicals in which the unpaired electron is localised in a p or π orbital show a measurable ^{17}O hyperfine interaction along one principal axis only.

Figure 3.3 shows the spectrum obtained when the MoO_3-SiO_2 catalyst was irradiated in H_2 and ^{17}O enriched O_2. The spectrum was analysed by first identifying all the ^{17}O hyperfine components of the known radical O_2^-. Since the ^{17}O enrichment was not 100%, these are a set of 6 lines due to the singly labelled radical ($^{17}O^{16}O^-$), and a set of 11 lines due to the doubly labelled radical in which the oxygen ncluei are equivalent ($^{17}O^{17}O$). The remaining features, due to the second signal, consist of 2 different sets of 6 lines centred on the same point, each set split further into doublets by the ^{1}H hyperfine interaction, plus a set of 36 lines due to the doubly labelled species. Not all of the lines are resolved, but it is clear that this species contains two oxygen atoms which are not magnetically equivalent; the hyperfine splitting for one oxygen is about twice that for the other.

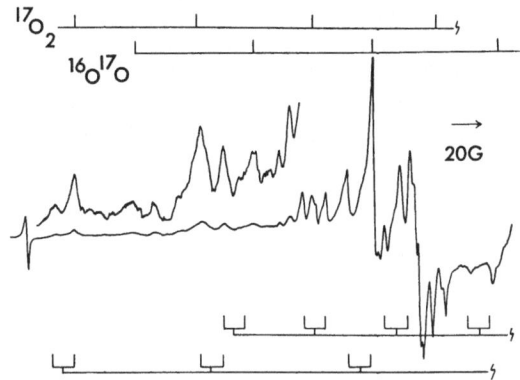

Fig.3.3. EPR spectrum obtained by irradiation of MoO_3-SiO_2 catalyst at 77 K with light of wavelength 340 nm in H_2 + ^{17}O enriched O_2

Table 3.1. EPR parameters of peroxy radicals

Radical	g-tensor components			Hyperfine tensor components (Gauss)			
HO_2 on MoO_3-SiO_2[a]	2.022	2.0078	2.0024	(^1H)	18	7	14
				(^{17}O)	0	0	54
					0	0	100
HO_2 (gas phase)[b]	2.042	2.0079	2.0015	(^1H)	12.8	2.8	13.9
HO_2 in ice[c]	2.035	2.0086	2.0042	(^1H)	12	4	12
HO_2 in $BaCl_2$[d]	2.020	2.007	2.007	(^1H)	16	8	8
RO_2 in PTFE[e]	2.038	2.0065	2.0020	(^{17}O)	0	0	46
					0	0	107

[a][3.31]; [b][3.33]; [c][3.34]; [d][3.35]; [e][3.36]

The second radical cannot therefore be HO·; it must be in fact be the hydroperoxyl radical HO_2·. Table 3.1 compares the spin Hamiltonian parameters of this radical with those of known peroxy radicals. The g tensor of HO_2· varies significantly with its environment, but the ^1H and ^{17}O hyperfine data allow unequivocal identification of the radical. Soria et al. [3.37] observed an EPR signal from irradiated TiO_2 surfaces which was attributed to HO_2·, but no ^1H or ^{17}O hyperfine splittings were reported. The MoO_3-SiO_2 catalyst remains the first surface on which this radical, known to be important in solution photochemistry, has been identified with certainty.

3.3.2 Transition Metal Ions on Surfaces

Many transition metal ions contain unpaired electrons, and the EPR spectrum
of a paramagnetic transition metal ion is strongly dependent on the environment of the ion and the particular ligands coordinated to it. The spectrum
of a paramagnetic metal ion on an oxide surface should therefore be a sensitive probe of surface chemistry occurring in the vicinity of the metal ion.
Many EPR studies of such supported transition metal ions have been reported;
there are, however, two major problems that limit the information obtainable.
The first is the existence of magnetic interactions between metal ions on a
surface. Depending on the distance between the metal ions, these may broaden
the EPR spectrum (typically at distances of 5-10 Å) or in extreme cases
(<4 Å) render the spectrum unobservable.

The EPR spectra of supported MoO_3 catalysts illustrate this problem.
Thermal reduction in hydrogen of silica- or alumina-supported MoO_3 produces
Mo^{+5} ions which have a single d electron and should be readily observed by
EPR. *Hall* and co-workers [3.38] first showed that the amount of Mo^{+5} in reduced MoO_3-Al_2O_3 catalysts detected by EPR spectroscopy is considerably less
than the amount measured by XPS or chemical methods. The discrepancy was attributed to magnetic coupling between Mo^{+5} ions in close proximity, render-
ing them invisible to EPR, i.e., the EPR measurements detected only those
Mo^{+5} ions which are magnetically isolated [3.39].

The coupling between Mo^{+5} ions on both alumina [3.40] and silica [3.41]
supports can be reduced in magnitude by treating the catalysts with gaseous
HCl. This is illustrated for silica supported Mo^{+5} in Fig.3.4. Exposure of
the reduced catalyst to HCl caused an approximately 20-fold increase in the
integrated intensity of the Mo^{+5} EPR signal which was completely reversed on
subsequent outgassing. The changes in the **g** and ^{95}Mo hyperfine tensors indicated that chloride ions replace oxide ions in the coordination sphere of
Mo^{+5}, and the uncoupling of magnetic interactions between adjacent Mo^{+5}
ions is attributed to replacement of bridging oxide ions:

$$Mo \overset{\diagup O \diagdown}{\quad} Mo + 2HCl \rightleftharpoons Mo \overset{\diagup Cl \quad Cl \diagdown}{\quad} Mo + H_2O \ . \qquad (3.5)$$

The second problem associated with transitional metal ions on oxide surfaces is that the surface environments are neither well defined nor uniform.
Thus the EPR spectrum, which may already be poorly resolved due to the macroscopic effect of random orientation of crystallites, is further broadened
due to microscopic effects of varying local environments. Subtle changes in

48

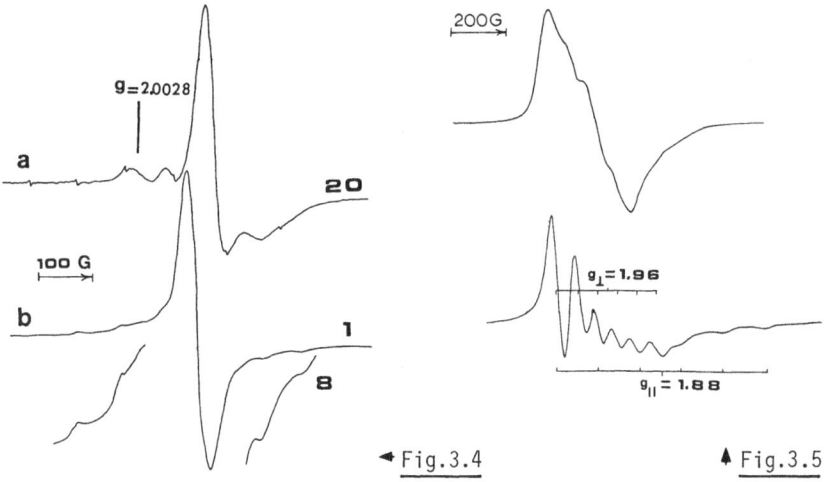

g=2.0028

Fig.3.4a,b. EPR spectra of a reduced MoO$_3$-SiO$_2$ catalyst before and after exposure to HCl (gas)

Fig.3.5. EPR spectra of ^{95}Mo^{+5} ions in: (a) reduced MoO$_3$-Al$_2$O$_3$ catalyst; (b) MoY zeolite

a spectrum due to adsorption or desorption of molecules on the surface may then be difficult to detect.

One successful approach to this problem has been to locate transition metal ions in the more uniform environments of molecular sieve zeolites. A zeolite structure may be considered as a well-defined three-dimensional surface within which metal ions can be located and small molecules adsorbed. The EPR spectra of transition metal ions in zeolites show better resolution than those of the same ions on oxide surfaces, as illustrated in Fig.3.5. This compares spectra of Mo^{+5} ions in a ^{95}Mo-enriched MoO$_3$-Al$_2$O$_3$ catalyst [3.42], and in zeolite Y [3.43]. Only in the zeolite environment are the 2 overlapping sets of 6 hyperfine lines due to the axially symmetric ^{95}Mo hyperfine tensor clearly resolved. Intrazeolite transition metal complexes may be regarded as models for the transition metal sites on "real" oxide catalysts, as first pointed out by *Lunsford* [3.44]. Depending on the zeolite structure, interactions between metal ions may also be eliminated in this way.

3.3.3 Mobility of Adsorbed Radicals

The effects of molecular motion on EPR spectra were referred to briefly above. Motion on the time scale between 10^{-7} and 10^{-10} s will cause variations in the normal EPR signal shapes due to modulation of the anisotropic terms in the spin Hamiltonian. This can be used in favourable cases to study

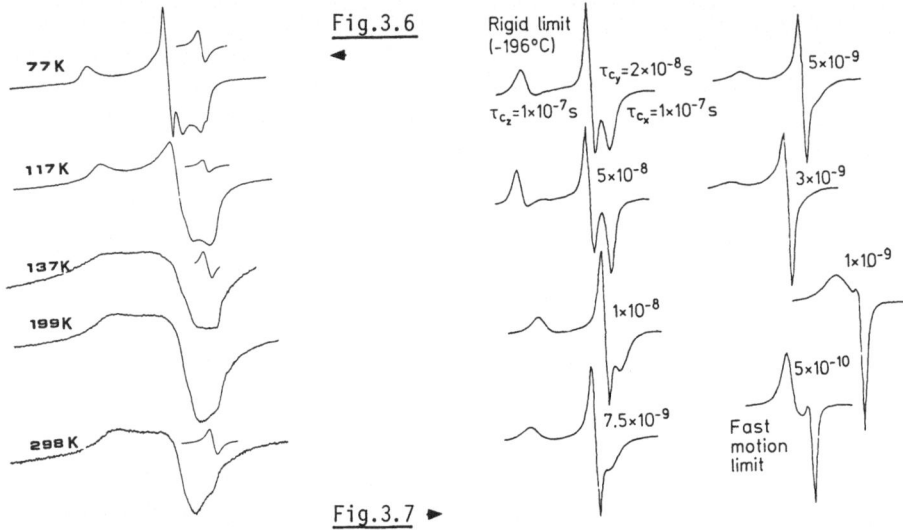

Fig.3.6

Fig.3.7 ►

Fig.3.6. EPR spectra of O_2^- on silica gel as a function of temperature

Fig.3.7. Computer-simulated spectra of O_2^- undergoing rotational diffusion about an axis perpendicular to the surface

the motion of species adsorbed on solid surface. For example, Fig.3.6 shows spectra of the O_2^- radical on a silica gel surface as a function of tempera-ture between 77 K and 298 K [3.45]. At 77 K, the powder spectrum shows the 3 features expected for an orthorhombic g tensor. On raising the temperature, the spectra broaden, and the low field turning point (associated with the largest g-tensor component) shifts to higher field. These changes, which were completely reversible, and the corresponding changes in the [17]O hyperfine pattern (not shown), indicate that partial averaging of the high and low field g-tensor components occurs due to the onset of rotational diffusion about an axis perpendicular to the surface. In the limit of rapid motion about that axis the g tensor would become $g_\perp = \frac{1}{2}(g_{zz} + g_{xx})$, and $g_\parallel = g_{yy}$.

Quantitative information about the rate of reorientation of the O_2^- can be obtained from computer simulation of the observed spectra. Figure 3.7 shows line shapes computed for O_2^- on silica gel as a function of the corre-lation time for Brownian rotational diffusion about an axis perpendicular to the surface (using the program of *Dalton* and *Robinson* [3.46]. The calculated spectra correctly reproduce the observed averaging of the g-tensor components, although not the broadening of the central component (which may be due to variations in the spin-spin relaxation time T_2). Estimates of correlation times can then be obtained by matching observed and simulated spectra. Very

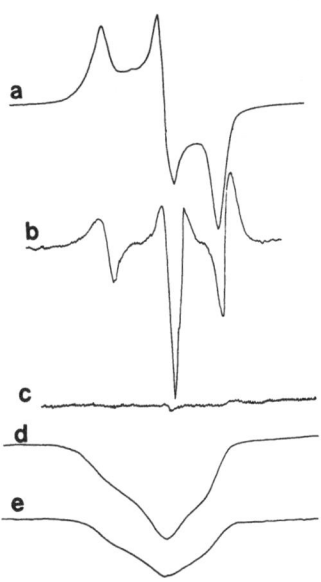

Fig.3.8a-e. EPR signals of O_2^- on MoO_3-SiO_2.
a) absorption in phase with first harmonic of
field modulation; b) absorption in phase with
second harmonic of field modulation; c) out of
phase with second harmonic of field modulation
at low microwave power; d) out of phase with
second harmonic of field modulation at high
power, 77 K; e) 298 K

similar results have been reported for O_2^- on silica-supported tungsten [3.47]
and titanium [3.48] surfaces. It appears likely that the mobile O_2^- species
is in fact located on the silica support in all cases.

The O_2^- radical on other oxide surfaces does not show evidence of mobility
on the EPR time scale. Evidence for surface motion on a longer time scale
can be obtained however from Saturation Transfer EPR measurements (STEPR).
The STEPR technique was developed to study the motion of spin labels attached
to large biomolecules and membranes [3.49], but can also be applied to simple
radicals adsorbed on solid surfaces [3.50]. The method involves measuring
absorption or dispersion signals out of phase with the magnetic field modulation, under conditions of saturation (high microwave power). Under non-sa-
turated conditions, when the populations of the spin states remain close to
their thermal equilibrium values, no signals will be detected out of phase
with the field modulation. When the spin system is saturated however, it can
no longer keep up with the field modulation (provided $\omega^{-1} \leq T_1$, where ω is
the modulation frequency and T_1 the spin lattice relaxation time), and out
of phase signals will be detected. This is illustrated in Fig.3.8, for O_2^-
on silica-supported MoO_3 [3.51]. Any reorientation of the radical on the
time scale of the magnetic field modulation ($\omega^{-1} \sim 10^{-5}$ s) will remove saturation from the spin system, since at its new orientation the radical will
no longer satisfy the resonance condition. The effect of motion on the time
scale of ω^{-1} is thus to remove intensity from the out-of-phase signals be-

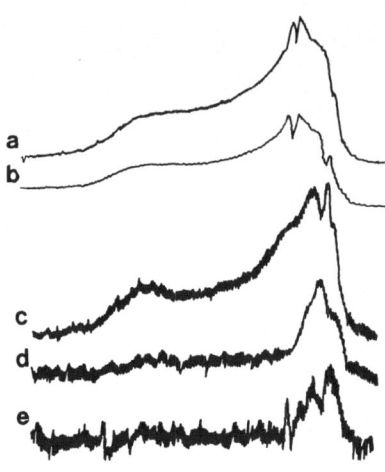

Fig.3.9. STEPR spectra (absorption out of phase with second harmonic of field modulation) of O_2^- on ZnO at temperatures between 77 K (a) and 298 K (e)

a

b

c

d

e

tween the turning points (those features corresponding to principal tensor components).

For O_2^- on MoO_3-SiO_2, no effects of motion are seen on the out-of-phase signals between 77 K and 298 K; thus on this surface O_2^- is immobile on the STEPR time scale. Figure 3.9 shows out-of-phase (STEPR) signals for O_2^- on ZnO between 77 K and 298 K. In this case, the effects of motion are seen on raising the temperature, as a loss of intensity in the central region of the spectrum.

The STEPR spectra of spin labels have been calibrated in terms of correlation times by examining the spin labels in solutions of varying viscosity [3.49]. Such calibrations are not possible for adsorbed O_2^-, but orders of magnitude for correlation times can be estimated from comparison between observed and simulated spectra. Table 3.2 summarises the available data on correlation times for rotational diffusion of O_2^- on different surfaces. The combination of EPR and STEPR covers a magnitude range of six orders in correlation times, although a high degree of accuracy is not possible. Rotational correlation times can in principle be obtained much more precisely from pulsed EPR measurements, but no such measurements on surface species have yet been reported.

3.3.4 Pulsed EPR Experiments

Time domain (pulsed) EPR spectroscopy has to date been confined to a few specialised laboratories able to construct the rapid response pulsed spectrometers required [3.52]; no commercial pulsed EPR instruments are presently available. One of the time domain methods which has been successfully applied

Table 3.2. Rotational correlation times for O_2^- on various surfaces

Surface	t_c (s)	
	77 K	298 K
SiO_2[a,b]	2×10^{-8}	2×10^{-9}
Ti/vycor[a,c]	5×10^{-8}	$<6 \times 10^{-9}$
Ag/vycor[d,e]	$>10^{-4}$	10^{-6}
ZnO[d,f]	10^{-3}	$\leq 10^{-5}$
MoO_3-SiO_2[d,f]	$\geq 10^{-3}$	$\geq 10^{-3}$
V_2O_5-SiO_2[d,f]	$\geq 10^{-3}$	$\geq 10^{-3}$

[a]EPR measurement; [b][3.45]; [c][3.48]; [d]STEPR measurement; [e][3.50]; [f][3.51]

to surfaces is the electron spin echo (ESE) technique. The ESE experiment is best described in the rotating frame, i.e., a coordinate system rotating with the precessing electron spin magnetisation vector about the static magnetic field direction. An initial pulse of high intensity microwave radiation (typically of 50 ns duration) rotates the magnetisation vector through 90° in the rotating frame, into a plane perpendicular to the field direction. Once the pulse is turned off the magnetisation vectors of the individual spins lose their coherence and precess apart. A second pulse of twice the duration of the first is then applied after a time interval τ, which refocuses the magnetisation vectors back into phase at a time τ following the second pulse, producing the so-called echo signal. As the time interval τ is increased, the amplitude of the echo signal decays with a time constant governed by the various relaxation processes that are possible in the spin system. However, the particular value of the ESE experiment is that a plot of echo amplitude versus delay time τ often displays periodic modulation due to hyperfine interactions of the unpaired electron with nearby nuclear spins. A nuclear spin in the vicinity of the unpaired electron will experience a magnetic field which results from the external field and the dipolar field from the unpaired electron. The nuclear spin will precess about this field, producing a modulation of the dipolar field at the electron due to the nuclear spin, which is observed in the spin echo display. The frequency of the modulation is determined by the magnitude of the dipolar coupling interaction between the electron and the nuclear spin.

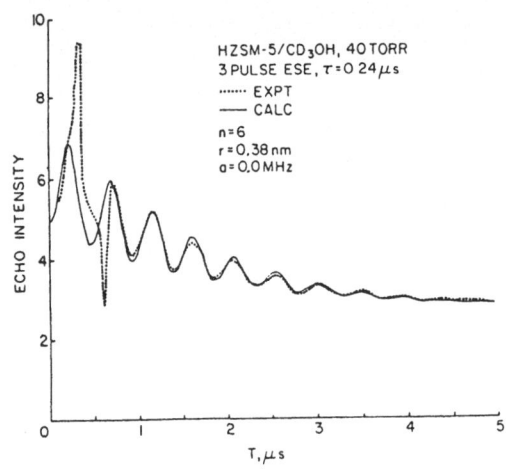

Fig.3.10. Comparisons of experimental (....) and calculated (——) 3-pulse ESE deuterium modulation for CD2OH in HZSM5 with 40 Torr of CD_3OH [3.53]

The ESE experiment can thus be used to measure hyperfine interactions which are too small to be resolved in the normal EPR spectrum. The largest hyperfine interaction that can be seen in ESE is that corresponding to a modulation period longer than the microwave pulse length, which can be experimentally varied. *Kevan* et al. [3.53] used ESE to examine the geometry of adsorbates on zeolites and other surfaces, e.g., their recent study of methanol in the zeolite HZSM5, where hydroxymethyl radicals (CH_2OH) were generated by γ irradiation of adsorbed methanol, and modulation of the ESE signals by nuclei in surrounding methanol molecules was measured.

Fourier transformation of the ESE decay function would give in principle the dipolar coupling constants directly. However, for instrumental reasons a complete decay function from $\tau = 0$ cannot be obtained, and Fourier transformation of a truncated decay function is fraught with difficulty. The alternative approach adopted by *Kevan* et al. [3.53] is to computer simulate the observed decay function, using the number and type of nuclei and their distances from the unpaired electron as adjustable parameters.

Figure 3.10 shows the observed and simulated ESE decay curves for the radical CD_2OH in HZSM5 exposed to 40 Torr of CD_3OH. In this case a 3-pulse sequence rather than the 2-pulse sequence described above was used to generate the echo, and the pulse lengths were chosen such that only deuteron modulation is present. The observed modulation pattern could be fitted to a model of 6 deuterium atoms at an average distance of 0.38 nm from the unpaired electron, which is located on the carbon atom of the CD_2OH radical (the measurements are made at 4.2 K, so that the adsorbed methanol molecules are frozen in position). The number of deuterium atoms at this distance increased

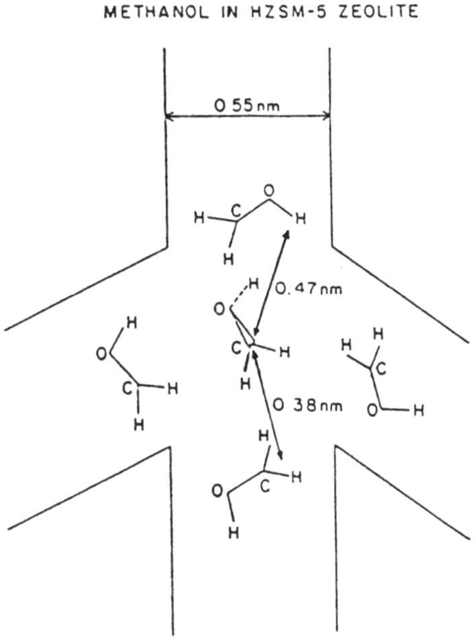

Fig.3.11. Model for the arrangement of methanol molecules in HZSM5 [3.53]

to 12 at a methanol pressure of 100 Torr, and fell to 3 at a pressure of 10 Torr. When the same experiment was repeated with CH_3OD, the echo modulation pattern could be fitted with 2 deuterium atoms at an average distance of 0.47 nm.

Figure 3.11 shows the structural model proposed by *Kevan* et al. to account for the ESE results. The hydroxymethyl radical is located at the zeolite channel intersections; the number of surrounding methanol molecules then depends on the pressure, with a maximum of 4 methyl groups at an average distance of 0.38 nm, and 2 hydroxyl groups at 0.47 nm.

The ESE experiment remains at the moment difficult to perform, but demonstrates potential as a surface technique capable of detecting longer range interactions than conventional EPR spectroscopy.

3.3.5 Well-Defined Surfaces

Most EPR experiments to date have been carried out on high surface area powders under conditions far removed from the single-crystal ultrahigh vacuum experiments of modern surface science. Mention has already been made of the use of zeolites as well-defined 3-dimensional surfaces in magnetic resonance experiments, but well-defined 2-dimensional surfaces have so far received little attention.

Several studies have been reported of single crystals cleaved under ultra-high vacuum to produce clean powders. For example, single crystals of silicon and germanium, when crushed under vacuum, give EPR signals which were originally attributed to "dangling bonds", or unpaired electrons in localised surface states characteristic of the clean surfaces [3.54]. A more recent investigation by *Lemke* and *Haneman* [3.55] established, however, that for silicon at least, the paramagnetic centres are associated with microcracks produced in the vacuum crushing or cleavage process, and cannot therefore be said to be characteristic of well-defined single-crystal surfaces.

Single crystals of metals cannot be placed in the sample cavity of an EPR spectrometer. An interesting approach to the problem of obtaining EPR spectra from clean metal surfaces has been described by *Freed* et al. [3.56]. An ultra-high vacuum microwave cavity was constructed from titanium and connected to a conventional pumping system. Clean metal films were then evaporated in vacuo onto the cavity walls, and EPR spectra measured of molecules subsequently adsorbed. The sensitivity (minimum detectable number of spins) of the UHV-EPR system was estimated to be 1×10^{10} cm^{-2} for a 1 Gauss wide signal.

The adsorption of paramagnetic molecules such as NO_2 [3.56] and di-t-butyl nitroxide (DTBN [3.57] on clean evaporated films of Cu or Ag gave no EPR signals. The authors attributed this either to coupling of the unpaired electron with conduction electrons in the metal (DTNB) or to chemisorption resulting in diamagnetic species (NO_2). Spectra of adsorbed DTNB were observed from oxidised surfaces, demonstrating the sensitivity of the technique.

An unexpected observation in these experiments was the appearance of an intense electron cyclotron resonance signal at g = 2 when the gas pressure in the cavity was held below 10^{-2} Torr. This phenomenon has been dubbed CREMSEE, or cyclotron resonance from microwave-induced secondary electron emission. Secondary electrons emitted from a clean metal surface above a certain threshold microwave power are induced into circular orbits in the external magnetic field. When the orbital frequency matches the microwave frequency, resonant absorption occurs. A feature of electron cyclotron resonance is its extremely high susceptibility, about 10^{13} larger than that for electron spin resonance. Thus a very small steady-state concentration of free electrons in the microwave cavity (typically 10^3 cm^{-3}) can give an intense signal.

A formal theory for the CREMSEE effect has not yet been developed, but the secondary electron emission and hence the intensity of the CREMSEE signal will clearly be sensitive to the state of cleanliness of the metal surface and the nature of any adsorbed species. The use of CREMSEE to measure the work func-

tion of oxidised silver films deposited in the UHV cavity has recently been
described [3.58].

The coupling of unpaired electrons in adsorbed radicals with conduction
electrons in clean metal surfaces will restrict possible applications of the
UHV-EPR technique developed by Freed, despite the novelty of the CREMSEE
phenomenon. However, the sensitivity of the EPR technique demonstrated in
this work for submonolayer ($\theta \sim 0.01$) quantities of adsorbed radicals should
be applicable to semiconductor and insulator surfaces. There appears to be
no reason why such EPR studies cannot be undertaken for paramagnetic species
at the surfaces of evaporated films or single crystals of semiconductors or
insulating oxides. Combination of EPR spectroscopy with other surface-science
techniques then becomes a viable proposition.

3.4 Applications of NMR Spectroscopy

3.4.1 ^{29}Si NMR of Zeolites

The high internal surface area of zeolites has made them attractive targets
for NMR spectroscopy. *Lechert* [3.59] has reviewed NMR studies of structure
and sorption problems in faujasite-type zeolites prior to 1976. The advent
of high-resolution solid-state NMR has revived interest in NMR spectroscopy
of zeolites, particularly the ^{29}Si NMR of the zeolite aluminosilicate lattice.

The ^{29}Si nucleus has a natural abundance of 4.7%, so that homonuclear dipolar interactions between ^{29}Si nuclei are negligible. Other magnetic nuclei
present in zeolites are ^{27}Al, and ^{1}H in adsorbed water or hydroxyl groups.
Dipolar coupling with ^{27}Al nuclei does not appear to broaden ^{29}Si signals
significantly, while the rapid motion of adsorbed water within zeolite pores
averages any dipolar coupling between water protons and ^{29}Si to zero. Only
in the case of hydroxyl groups bound to silicon is a significant heteronuclear dipolar coupling present, and this can be removed by proton decoupling
and cross polarisation. Magic angle spinning removes ^{29}Si chemical shift
antisotropy, allowing silicon atoms in different sites within the zeolite
lattice to be resolved. The isotropic chemical shift for ^{29}Si in solid alu-
minosilicates varies with the number of aluminium atoms connected to the silicon via bridging oxide ions. Five distinct chemical shift ranges have been
identified which depend on the number of AlO_4 tetrahedra connected to an
SiO_4 tetrahedron [3.60]. The ^{29}Si NMR spectrum can thus provide direct information about the distribution of silicon and aluminium atoms amongst the
zeolite lattice sites, and such information has been obtained for many different zeolite structures [3.60-63].

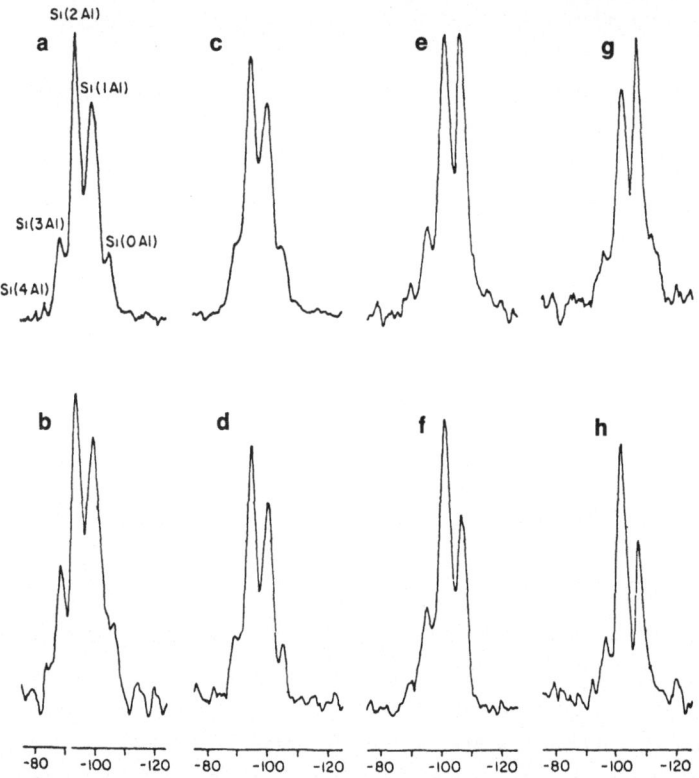

Si(2 Al)

Si(1 Al)

Si(3 Al)

Si(0 Al)

Si(4 Al)

a   c   e   g

b   d   f   h

-80 -100 -120   -80 -100 -120   -80 -100 -120   -80 -100 -120

Fig.3.12a-h. ^{29}Si NMR spectra of: a,b) NaY; c,d) NH$_4$NaY; e,f) steamed NH$_4$NaY; g,h) acid washed. Upper spectra were recorded without and lower spectra with ^{1}H cross polarisation [3.64]

Figure 3.12 shows ^{29}Si NMR spectra reported recently by *Engelhardt* et al. [3.64] in a study of the dealumination of zeolite Y, the processes by which aluminium is removed from the zeolite lattice. The initial spectrum of the sodium ion exchanged zeolite (a) shows 4 distinct signals due to silicon atoms surrounded by 0, 1, 2 and 3 aluminium atoms. No change in the spectrum occurred when proton cross polarisation was employed (b), indicating that none of the silicon atoms contained hydroxyl groups. Exchange of 50% of the sodium ions with NH$_4^+$ caused no change (c,d), but subsequent heating in water vapour at 540°C modified the spectrum (e,f). The hydrothermal treatment increases the amount of silicon connected to 0 and 1 aluminium atoms, and decreases the amount of silicon having 2 and 3 aluminium neighbours. There is now also a significant difference between spectra recorded with and without cross polarisation. The signal at -100 ppm which is enhanced by proton cross polarisation is attributed to SiOH groups. The intensity changes indicate that the zeolite has lost about 12 aluminium atoms per unit cell from the lattice, with new SiOH groups being formed at the resulting lattice defects. Subsequent treatment with dilute hydrochloric acid caused further loss of aluminium (g,h).

These studies have demonstrated the usefulness of ^{29}Si NMR for studying zeolite structures. The obvious extension to examine the interaction of adsorbed molecules with the zeolite lattice, possibly under catalytic conditions, has been inhibited by the experimental difficulty of magic angle spinning of samples in a controlled atmosphere at elevated temperatures.

3.4.2 ^{13}C NMR of Adsorbed Molecules

The chemisorption of CO on metal clusters is a topic of considerable current interest. ^{13}C NMR spectroscopy has been successfully applied to this problem by *Duncan* et al. [3.65,66] in a study of CO chemisorbed on alumina-supported rhodium catalysts. This work did not, strictly speaking, involve high-resolution NMR, since magic angle spinning was not employed. The spectra are thus broadened due to chemical shift anisotropy. Homonuclear dipolar broadening was less in magnitude than the chemical shift anisotropy at the ^{13}CO concentrations used, while any heteronuclear dipolar broadening from ^{1}H, ^{27}Al or ^{163}Rh was negligible. Figure 3.13 shows the NMR spectrum of the complex $Rh_2Cl_2(CO)_4$, which may be regarded as a model for chemisorbed CO. The chemical shift tensor is almost axially symmetric, and the absorption spectrum of the powdered sample is the expected envelope of spectra from all orientations.

The signal shapes observed for CO chemisorbed on rhodium were more complex than that of the model, as shown in Fig.3.14. These spectra contain 2 overlapping signals which could be separated from the difference in their spin-lattice relaxation times. In a 2-pulse NMR experiment, a 180° pulse followed after a time τ by a 90° pulse, the measured signal (free induction decay after the second pulse) will contain contributions only from those spins having $T_1 \leq \tau$. Figure 3.15 shows spectra measured as a function of τ. The authors thus separate signals having average chemical shifts $[\sigma_{av} = \frac{1}{3}(\sigma_{xx} + \sigma_{yy} + \sigma_{zz})]$ of -177 and -199 ppm, and T_1 values of 5.6 ms and 64 ms, respectively.

Infrared studies of the adsorption of CO have established 3 different states of CO on dispersed Rh:

```
   O   O          O              O
    \ /           ‖              ‖
     C  C         C              C
    \ /           ‖            /   \
    Rh          — RH —      — Rh — Rh —
   (I)           (II)          (III)
```

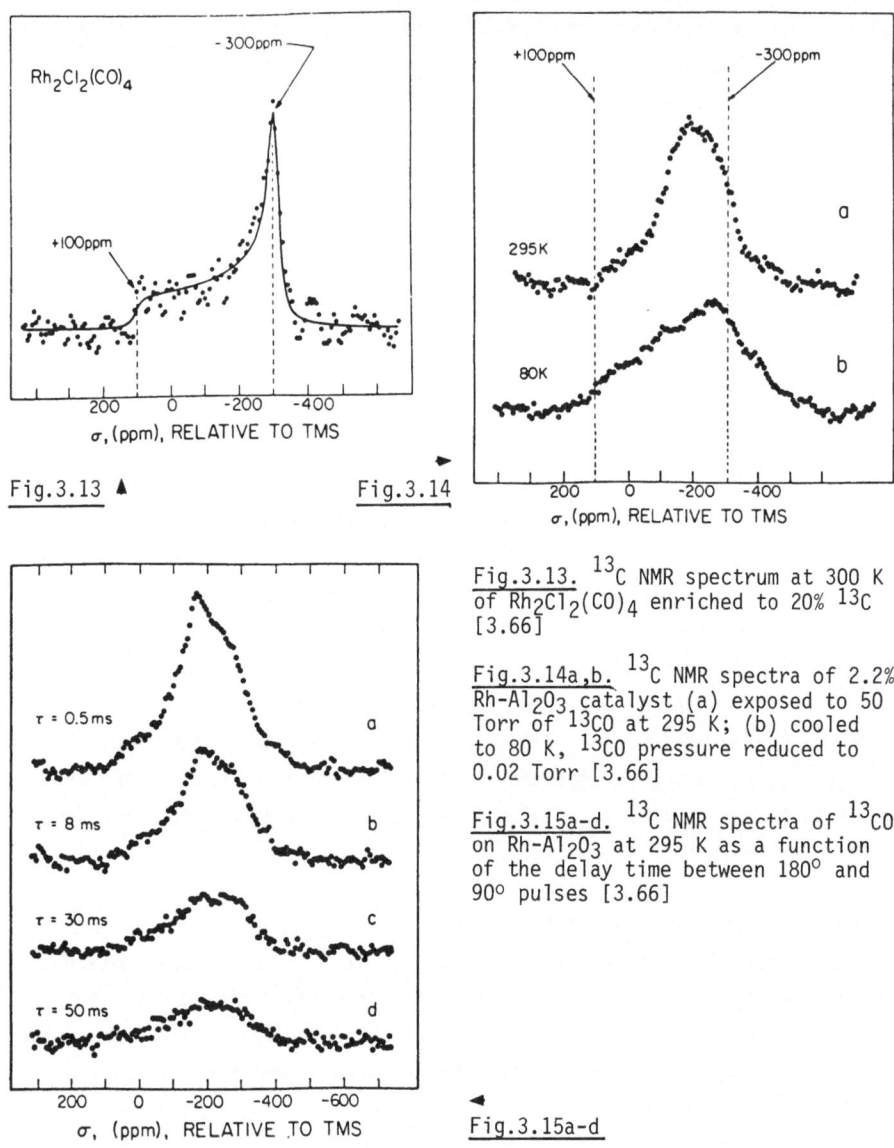

Fig.3.13. ^{13}C NMR spectrum at 300 K of $Rh_2Cl_2(CO)_4$ enriched to 20% ^{13}C [3.66]

Fig.3.14a,b. ^{13}C NMR spectra of 2.2% $Rh-Al_2O_3$ catalyst (a) exposed to 50 Torr of ^{13}CO at 295 K; (b) cooled to 80 K, ^{13}CO pressure reduced to 0.02 Torr [3.66]

Fig.3.15a-d. ^{13}C NMR spectra of ^{13}CO on $Rh-Al_2O_3$ at 295 K as a function of the delay time between 180° and 90° pulses [3.66]

Fig.3.13 ▲   Fig.3.14 ➤

Fig.3.15a-d ◄

with species (II) and (III) being formed on small rafts or clusters of rhodium [3.67,68] and species (I) on isolated rhodium atoms [3.67] or at the edges of rafts [3.68]. Species (II) and (III) are expected to exchange rapidly on the NMR time scale, and should thus have the same ^{13}C spin-lattice relaxation time. *Duncan* et al. thus attribute the rapidly relaxing signal ($\sigma = -177$, $T_1 = 5.6$ ms) to species I on isolated Rh atoms (the rapid relaxation caused by paramagnetic impurities in the alumina support), and the slowly

60

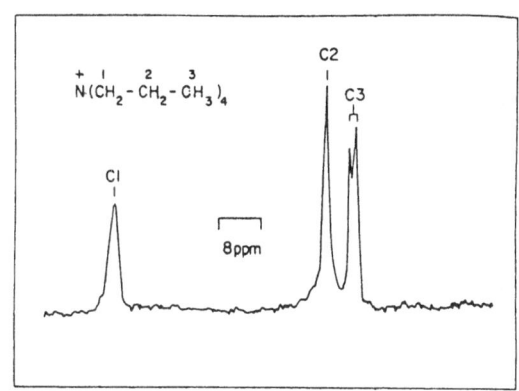

Fig.3.16. ^{13}C NMR spectrum (^{1}H cross polarisation, MAS) of TPA in HZSM5 [3.69]

relaxing signal to an average of species II and III. The chemical shift of the second signal (σ_{av} = -199 ppm) lies between that expected from model compounds for linearly bonded (-184 ppm) and bridge bonded (-228 ppm) carbonyl ligands.

From the ^{13}C NMR data it is possible to determine the amounts present of each of the 3 types of chemisorbed CO. The intensity of the rapidly relaxing signal gives the amount of species I; the intensity of the slowly relaxing signal gives the total of II and III, and the relative amounts of II and III are obtained from the average chemical shift value. In this way Duncan et al. were able to calibrate in turn the extinction coefficients of the infrared bands due to each species. This combination of NMR and infrared spectroscopy is a particularly interesting illustration of the potential of th NMR technique.

The resolution of ^{13}C NMR spectra from chemisorbed species can be dramatically improved by magic angle spinning. As a final example, Fig.3.16 shows a ^{13}C spectrum reported for the tetrapropylammonium cation (TPA) in the zeolite ZSM5 [3.69]. In this case the cation is incorporated into the zeolite during its synthesis, and the interest lies in determining its location. Proton decoupling removes the heteronuclear dipolar broadening, and magic angle spinning the chemical shift anisotropy. The resulting well-resolved spectrum shows the expected 3 different carbon signals for TPA. The CH$_3$ signal shows an additional splitting due to the existence of methyl groups in two different environments within the zeolite. *Nagy* et al. [3.69] suggest, from consideration of the known pore geometry in ZSM5, that cationic nitrogen of TPA must be located at the intersection of zeolite channels, with 2 propyl groups extending into the linear channels, and 2 into the zigzag channels (the channel structure of ZSM5 was represented schematically in Fig.3.11).

3.5 Concluding Remarks

This review has attempted to outline the present scope of magnetic resonance techniques in surface science, by reference to selected examples. By now EPR spectroscopy is well established as a technique for studying radicals and transition metal ions on high area surfaces. Although spectra are often broad and poorly resolved, appropriate use of isotopic substitution and computer simulation will usually permit unambiguous identification of the species observed. However, in situ studies of catalyst surfaces under high temperature conditions are not possible. The potential of EPR spectroscopy for studies of well-defined low area surfaces has now been demonstrated, and future progress in this area may be anticipated. Future development of commercial pulsed EPR spectrometers will stimulate further applications of the electron spin echo technique. Measurements of the surface mobility of adsorbed radicals should also be possible to a greater degree of accuracy with pulsed spectrometers than can be achieved at present with conventional EPR and STEPR.

The NMR spectroscopy cannot achieve the sensitivity of the EPR technique, and will remain restricted to high area surfaces. The development of pulsed spectrometers for high resolution measurements on solids has, however, increased the potential scope of the NMR technique enormously. Finally, the work of Duncan, Yates and Vaughan on chemisorbed carbon monoxide has demonstrated yet again the value of combining spectroscopic techniques when investigating any problem in surface science.

Acknowledgements. The work cited in this review from the author's research group at the University of Wisconsin-Milwaukee involved the following post-doctoral fellows and students: Dr. S. Abdo, Dr. A. Kazusaka, Dr. S. Seyed-monir, S. Balistreri, J. Lopata, D. Van Galen and W. Timmer. Support from the Laboratory for Surface Studies, the Petroleum Research Fund of the American Chemical Society and the Research Corporation is gratefully acknowledged.

References

3.1 J.H. Lunsford : In *Chemistry and Physics of Solid Surfaces*, Vol. I, ed. by R. Vanselow and S.Y. Tong (CRC Boca Raton 1976) p.255
3.2 P.H. Kasai, R.J. Bishop: ACS Monographs **171**, 350 (1976)
3.3 J.H. Lunsford: In *Spectroscopy in Heterogeneous Catalysis*, ed. by W.N. Delgass, G. Haller, R. Kellerman, J.H. Lunsford (Academic, London 1979) p.183
3.4 M. Che: In *Magnetic Resonance in Colloid and Interface Science*, ed. by J.P. Fraissard, H.A. Resing (NATO Advanced Study Institutes Ser.C.61, Riedel 1980) p.79
3.5 R.F. Howe: Adv. Colloid Interface Sci. **18**, 1 (1982)
3.6 H. Pfeiffer: Phys. Rep. **26**, 293 (1976)

3.7 J.H. Lunsford: See Ref. [3.3, p.326]
3.8 J. Tabony: Progr. Nucl. Magn. Spectros. Reson. **14**, 1 (1980)
3.9 T.M. Duncan, C. Dybowski: Surf. Sci. Rep. **1**, 157 (1981)
3.10 W. Derbyshire: Royal Society of Chemistry Specialist Periodical Report. Nucl. Mag. Res. **11**, 264 (1981)
3.11 H. Resing, C.G. Wade (eds.) *Magnetic Resonance in Colloid and Interface Science* (ACS Symp. Ser. **34**, 1976)
3.12 J.P. Fraissard, H.A. Resing (eds.): *Magnetic Resonance in Colloid and Interface Science* (NATO Advanced Study Institutes Ser.C.61, Riedel 1980)
3.13 J.E. Wertz, J.R. Bolton: *Electron Spin Resonance* (McGraw-Hill, New York 1972)
3.14 W. Gordy: *Theory and Applications of Electron Spin Resonance* (Wiley, New York 1980)
3.15 N. Atherton: *Electron Spin Resonance* (Ellis Horwood, 1973)
3.16 A. Carrington, A.D. McLachlan: *Introduction to Magnetic Resonance* (Harper and Row, New York 1967)
3.17 C.P. Slichter: *Principles of Magnetic Resonance*, 2nd ed. (Springer, Berlin, Heidelberg, New York 1978)
3.18 F.A. Rushworth, D.P. Turnstall: *Nuclear Magnetic Resonance* (Gordon and Breach, New York 1973)
3.19 T.C. Farrar, E.D. Becker: *Pulse and Fourier Transform NMR* (Academic, New York 1971)
3.20 For example, G.P. Lozos, B.M. Hoffman, C.G. Franz: Quantum Chemistry Program Exchange **265** (1973)
3.21 A.C. Cirillo, J.M. Dereppe, W.K. Hall: J. Catal. **61**, 170 (1980)
3.22 V.B. Kazansky, V.Y. Borovkov, G.M. Zhidomirov: J. Catal. **39**, 205 (1975)
3.23 R.W. Vaughan: Ann. Rev. Phys. Chem. **29**, 397 (1978)
3.24 C.S. Yannoni: Acc. Chem. Res. **15**, 201 (1982)
3.25 C.A. Fyfe: Ann. Rep. NMR Spectrosc. **12**, 1 (1982)
3.26 D. Freude, M. Hunger, H. Pfeiffer: Chem. Phys. Lett. **91**, 307 (1982)
3.27 U. Haeberlen: *High Resolution NMR in Solids. Selective Averaging* (Academic, New York 1976)
3.28 M. Mehring: *High Resolution NMR in Solids* (Springer, Berlin, Heidelberg, New York 1976)
3.29 J.H. Lunsford: Catal. Rev. **8**, 135 (1973)
3.30 M. Che, A.J. Tench: Adv. Catal. 31, 78 (1982)
3.31 S.R. Seyedmonir, R.F. Howe: J. Chem. Soc. Farad. Trans. I, in press
3.32 S. Balistreri, R.F. Howe: In *Magnetic Resonance in Colloid and Interface Science*, ed. by J.P. Fraissard, H.A. Resing (NATO Advanced Study Institutes Ser.C.61, Riedel 1980) p.489
3.33 C.E. Barnes, J.M. Brown, A. Carrington, J. Pinkstone, T.J. Sears: J. Molec. Spectrosc. **72**, 86 (1978)
3.34 S.J. Wyard, R.C. Smith, F.J. Adrian: J. Chem. Phys. **49**, 2780 (1968)
3.35 S. Radhakrishna, B.V.R. Chowdari, A. Kasi Viswanath: Chem. Phys. Lett. **40**, 134 (1976)
3.36 M. Che, A.J. Tench: J. Chem. Phys. **64**, 2370 (1976)
3.37 A.R. Gonzalez-Elipe, G. Munuera, J. Soria: J. Chem. Soc. Farad. Trans. I **74**, 748 (1978)
3.38 W.K. Hall, M. LoJacono: In *Proceedings of the Sixth International Congress on Catalysis*, ed. by G.C. Bond, P.B. Wells, F.C. Tompkins (Chemical Society, London 1977) p.246
3.39 S. Abdo, R.B. Clarkson, W.K. Hall: J. Phys. Chem. **80**, 2431 (1976)
3.40 S. Abdo, A. Kazusaka, R.F. Howe: J. Phys. Chem. **85**, 1380 (1981)
3.41 S.R. Seyedmonir, R.F. Howe: J. Chem. Soc. Farad. Trans. I **80**, 87 (1984)
3.42 S.R. Seyedmonir: Ph.D. Thesis (University of Wisconsin-Milwaukee, 1982)
3.43 S. Abdo, R.F. Howe: J. Phys. Chem. **87**, 1722 (1983)
3.44 J.H. Lunsford: Catal. Rev. **12**, 137 (1975)
3.45 W. Timmer: M.Sc. Thesis (University of Wisconsin-Milwaukee, 1982)
3.46 B.H. Robinson, L.R. Dalton: J. Chem. Phys. **72**, 1312 (1980)

3.47 A. Kazusaka, L.K. Yong, R.F. Howe: Chem. Phys. Lett. **57**, 592 (1978)
3.48 M. Shiotani, G. Moro, J.H. Freed: J. Chem. Phys. **74**, 2616 (1981)
3.49 J.S. Hyde, L.R. Dalton: In *Spin Labelling II*, ed. by L.J. Berliner (Academic, New York 1979) p.1
3.50 R.B. Clarkson, R.G. Kooser: Surf. Sci. **74**, 325 (1978)
3.51 D. Van Galen, S. Abdo, J. Lopata, R.F. Howe: To be published
3.52 L. Kevan, R.N. Schwartz (eds.): *Time Domain Electron Spin Resonance* (Wiley, London 1979)
3.53 C.S. Narashimhan, M. Narayana, L. Kevan: J. Phys. Chem. **87**, 984 (1983)
3.54 D. Haneman: Jap. J. Appl. Phys. Suppl. 2, **2**, 371 (1974)
3.55 B.P. Lemke, D. Haneman: Phys. Rev. B**17**, 1893 (1978)
3.56 M. Nilges, G. Barkley, M. Shiotani, J.H. Freed: J. de Chim Physique **78**, 909 (1981)
3.57 M. Nilges, J.H. Freed: Chem. Phys. Lett. **82**, 203 (1981)
3.58 M. Nilges, J.H. Freed: Chem. Phys. Lett. **85**, 499 (1982)
3.59 H. Lechert: Catal. Rev. Sci. Eng. **14**, 1 (1976)
3.60 E. Lipmaa, M. Magi, A. Samosan, G. Engelhardt, A.R. Grimmer: J. Am. Chem. Soc. **102**, 4889 (1980)
3.61 S. Ramdas, J.M. Thomas, J. Klinowski, C.A. Fyfe, J.S. Hartman: Nature **292**, 228 (1981)
3.62 J. Klinowski, S. Ramdas, J.M. Thomas, C.A. Fyfe, J.S. Hartman: J. Chem. Soc. Faraday Trans 2, **78**, 1025 (1982)
3.63 M.T. Melchior, D.E.W. Vaughan, A.J. Jacobson: J. Am. Chem. Soc. **104**, 4859 (1982)
3.64 G. Engelhardt, U. Lohse, A. Samosan, M. Magi, M. Tarmark, E. Lipmaa: Zeolites **2**, 59 (1982)
3.65 T.M. Duncan, J.T. Yates, R.W. Vaughan: J. Chem. Phys. **71**, 3129 (1979)
3.66 T.M. Duncan, J.T. Yates, R.W. Vaughan: J. Chem. Phys. **73**, 975 (1980)
3.67 J.T. Yates, T.M. Duncan, S.D. Worley, R.W. Vaughan: J. Chem. Phys. **70**, 1219 (1979)
3.68 D.J.C. Yates, L.L. Murrell, E.B. Prestridge: J. Catal. **57**, 41 (1979)
3.69 J.B. Nagy, Z. Gabelica, E.G. Derouane: Zeolites **3**, 43 (1983)

4. Mössbauer Spectroscopy: Applications to Surface and Catalytic Phenomena

B. J. Tatarchuk and J. A. Dumesic

With 14 Figures

4.1 Introduction

Since the observation of recoilless nuclear resonant absorption by Mössbauer in 1957, Mössbauer spectroscopy has received considerable attention. Because of a wealth of electronic, geometric, magnetic, and structural information provided by this technique, Mössbauer spectroscopy has been used as a powerful diagnostic tool in many physical and biological disciplines, as evidenced by a number of excellent reviews and texts in this area [4.1-12].

In the normally employed transmission geometry, specimen thicknesses between 10^{-1} and 10^{-3} mm are required so that for most purposes transmission Mössbauer spectroscopy has been regarded as a bulk characterization tool. More recently, however, methods for applying transmission Mössbauer spectroscopy to surface and catalytic phenomena have been devised and a number of recent reviews have been devoted to this topic [4.13-19]. The majority of these applications make use of highly dispersed Mössbauer isotopes (absorbers or emitters) in the form of small crystallites (circa <10 nm) or thin foils (circa <2.5 nm thick). Thus, while a "bulk" technique is employed, a significant fraction of the absorber/emitter atoms lies on or near the surface. In this manner surface and catalytic information can be obtained in situ using liquid/gaseous reaction conditions or ultrahigh vacuum environments [4.13-19]. Although these capabilities are most important for applied efforts, it is clear that transmission Mössbauer spectroscopy is not an inherently surface-sensitive technique, but requires careful sample preparation and a knowledge of the surface-to-volume ratio of the specimen (e.g., particle size) in order to estimate the surface contribution to the spectrum.

Backscattered electron and photon signals, emitted following decay of an excited absorber nucleus, may be used to obtain "surface-specific" Mössbauer spectra. Indeed, conversion electrons, Auger electrons, and fluorescent X-rays are all produced in sufficient quantities and at low enough energies to provide a backscattered surface-specific Mössbauer spectrum. Because of the inherent advantages in this approach, backscattered photon and electron spectroscopies are becoming increasingly more popular for surface and catalytic studies. Unfortunately, backscattered conversion electron Mössbauer spectroscopy (BSCEMS), a cross between nuclear and electron spectroscopies, is still in a developmental stage. Due to the short path length of low-energy electrons (circa <10 kV) in solids, BSCEMS holds promise for performing nondestructive depth profiling in the topmost 100 nm of a specimen surface while possessing sufficient sensitivity to observe a single monolayer of resonant nuclei. Backscattered photons of similar energy (<10 kV) may originate from deeper in the specimen (circa 10^4 nm) so that combined detection of backscattered photon and electron spectra may permit nondestructive depth profiling over the outer 0.25 to 10^4 nm of the sample. Furthermore, this type of electron/photon spectroscopy can be added to existing UHV chambers and used as a complementary characterization tool with existing surface analysis techniques. Since the capabilities noted above are most suitable for studying problems in adhesion, corrosion, surface chemistry, catalysis, surface science, and thin-film coating procedures, it is clear that backscattered surface-specific Mössbauer spectroscopy is a technique with significant future potential.

The intent of the following discussion is to review briefly recent efforts and advances in surface science and catalysis using backscattered conversion electron and photon Mössbauer spectroscopies. General knowledge of the Mössbauer effect and its uses in determining chemical, electronic, and magnetic insight for surface studies is assumed [4.13-19]. Specifically, the role of this review will be: (i) to present an overview of the physical principles behind backscattered conversion electron and photon Mössbauer spectroscopies; (ii) to summarize past experimental and theoretical efforts in these areas; (iii) to present recent advances in surface-specific Mössbauer equipment and theory; and finally (iv) to speculate on the potential for future developments in the theory, equipment or applications of this technique.

4.2 Physical Principles Important to Surface-Specific Mössbauer Spectroscopy

4.2.1 Background

Mössbauer spectroscopy is normally conducted in a transmission geometry, wherein the resonant absorption of a γ-ray by the nucleus is monitored as a function of incident γ-ray energy (i.e., the Mössbauer spectrum). Since nuclear energy levels are extremely sensitive to perturbations in the shape and density of the surrounding electron cloud, the absorption spectrum provides detailed information regarding the electronic, magnetic, geometric, and dynamic nature of the solid state. Following absorption of a γ-ray, an excited nucleus quickly decays (circa 10 ns) to the ground state. Because this relaxation usually occurs via emission of a conversion electron, and subsequent Auger electrons or X-ray photons, detection of these backscattered species greatly enhances the inherent sensitivity (i.e., signal-to-background ratio) of the Mössbauer experiment. In the case of a backscattered electron spectrum, as little as one monolayer of resonant nuclei is observable [4.20], whereas in transmission geometry, a specimen ~1 μm thick is needed. Moreover, backscattered electrons and photons typically have low energies (circa <10 kV) and are emitted at a number of discrete energies. Therefore, detection at a particular electron or photon energy not only permits a Mössbauer spectrum to be collected but further requires that the signal originate from a depth less than or equal to the escape depth of that electron/photon. Collection of backscattered conversion electron and photon spectra allows specimen surfaces to be nondestructively depth profiled (i.e., individual Mössbauer spectra recorded from differential volume elements) within ~1.0 nm to 20 μm of the surface. Detection of these signals is particularly suited for the study of thin films (i.e., corrosion products) and the correlation and understanding of physical processes which occur at both the external surface (circa <1.0 nm) and further into the bulk (circa 20 μm). Backscattered conversion electron and photon spectra are uniquely able to link surface phenomena such as adsorption and reaction with behavior deeper in the solid, such as diffusion, phase behavior, and bulk reactivity.

Since these techniques are still in an early stage of development, the following discussion outlines a number of physical factors which form the basis for surface-specific Mössbauer studies.

4.2.2 Feasibility and Limitations of Mössbauer Studies

Figure 4.1 details the manner in which transmission Mössbauer spectroscopy (TMS) is usually performed. While Mössbauer spectra have been observed from over 83 isotopes [4.21], a recent review by *Dumesic* and *Topsøe* [4.14] notes that only about 38 of these resonances can yield detailed chemical, electronic, and magnetic information from the solid state. If it is further required that the source matrix have a half-life >30 days and the γ-ray energy be small enough to allow significant recoilless emission/absorption at room temperature, then only 8 nuclei remain, consisting of: ^{57}Fe, ^{119}Sn, ^{121}Sb,

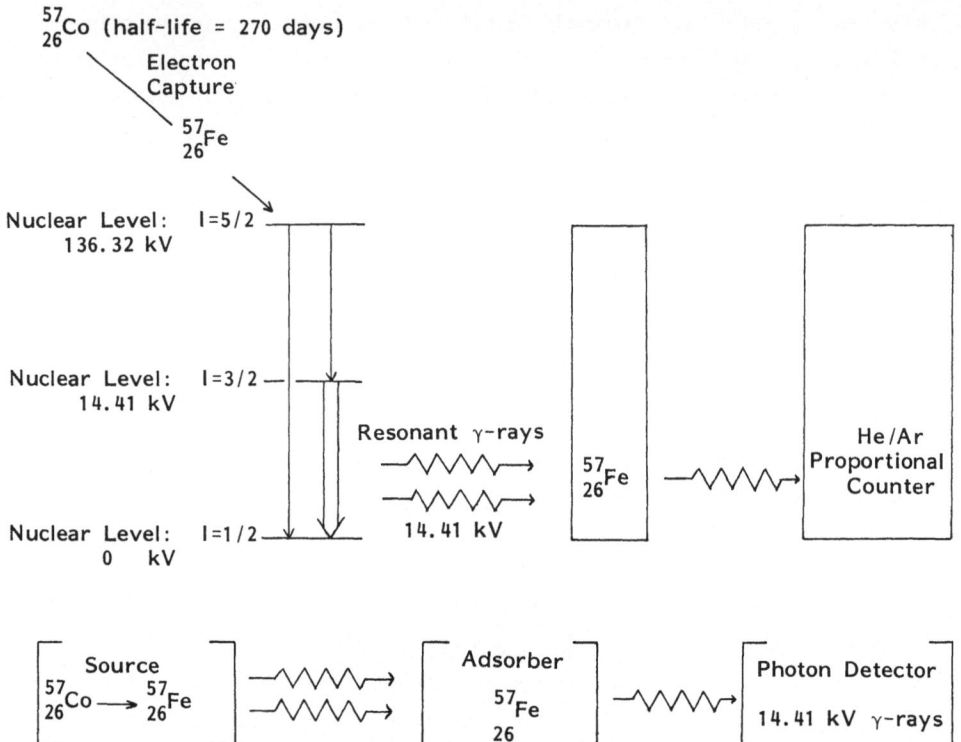

$^{57}_{26}$Co (half-life = 270 days)

Electron Capture

$^{57}_{26}$Fe

Nuclear Level: I=5/2
136.32 kV

Nuclear Level: I=3/2
14.41 kV

Resonant γ-rays

$^{57}_{26}$Fe

He/Ar Proportional Counter

Nuclear Level: I=1/2
0 kV

14.41 kV

Source
$^{57}_{26}$Co → $^{57}_{26}$Fe

Adsorber
$^{57}_{26}$Fe

Photon Detector

14.41 kV γ-rays

Fig.4.1. Schematic diagram of transmission geometry typically used for Möss-bauer spectroscopy: the ^{57}Fe example

^{125}Te, ^{149}Sm, ^{151}Eu, ^{169}Tm, and ^{181}Ta. These severe restrictions placed on half-life and γ-ray energy are sufficient to ensure that the above resonances can be routinely observed in almost any Mössbauer laboratory, without neces-sitating use of cryogenic liquids to increase the fraction of resonant emis-sions/absorptions. Therefore, it is apparent that these 8 nuclei form a reasonable working basis for surface-specific Mössbauer experiments. However, materials which can be analyzed by Mössbauer spectroscopy are not restricted to just these metals. Rather, the sensitivity of Mössbauer spectroscopy is such that useful spectra can be recorded from almost any solid provided it contains just a few percent of one of these metals. In this fashion, the de-tailed information provided by the Mössbauer resonance can be used to probe electronic, magnetic, geometric, and solid-state properties of many host materials.

It is important to note that BSCEMS or backscattered photon Mössbauer spectroscopy (BSPMS) are decidedly more sensitive than TMS. A conversion electron spectrometer can detect a ^{57}Fe film about a monolayer thick [4.20]

while TMS requires a minimum of about 10^3 nm. Thus, BSCEMS appears to be at least 10^4 times more sensitive than TMS.

4.2.3 Physical Bases for Backscattered Mössbauer Spectroscopies

The key aspect of surface-specific Mössbauer spectroscopy is the collection of backscattered photons and electrons which are emitted when the nucleus decays from the excited state, Fig.4.2. The depth-profiling capability of these techniques results because backscattered electrons and photons have limited path lengths in solids, making these spectra inherently surface sensitive.

In order to understand these physical bases better, Fig.4.3 depicts a simplified excitation/decay scheme. A more detailed account of a specific nucleus (e.g., ^{57}Fe) will be given below. The salient point of Fig.4.3 is that nuclear relaxation does not generally result in reemission of the original γ-ray. Rather, the nucleus transfers excess energy to an s electron in order to eject it from the atom with a kinetic energy equal to $E_\gamma - E_{B.E.}$, where $E_{B.E.}$ is the binding energy of the electron to the atom and E_γ is the energy of the original γ-ray. This phenomenon is known as the internal con-

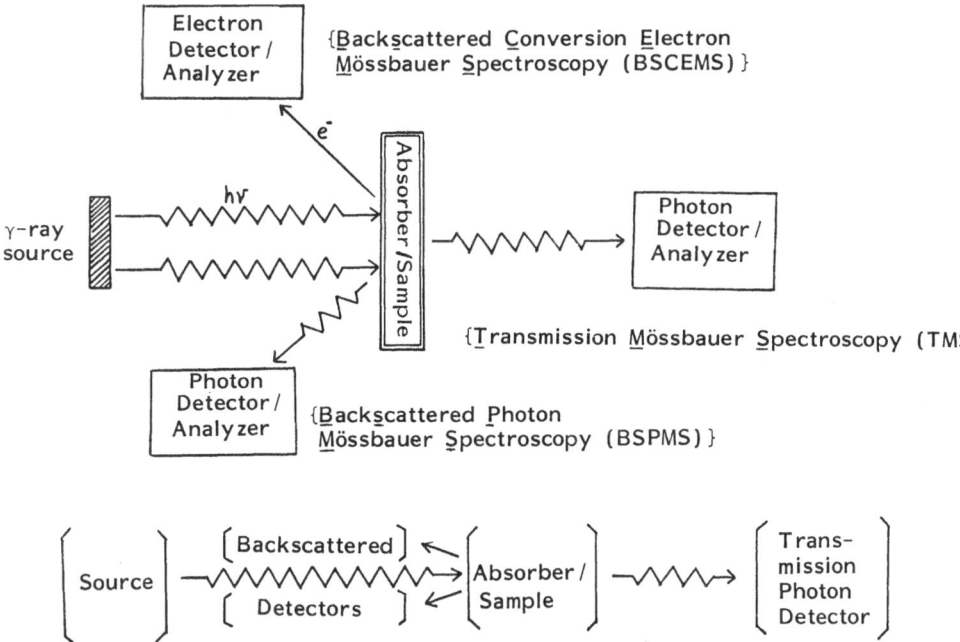

Fig.4.2. Backscatter geometry used for surface-specific Mössbauer spectroscopy

1. Nuclear Excitation:

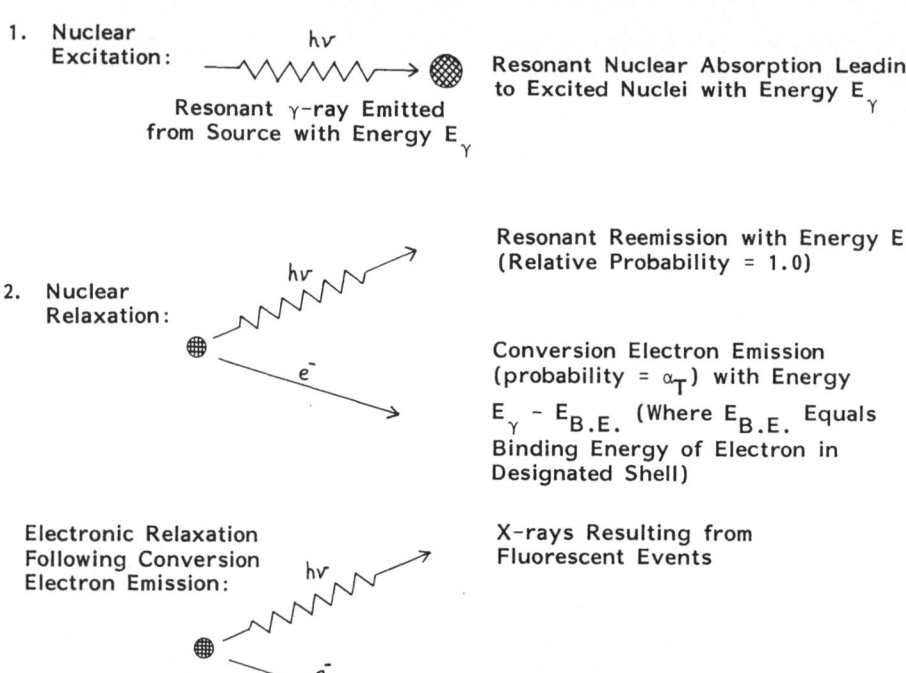

Resonant γ-ray Emitted from Source with Energy E_γ

Resonant Nuclear Absorption Leading to Excited Nuclei with Energy E_γ

2. Nuclear Relaxation:

Resonant Reemission with Energy E_γ (Relative Probability = 1.0)

Conversion Electron Emission (probability = α_T) with Energy $E_\gamma - E_{B.E.}$ (Where $E_{B.E.}$ Equals Binding Energy of Electron in Designated Shell)

3. Electronic Relaxation Following Conversion Electron Emission:

X-rays Resulting from Fluorescent Events

Electrons Resulting from Auger Cascades

Fig.4.3. Excitation/decay mechanisms during Mössbauer spectroscopy

version process and accounts for a portion of the backscattered radiation. The remainder of the backscattered signal is produced when the core level electron hole, formed by internal conversion, relaxes via: (i) Auger cascades which release additional electrons and also create more electron holes, or (ii) electron cascades which produce X-ray fluorescent events. The exact number of electrons and photons produced during relaxation is a complex problem treated in some detail below. The ratio of conversion electron events to γ-ray reemissions, known as the total internal conversion coefficient α_T, has been measured for most Mössbauer isotopes. Table 4.1 lists these ratios and shows that for the 8 nuclei mentioned above, electron production is greatly favored. The preferential generation of electrons, the multiplication of the electron signal by repeated Auger events, and the inherently higher signal-to-background ratio of the backscatter experiment all account for the superior sensitivity and surface specificity of BSCEMS.

To elucidate better the surface sensitivity and nondestructive depth-profiling capability of BSCEMS and BSPMS, Table 4.2 indicates the types of electrons and photons generated during a typical nuclear relaxation. In this in-

Table 4.1. Total internal conversion coefficients and resonant gamma ray energies for selected Mössbauer nuclei

Nuclei	Total internal conversion coefficient[a] $[\alpha_T]$	Resonant gamma ray energy $[E_\gamma/keV]$	Reference
^{57}Fe	10.11	14.41	[4.11]
^{119}Sn	5.12	23.87	[4.11]
^{121}Sb	~ 10.0	37.15	[4.11]
^{125}Te	12.7	35.48	[4.11]
^{149}Sm	~ 12.0	22.50	[4.11]
^{151}Eu	29.0	21.60	[4.11]
^{169}Tm	220	8.40	[4.11]
^{181}Ta	46.0	6.24	[4.12]

[a]Total internal conversion coefficient defined as the ratio of nuclear re-laxations initiated by conversion electron events compared to the number of resonant photon reemissions.

stance ^{57}Fe is used as an example, although any of the other 7 nuclei yield similar results. By using the 3 detectors pictured in Fig.4.2, at least 9 different Mössbauer signals can be monitored from an ^{57}Fe absorber. As shown in Table 4.2, the transmitted Mössbauer spectrum collects information over the entire sample depth provided a thin specimen is used (circa 1-100 μm thick). If very thick specimens are examined only the backscattered geometry may be employed. Backscattered γ and X-rays yield signals from the topmost 14 μm and 11 μm, respectively, facing the photon detector. Six different electron signals, each emitted at a discrete energy, can also be recorded using a backscatter electron detector. Since each of these 9 signals is gen-erated in the solid at a discrete energy, determined by the resonant atom nuclear and electronic levels, comparison of these various Mössbauer spectra will yield depth profiled information. Indeed, physical and mathematical treat-ments discussed below have been used to depth profile the topmost 300 nm of an iron specimen with a resolution of ~5.0 nm.

4.3 Surface-Specific Mössbauer Studies: Equipment and Applications

4.3.1 Backscattered Photon Detection

The first studies directed toward measurement of surface-specific Mössbauer spectra were recorded utilizing existing Ar/He proportional counters in a backscatter configuration, as shown in Fig.4.4a. Using this approach, *Terrell*

Table 4.2. Photon and electron signals generated by a typical Mössbauer nucleus: ^{57}Fe

Detection geometry	Species detected	Sampling depth[c]	Probability of species generation[d]
Transmission	14.41 keV γ-ray	0.002-0.1 mm	-
Backscatter[a]	14.41 keV γ-ray	~ 14 μm	$\dfrac{1}{1+\alpha_T} = 0.09$
Backscatter[b]	6.4 keV X-ray	~ 11 μm	$\dfrac{0.3\alpha_K}{1+\alpha_T} = 0.24$
Backscatter[b]	14.3 keV M-shell conversion electron	~ 80 nm	$\dfrac{\alpha_M}{1+\alpha_T} = 0.009$
Backscatter[b]	13.6 keV L-shell conversion electron	~ 73 nm	$\dfrac{\alpha_L}{1+\alpha_T} = 0.09$
Backscatter[b]	7.3 keV K-shell conversion electron	~ 24 nm	$\dfrac{\alpha_K}{1+\alpha_T} = 0.81$
Backscatter[b]	5.4 keV KLL Auger electron	~ 13 nm	$\dfrac{0.7\alpha_K}{1+\alpha_T} = 0.57$
Backscatter[b]	0.65 keV LMM Auger electron	~ 5 nm	$\dfrac{1.7\alpha_K+\alpha_L}{1+\alpha_T} \cong 1.5$
Backscatter[b]	0.05 keV MNN Auger electron	~ 2 nm	$\dfrac{4(0.7\alpha_K)+2(0.3\alpha_K)+2\alpha_L}{1+\alpha_T} \cong 2.9$

[a]Only specimens <0.1 mm may be used for TMS. [b]Using backscatter geometry there is no limit to specimen thickness. [c]Sampling depth defined as sample thickness necessary to produce a signal one-half as intense as produced by an infinitely thick absorber (photons assumed to escape with full energy, electrons with energies ≥ 0 eV). Calculations performed assuming that: (i) sample/adsorber contains only 2.19 atomic % ^{57}Fe, (ii) recoil-free fraction of the source and absorber equals 0.7, and (iii) electron/photon attenuation follows the formalism described in [4.22]. [d]Probabilities for species generation total more than unity since this table is based on unit relaxation of the excited nuclear state.

and *Spijkerman* [4.24] recorded 6.4 kV X-rays which were produced following K-shell conversion (Fig.4.4b). These spectra, Fig.4.4b, were used to identify a ~20 μm overlayer of β-FeOOH which formed on a thick iron plate following exposure to a HCl/H_2O mixture. Similar work by *Herskowitz* et al. [4.25] and *Ord* et al. [4.23] also utilized the 6.4 kV K_α X-ray, because it was pro-

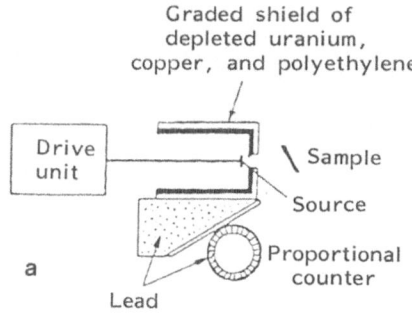

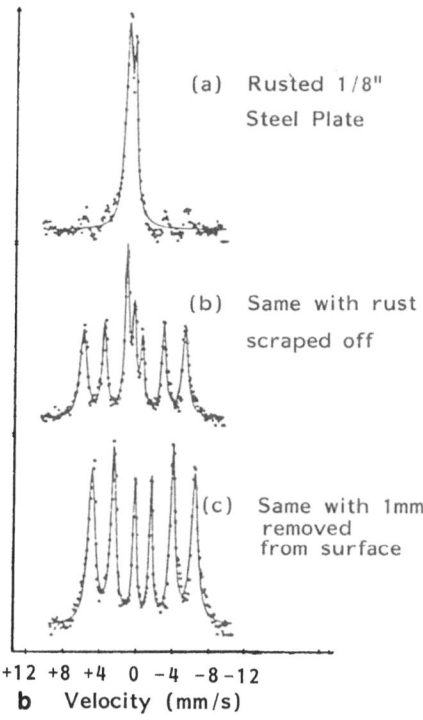

(a) Rusted 1/8" Steel Plate

(b) Same with rust scraped off

(c) Same with 1mm removed from surface

+12 +8 +4 0 -4 -8 -12

b Velocity (mm/s)

Fig.4.4a. Apparatus used for back-scattered photon Mössbauer spectroscopy [4.23], reproduced with permission. b) Backscattered Mössbauer spectra using the 6.4 kV X-ray emission, following treatment of a steel plate in an HCl/air mixture to produce β-FeOOH surface layer [4.24], reproduced from Applied Physics Letters with permission

duced more frequently than reemitted 14.41 kV γ-rays, and these workers [4.25] were able to record good spectra from films as thin as 1 nm. While this result certainly demonstrates the sensitivity of backscatter geometry, the ability to record spectra from films which are 2×10^4 nm thick clearly shows that this technique is not strongly surface sensitive.

4.3.2 Backscattered Electron Detection: Gas-Filled Electron Counters

One of the earliest and still most popular backscatter conversion electron detectors was constructed by *Fenger* [4.26] in 1968 and is schematically shown in Fig.4.5a. To permit easy and relatively unattenuated passage of photons, the counter was made as thin as possible with entrance and exit windows constructed from gamma-ray-transparent materials such as Lucite, Kapton, or beryllium. A counting gas is then chosen to measure selectively scattered electrons (He/10% CH_4). By placing the sample within the detector the majority of the detected signal results from backscattered conversion and Auger electrons, and the composition of the counting gas and anode voltage can be adjusted to optimize the spectrum collection. If thin samples are used it is also possible to record simultaneously the transmission Mössbauer spectrum using an additional detector located behind the exit window. More recent de-

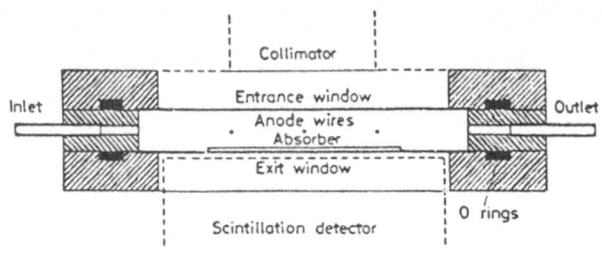

Collimator

Inlet | Entrance window | Outlet
Anode wires
Absorber

Exit window

Scintillation detector

O rings

a

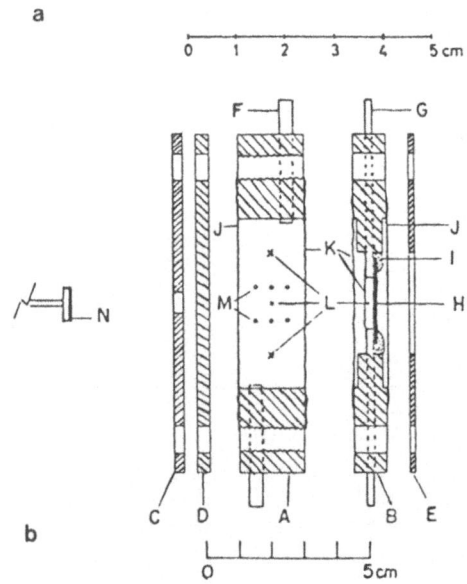

0 1 2 3 4 5 cm

F — ⌐ ⌐ — G

J — — J
— K — I
M — L — H

/ — N

C D   A   B E

0   5 cm

b

Fig.4.5a. Cut view of He/10% CH$_4$ gas-filled proportional counter used by Fenger [4.26]. Scintillation detector used only for transmission and coincidence counting, reproduced with permission. b) Detector assembly, with partioned counting chamber, used in [4.27] for measuring separate backscattered electron and X-ray signals, reproduced with permission. (A) Lucite frame for X-ray counter; (B) Lucite frame for electron counter; (C) 3 mm thick lead plate; (D) 4 mm thick Lucite plate; (E) 1 mm thick Al plate; (F) inlet for the counting gas of X-ray counter; (G) inlet for the counting gas of electron counter; (H) sample; (I) sealing compound (Apiezon); (J) 20 μm thick Al foil; (K) Al evaporated on Mylar film; (L) 50 μm diam. W wire; (M) 100 μm diam. W wire; (N) Mössbauer source 10 mCi ^{57}Co in Cu

signs (Fig.4.5b) have divided the counting chamber with thin aluminum/mylar films [4.27] so that both backscattered photon and electron spectra can be recorded simultaneously in different parts of the detector.

Using a design similar to that of *Fenger* [4.26], *Spijkerman* and *Swanson* [4.28] recorded the backscattered conversion electron spectrum from the topmost 5-300 nm of an iron foil using a He/10% CH$_4$ counting gas and then switched this miture to Ar/10% CH$_4$ to record a spectrum from the outer 10-20 μm of the foil using backscattered 6.4 kV X-rays. By vacuum evaporating various thicknesses of metallic iron onto a stainless steel substrate and comparing the spectral contribution from each phase in the conversion electron spectrum (Fig.4.6), *Spijkerman* and *Swanson* were able to estimate the maximum sampling depth of ^{57}Fe-BSCEMS at ~300 nm.

Since these early experiments [4.26,28,29], gas-filled electron counters have seen continued use because of their simple design and operation. Appli-

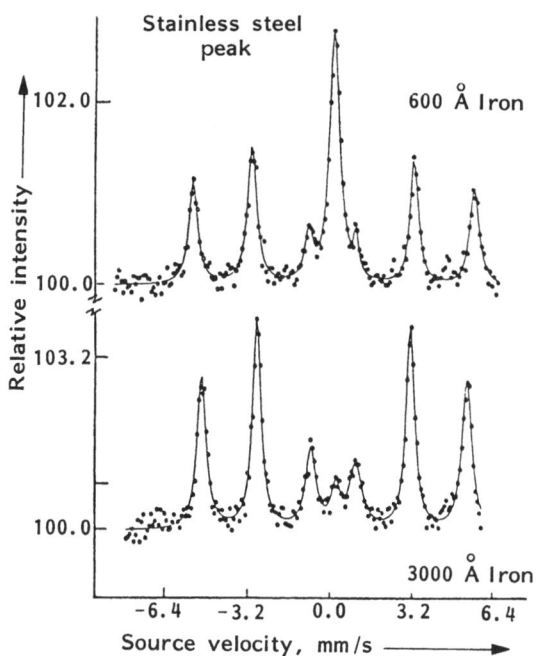

Fig.4.6. Backscattered conversion electron Mössbauer spectra obtained by Spijkerman [4.29] for two thicknesses of metallic Fe evaporated on a stainless steel substrate

cations include: (i) the analysis and identification of corrosion and oxidation products on iron foils and low carbon steels [4.30-34]; (ii) the kinetics, stoichiometry, and morphology of multiphasic, layered iron oxides [4.31,34-37]; (iii) iron implanted into graphite [4.38], aluminum [4.39], beryllium [4.39] and the 3d, 4d, and 5d transition metals [4.40]; (iv) surface stress measurements from Fe and Sn substrates [4.41,42]; (v) the orientation of the electric-field gradient and the oxidation of biotite [4.43,44]; (vi) the formation of surface iron carbides and their decarburization using H_2 [4.45,46]; (vii) the formation of surface austenite during surface grinding of carbon steel [4.47]; (viii) the oxidation of iron phosphates [4.48]; (ix) the determination of the location and oxidation state of iron in phosphate glasses [4.49]; (x) the phosphiding of iron to assess corrosion resistant finishes [4.50,51]; (xi) the surface nitriding and and implantation of steel to improve hardness and wear properties [4.52]; (xii) the characterization of iron/germanium-amorphous films [4.53]; (xiii) the determination of iron and iron-cobalt phases present in Fe-Co ammonia synthesis catalysts [4.54]; and (xiv) the surface monitoring of β-Sn, $CaSnO_3$, and SnO_2 substrates [4.55].

Additional efforts and refinements in detector design have been made in many of the above-noted studies. Helium/~10% butane mixtures have been shown to be more sensitive for electron counting than He/~10% CH_4 [4.27,30,56], and this gas mixture also appears to offer greater energy resolution and discrimination in the pulse-height spectrum [4.30]. Using carefully prepared standards of iron evaporated onto stainless steel [4.33] or copper substrates [4.20], the depth-profiling capabilities of BSCEMS have been determined with the inherent sensitivity of BSCEMS established at less than one monolayer of [57]Fe [4.20]. Gas-filled electron detectors using He/∼10% CH_4 mixtures have also been operated at temperatures from 77 to 560 K by *Sawicki* et al. [4.58] and *Isozumi* et al. [4.57], respectively.

In summary, it can be seen that gas-filled proportional counters can be used to obtain either backscattered conversion electron or photon spectra. These detectors are inexpensive, easy to assemble, and can be readily added to existing transmission Mössbauer spectrometers. Moreover, because the spectrum is recorded in a backscatter geometry, thick specimens can be used without special preparation, greatly increasing the impact of this technique for applied studies. A disadvantage, however, is the fact that samples used for BSCEMS and BSPMS must be exposed to both counting and quench gases during data collection. Gas-filled conversion electron detectors are therefore difficult to combine with modern surface-analysis techniques employing ultra-high vacuum environments. Little temperature variation of the sample (i.e., detector) has been demonstrated and only modest energy resolution in the electron spectrum has been observed. Without better resolving capabilities these devices must be viewed as "integral" detectors which collect any electron with energy above some threshold close to zero eV.

This last trait in particular greatly reduces the utility of this counting device, as it will be shown in subsequent discussions that discrimination between the various electrons which are generated during nuclear or electronic relaxation can be used to obtain depth-profiled structural, electronic, magnetic, and stoichiometric information.

4.3.3 Backscattered Electron Detection: Magnetic Spectrometers

To take advantage of the inherent depth-profiling capabilities offered by conversion and Auger electrons, a number of electron spectrometers have been developed for surface-specific Mössbauer investigations. These analyzers include magnetic spectrometers of the β-ray design, electron spectrometers of the cylindrical mirror variety and retarding field optics in conjunction with continuous dynode or more conventional Cu/Be electron multipliers. Schematically, these various designs are shown in Figs.4.7-9.

Beta-ray spectrometers were first used for [119]Sn-BSCEMS in 1961 by *Mitrofanoff* and *Shpinel* [4.62] and later by *Bonchev* et al. [4.63]. Indeed, Bonchev

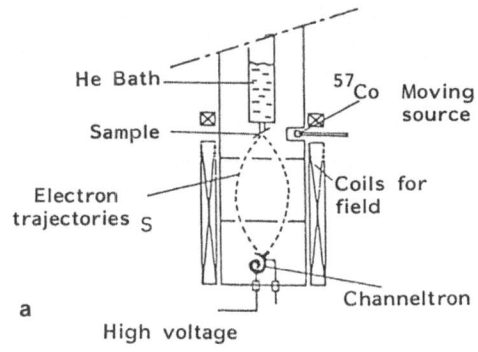

He Bath

Sample

Electron
trajectories S

a

High voltage

^{57}Co Moving
source

Coils for
field

Channeltron

Fig.4.7a. Low-temperature β-ray spectrometer used in [4.59] for collecting BSCEMS spectra from Fe_xGe_{1-x} amorphous alloys, reprinted with permission from [Solid State Communications, Vol.25, Massenet O. and Daver H. "Low Temperature Conversion Electron Mössbauer Spectroscopy on Fe_xGe_{1-x} Amorphous Thin Films"] (Pergamon Ltd. 1968). Design features external source and Doppler velocity transducer. b) β-ray spectrometer used in [4.60] to study oxidation products on Fe. Note that in this design the source is stationary, reprinted with permission from "Applications of Surface Science"

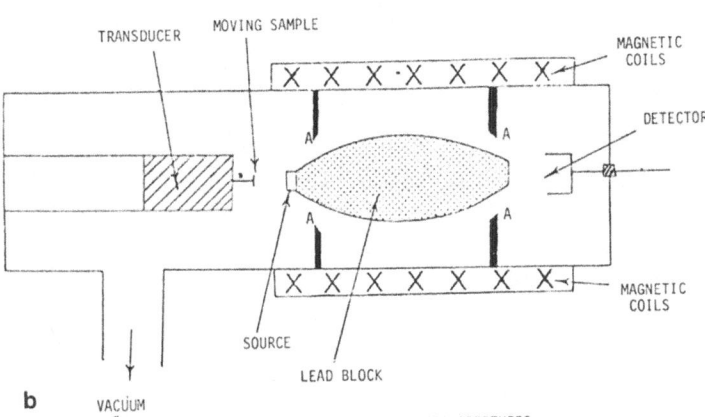

TRANSDUCER   MOVING SAMPLE

MAGNETIC
COILS

DETECTOR

A   A

A   A

MAGNETIC
COILS

SOURCE

LEAD BLOCK

b   VACUUM
($<10^{-5}$torr)

A = LEAD APERTURES

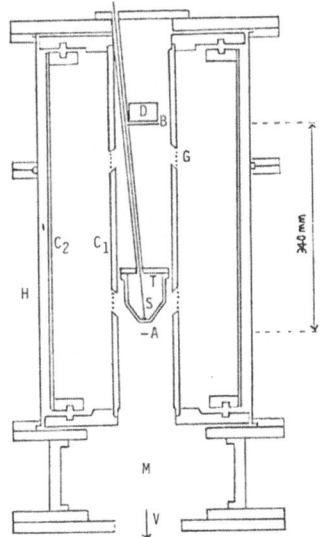

D B

G

C$_2$ C$_1$

H

340 mm

T
S
A

M

V

Fig.4.8. Cylindrical mirror analyzer with ex situ sample/absorber treatment chamber, reproduced with permission [4.61]. $C1 \equiv$ inner cylinder, $C2 \equiv$ outer cylinder, $G \equiv$ thin grids, $A \equiv$ absorber, $B \equiv$ detector baffle, $D \equiv$ detector, $T \equiv$ transducer rod, $S \equiv$ source, $H \equiv$ vacuum tank, $M \equiv$ sample/absorber treatment chamber, $V \equiv$ outlet to vacuum system, reproduced with permission [4.61]

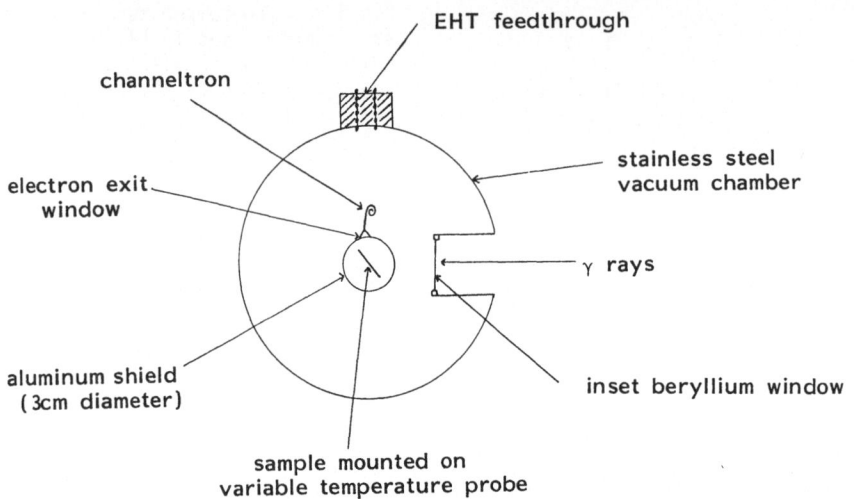

EHT feedthrough

channeltron

electron exit window

aluminum shield (3cm diameter)

stainless steel vacuum chamber

γ rays

inset beryllium window

sample mounted on variable temperature probe

<u>Fig.4.9.</u> Apparatus for detecting backscattered conversion electrons in vacuo using a channeltron electron detector. System may be operated as a vacuum compatible-integral detector [4.60], reproduced with permission from "Applications of Surface Science"

was one of the first workers to exploit the depth-profiling capability of BSCEMS by utilizing a β-ray design in conjunction with a Bethe-Block expression suitable for low-energy electrons [4.64]. Since electrons generated at specific energies but different depths of the surface will emerge from the sample with different energies [4.64], Bonchev tuned the β-ray spectrometer to progressively lower electron energies in order to probe more deeply into the specimen. In this same vein *Liljequist* et al. [4.61,65-69], in an extensive theoretical and experimental program concerning ^{57}Fe-BSCEMS, developed procedures capable of obtaining an individual conversion electron Mössbauer spectrum from a 5.0 nm differential volume element anywhere within the topmost 300 nm of the specimen surface. In this method several BSCEMS spectra are recorded at spectrometer pass energies slightly below the 7.3 kV K-shell conversion electron. Each of these spectra is then considered to represent a large number of superimposed "depth differential" spectra built from ejected electrons originating at different depths of the sample. If the probability is known for a given electron (from a certain depth) to appear at the spectrometer energy setting, then the weights of the various differential spectra are also known. Thus, a number of BSCEMS spectra measured at different energy settings can be used to separate statistically the data into differential spectra containing distinct contributions from specified depths of the surface [4.61,65-69].

Since the extensive work of *Liljequist*, a number of other authors have also employed the β-ray spectrometer, including *Schunck* et al. [4.70] for the depth profiling of ^{57}Fe and ^{119}Sn compounds, *Gruzin* et al. [4.71] who depth profiled the oxygen concentration in plasma anodized iron films, and *Shinohara* et al. [4.72] for measuring the 4s electron density and isomer shift of Fe atoms in Cr, Ni, Cu, Pt and Au hosts. *Massenet* [4.59,73] characterized the magnetic properties of amorphous Fe_xGe_{1-x} and FeCoTi alloy films at liquid helium temperatures, and *Jones* et al. [4.60,74] studied the kinetics and composition of iron oxide growth [4.60] and also iron implantation into aluminum foils [4.74].

β-ray spectrometers used in these studies have resolved electrons from the individual K and L conversion shells of iron as well as from KLL and LMM Auger events which follow these initial electron emissions [4.60,70]. However, in spite of this superior resolution, β-ray spectrometers are usually "tuned in" only in the vicinity of the 7.3 kV K-shell conversion electron so as to maximize the count rate of resonant electrons and increase the signal-to-background ratio. This particular energy window is chosen because (i) the probability of emission is high; (ii) the electron energy, which is higher than for a KLL or LMM Auger event, permits a larger sample volume to be measured (i.e., more signal); (iii) the β-ray spectrometers are usually designed for best transmission characteristics in this range; and (iv) the electron energy is high enough to minimize interferences by low-energy, non-resonant secondary electrons. Unfortunately, while the 7.3 kV energy window provides for a good signal-to-background ratio, the count rates for these studies are so low that it may require on the order of 24 hours to obtain a spectrum with good statistics (i.e., signal-to-noise ratio, proportional to the square root of the number of counts). In view of this fact, it is clear that large amounts of experimental time may be necessary to generate the requisite number of spectra to permit 5.0 nm "depth-differential" resolution according to the method of *Liljequist* [4.61,65-69]. Other disadvantages of this technique include: (i) the use of complex detectors and electronics; (ii) the requirement of extensive computer hardware and software; and (iii) the necessity of a good background vacuum to permit electron transport through the analyzer. More involved sample handling is required to reproducibly place flat sample surfaces at a fixed focal point in the spectrometer. Since it is known that lower energy Auger electrons (LMM and below) are produced more frequently than other types of resonant electrons (see Table 4.2 for the ^{57}Fe example), spectrometers which are capable of focusing these low-energy signals could be used to decrease the time required to collect a backscattered electron spectrum.

4.3.4 Backscattered Electron Detection: Retarding Field and Electrostatic Analyzers

In addition to the β-ray spectrometers discussed above, other analyzer types have been utilized for BSCEMS, as shown in Figs.4.8,9. A cylindrical mirror analyzer (CMA) was employed by *Bäverstam* et al. [4.61] while *Keune* et al. [4.75] and *Toriyama* et al. [4.76] used spherical analyzers. Because these analyzers provide 2% to 3% resolution, individual electron emission levels are easily resolved, yet count rates are low and data collection takes considerable time. To circumvent this problem, other workers [4.22,60,77-82] have employed Retarding Field Analyzers (RFA). These systems offer the advantage that while differential analyses may still be obtained by subtracting spectra obtained at different retarding voltages, these detectors may also be operated in the integral mode to monitor essentially all electrons reaching the detection apparatus with energy above zero eV. This feature significantly increases resonant count rates, permitting more rapid spectrum collection.

Furthermore, BSCEMS spectra can be obtained in minutes and samples do not have to be exposed to counting or quenching gases. As shown below in Fig.4.11. these analyzers/detectors may be fitted to existing surface analysis instruments or combined with other UHV techniques. Alternatively, combined UHV-high-pressure systems, similar to those developed by *Somorjai* et al. [4.83] could be used which permit high-pressure/high-temperature pretreatment followed by analysis using UHV techniques in the same chamber. Retarding field analyzers offer a favorable compromise by allowing collection of both integral and differential spectra following exposure to UHV or gas-filled environments at various temperatures and pressures. *Carbucicchio* [4.80] has recently constructed a RFA-type instrument which allows the incidence and scattering angles to be varied as well as permitting sample temperature variations from 78 to 800 K.

In addition to studies on ^{57}Fe-enriched metallic iron and stainless steel foils, retarding field analyzers have been used for unenriched (i) Fe and Sn samples [4.84], (ii) Fe and Fe-Si thin films [4.77,79], (iii) surfaces of TiFe intermetallics and intermetallic hydrides [4.78], (iv) metallic iron and multiphasic layered iron oxides [4.80], (v) FeOOH layers formed by HCl/H_2O treatment of Fe [4.60], (vi) the surface composition, and phase identification of FeNiMo alloys [4.81], (vii) ^{57}Fe implanted into Si [4.85],

(viii) and Fe/TiO$_2$ model supported catalysts consisting of ~5.0 nm overlayers of ^{57}Fe nucleated into circa 20 nm crystallites on the surface of a TiO$_2$ substrate [4.22].

The flexibility of retarding field analyzers for performing various types of BSCEMS experiments is a decided advantage of this analyzer type. Integral and differential analyses can be performed and the analyzer/detector readily integrated into existing vacuum systems or reaction chambers. The potential for future developments in this area appears good and subsequent discussions will treat the applications of this technique in more detail.

4.4 Theoretical Description Pertinent to Surface–Specific Mössbauer Techniques

To model surface-specific Mössbauer studies successfully the individual physical processes involved must first be clearly identified. Figure 4.10 schematically denotes the locations of these events. Because the five processes shown in Fig.4.10 are common to any surface-specific Mössbauer study, they will be briefly described and then used as a framework to review existing literature.

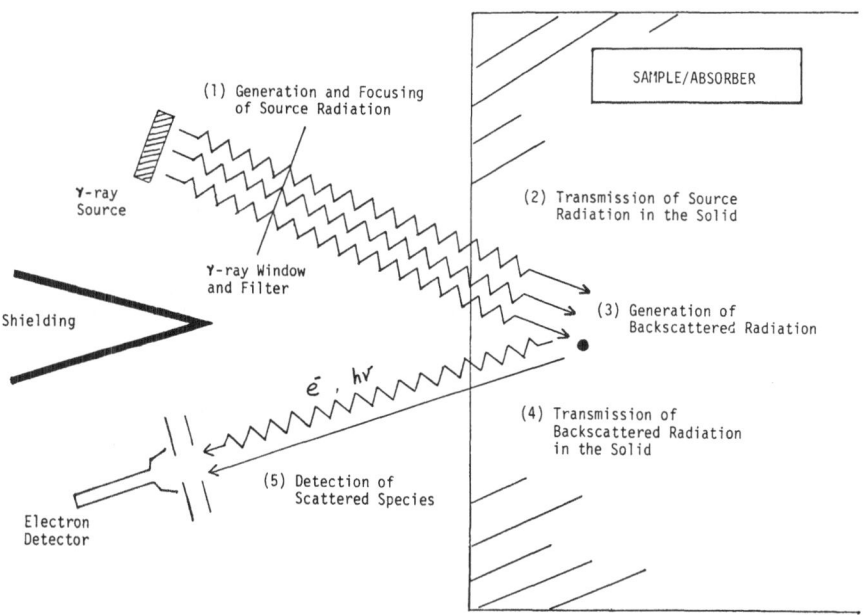

Fig.4.10. Physical processes involved in surface-specific Mössbauer studies

4.4.1 Generation and Focusing of Source Radiation

Gamma rays capable of causing resonant absorption in the sample are generated during a complex nuclear decay scheme leading to the production of a
number of different photons and electrons [4.8,86]. Details of the incident
energy spectrum must be obtained before the ratio of resonant to nonresonant
processes in the absorber can be evaluated. Accordingly, if estimates of
the count rate are desired, then the activity (viz., number of nuclear decays per second) of the source and the exact yield of photons and electrons
per decay must be known. In practice these data are difficult to estimate
and sample windows of various thickness and composition which will scatter
some photons are usually employed to introduce source radiation into the
sample cell. Thus, it appears that the most straightforward means of discern-
ing the effective energy distribution from the source is by direct measurement.

4.4.2 Transmission of Source Radiation in the Solid

Because absorption cross sections for resonant (i.e., Mössbauer) and nonresonant (viz., Compton and photoelectric) events are known, the transmission
of source photons into the solid can be described by fairly simple, exponen-
tial-type attenuation functions [4.86,87]. In most instances the transmission
and scattering of source electrons is not considered important as they are
easily removed from the source beam by thin metal or polymer films which
cause little attenuation in the photon flux.

4.4.3 Generation of Backscattered Radiation

Backscattered radiation generated by resonant and nonresonant events produces
discrete electron and photon species at various depths in the solid. While
momentum must be conserved during the scattering process, to simplify the
ensuing calculations, scattered electrons and photons are usually assumed to
be generated with spherical symmetry from the interacting atoms or nuclei.
Indeed, because incident photon energies are generally ≤ 100 kV this assumption introduces relatively little error.

Resonant absorption of a source γ-ray is followed by (i) reemission of
another γ-ray of comparable energy or (ii) emission of a conversion electron
and subsequent Auger electrons or fluorescent X-rays. The latter possibility,
known as the internal conversion process, has been quantified for most Mössbauer nuclei. However, because these core level Auger and fluorescent events
do not consume electron vacancies, continued relaxation via lower energy

electron and photon emissions will occur. To date, details regarding this low-energy cascade process have not been determined. Further work is needed if the measurement of low-energy resonant electrons is to provide additional insight into BSCEMS. Indeed, this area will be treated in more detail subsequently.

4.4.4 Transmission of Backscattered Radiation in the Solid

Once a photon or electron has been generated in the solid, the probability of that particle (or a related particle) reaching the surface must be considered. This scattering/transmission problem is quite complex due to the many scattering mechanisms which exist. However, at least a rudimentary understanding of these processes is required for proper understanding and interpretation of backscatter Mössbauer spectroscopy. For this reason, most depth-selective studies utilize electron/photon signals which are generated close to a particular emission energy. Simple exponential scattering expressions can then be used in conjunction with estimates of the inelastic mean free path for a particular electron energy, or alternatively, known Compton and photoelectric cross sections can be used if photons are being considered.

Further advances in this area are needed if the transmission of low-energy electron or photons, produced by multiple scattering events, are to be used in BSCEMS and BSPMS. To allow quantitative interpretation of BSCEMS spectra, using integral detectors for short counting times, a better understanding of low energy resonant particle transport mechanisms is required.

4.4.5 Detection of Scattered Species

The previously described phenomena (processes 1-4) produce a resonant and nonresonant flux of electrons and photons at the surface which vary in their angular and energy distributions. By summing some portion of these distributions in a synchronous fashion with the source velocity the Mössbauer spectrum is obtained. The information contained in the spectrum will depend on: (i) the electron or photon energy windows which are used to record the spectrum, (ii) the solid angle and take-off angles which are viewed by the detector(s), (iii) the sensitivity of the detector to electrons and photons of a particular energy, and finally (iv) the chemical composition and physical properties of the sample itself.

To date, consideration of the five areas described above has not been attempted in a comprehensive or unified manner. A quantitative understanding of these phenomena is required, however, for the interpretation of depth-selective and surface-specific Mössbauer spectroscopies. The following discussion briefly reviews the past progress and efforts that have been made in each of these areas.

One of the earliest and most extensive theoretical examinations of backscatter Mössbauer spectroscopy was developed by *Krakowski* and *Miller* [4.88] in 1972. In their mathematical framework they included photon attenuation by both resonant and nonresonant means and then accounted for subsequent electron transport using either a simple exponential attenuation function or a Fermi age diffusion model. While this effort was most ambitious in detail and the first of its kind in terms of scope and concept, some physical factors were not considered, including: (i) different types of photons emitted by the source; (ii) complete decay of the excited atom by Auger transitions and fluorescent events below the KLL energy; and (iii) detection of scattered electrons and photons.

Following the work of *Krakowski* and *Miller*, similar efforts by *Bainbridge* [4.89] and *Huffman* [4.90] considered the analysis of multilayer iron oxide films. The work of *Huffman* [4.90] was also combined with further experimental efforts [4.35]. These models for ^{57}Fe-BSCEMS included only K-shell conversion and KLL Auger contributions in the resonant spectrum and did not attempt to evaluate background count rates caused by nonresonant photons (122 and 136 keV γ-rays) emitted from the source.

Bäverstam et al. [4.91], to support their accompanying experimental program [4.61,65-69] for β-ray detectors, modeled the transmission of resonant electrons and calibrated their "depth-differential interpolation technique" to the backscattered electron spectrum from a foil of known thickness and composition. The goal of this effort was not to model the entire backscattered conversion electron process but rather to develop a method for empirically determining depth-profiled data based on the intensity and energy loss of electrons emitted close to a strong resonance peak. This approach was shown to be quite successful and "depth-differential" spectra from volume elements as thin as 5.0 nm were obtained using this model.

In a similar empirical fashion, *Bonchev* et al. [4.92-94] and *Proykova* [4.95] devised means for measuring the depth distribution of resonant matter in a thin surface layer. Active thin films (^{119}Sn) were prepared and backscattered electron spectra were measured from the emitting film after it was covered by known thicknesses of different metallic overlayers. In this fashion the distribution of resonant material in an unknown sample could be determined by detailed comparison with the "standard spectra".

Dynamic aspects of the conversion electron process have been considered by *Chugunova* and *Mitrin* [4.96] who used the electron-nuclear density matrix to investigate the potential for using BSCEMS to study ultrasonic modulations in Rayleigh surface waves and details regarding γ-ray magnetic resonance effects at the surface.

In addition to the above-noted studies, which are primarily theoretical in scope, a number of other physical details concerning BSCEMS have been gleaned during primarily experimental efforts. *Simmons* et al. [4.31] and *Tricker* et al. [4.32-34] have modeled the BSCEMS process in attempts to rationalize and calibrate spectra obtained from multilayer foils of known composition and thickness.

One interesting insight provided by *Tricker* et al. [4.97,98] is the observation that evaporated overlayers of nonresonant metals on top of resonant iron containing films preferentially increase the relative spectral contribution from the component which lies deepest within the resonant layer. In these studies a 10 nm metallic iron overlayer was evaporated on top of a stainless steel foil and then coated with various thicknesses of Al, Cr, and Au. It was observed that the spectral area ratio (Fe/stainless steel) decreased with increasing thickness of the nonresonant overlayer when a He/CH_4 proportional counter was used. *Tricker* et al. [4.97,98] attributed this phenomenon to the production of photoelectrons in the nonresonant overlayers by a large flux of resonant backscattered X-rays from the massive stainless steel substrate.

More recently, *Deeney* et al. [4.99] and *Tatarchuk* [4.22] suggested a different interpretation and demonstrated that the magnitude of such backscattered signals cannot account for the observed effect. It appears more likely that these nonresonant overlayers strongly attenuate low-energy (circa ≤ 10 eV) resonant electrons from the iron film but have little effect on low-energy electrons from the stainless steel substrate as these electrons have already been "filtered" from the backscattered electron spectrum by the iron overlayer. Indeed, the BSCEMS formalism developed by *Tatarchuk* [4.22] suggests the importance of these species and further experimental efforts by *Tatarchuk* [4.22] and *Tyliszczak* et al. [4.83] confirm the existence of strong resonance signals below 10 eV. *Tyliszczak* et al. also demonstrated that the electron spectrum collected with electrons below 10 eV possessed broadened spectral features attributable to inhomogeneities in the surface hyperfine field in the topmost atomic layers of the surface. These results [4.22,83] suggest that low-energy electron contributions to the backscattered electron spectrum may be significant, and that these signals may result from electronic relaxations of an excited atom at energies below the LMM and MNN Auger transitions.

4.4.6 Summary

A number of advances and modeling procedures have been developed for backscattered photon and electron Mössbauer spectroscopies. Further theoretical and experimental work is needed, however, before these treatments can be used to interpret spectra readily for depth-profiled chemical and structural information. Of the five physical processes shown in Fig.4.10, it is clear that the detailed nature of electronic relaxation processes below 10 eV (process 3) and the transport of electron signals in the solid (process 4) are most crucial to the further development of surface-specific Mössbauer studies.

4.5 New Developments in Surface-Specific Mössbauer Spectroscopy

4.5.1 Experimental

As noted earlier, the possibility exists for backscatter Mössbauer spectrometers of appropriate design to be added as complementary characterization tools for existing UHV surface analysis systems. *Tatarchuk* and *Dumesic* [4.100-102] have constructed such a device as shown in Fig.4.11 which combines molecular beam reactive scattering, Auger electron spectroscopy, temperature-programmed desorption, and BSCEMS. Iron/titania model supported catalysts were examined using these techniques and in a parallel set of experiments TEM and X-ray photoelectron spectroscopy (XPS) were also employed to

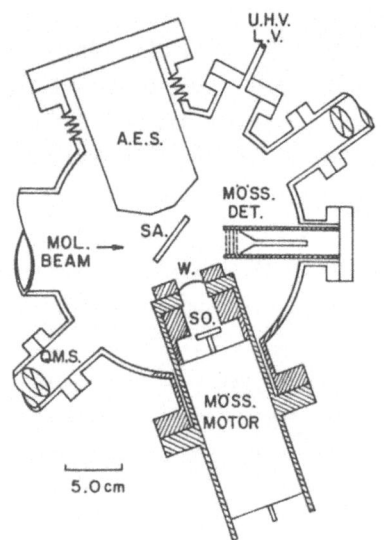

Fig.4.11. Retarding field-conversion electron Mössbauer analyzer located in a vacuum system with other physical and chemical characterization probes: MÖSS MOTOR ≡ Doppler velocity transducer, SO. ≡ ^{57}Co/Pd γ-ray source, W. ≡ Teflon-coated Kapton γ-ray window, S.A. ≡ sample, MÖSS DET. ≡ electron detector with retarding grids, U.H.V. L.V. ≡ ultrahigh vacuum leak valve, A.E.S. ≡ Auger electron spectrometer, MOL. BEAM ≡ molecular beam axis, Q.M.S. ≡ quadrupole mass spectrometer, with permission [4.22,101]

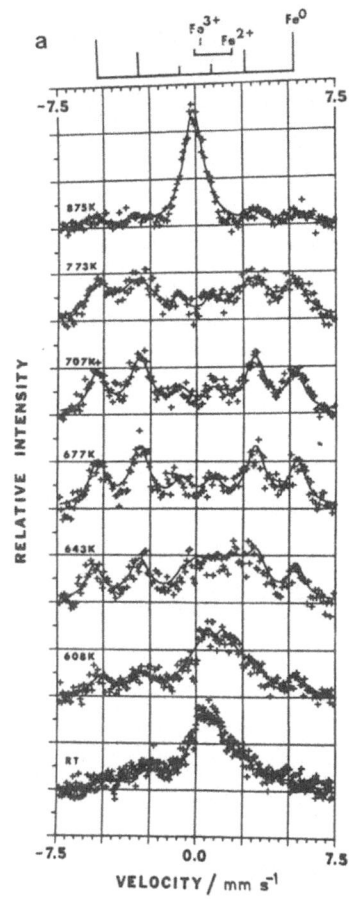

Fig.4.12a. BSCEMS spectra recorded in integral mode from 5 nm Fe/TiO_2 model catalyst following reduction at the indicated temperatures. One vertical square equivalent to 2% effect (RT ≡ initial iron overlayer) [4.101], reproduced with permission. b) Fe-$2p_{3/2}$ X-ray photoelectron spectra recorded from 5 nm Fe/TiO_2 specimen after reduction at the indicated temperatures. Specimen preparation and reduction treatments are chosen to mimic the conditions of Fig.4.12a. Vertical sensitivities for each spectrum are shown adjacent to the reduction temperature. Dashed spectra were obtained after subsequent oxidation (RT ≡ initial iron overlayer), reproduced with permission [4.101]

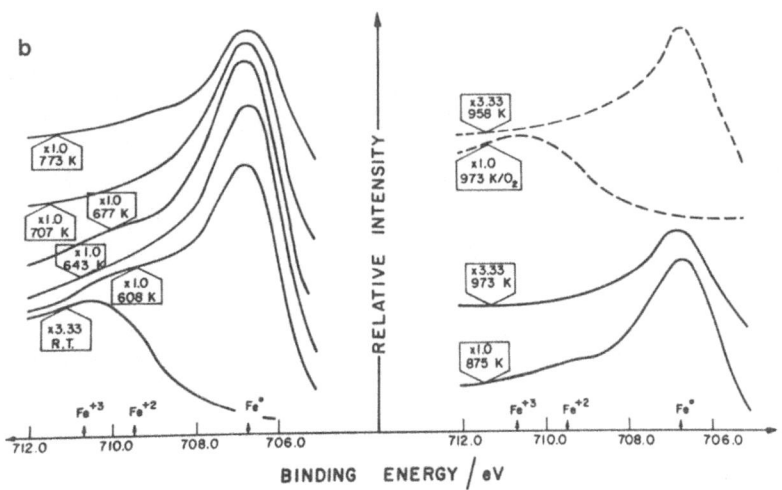

examine the morphology and oxidation state, respectively, of the iron overlayer. Figure 4.12a shows a set of BSCEMS spectra recorded from one such
TiO_2 film onto which ~5.0 nm of ^{57}Fe was evaporated and subsequently reduced
in H_2 at increasingly higher temperatures. For comparison, Figs.4.12b,c show
identically prepared specimens using XPS and TEM, respectively, which have
been reduced at identical conditions. As expected, these results show an
increased tendency toward reduction of the iron overlayer with increasing
reduction temperature until at 773 K a broadening in the Mössbauer spectrum
is observed, consistent with a change in the morphology of the iron overlayer
(also noted with TEM). Finally, reduction at 875 K and above caused a decrease in the areas under both BSCEMS and XPS spectra, which was attributed
to an attenuation of the resonant electron signal caused by a diffusion of
iron into the TiO_2 support. At these same conditions BSCEMS gave evidence
of a new spectral component which could not be attributed to a superparamag-
netic α-iron phase but rather to the formation of either a γ-iron or Fe_xTi
$(1 \leq x \leq 2)$ intermetallic phase.

Over the same reduction regime, transmission electron micrographs demonstrated that reduction of the initially contiguous and oxidized iron overlayer

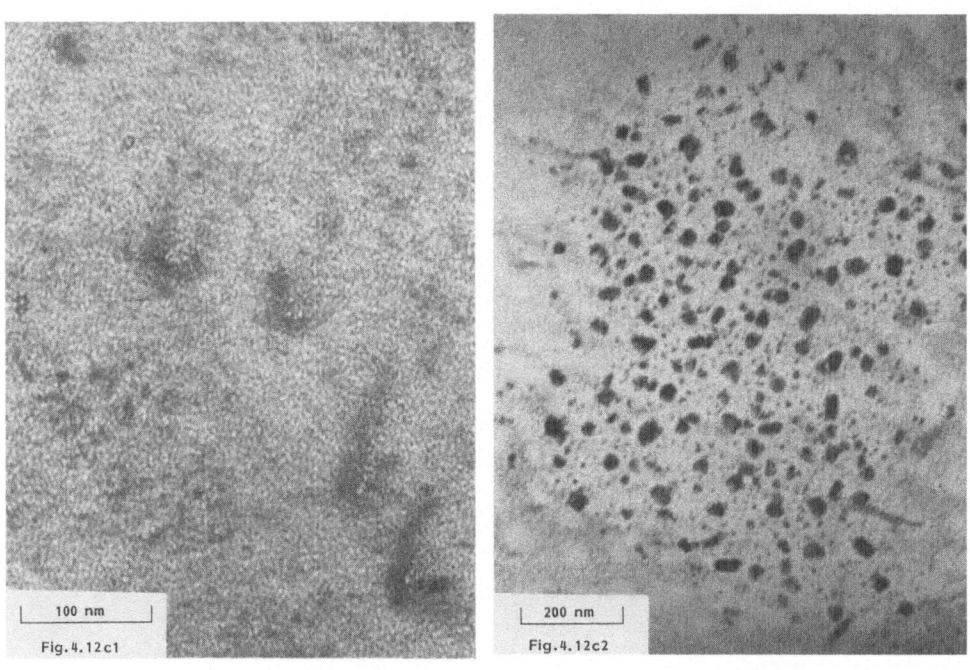

Fig.4.12c1,2   (Figure caption see opposite page)

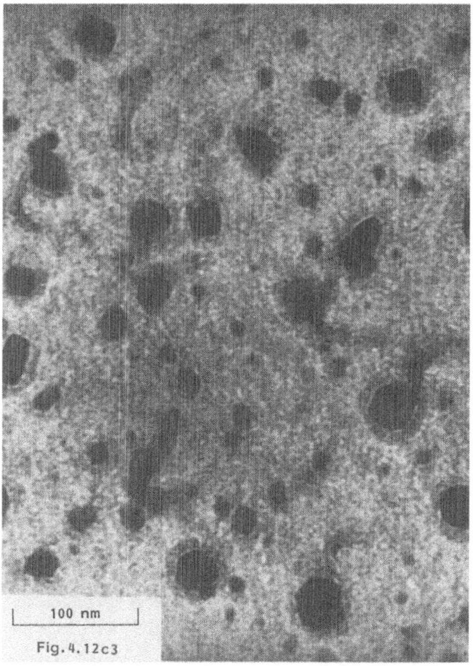

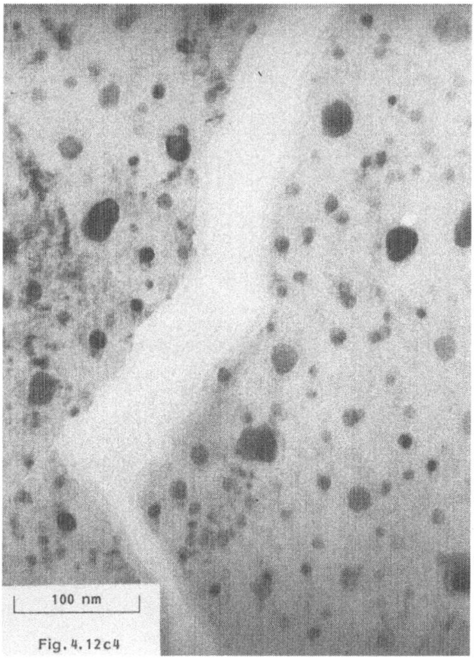

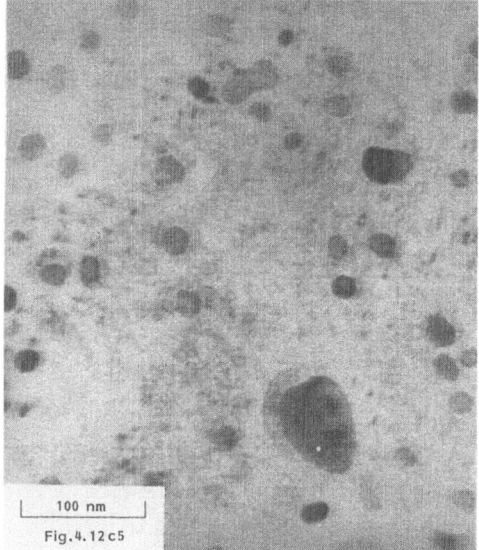

Fig.4.12c1-5. Transmission electron micrographs of Fe/TiO$_2$ specimens prepared by vacuum evaporating a 5 nm Fe overlayer onto a TiO$_2$ substrate. Specimen preparation and reduction treatments are chosen to mimic the conditions
of Figs.4.12a,b. 1 ≡ initial Fe overlayer, 2 ≡ reduced at 643 K, 3 ≡ reduced
at 707 K, 4 ≡ reduced at 773 K, 5 ≡ reduced at 875 K, reproduced with permission [4.100]

Table 4.3. TEM, XPS, BSCEMS reduction summary (5.0 nm Fe)

Reduction temp. (K)	Observed TEM trends	Observed XPS trends	Observed BSCEMS trends	Reduction regimes
Initial iron overlayer	Contiguous Fe film	Fe^{3+}/Ti^{4+}	55% Fe^{2+}, 32% Fe^0, 13% Fe^{3+}	
608	Nonuniform nucleation, most probable particle size,	Fe^0 and Fe^{2+}/Ti^{4+}	59% Fe^{2+}, 41% Fe^0	
643	<D> <10 nm	Fe^0/Ti^{4+}	57% Fe^0, 43% Fe^{2+}	Low-temperature, reduction nucleation, and growth
677	Uniform nucleation: three-dimensional crystallites,	Fe^0/Ti^{4+}	100% Fe^0	
707	<D> >10 nm			
773	Contrast between crystallites, <D> ≅ 10 nm	Fe^0/Ti^{4+}	Fe^0, broadened spectrum	Spreading of iron over support
875	Contrast between crystallites, decrease in number, density, and volume of crystallites,	Decrease in Fe spectral area, increase in Ti spectral area, reduction of Ti^{4+}	Spectral singlet (e.g., γ-Fe, Fe_XTi), ~32% decrease in spectral area	Diffusion of iron into support
973	<D> ≤10 nm			

corresponded to the nucleation and growth of discrete metallic iron crystallites. Furthermore, a change in crystallite morphology (viz., TEM particle contrast) following reduction at 773 K and a loss of surface iron following reduction at ≥875 K were both consistent with BSCEMS results [4.22,100-102].

Agreement between XPS, TEM, and BSCEMS was excellent, as shown in Table 4.3. Moreover, comparison of these results provided complementary insights into the surface properties and phase behavior of Fe/TiO$_2$ thin films which were not readily observable by other means. The ability to obtain Mössbauer spectra from samples containing relatively little iron and located within a multipurpose surface analysis system demonstrates the capability, sensitivity, and versatility of BSCEMS.

4.5.2 Theoretical

The purpose of this section is to present a physical/mathematical framework amenable to quantitative interpretation of BSCEMS spectra collected in an integral fashion with electrons above zero eV. In particular, a formalism is presented which can be used to account for intensities of signals produced by (i) backscattered electrons, (ii) backscattered photons, and (iii) forward-attenuated photons (i.e., transmission Mössbauer spectra). A framework which simultaneously accounts for all these effects is necessary to interpret BSCEMS spectra quantitatively for depth-profiled information. For exemplary purposes, ^{57}Fe is considered to be the absorbing nucleus in this treatment, although the formalism presented here can be generalized to account for any Mössbauer resonance. Since the transmission/attenuation of γ-rays has been treated in a number of discussions [4.8,9,11,22] dealing with normal-transmission Mössbauer spectroscopy, it is essential that this development of ^{57}Fe BSCEMS begin with a consideration of the ^{57}Fe excitation/decay scheme.

a) Excitation/Decay Scheme of an Excited $^{57}_{26}$Fe Nucleus

The ^{57}Fe decay process is encountered twice during the course of a BSCEMS experiment: once in the source matrix where photons are considered, and again in the absorber where both the photon and electron yields are of interest.

Table 4.4 is an energy-level diagram detailing the manner in which 100 excited $^{57}_{26}$Fe nuclei decay from the $I = 5/2$ state. Nine nuclei relax directly to the ground state by emitting a 136.32 kV γ-ray [4.9] while the remaining 91 nuclei decay to the $I = 3/2$ level releasing a 121.91 kV γ-ray. From the 91 excited nuclei that remain ($I = 3/2$) 8.2 decay to the ground state by emission of a 14.41 kV γ-ray, 0.9 generate a 14.3 kV M-shell conversion electron, 8.2 release a 13.6 kV L-shell conversion electron, and 73.7 relax by emitting a 7.3 kV K-shell conversion electron. Because of the core level vacancies produced by internal conversion, the ^{57}Fe atom is not in the electronic ground state and will decay further by one of two competing processes involving either an Auger cascade or X-ray fluorescence. In the case of the 73.7 K-shell holes, 51.9 will relax further by emitting a KLL Auger electron (resulting in an Fe^{2+} ion with two L-shell holes), and the remainder (i.e., 21.8) relax by emission of a 6.4 kV K_α X-ray (resulting in a Fe^{1+} ion with one L-shell hole). These electron holes are also accompanied by the L-shell and M-shell holes produced by L and M internal conversion, respectively, and it is clear that the atom will continue to relax by a sequence of Auger cas-

Table 4.4. Decay scheme for an excited $^{57}_{26}$Fe nucleus. Based on 100 initial I = 5/2 states, solid lines represent photons, dashed lines represent electrons

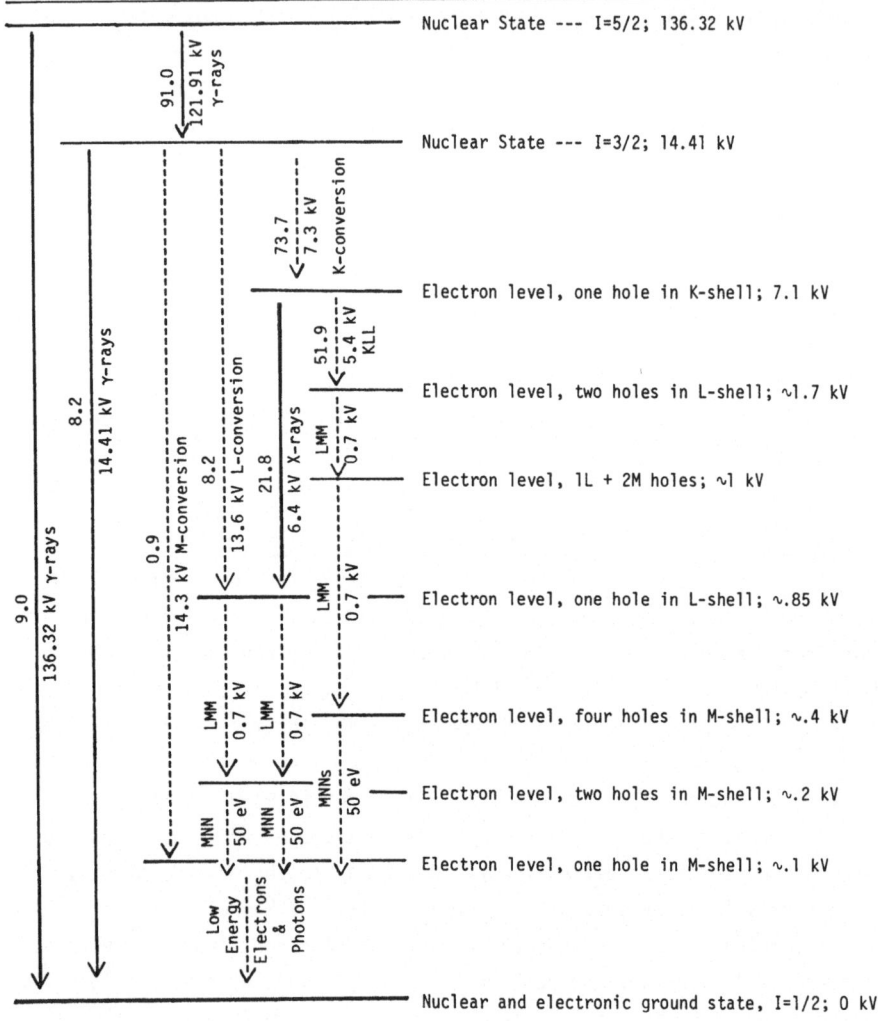

cades and X-ray fluorescences. LMM events are known to occur during BSCEMS and *Carbucicchio* [4.80] cites a measured LMM intensity of 96% of the KLL value.

As seen in Fig.4.13, the possibilities for continued relaxations via LMM and MNN Auger emissions increase dramatically, making an exact calculation of the electron/photon cascade a difficult task. *Krakowski* and *Miller* [4.88] suggest that these calculations be undertaken utilizing tabulated values of atomic binding energies, internal conversion coefficients, and X-ray fluor-

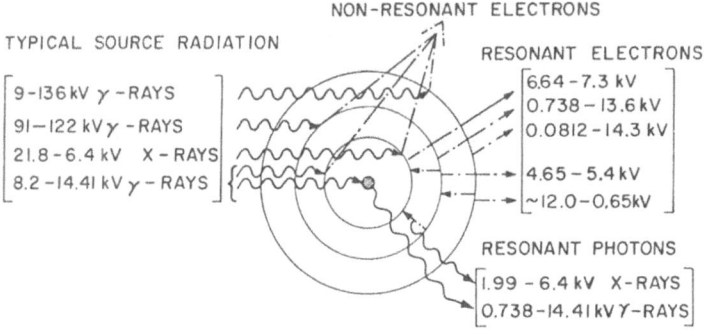

Fig.4.13. Excitation/decay scheme of a ^{57}Fe atom. Resonant yields are calculated assuming 8.2 resonant absorptions

escences. However, these calculations may be of limited value since repeated ionization of the iron atom to a higher oxidation state may invalidate the use of X-ray fluorescence and binding energy data based on a lesser number of electron holes.

Having outlined the relaxation effects that occur at an excited $^{57}_{26}$Fe atom (Table 4.4), it is important to consider the combination of excitation and decay processes that are involved at an *absorber atom* during BSCEMS. Thus, Fig.4.13 depicts a situation where the expected radiation from an ideal source matrix is allowed to impinge on an isolated $^{57}_{26}$Fe atom. Radiation from the source can be resonantly or nonresonantly scattered. In the case of 136, 122, and 6.4 kV photons, only nonresonant processes are important and produce electrons via Compton and photoelectric processes with an efficiency that depends on their cross sections. (Note: photons produced by Compton scattering are neglected in this diagram and will be discussed later.) The 14.41 kV γ-ray can undergo these same nonresonant processes or be resonantly absorbed at the nucleus (i.e., Mössbauer effect). Since the resonant cross section is ∼2 orders of magnitude greater than the sum of the nonresonant cross sections, it is clear that resonant absorption is favored. If it is assumed that all 8.2 incoming 14.41 kV γ-rays are resonantly absorbed, the eventual decay of this atom will produce a number of resonant electrons and photons with the intensities and energies shown in Fig.4.13. Detection of these electrons and photons versus the velocity of the source matrix constitutes the basis of BSCEMS. The signal-to-background ratio inherent to the spectrum, detecting either photons or electrons, depends on the ratio of resonant to nonresonant events and is of vital interest to experimentalists concerned with interpretation of Mössbauer spectra. For this reason, a major part of the remaining discussion will revolve around efforts to model the excitation/

decay scheme shown in Fig.4.13. Specifically, it is of interest to examine both physically and mathematically how the signal-to-background ratio (S/B) depends on factors such as isotopic enrichment and sample configuration. Knowledge of the sample configuration at various depths into the specimen (i.e., comparison of backscattered electron and photon spectra) can be used to glean depth-profiled chemical and physical information important to studies in catalysis, corrosion, semiconductor fabrication, and ion implantation.

b) Expected Signal-to-Background Ratio of the Electron Spectrum: The Isolated ^{57}Fe Atom

Estimates for the percent effect (E_e) of the electronic spectrum, which is equal to the signal-to-background ratio times 100% [see (4.1)], can be calculated using (4.2) [4.8,9,11,84].

$$E_e = \frac{(N_0 - N_\infty)}{N_\infty}(100\%) \quad , \tag{4.1}$$

where $E_e \equiv$ percent effect of the electronic spectrum (%), $N_0 \equiv$ signal at maximum resonance, single-peak spectrum assumed, $N_\infty \equiv$ background signal far from resonance.

Furthermore,

$$E_e = \frac{100\% \ a_n f_a f_s N_{14.41 \ kV} \sigma_0 \ (\alpha_T/1+\alpha_T)}{2 \sum_j a_j \sum_i N_i (\sigma_{c,i,j} + \sigma_{p,i,j})} \quad , \tag{4.2}$$

where $a_n \equiv$ fractional abundance of ^{57}Fe; f_a;$f_s \equiv$ recoil-free fraction of absorber or source, respectively; $N_i \equiv$ fraction of incident photon flux at energy i; $\sigma_0 \equiv$ Mössbauer absorption cross section for a 14.41 kV resonant γ-ray (Barns/atom); $\alpha_T \equiv$ total internal conversion coefficient for ^{57}Fe in the I = 3/2 state (10.11); $a_j \equiv$ fractional abundance of species j; $\sigma_{c,i,j}$; $\sigma_{p,i,j} \equiv$ Compton or photoelectric cross section, respectively, for species j using incident photons of energy i (Barns/atom).

Inherent to (4.2) are the following conditions/assumptions: (i) 136 kV resonant absorption is neglected; (ii) electron generation is isotropic; (iii) Auger electrons are not considered; and (iv) the sample must exist as a film only a monolayer thick since photon and electron attenuation by a sample/support matrix are not considered.

Using the cross sections found in Table 4.5 and the distribution of incoming photons shown in Fig.4.13, it is possible to evaluate (4.2). A value of 4230% is obtained when the sample consists of a monolayer of ^{57}Fe atoms

Table 4.5. Cross section data for Fe [4.84]. Units of cross sections are Barns/atom (1 Barns = 10^{-24} cm^2)

Event/species	Photon Energies (kV)			
	6.4	14.41	122.91	136.4
Resonant absorption $^{56}_{26}$Fe	0	0	0	0
Photoelectric event $^{56}_{26}$Fe	7.3×10^3	5.9×10^3	10.0	7.4
Compton scattering $^{56}_{26}$Fe	5.5	9.7	12.0	11.6
Resonant absorption $^{57}_{26}$Fe	0	2.4×10^6	0	0
Photoelectric event $^{57}_{26}$Fe	7.3×10^3	5.9×10^3	10.0	7.4
Compton scattering $^{57}_{26}$Fe	5.5	9.7	12.0	11.6

and both the source and absorber are assumed to have recoil-free fractions of unity. Experimentally measured values of E_e are usually much less than this value so that circa 4000% must be considered an upper limit subject to the assumptions of (4.2) (i.e., isolated atom).

c) Expected Signal-to-Background Ratio of the Electron Spectrum: The Semi-Infinite Absorber

Figure 4.14 illustrates a situation commonly encountered during electron detection for ^{57}Fe-BSCEMS. In this experiment the complement of photons that originates from the source matrix is collimated and allowed to intercept the surface of a ^{57}Fe-containing specimen at some angle θ_1 to the surface normal. After proceeding an arbitrary distance t_a into the sample (t_a measured along the direction of travel), the photons may be absorbed either resonantly or nonresonantly generating at least one electron per interaction. These electrons may or may not leave the solid depending on their energy and the direction in which they are ejected from the atom. A finite number of electrons do escape the solid, however, and some fraction of these which emanate at an angle θ_2 from the normal are detected by an electron detector. Because the electrons (resonant and nonresonant) are assumed to be generated isotropically, an expression of the same form as (4.2) can be used to calculate an expected value for the percent effect (E_e). If θ_1 equals θ_2, this expression takes the following form:

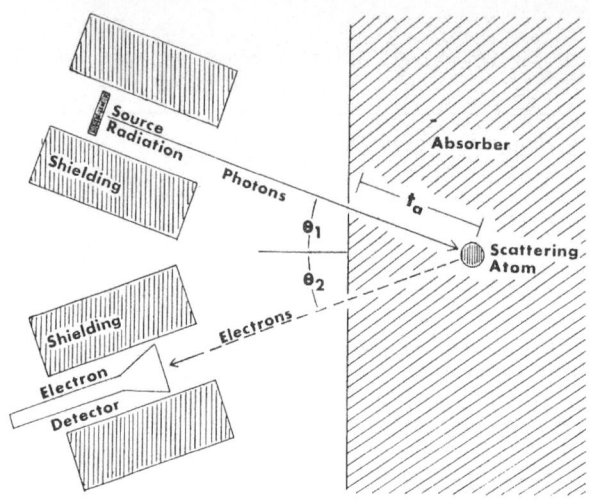

$$E_e = \frac{100\% a_n f_a f_s N_{14.41kV} \sigma_0 \left(\frac{\alpha_T}{1+\alpha_T}\right) \int\limits_0^\infty \Psi_{Res,t_a} P_{Res,t_a} \, dt_a}{2 \sum\limits_j a_j \sum\limits_i \left[N_i (\sigma_{c,i,j} + \sigma_{p,i,j}) \int\limits_0^\infty \Psi_{i,t_a} P_{i,t_a} \, dt_a \right]}, \qquad (4.3)$$

where $t_a \equiv$ distance or thickness in units of mg/cm^2, (e.g., for $^{56}_{26}$Fe: 1 mg/cm^2 = 1272 nm); $\Psi_{Res,t_a} \equiv$ the probability that a resonant 14.41 kV γ-ray is transmitted a distance t_a through the absorber matrix; $P_{Res,t_a} \equiv$ the probability that an electron generated by a resonant process is transmitted a distance t_a through the absorber matrix; Ψ_{i,t_a}; $P_{i,t_a} \equiv$ the probability that a photon or electron, respectively, of energy i is transmitted a distance t_a through the absorber matrix.

Equation (4.3) is identical to (4.2) except for the addition of an integral to the numerator and denominator. Phenomenologically, (4.3) states that the ratio of resonant to nonresonant processes, occurring at any depth t_a in the absorber, can still be described by (4.2) provided that corrections are made to account for attenuation of both photons and electrons. However, since electron attenuation depends on electron energy, a further expansion of (4.3) is necessary to account for: (i) K, L, and M-shell conversion electrons, (ii) generation of nonresonant electrons, and (iii) Auger electrons emitted following (i) and (ii). Thus, (4.3) becomes

$$E_e = \frac{C \sum\limits_{h=K,L,M} \left[\left(\frac{\alpha_h}{1+\alpha_T}\right) \int\limits_0^\infty \Psi_{Res,t_a} (P_{h,t_a} + e_{KLL,h}P_{KLL,t_a} + e_{LMM,h}P_{LMM,t_a}) dt_a \right]}{\sum\limits_j a_j \sum\limits_i \left[N_i(\sigma_{c,i,j} + \sigma_{p,i,j}) \int\limits_0^\infty \Psi_{i,t_a} (P_{i,t_a} + e_{KLL,i}P_{KLL,t_a} + e_{LMM,i}P_{LMM,t_a}) dt_a \right]}$$

(4.4)

where

$$C \equiv \frac{100\% a_n f_a f_s N 14.41 \sigma_0}{2} \quad ,$$

and $\alpha_{K,L,M} \equiv$ K,L,M-shell internal conversion coefficient, respectively, $\alpha_K = 9.0$, $\alpha_L = 1.0$, $\alpha_M = 0.11$, $\alpha_T = \alpha_K + \alpha_L + \alpha_M = 10.11$. Furthermore, P_{h,t_a}; P_{KLL,t_a}; $P_{LMM,t_a} \equiv$ the probability that an h-shell conversion electron (h = K,L,M) or respectively that a KLL or LMM Auger electron is transmitted a distance t_a through the absorber matrix. In addition, $e_{KLL,h}$; $e_{LMM,h} \equiv$ the number of KLL or LMM Auger electrons, respectively, generated by a single Fe atom after removal of a h-shell electron (h = K,L,M) by an internal conversion process; $e_{KLL,i}$; $e_{LMM,i} \equiv$ the number of KLL or LMM Auger electrons generated by a single Fe atom, respectively, after Compton or photoionization by an incident photon of energy i; and $P_{i,t_a} \equiv$ the probability that a Compton or photoelectron, generated by an incident photon of energy i, is transmitted a distance t_a through the absorber matrix.

Equation (4.4) is now in a form suitable for evaluation; however, it still retains a number of assumptions. (i) Compton events are treated as if they were actually photoelectric processes. Since photoelectric cross sections for 6.4 and 14.4 kV photons are much larger than Compton cross sections (Table 4.5), this assumption has little effect on the intensity of nonresonant (i.e., background) radiation produced by these photons. In the case of 122 and 136 kV photons, where Compton and photoelectric effects are comparable, this assumption is somewhat more significant. Indeed, a 50% reduction in the nonresonant contribution produced by 136 and 122 kV photons would result if Compton electrons are not detected. However, it can be estimated [4.105] that a Compton electron scattered in the opposite direction of an incoming 136 kV photon (i.e., 180° change in direction) possesses about 47 kV of kinetic energy. (ii) Auger electrons with energies <LMM transitions have not been included in (4.4) because the frequencies of these events are difficult to estimate, as noted earlier. It is also expected that since they have relatively low energies (circa 0.65 kV, and thus short path lengths in solids), they will contribute little to the resonant spectrum in comparison to higher energy electrons when semi-infinite absorbers are used. (iii) The

generation of all electrons is assumed to be isotropic. As pointed out by
Huffman [4.90], this is an excellent assumption for the conversion and Auger
electrons, but it will break down somewhat for the higher energy (e.g., 122
and 136 kV) Compton and photoelectrons.

Now that the assumptions implicit in (4.4) have been noted, the various
terms of this equation can be expanded prior to its evaluation.

Expansion of Terms in (4.4): $e_{KLL,h}$; $e_{LMM,h}$; $e_{KLL,i}$; $e_{LMM,i}$

Table 4.6 contains values for the "e-type" parameters encountered in (4.4).
For example, Table 4.4 shows that 0.7 KLL Auger electrons are generated for
each K-shell conversion; thus, $e_{KLL,K} = 0.7$. In a similar fashion ~1.7 and
~1.0 LMM Auger electrons are generated for each K- and L-shell conversion,
respectively (assuming an X-ray fluorescence yield near zero), so that
$e_{LMM,K} \cong 1.7$ and $e_{LMM,L} \cong 1.0$. The values of $e_{KLL,L}$, $e_{KLL,M}$, and $e_{LMM,M}$ are
all zero as these terms violate Auger transition rules while $e_{KLL,6.4}$ must
be zero since a photon of 6.4 kV possesses insufficient energy to cause a
K-shell ionization. Since nonresonant ionization caused by 14.41 kV photons
occurs primarily in the K-shell (largest total cross section), $e_{KLL,14.41}$
can be estimated to be ~0.7 with $e_{LMM,14.41}$ then equal to ~1.7 due to the
relaxation events which follow the KLL Auger electron and/or 6.4 kV X-ray.
Similarly, since nonresonant ionization by 6.4 kV photons occurs primarily
in the L shell, $e_{LMM,6.4} \cong 1.0$ (Table 4.4).

Table 4.6. Estimated values of $e_{KLL,h}$; $e_{LMM,h}$; $e_{KLL,i}$; and $e_{LMM,i}$

Parameter	Value
$e_{KLL,K}$	0.7
$e_{KLL,L}$	0
$e_{KLL,M}$	0
$e_{LMM,K}$	~1.7
$e_{LMM,L}$	~1.0
$e_{LMM,M}$	0
$e_{KLL,6.4}$	0
$e_{KLL,14.41}$	~0.7
$e_{KLL,122}$	~0
$e_{KLL,136}$	~0
$e_{LMM,6.4}$	~1.0
$e_{LMM,14.41}$	~1.7
$e_{LMM,122}$	~0
$e_{LMM,136}$	~0

98

Since the relatively large penetration depth of 122 and 136 kV photons produces ionizations that are generally at depths greater than the Auger escape depth, it may be estimated that $e_{KLL,122} = e_{KLL,136} = e_{LMM,122} = e_{LMM,136} \cong 0$.

Expansion of Terms in (4.4): Ψ_{i,t_a} and Ψ_{Res,t_a}

Nonresonant photon attenuation can be described by:

$$\Psi_{i,t_a} = \exp(-\mu_{i,j} t_a) \quad , \tag{4.5}$$

where $\mu_{i,j} \equiv$ mass attenuation coefficient of photons with energy i by species j (cm^2/mg), $i = 6.4$, 122, or 136 kV.

Since photon attenuation by pair production as well as Rayleigh and Thomson scattering are insignificant in the energy range of interest [4.86] (i.e., ≤ 136 kV), the mass attenuation coefficient depends on the sum of the Compton and photoelectric cross sections. Thus:

$$\mu_{i,j} = 1.079 \times 10^{-5} \left(\frac{cm^2}{mg} \Big/ \frac{Barns}{atom}\right)\left[\sigma_{c,i,j} + \sigma_{p,i,j}\right] \quad , \tag{4.6}$$

and the values of $\mu_{i,j}$ can be determined using Table 4.5.

For 14.41 kV γ-rays, both resonant and nonresonant attenuations may take place, and the following equation has been developed by *Margulies* and *Ehrman* [4.87] to account for this process:

$$\Psi_{14.41\ kV,t_a} = [\exp(-\mu_{i,j} t_a)(1 - f_s)] + [(\exp(-\mu_{i,j} t_a))f_s \exp(-1/2\ T_a)$$

$$J_0(1/2\ iT_a)] \quad , \tag{4.7}$$

where $J_0 \equiv$ the zero-order Bessel function, and

$$T_a = f_a n_a a_n \sigma_0 t \quad , \tag{4.8}$$

with $T_a \equiv$ effective thickness; $n_a \equiv$ atomic density $(atoms/cm^2)$; and $t \equiv$ thickness or distance (cm).

For iron, T_a may be conveniently expressed as:

$$T_a = f_a a_n c t_a \quad , \tag{4.9}$$

where $c = 25.663$ cm^2/mg.

Examination of (4.7) reveals that the first term represents nonresonant attenuation of that fraction $(1-f_s)$ of the 14.14 kV γ-rays which are nonresonant (i.e., cannot cause Mössbauer absorption), while the second term considers both resonant and nonresonant absorption of the resonant beam (f_s).

Thus, the attenuation of the resonant γ-ray flux may be written as:

$$\Psi_{Res,t_a} = \exp[-(\mu_{i,j} + 1/2(f_a\text{a}_nc))t_a]J_0(1/2i(f_a\text{a}_nct_a)) \quad , \qquad (4.10)$$

with the Bessel function represented in the following manner:

$$J_0(ix) = 1 + (x/2)^2 + \frac{(x/2)^4}{1^2 \cdot 2^2} + \frac{(x/2)^6}{1^2 \cdot 2^2 \cdot 3^2} + \dots \quad . \qquad (4.11)$$

Expansion of Terms in (4.4): P_{h,t_a} ; P_{i,t_a} ; P_{KLL,t_a} ; P_{LMM,t_a}

The fraction of a monoenergetic electron beam that is transmitted through a solid to emerge with energy greater than zero eV has been measured by a number of workers. *Subba Rao* [4.103] recently fitted these data to a generalized expression (of the Fermi-function type) which appears to give good agreement with both transmission and absorption curves for electrons with incident energies between 10 and 3000 kV. Since electrons generated during BSCEMS are assumed to be emitted at a number of discrete energies, it is expected that this expression can be used to account for their transmission. Thus,

$$P_{n,t_a} = [1 + \exp(-\mu_e x_0)]/\{1 + \exp[\mu_e(t_a - x_0)]\} \quad , \qquad (4.12)$$

with

$$\mu_e = [9.2Z^{-0.2} + 16Z^{-2.2}][(1.66 \times 10^{-3}/Z^{0.33})n^{(2.579-0.219 \log_{10}n)}]^{-1} \quad , \qquad (4.13)$$

$$x_0 = [0.63(Z/A) + 0.27][(1.66 \times 10^3/Z^{0.33})n^{2.579-0.219 \log_{10}n}] \quad , \qquad (4.14)$$

where $n \equiv$ electron energy (kV); $Z \equiv$ atomic number; $A \equiv$ molecular weight (g/mole); $P_{n,t_a} \equiv$ the probability that an electron with energy n is transmitted a distance t_a through the absorber matrix.

The resemblance between P_{n,t_a} and the other quantities (e.g., P_{h,t_a}) is clear, so that

$$P_{h,t_a} = P_{n,t_a} \quad , \qquad (4.15)$$

where

$$\begin{bmatrix} h = K \rightarrow n = 7.3 \text{ kV} \\ h = L \rightarrow n = 13.6 \text{ kV} \\ h = M \rightarrow n = 14.3 \text{ kV} \end{bmatrix}$$

$$P_{KLL,t_a} = P_{n,t_a} \quad ; \quad n = 5.4 \text{ kV} \quad , \qquad (4.16)$$

and

$$P_{LMM,t_a} = P_{n,t_a} \; ; \; n = 0.65 \text{ kV} \; . \tag{4.17}$$

It should be noted that since LMM Auger electrons possess energies well below 10 kV, they may not be adequately described by (4.12). In fact, $P_{0.65,t_a}$ = 0.05 when t_a is ~2.18×10^{-4} mg/cm^2, which corresponds to about a monolayer of iron. This represents an unusually strong attenuation of these electrons, while $P_{5.4,t_a}$ = 0.05 when t_a is equivalent to ~50 nm (which appears reasonable). Therefore, future modeling efforts may wish to use a different form of the electron attenuation function specifically to account for the behavior of these low-energy electrons.

In the case of P_{i,t_a}, additional assumptions must be made since electrons generated nonresonantly by photons of energy i can have any one of a number of discrete energies between i and i-E_K, where E_K is the binding energy of an electron in the K-shell. For 122 and 136 kV photons, this energy interval is *relatively* small so that $n \cong i$ and

$$P_{i,t_a} = P_{n,t_a} : \text{for i = 122 or 136 kV} \; . \tag{4.18}$$

For 14.41 kV γ-rays, the major ionization takes place in the K-shell, so that

$$P_{i,t_a} = P_{n,t_a} \; ; \; n = 7.1 \text{ kV when i = 14.41 kV} \; , \tag{4.19}$$

and in a similar fashion for the 6.4 kV X-rays:

$$P_{i,t_a} = P_{n,t_a} \; ; \; n = 5.6 \text{ kV when i = 6.4 kV} \; . \tag{4.20}$$

Evaluation of (4.4)

Table 4.7 contains tabulated values for the integrals encountered during expansion of (4.4). Expansion and evaluation of (4.4) yields a value of 757% effect (Table 4.8), which is a factor of ~5 less than that obtained for the isolated atom. As observed from Table 4.8, the large decrease results primarily from the contribution of 122 kV photons. Since these photons have relatively small total cross sections (compared with 14.41 and 6.4 kV photons), it is clear that the large background contribution results because (i) 122 kV photons comprise a large fraction of the incident radiation (circa 70%) and (ii) 122 kV electrons have relatively large transmission lengths in the solid matrix.

By changing limits on the integrals shown in Table 4.7, and altering $\Psi_{14.41,t_a}$ and Ψ_{Res,t_a} to account for predominantly nonresonant attenuation, it is possible to calculate the percent effect of thinner films. In this

Table 4.7. Numerical evaluation of the integrals $\int_0^\infty \Psi_{i,t_a} P_{i,t_a}\, dt_a$, at $a_n = 0.91$ and $f_s = f_a = 0.7$

Photon energy (kV)	Integral form	Equivalent form	Integral value (arbitrary units)
14.41 Res	$\int_0^\infty \Psi_{Res,t_a} P_{K,t_a}\, dt_a$	$\int_0^\infty \Psi_{Res,t_a} P_{7.3,t_a}\, dt_a$	3.33×10^{-2}
14.41 Res	$\int_0^\infty \Psi_{Res,t_a} P_{L,t_a}\, dt_a$	$\int_0^\infty \Psi_{Res,t_a} P_{13.6,t_a}\, dt_a$	8.76×10^{-2}
14.41 Res	$\int_0^\infty \Psi_{Res,t_a} P_{M,t_a}\, dt_a$	$\int_0^\infty \Psi_{Res,t_a} P_{14.3,t_a}\, dt_a$	9.35×10^{-2}
14.41 Res	$\int_0^\infty \Psi_{Res,t_a} P_{KLL,t_a}\, dt_a$	$\int_0^\infty \Psi_{Res,t_a} P_{5.4,t_a}\, dt_a$	1.84×10^{-2}
14.41 Res	$\int_0^\infty \Psi_{Res,t_a} P_{LMM,t_a}\, dt_a$	$\int_0^\infty \Psi_{Res,t_a} P_{0.65,t_a}\, dt_a$	1.13×10^{-4}
136	$\int_0^\infty \Psi_{i,t_a} P_{i,t_a}\, dt_a$	$\int_0^\infty \Psi_{136,t_a} P_{136,t_a}\, dt_a$	11.15
122	$\int_0^\infty \Psi_{i,t_a} P_{i,t_a}\, dt_a$	$\int_0^\infty \Psi_{122,t_a} P_{122,t_a}\, dt_a$	9.314
14.41	$\int_0^\infty \Psi_{i,t_a} P_{i,t_a}\, dt_a$	$\int_0^\infty \Psi_{14.41,t_a} P_{7.1,t_a}\, dt_a$	3.53×10^{-2}
14.41	$\int_0^\infty \Psi_{i,t_a} P_{KLL,t_a}\, dt_a$	$\int_0^\infty \Psi_{14.41,t_a} P_{5.4,t_a}\, dt_a$	1.95×10^{-2}
14.41	$\int_0^\infty \Psi_{i,t_a} P_{LMM,t_a}\, dt_a$	$\int_0^\infty \Psi_{14.41,t_a} P_{0.65,t_a}\, dt_a$	1.13×10^{-4}
6.4	$\int_0^\infty \Psi_{i,t_a} P_{i,t_a}\, dt_a$	$\int_0^\infty \Psi_{6.4,t_a} P_{5.6,t_a}\, dt_a$	2.25×10^{-2}
6.4	$\int_0^\infty \Psi_{i,t_a} P_{LMM,t_a}\, dt_a$	$\int_0^\infty \Psi_{6.4,t_a} P_{0.65,t_a}\, dt_a$	1.13×10^{-4}

fashion, Table 4.8 also contains pertinent quantities for a 5.0 nm overlayer of 91% $^{57}_{26}$Fe covering a semi-infinite $^{56}_{26}$Fe foil. This calculation shows that ~12.5% of the resonant signal results from the topmost 5.0 nm and that the contribution of low-energy electrons (i.e., $\leq$5.4 kV) increases significantly. Indeed, it is suspected that if the Subba Rao expression for electron transmission were replaced by one better able to describe the behavior of LMM Auger electrons, the relative importance of these low-energy electrons would be enhanced.

Table 4.8. Evaluation of (4.4), where $a_n = 0.91$, $f_a = f_s = 0.7$

	Semi-infinite foil of $0.91 - {}^{57}_{26}Fe$	5.0 nm overlayer of $0.91 - {}^{57}_{26}Fe$ on semi-infinite foil of ${}^{56}_{26}Fe$
Percent effect (E_e) with 6.4 kV X-rays	757.%	95.%
Percent effect (E_e) without 6.4 kV X-rays	876.%	110.%
(Photon type)	(% of background)	(% of background)
136	7.2	6.9
122	70.3	67.5
14.41	9.0	12.6
6.4	13.5	13.0
(Electron type)	(% of signal)	(% of signal)
K-shell conversion	57.9	53.5
KLL Auger	22.7	37.0
LMM Auger	0.3	2.7
L-shell conversion	17.1	6.0
LMM Auger	0.02	0.2
M-shell conversion	2.0	0.7

d) Expected Signal-to-Background Ratio of the Photon Spectrum: The Semi-Infinite Absorber

For a semi-infinite absorber the magnitude of the backscattered X-ray intensity (I_x) which arises from resonant absorption is proportional to

$$I_x \approx \left(\frac{0.3\alpha_K}{1 + \alpha_T}\right) \int_0^\infty \Psi_{Res,t_a} \Psi_{6.4,t_a} dt_a \quad , \tag{4.21}$$

where $I_x \equiv$ backscattered X-ray intensity from resonant absorption.

By evaluating the rhs of (4.21) and comparing it with the summation in the numerator of (4.4), it is observed that the backscattered X-ray flux that results from resonant absorption is ~150% of the resonant electron signal ($f_a = f_s = 0.7$, $a_n = 0.91$). This large effect reflects the milder attenuation experienced by X-rays and underlines the necessity of BSCEMS spectrometers being able to discriminate between electrons and photons. Similarly, the backscattered 14.41 kV photon intensity (I_γ) can be approximated from

$$I_\gamma \approx \left(\frac{1}{1 + \alpha_T}\right) \int_0^\infty \Psi_{Res,t_a} [(1 - f_a)\exp(-\mu_{14.41}t_a) + f_a\Psi_{Res,t_a}]dt_a \quad , \tag{4.22}$$

which yields a value of 30.3% of the resonant electron signal when $f_a = f_s$ = 0.7 and $a_n = 0.91$.

Of interest to the present discussion then is whether or not these back-scattered photon fluxes can produce significant electron signals.

For γ-rays, the number of reemitted photons will be $(1/1 + \alpha_T)$ or ~9% of the initial resonant flux. If these γ-rays could be removed and refocused (i.e., collimated) onto the absorber, as is the incident radiation, it would correspond to using a source with $(0.09\ f_a)$ additional activity. In actuality, these reemitted γ-rays are not collimated in the direction of the source and their maximum intensity is not at the surface of the absorber (where electron attenuation is at a minimum), so that the relative intensity of electrons produced by backscattered γ-rays must be $<<0.09\ f_a$.

An upper limit on the electron signal generated by resonant backscattered X-rays (I_{xe}) can be estimated from

$$I_{xe} \leq \left(\frac{0.3\alpha_K}{1 + \alpha_T}\right)\left[\int_0^\infty {}^\Psi Res,t_a\ dt_a - \int_0^\infty {}^\Psi Res,t_a {}^\Psi 6.4,t_a\ dt_a\right] \quad . \tag{4.23}$$

The first integral in this expression represents the resonant backscattered X-ray intensity assuming no attenuation by the solid, while the second integral determines this same value including attenuation by the solid matrix. The difference of these two terms sums the number of attenuations, which corresponds to electron production. If all of these electrons (which are generated isotropically) were directed at the detector and experienced no attenuation, then the value of I_{xe} would be 5.7% of the total resonant electron signal. Because of the assumptions involved in calculating I_{xe}, it is clear that I_{xe} accounts for $<<5.7\%$ of the conversion electron spectrum.

4.5.3 Experimental Measurements of Percent Effect: Verification of Proposed Model

As observed by *Tatarchuk* and *Dumesic* [4.101,102], BSCEMS spectra of 5.0 nm of 91% $^{57}_{26}$Fe on TiO_2 produced resonant effects equivalent to circa 20% mm/s with 6.4 kV X-rays not removed. It is noteworthy that the evaluation of (4.4) produced a value of 95% effect (6.4 kV X-rays unfiltered) for 5.0 nm of 91% $^{57}_{26}$Fe on $^{56}_{26}$Fe. Converting this effect to a spectral area using

$$\text{Spectral Area (\% mm/s)} = 2E_e\Gamma_n\ \pi/2 \quad , \tag{4.24}$$

where $\Gamma_n \equiv$ natural or Heisenberg linewidth $\cong 0.096$ mm/s, yields a value of 28.7% mm/s. Thus, there is good agreement between (4.4) and experimental data.

It was also observed [4.22] that an iron foil (0.1 mil thick, enriched to 91% $^{57}_{26}$Fe) mounted on a copper/titanium holder produced a resonant effect of ~36% mm/s. In this configuration the iron foil covered only a 0.14 fraction of the area viewed by the electron detector. If the result of Table 4.8 (semi-infinite foil of 91% $^{57}_{26}$Fe) is normalized by this view factor [i.e., (0.14) × (757%)], a value of 32% mm/s is obtained. Since fair agreement is obtained for ^{57}Fe absorbers at both extremes of sample thickness, it appears that (4.4) may be used to represent adequately the conversion electron process for integral detectors accepting electrons above zero eV.

4.6 Potential for Future Developments in Surface-Specific Mössbauer Spectroscopy

Surface-specific Mössbauer spectroscopies in the form of BSCEMS and BSPMS have direct application to a number of important research fields, including catalysis, surface science, corrosion, thin-film coatings, adhesion, ion implantation/surface hardening, laser/electron glazing, and assorted biological, metallurgical and geological problems. Because of the detailed structural, chemical, electronic, and magnetic information which is provided by the Mössbauer effect, and the surface specific and nondestructive depth-profiling capabilities of BSCEMS and BSPMS in particular, it is expected that these two techniques will see a growing implementation in these areas. In particular, combined BSCEMS and BSPMS studies have the unique ability to examine and relate phenomena which occur simultaneously at both the outermost surface of a specimen (circa 0.25 nm) and further into the bulk (circa 20 μm).

At present, the theoretical understanding and available equipment for BSCEMS or BSPMS are limited. Refinements in these techniques, however, are likely to be along a number of different avenues. One such approach may be the development of dedicated Mössbauer spectrometers capable of performing integrated BSCEMS and BSPMS studies for detailed and nondestructive depth profiling over the range 0.25 nm - 20 μm. A second goal of surface-specific Mössbauer spectroscopy would be the refinement of retarding field and cylindrical mirror-type analyzers to add to existing surface analysis instrumentation. And finally, a third approach would be the development of improved proportional counters for integral detection of BSCEMS and/or measurements of BSPMS using resonant γ-rays or fluorescent X-rays. These latter devices are inexpensive, portable, and can collect spectra in a matter of minutes. Since the surfaces of large objects can be readily examined, it is possible that such detectors may be particularly suited for commercial production monitoring or other on site testing/characterization procedures. Indeed, simultaneous collection of an integral electron spectrum, a resonant γ-ray spectrum, and a fluorescent X-ray spectrum allows three different depth regions to be rapidly examined.

While it was suggested earlier that BSCEMS and BSPMS can be easily obtained from eight different metals including ^{57}Fe, ^{119}Sn, ^{121}Sb, ^{125}Te, ^{149}Sm, ^{151}Eu, ^{169}Tm, and ^{181}Ta, it should be noted that Mössbauer spectroscopy studies are not restricted to just these elements. In fact, the sensitivity of BSCEMS and BSPMS is such that they can examine almost any solid material provided it contains just a few percent or less of one of these metals. To date, only Sn and Fe have seen significant use of BSCEMS and BSPMS; however, other nuclei such as ^{181}Ta are so sensitive to conversion electron studies that the effects of an adsorbed gas can be observed [4.104].

To enhance the capabilities and impact of BSCEMS or BSPMS, further work is needed to understand the nuclear/electronic relaxation process more fully and to gain better insight into the transport of low-energy resonant electrons in the solid. Indeed, the high sensitivity of BSCEMS and BSPMS, combined with the ability to make in situ measurements, the lack of limitation on sample size, and the capability to make nondestructive depth profiles make these techniques good choices for further study and development.

References

4.1 H. Frauenfelder: *The Mössbauer Effect* (Benjamin, New York 1962)
4.2 G.K. Wertheim: *Mössbauer Effect: Principles and Applications* (Academic, New York 1964)
4.3 I.J. Gruverman (ed.): *Mössbauer Effect Methodology* (Plenum, New York 1965, to present
4.4 A.H. Muir, K.J. Ando, H.M. Coogan: *Mössbauer Effect Data Index, 1958-1965* (Interscience, New York 1966)
4.5 J.G. Stevens, V.E. Stevens: *Mössbauer Effect Data Index 1965-1975* (Adam Hilger, London 1976)
4.6 V.I. Goldanskii, R. Herber (eds.): *Chemical Applications of Mössbauer Spectroscopy* (Academic, New York 1968)
4.7 L. May (ed.): *An Introduction to Mössbauer Spectroscopy* (Plenum, New York 1971)
4.8 N.N. Greenwood, T.C. Gibb: *Mössbauer Spectroscopy* (Chapman and Hall, London 1971)
4.9 G.M. Bancroft: *Mössbauer Spectroscopy* (Wiley, New York 1973)
4.10 S.G. Cohen, M. Pasternak (eds.): *Perspectives in Mössbauer Spectroscopy* (Plenum, New York 1973)
4.11 T.C. Gibb: *Principles of Mössbauer Spectroscopy* (Chapman and Hall, London 1976)

4.12   P. Gutlich, R. Link, A. Trautwein: *Mössbauer Spectroscopy and Transition Metal Chemistry* (Springer, Berlin, Heidelberg, New York 1978)
4.13   M.C. Hobson: "The Mössbauer Effect in Surface Science", in *Progress in Surface and Membrane Science*, Vol.5 (1972) p.1
4.14   J.A. Dumesic, H. Topsøe: "Mössbauer Spectroscopy Applications to Heterogeneous Catalysis", in *Advances in Catalysis*, Vol.26 (1977) p.121
4.15   G.W. Simmons, H. Leidheiser: "Corrosion and Interfacial Reactions", in *Applications of Mössbauer Spectroscopy*, Vol.1, ed. by R.L. Cohen (Academic, New York 1976)
4.16   W.N. Delgass, G.L. Haller, R. Kellerman, J.H. Lunsford: *Spectroscopy in Heterogeneous Catalysis* (Academic, New York 1979)
4.17   S. Mørup, J.A. Dumesic, H. Topsøe: "Magnetic Microcrystals", in *Applications of Mössbauer Spectroscopy*, Vol.2, ed. by R.L. Cohen (Academic, New York 1980)
4.18   H. Topsøe, J.A. Dumesic, S. Mørup: "Catalysis and Surface Science", in *Applications of Mössbauer Spectroscopy*, Vol.2, ed. by R.L. Cohen (Academic, New York 1980)
4.19   G.P. Huffman: "Mössbauer Studies of Surface-Treated Steels", in *Applications of Mössbauer Spectroscopy*, Vol.2, ed. by R.L. Cohen (Academic, New York 1980)
4.20   M. Petrera, U. Gonzer, U. Hansmann, W. Keune, J. Lauer: J. Phys. Coll. C6, 295 (1976)
4.21   G.K. Shenoy, F.E. Wagner (eds.): *Mössbauer Isomer Shifts* (North-Holland, New York 1978)
4.22   B.J. Tatarchuk: Ph.D. Thesis, University of Wisconsin, Madison (1981)
4.23   R.N. Ord, C.L. Christensen: Nucl. Instrum. Methods 91, 293 (1971)
4.24   J.H. Terrell, J.J. Spijkerman: Appl. Phys. Lett. 13, 11 (1968); R.H. Forsyth, J.H. Terrell: Bull. Am. Phys. Soc. 13, 61 (1968)
4.25   N. Herskowitz, J.C. Walker: Nucl. Instrum. and Methods 53, 273 (1967)
4.26   J. Fenger: Nucl. Instrum. Methods 69, 268 (1969)
4.27   Y. Isozumi, D.I. Lee, I. Kádár: Nucl. Instrum. Methods 120, 23 (1974)
4.28   K.R. Swanson, J.J. Spijkerman: J. Appl. Phys. 41, 3155 (1970)
4.29   J.J. Spijkerman: Mössbauer Eff. Methodol. 7, 85 (1971)
4.30   H. Onodera, H. Yamamoto, H. Watanabe, H. Ebiko: Jpn. J. Appl. Phys. 11, 1380 (1972)
4.31   G.W. Simmons, E. Kellerman, H. Leidheiser: Corrosion-NACE 29, 227 (1973)
4.32   M.J. Tricker, J.M. Thomas, A.P. Winterbottom: Surf. Sci. 45, 601 (1974)
4.33   J.M. Thomas, M.J. Tricker, A.P. Winterbottom: J. Chem. Soc. Faraday Trans. 2, 71, 1708 (1975)
4.34   M.J. Tricker, A.G. Freeman, A.P. Winterbottom, J.M. Thomas: Nucl. Instrum. Methods 135, 117 (1976)
4.35   G.P. Huffman, H.H. Podgurski: Oxid. Metals 10, 377 (1976)
4.36   A. Sette Camara, W. Keune: Corros. Sci. 15, 441 (1975)
4.37   M.J. Graham, D.F. Mitchell, D.A. Channing: Oxid. Metals 12, 247 (1978)
4.38   M.J. Tricker, R.K. Thorpe, J.H. Freeman, G.A. Gard: Phys. Status Solidi A: 33, K97 (1976)
4.39   J. Stanek, J. Sawicki, B. Sawicka: Nucl. Instrum. Methods 130, 613 (1975)
4.40   B.D. Sawicka, J. Sawicki, J. Stanek: Phys. Lett. 59A, 59 (1976)
4.41   R.L. Collins, R.A. Mazak, C.M. Yagnik: Mössbauer Eff. Methodol. 8, 191 (1973)
4.42   R.C. Mercader, T.E. Cranshaw: J. Phys. F 5, L124 (1975)
4.43   M.J. Tricker, A.G. Freeman: Surf. Sci. 52, 549 (1975)
4.44   M.J. Tricker, A.P. Winterbottom, A.G. Freeman: J. Chem. Soc. Dalton Trans., 1289 (1976)
4.45   J.G. Goodwin, G. Parravano: J. Vac. Sci. Technol. 14, 1157 (1977)

4.46 J.G. Goodwin, G. Parravano: J. Phys. Chem. **82**, 1040 (1978)
4.47 L.J. Swartzendruber, L.H. Bennett: Scr. Metall. **6**, 737 (1972)
4.48 M.J. Tricker, L.A. Ash, W. Jones: J. Inorg. Nucl. Chem. **41**, 891 (1979)
4.49 J. Sawicki, B. Sawicka, O. Gzowski: Phys. Status Solidi A **41**, 173 (1977)
4.50 F.J. Berry: Trans. Met. Chem. **4**, 209 (1979)
4.51 F.J. Berry: J. Chem. Soc. Dalton Trans., 1736 (1979)
4.52 Y. Ujihira, A. Handa: J. Phys. Coll. C2, 586 (1979)
4.53 O. Massenet, H. Daver: Solid State Commun. **21**, 37 (1977)
4.54 M.J. Tricker, P.P. Vaishnava, D.A. Whan: Applied Catalysis **3**, 283 (1982)
4.55 F.A. Deeney, P.J. McCarthy: Nucl. Instrum. Methods **159**, 381 (1979)
4.56 M. Inaba, K. Nomura, Y. Ujihira: J. Phys. Coll. C1, 115 (1980)
4.57 Y. Isozumi, M. Kurakado, R. Katano: Nucl. Instrum. Methods **166**, 407 (1979)
4.58 J.A. Sawicki, B.D. Sawicka, J. Stanek: Nucl. Instrum. Methods **138**, 565 (1976)
4.59 O. Massenet, H. Daver: Solid State Commun. **25**, 917 (1978)
4.60 W. Jones, J.M. Thomas, R.K. Thorpe, M.J. Tricker: Appl. Surf. Sci. **1**, 388 (1978)
4.61 U. Bäverstam, B. Rodlund-Ringström, C. Bohm, T. Ekdahl, D. Liljequist: Nucl. Instrum. Methods **154**, 401 (1978)
4.62 K.P. Mitrofanoff, V.S. Shpinel: Zh. Eksp. Teor. Fiz. **40**, 983 (1961)
4.63 Z. Bonchev, A. Jordanov, A. Minkova: Nucl. Instrum. Methods **70**, 36 (1969)
4.64 K. Siegbahn: *Beta and Gamma-Ray Spectroscopy* (North-Holland, Amsterdam 1955) p.85
4.65 U. Bäverstam, T. Ekdahl, C. Bohm, B. Ringström, V. Stefansson, D. Liljequist: Nucl. Instrum. Methods **115**, 373 (1974)
4.66 U. Bäverstam, C. Bohm, T. Ekdahl, D. Liljequist, B. Ringström: Nucl. Instrum. Methods **118**, 313 (1974)
4.67 U. Bäverstam, C. Bohm, T. Ekdahl, D. Liljequist, B. Ringström: Nucl. Instrum. Methods **9**, 259 (1974)
4.68 D. Liljequist, T. Ekdahl, U. Bäverstam: Nucl. Instrum. Methods **155**, 529 (1978)
4.69 D. Liljequist, B. Bodlund-Ringström: Nucl. Instrum. Methods **160**, 131 (1979)
4.70 J.P. Schunck, J.M. Friedt, Y. Llabador: J. Physique Appl. **10**, 121 (1975)
4.71 P.L. Gruzin, V.N. Gorokhov, Y.V. Petrikin: Zavodskaya Laboratoriya **41**, 984 (1975)
4.72 T. Shinohara, M. Fujioka, H. Onodera, K. Hisatake, H. Yamamoto, H. Watanabe: Hyperfine Interactions **1**, 345 (1976)
4.73 O. Massenet: Nucl. Instrum. Methods **153**, 419 (1978)
4.74 W. Jones, M.J. Tricker, G.A. Gard: J. Mater. Sci. **14**, 751 (1979)
4.75 T. Shigemastsu, H.D. Pfannes, W. Keune: "Symposium on Recent Chemical Applications of Mössbauer Spectroscopy", 179th National Meeting of the American Chemical Society, Divisions of Nuclear Chemistry and Technology and Inorganic Chemistry, Houston, Texas, March 1980, Paper 58
4.76 T. Toriyama, M. Kigawa, M. Fujioka, K. Hisatake: Jpn. J. Appl. Phys. Suppl. **2**, Pt. 1, 733 (1974)
4.77 T. Oswald, M. Ohring: J. Vac. Sci. Technol. **13**, 40 (1976)
4.78 M. Ron, R.S. Oswald, M. Ohring, G.M. Rothberg, M.R. Polcari: Bull. Am. Phys. Soc. **21**, 273 (1976)
4.79 R.S. Oswald, M. Ron, M. Ohring: Solid State Commun. **26**, 883 (1978)
4.80 M. Carbucicchio: Nucl. Instrum. Methods **144**, 225 (1977)
4.81 V.A. Bychkov, P.L. Gruzin, Y.V. Petrikin, L.A. Sharova: Zavodskaya Laboratoriya **44**, 970 (1978)

4.82 T. Toriyama, K. Saneyoshi, K. Hisatake: J. Phys. Coll. C2, 40, 15
      (1979)
4.83 D.R. Kahn, E.E. Petersen, G.A. Somorjai: J. Catal. 34, 294 (1974)
4.84 B. Davis: Ph.D. Thesis, Stevens Inst. of Technology (1973)
4.85 T. Tyliszczak, J.A. Sawicki, J. Stanek, B.D. Sawicka: J. Phys. Coll.
      C1, 41, 117 (1980)
4.86 P. Marmier, E. Sheldon: *Physics of Nuclei and Particles*, Vol.1 (Aca-
      demic, New York 1969)
4.87 S. Margulies, J.R. Ehrman: Nucl. Instrum. Methods 12, 131 (1961)
4.88 R.A. Krakowski, R.B. Miller: Nucl. Instrum. Methods 100, 93 (1972)
4.89 J. Bainbridge: Nucl. Instrum. Methods 128, 531 (1975)
4.90 G.P. Huffman: Nucl. Instrum. Methods 137, 267 (1976)
4.91 U. Bäverstam, C. Bohm, B. Ringström, T. Ekdahl: Nucl. Instrum. Methods
      108, 439 (1973)
4.92 T. Bonchev, A. Minkova, G. Kushev, M. Grozdanov: Nucl. Instrum. Methods
      147, 481 (1977)
4.93 M. Grozdanov, T. Bonchev, V. Lilkov: Nucl. Instrum. Methods 165, 231
      (1979)
4.94 T. Bonchev, M. Grozdanov, L. Stoev: Nucl. Instrum. Methods 165, 237
      (1979)
4.95 A. Proykova: Nucl. Instrum. Methods 160, 321 (1979)
4.96 G.P. Chugunova, A.V. Mitrin: Phys. Status Solidi B 81, 69 (1977)
4.97 M.J. Tricker, L.A. Ash, T.E. Cranshaw: Nucl. Instrum. Methods 143,
      307 (1977)
4.98 M.J. Tricker, L. Ash, W. Jones: Surf. Sci. 79, L333 (1979)
4.99 F.A. Denney, P.J. McCarthy: Nucl. Instrum. Methods 166, 491 (1979)
4.100 B.J. Tatarchuk, J.A. Dumesic: J. Catal. 70, 308 (1981)
4.101 B.J. Tatarchuk, J.A. Dumesic: J. Catal. 70, 323 (1981)
4.102 B.J. Tatarchuk, J.A. Dumesic: J. Catal. 70, 335 (1981)
4.103 B.N. Subba Rao: Nucl. Instrum. Methods 44, 155 (1966)
4.104 D. Salomon, P.J. West, G. Weyer: Hyperfine Interactions 5, 61 (1977)
4.105 D. Halliday, R. Resnik: *Fundamentals of Physics* (Wiley, New York 1970)

5. Heterogeneous Photocatalysis with Semiconductor Particulate Systems

K. Kalyanasundaram and M. Grätzel

With 10 Figures

5.1 Introduction

The field of photochemical conversion of solar energy has become an exciting
and rapidly growing area of research over the last few years. Light-driven
redox reactions coupled with redox catalysts are being investigated as a possible route for the generation of fuels by visible light [5.1-4]. With the
limitations imposed by the near diffusion-controlled rates for the back reactions that follow endergonic (uphill) photoredox reactions with organic
and inorganic dye-based systems, attention is being directed towards semiconductor particulate and colloidal systems as light harvesting units. Heterogeneous photocatalysis with semiconductor particulate systems offer several
advantages. Colloidal semiconductors combine a number of desirable properties
such as high extinction coefficients, fast carrier diffusion to the interface
and suitable positioning of valence and conduction bands to achieve high efficiencies in light-energy conversion processes. The transparent nature of
these sols allows ready detection of short-lived intermediates by fast kinetic spectroscopy. Particularly attractive is the added possibility of modi-
fying the surface of the semiconductor particles by chemisorption, chemical
derivatization and/or catalyst deposition assisting the light-induced charge
separation and subsequent fuel-generating dark reaction.

Various semiconductor particulate systems have been examined for different photocatalytic ($\Delta G > 0$) and photosynthetic ($\Delta G < 0$) reactions [5.5-80].
Table 5.1 lists such studies with a few selected recent references. These include studies of 'naked' as well as 'catalyst-loaded' semiconductor dispersions, and reactions either at the solid-gas or solid-liquid interface. Semiconductor particles are also used as inert (non-light-absorbing) carriers of
organic and inorganic sensitizers, relay species and catalysts in various
dye-sensitization studies. In this review, we have attempted to highlight
two specific areas: features of interfacial electron transfer reactions that
emerge from pulsed laser studies of colloidal semiconductors and the novel

photochemistry that is being explored with metal and metal-oxide loaded semiconductor particulate systems. Recently, a few other reviews dealing with 'heterogeneous photocatalysis' have appeared [5.5-12], of which [5.12] in particular gives a comprehensive overview of recent work.

An increasing number of physical methods are being brought into use to characterize semiconductor dispersions and for the quantitative analysis of the various photoprocesses. Table 5.2 summarizes some of these methods presently utilized in various laboratories.

Table 5.2. Physical methods used to characterize semiconductor dispersions and to study their photoreaction

Property/reaction	Technique
Characterization	
Size and polydispersity	Low-angle light scattering, X-ray and electron microscopy
Concentration	Atomic absorption spectroscopy
Surface area, porosity	Gas Adsorption, such as BET
Surface charge	Micro- and photoelectrophoresis, slurry electrodes
Band gap	Photoacoustic, diffuse reflectance spectroscopy
Catalyst deposits	Electron spectroscopy (ESCA, Auger), electron microscopy (TEM, STEM)
Surface states, impurities lattice defects	Luminescence (time-resolved steady state)
Reactions	
Adsorption, desorption	Photoconductivity, electron spin resonance, dynamic mass spectrometry, Hall effect
Intermediate detection and mechanism	Laser flash photolysis, laser flash conductance, ESR spin traps, product analysis in a static or flow reactor, photoelectrochemical cells with macro-semiconducotor electrodes, resonance Raman spectroscopy

The band theory of solids and the electronic theory of catalysis have proved to be very useful in the understanding and rationalizing various photoprocesses studied with semiconductor electrode and powder systems. According to the band model, the occurrence and the efficiency of various photoredox processes are intimately related to the location of the valence and conduction bands. Absorption of light of energy $E > E_{bandgap}$ leads to generation of electron-hole pairs and these, under the influence of the electric field, move in the conduction and valence bands, respectively. The resulting nonequilibrium distribution of electrons (e^-) and holes (h^+) gives rise to reduction or oxidation processes with adsorbed species, surface groups and/or with the bulk semiconductor itself (Fig.5.1).

The following section deals with charge transfer processes involving conduction band electrons (e_{cb}^-) or valence band holes (h^+) in colloidal semiconductor particles with an acceptor or donor present in the solution. The small size of the particles renders scattering of light negligibly small, enabling

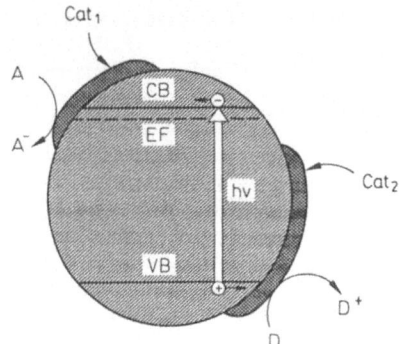

ready kinetic analysis of interfacial electron transfer processes by laser flash photolysis. The reaction of conduction-band electrons and valence-band holes with reactants is followed by fast kinetic spectroscopy or pulse con-ductometry. The competitive trapping of electrons by noble metal catalysts deposited onto the semiconductor can also be investigated. Finally the study also establishes a technique for determining the Fermi potential of semicon-ductor particles [5.81-97].

5.2 Studies with Colloidal Semiconductor Dispersions

5.2.1 General Considerations

We shall first consider the question of light absorption by ultrafine par-ticles. The maximum absorption coefficients for the semiconductor materials investigated are of the order of 10^5cm^{-1}, corresponding to an absorption length of at least 1000 Å. In view of the small dimensions (50-100 Å), light traverses many particles before complete extinction occurs, thus producing electron-hole pairs spatially throughout the particles along the optical path. This distinguishes the colloidal particles from semiconductor powders or elec-trodes where charge carriers are created mainly near the surface.

Consider a situation where a colloidal semiconductor particle is excited by a short (approximately 10 ns) laser pulse resulting in the generation of electron-hole pairs. The reaction of conduction band electrons with a relay compound R present in the bulk solution is observed subsequently by fast kine-tic spectroscopy.

One can envisage three elementary steps involved in the charge transfer event [5.81]. i) The charge carriers diffuse from the particle interior to the interface. This is a very rapid process, the average transit time τ being given by

$$\tau = (r_0^2/\pi^2 D) \quad , \tag{5.1}$$

where r_0 is the radius of the particle and D the diffusion coefficient of e_{cb}^- or h^+. For TiO_2 particles with $r_0 = 50$ Å and an electron mobility of 0.5 cm^2/Vs, $\tau = 2.5$ ps. This transit time is much faster than the estimated value of 100 ns for recombination, calculated for bulk recombination in a direct band-gap semiconductor with a majority carrier density of 10^{17}cm^{-3}. A similar estimation yields a recombination time of 100 ps at a majority carrier density of 2.10^{19}cm^{-3}. Note that τ increases with the square of the particle radius. For TiO_2 powders, a typical particle size would be 1 µm, which corresponds to $\tau = 100$ ns.

ii) Encounter complexes form between the electron (or hole) acceptors present in solution. The rate of this process is diffusion limited and hence determined by the viscosity of the medium and the radius of the reactants. Note that this diffusional displacement plays no role in systems where the relay adheres to the particle surface.

iii) Interfacial electron transfer is the third step. This involves movement of charges from the semiconductor particle surface across the Helmholtz layer to the relay species in solution. The rate constant k_{et} of this process is measured in cm/s. The sequence of encounter complex formation between the semiconductor particle and relay and subsequent electron transfer can be treated kinetically by solving Fick's second law of diffusion. For the observed bimolecular rate constant for electron transfer one obtains

$$\frac{1}{k_{obs}} = \frac{1}{4\pi r^2}\left(\frac{1}{k_{et}} + \frac{r}{D}\right) \quad , \tag{5.2}$$

where r is the reaction radius corresponding to the sum of the radii of the semiconductor particle and electron relay and D the sum of their respective diffusion coefficients.

5.2.2 Dynamics of Reduction of Methyl Viologen by e_{cb}^- of Colloidal TiO_2

The dynamics of reduction of the popular one-electron relay methyl viologen (MV^{2+}) upon excitation of the semiconductor in the band-gap (347 nm laser pulses) has been investigated on colloidal TiO_2 sols [5.82-84]. Particles were prepared by hydrolysis of titanium isopropoxide or $TiCl_4$ in acidic aqueous solution. They had a radius of 50-100 Å and consisted of amorphous material mixed with anatase. At pH > 3 a solution of polyvinyl alcohol is used to stabilize the particles. The yields of MV^+ as well as its growth kinetics were found to be strongly pH dependent. The variation of MV^+ (after completion

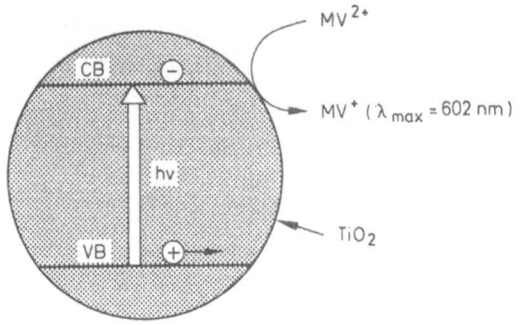
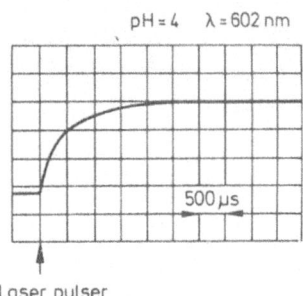

Fig.5.2. Light-induced electron transfer from the conduction band of a colloidal TiO$_2$ particle to MV^{2+} in aqueous solution. Oscillogram shows absorption growth of MV$^+$ at 602 nm [MV^{2+}] = 10^{-3} M

of electron transfer) with pH exhibits sigmoidal behavior. No reduction of
MV^{2+} occurs at pH $\leq$ 2.0. The yields of MV$^{\cdot+}$ increase sharply between pH 2.5
and 5.0 and attain a plateau for neutral and basic solutions. The experimental data in Fig.5.2 can be used to derive the conduction-band position or
Fermi potential of the semiconductor particles using the following simple
model. Laser excitation of semiconductor particles leads to a nonequilibrium
population of e_{cb}^- and h$^+$ and, therefore, the splitting of Fermi potential
into two quasilevels: one for h$^+$ and one for the electrons. The latter practically merges with the conduction band as the carrier density produced by
the laser pulse in the TiO$_2$ particles exceeds 10^{19} cm^{-3}. After equilibrium
with the redox couples in solution, i.e., completion of electron transfer,

$$E_f(e^-) = E_f(MV^{2+}/MV^{\cdot+}) = E^0(MV^{2+}/MV^{\cdot+}) + 0.059 \log \frac{[MV^{2+}]}{[MV^{\cdot+}]} . \qquad (5.3)$$

If equilibrium takes place under the pH conditions where only few electrons
leave the particles, then

$$E_f(e^-) \simeq E_{cb}(pH\ 0) - 0.059\ pH \qquad (5.4)$$

remains valid during the whole electron transfer process. From (5.3,4) one
expects a plot of log [MV^{2+}]/[MV$^{\cdot+}$] versus pH to yield a straigth line with
a slope of unity, which has indeed been observed for two viologen derivatives.
Using this method the conduction-band position $E_{cb}(pH\ 0)$ of colloidal TiO$_2$
prepared from the hydrolysis of titanium isopropoxide and TiCl$_4$ has been determined to be -0.13 and -0.11V (NHE), respectively.

The position of the conduction band edge of the colloidal TiO$_2$ particle
influences greatly the rate of MV$^+$ formation. Figure 5.3 shows data obtained

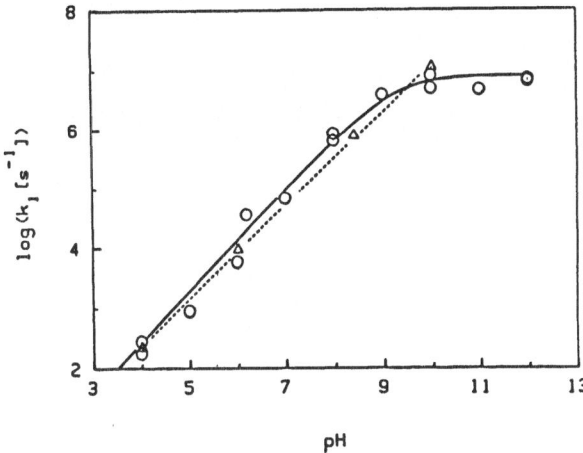

Fig.5.3. Reduction of MV^{2+} (⊖, 2×10^{-4} M) and $C_{14}MV^{2+}$ (▲, 2×10^{-4} M) by e_{CB}^{-} (TiO_2). The observed rate constant for MV^+ and $C_{14}MV^+$ formation is plotted as a function of pH. The solid line represents a computer fit for MV^{2+} reduction using $\alpha = 0.84$ and $k_{et}^{o} = 10^{-2}$ cm/s; the dashed line, with slope 0.78, was drawn through the $C_{14}MV^{2+}$ points. $[TiO_2]$ = 0.5 g/l protected by 1 g/l PVA

from the laser photolysis of colloidal TiO_2 (500 mg/l) in the presence of 2×10^{-14} M viologen. The logarithm of the observed rate constant (k_1) for the reduction of viologens MV^{2+} and the surfactant derivative

$$CH_3-(CH_2)_{13}-\overset{+}{N}\langle\bigcirc\rangle-\langle\bigcirc\rangle\overset{+}{N}-CH_3$$

$$(C_{14}MV^{2+})$$

$C_{14}MV^{2+}$ is plotted as a function of solution pH. The k_1 values were determined by monitoring the growth of the 602 nm absorption of the viologen cation radicals after exciting the TiO_2 colloid by a 20 ns laser pulse. For $C_{14}MV^{2+}$ a precisely linear relation is obtained between log k_1 and pH over a domain of at least 7 units, the slope of the line being 0.78. A straight line with similar slope is also obtained when MV^{2+} is used as an electron acceptor. However, in this case linearity of the log k_1 (pH) function is restricted to pH < 10. At higher alkalinity the curve bends sharply, k_1 attaining a limit of ~10^7 s^{-1}. Another noteworthy difference in the kinetic behavior of MV^{2+} concerns the effect of concentration on k_1. While k_1 increases with MV^{2+} concentration, it is not affected when $[C_{14}MV^{2+}]$ is varied from 2×10^{-4} to 10^{-3} M. The solid line in Fig.5.3 is a computer plot of

117

$$\frac{1}{k_1^I} = \frac{1}{4\pi r^2} \frac{1}{k_{et}^0 \exp(\alpha\ pH - \alpha\ 5.1)} + \frac{r}{D} \quad, \tag{5.5}$$

which is obtained from (5.2) by expressing k_{et} according to the Tafel relation:

$$k_{et} = k_{et}^0 \exp(-\frac{\alpha}{0.059} \mu) \quad, \tag{5.6}$$

where k_{et}^0 is the heterogeneous electron transfer rate at zero driving force, the transfer coefficient and α the overvoltage, and substituting:

$$\mu = 0.3 - 0.059\ pH \quad. \tag{5.7}$$

The predictions of (5.5) (using the experimentally determined parameter $r = 55$ Å and $D = 10^{-5} cm^2/s$) are in excellent agreement with the results supporting the validity of the kinetic model applied. From this curve, α is evaluated as 0.85 and $k_{et}^0 = 0.01$ cm/s. For comparison, TiO_2 colloids prepared via hydrolysis of titanium isopropoxide give an α value of 0.5 [5.82]. The difference between the two preparations has been attributed to the participation of surface states in the conduction-band process [5.83].

Substitution of a methyl group of MV^{2+} by a tetradecyl chain strongly enhances its adsorption to the surface of TiO_2 particles. Charge transfer from a semiconductor particle to a surface adsorbed species cannot be treated by (5.5). The correct interpretation of the k_1 values listed for $C_{14}MV^{2+}$ in Fig.5.3 is that of a reciprocal average time for electron transfer from the conduction band of the particle to the adsorbed acceptor molecule. A simple consideration shows that k_1 is related to the electrochemical rate constant k_{et} via

$$k_{et} = k_1 \cdot d \quad, \tag{5.8}$$

where d is the average distance over which the electron jump occurs. Assuming that the C_{14} chain of $C_{14}MV^{2+}$ extends radially away from the TiO_2 surface, the viologen moiety facing the aqueous phase, d, is calculated as circa 25 Å and $k_{et}^0 = 10^{-3}$ cm/s, which is about 10 times smaller than the corresponding value for methyl viologen. This difference is likely to arise from the closer contact of the latter acceptor to the TiO_2 surface. The α value for $C_{14}MV^{2+}$ reduction derived from Fig.5.3 is 0.78. A transfer coefficient of 0.5 is predicted from a Marcus-type free-energy relation, i.e.,

$$\Delta G^{\ddagger} = \lambda\left(1 + \frac{\Delta G^0}{4\lambda}\right)^2 \quad, \tag{5.9}$$

where $\Delta G^{\ddagger}$, $\Delta G^{O}(\simeq n)$ and λ are the free energy of activation, the free energy
of reaction and the reorganization energy, respectively. In the normal region
where $\Delta G^{O}/< 4\lambda$, $d\Delta G^{O} = 0.5$, such a relation seems to apply to TiO_2 particles
prepared from isopropoxide, but not to those produced via hydrolysis of $TiCl_4$.
The relatively large α value found for the latter is, however, compatible
with other free-energy relations derived empirically.

5.2.3 Consecutive and Simultaneous Two-Electron Reduction of Viologens on Colloidal TiO_2

Irradiation of TiO_2 sols in alkaline solutions in the presence of $C_{14}MV^{2+}$
leads to the formation of doubly reduced viologen ($C_{14}MV^{O}$). Pulsed laser
studies clearly indicate the mechanism of formation of $C_{14}MV^{O}$ to be consecutive electron transfer [5.84]. The rate constants for these single-electron
transfer events at $2 \times 10^{-4}M$ $C_{14}MV^{2+}$ and pH 11 are $10^8 s^{-1}$ and $5 \times 10^4 s^{-1}$, respectively. With a cofacial viologen dimer DV^{4+},

the mechanism of the e_{CB}^- reaction depends on pH. In acidic solution consecutive two-electron transfer is observed, Fig.5.4. However, laser photolysis
studies show that simultaneous two-electron reduction occurs at higher pH
(Fig.5.5)

$$2e_{CB}^- + DV^{4+} \longrightarrow DV^{2+} \tag{5.10}$$

and this is followed by comproportionation, i.e.,

$$DV^{2+} + DV^{4+} \longrightarrow 2DV^{3+} \quad . \tag{5.11}$$

(The end of laser pulse spectrum is that of DV^{2+} which has an absorption maximum at 536 nm. After completion of the slower process when the transient absorption has reached a plateau, the species present has the spectral characteristics of DV^{3+} with a maximum at 636 nm.) The rate constant for the latter
reaction has been evaluated to be $1.5 \times 10^7 M^{-1} s^{-1}$.

5.2.4 Reduction of $Rh(bipy)_3^{3+}$ on Colloidal TiO_2

Reduction of $Rh(bipy)_3^{3+}$ by the conduction-band electrons of TiO_2 has been investigated by continuous and laser photolysis techniques [5.83]. The occur-

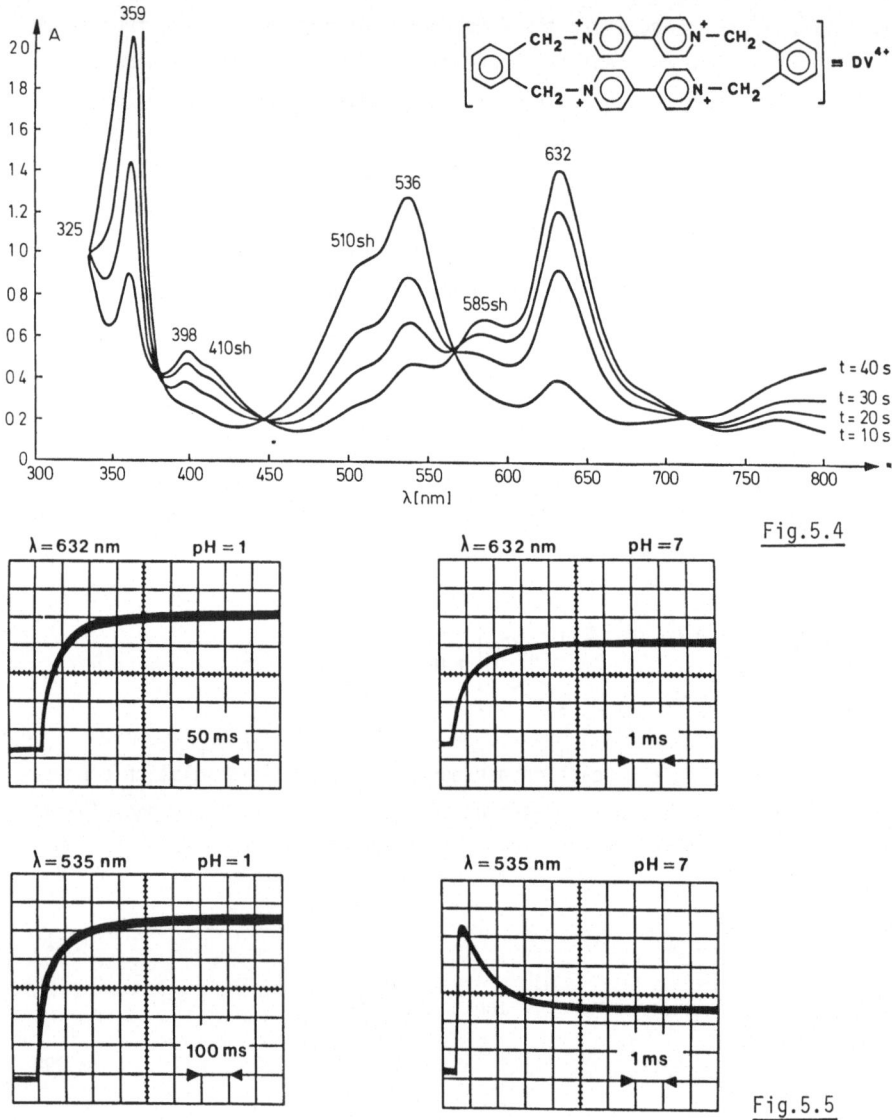

Fig.5.4

Fig.5.5

Fig.5.4. Spectral changes observed under irradiation of deaerated TiO_2 solutions (500 mg/l) by $\lambda > 330$ nm light in the presence of 2×10^{-4}M DV^{4+}, after 10, 20, 30 and 40 s irradiation. The spectra were measured against colloidal TiO_2 (500 mg/l) as reference solution. Optical pathlength 1 cm obtained. Note the clean isosbestic points at 714, 567, 447, 380 and 325 nm

Fig.5.5. Oscillograms from the laser photolysis of deaerated aqueous colloidal dispersions of TiO_2 (0.5 g/l) in the presence of 2×10^{-4}M DV^{4+}. The 632 nm absorbance indicates the temporal behavior of DV^{3+} while that at 535 nm reflects the temporal behavior of DV^{2+}. The oscillogram in the lower left corner was obtained with preirradiated solutions where most of the DV^{4+} had been converted to DV^{3+} prior to laser exposure

rence of one-electron reduction step

$$Rh(bipy)_3^{3+} + e_{cb}^- \longrightarrow Rh(bipy)_3^{2+} \qquad (5.12)$$

has been confirmed by the formation of a transient product with a characteristic absorption spectrum of $Rh(bipy)_3^{2+}$. Kinetic studies show that there is
a drastic pH effect on the reaction rate which increases more than 1000 times
when the pH is increased by only 4.8 units. Detailed analysis showed that
the kinetics of this reaction can be treated as for MV^{2+}. The values of transfer coefficient and k_{et}^0 derived are 0.64 and 0.4 cm/s, respectively. Furthermore, $Rh(bipy)_3^{2+}$ reacts to give red colored $Rh(bipy)_2^+$.

5.2.5 Dynamics of Hole Transfer Reactions on Colloidal TiO_2

Reactions of valence band holes with electron donors such as halide or thio-
cyanate (SCN^-) are also readily monitored by pulsed laser techniques [5.82,
85,86]. The oxidation of these species follows the sequence

$$S^- \xrightarrow{\;h^+\;} X. \xrightarrow{\;X^-\;} X_2^-. \quad , \qquad (5.13)$$

and results in the formation of X_2^-. ions which are readily monitored by their
characteristic absorption spectra. Kinetic studies show that the hole transfer takes place within the 10 ns duration of the laser pulse (Fig.5.6), indicating that the reaction mainly involves species adsorbed to the semiconductor particle. The efficiency of the process followed the sequence $Cl^- > Br^-$
$> SCN^- \simeq I^-$ and hence is closely related to the redox potential fo the X^-/X_2^-
couple. The yields also decrease sharply with increasing pH (yields are negligibly small at pH > 2.5) due to the competitive reaction of h^+ with water:

$$4\;h^+ + 2\;H_2O \longrightarrow 4\;H^+ + O_2 \quad . \qquad (5.14)$$

At pH 1.0, the quantum yields are in the range of 0.08 (Cl_2^-.) to 0.8 (I_2^-.).
Essentially similar results have been reported by *Henglein* [5.86], though
the quantum yields reported are significantly lower. The yields of Cl_2^-. and
Br_2^-. are greatly improved when RuO_2 is deposited onto the TiO_2 particles.
Presumably the role of RuO_2 is that of a hole scavenger, mediating the electron transfer. Tetranitromethane has also been shown to be a mediator for
similar hole transfer processes.

An interesting experiment with solutions of colloidal TiO_2 containing
Na_2CO_3 has recently been described by *Chandrasekaran* and *Thomas* [5.89]. Formation of CO_3^- radicals was analyzed by laser photolysis, monitoring their
characteristic absorption at 600 nm. Under steady-state UV light illumination

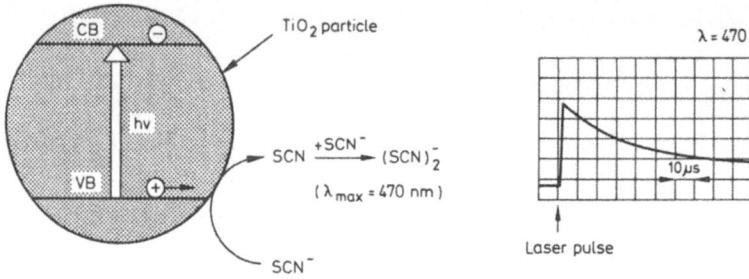

Fig.5.6. Hole transfer from the valence band of colloidal TiO_2 to SCN^- ion adsorbed on the particle surface. The oscillogram shows the temporal behavior of the absorption of $(SCN)_2^-$ radical anions at 470 nm

formaldehyde is produced, attributable to the reaction sequence

$$CO_3^- + h^+ \longrightarrow CO_3 \longrightarrow CO + O_2 \qquad (5.15)$$

$$CO + 2e^- + 2H^+ \longrightarrow H_2CO \quad . \qquad (5.16)$$

5.2.6 Halide Oxidation of Colloidal Fe_2O_3 Sols

Halide oxidation of colloidal Fe_2O_3 sols (particle radius 600 Å) has been examined subsequent to 347.2 nm laser pulse excitation [5.85]. With iodide, as for TiO_2 colloids, the formation of I_2^- appears promptly (within the laser pulse) with high quantum yield (0.8). In subsequent reactions I_2^- disappears according to a second-order law with a specific rate of $k = 7 \times 10^9 M^{-1}s^{-1}$. The high quantum yields observed with these finely divided sols (as compared to disappointingly low efficiencies on single-crystal, polycrystalline electrodes) again demonstrate the advantages of colloidal sols in solar energy conversion devices.

5.2.7 Colloidal TiO_2 in the Visible Light Induces Cleavage of Water

In these experiments, colloidal TiO_2 (prepared from $TiCl_4$) was loaded with 1% Pt via photoplatinization and used in conjunction with RuO_2 deposited on TiO_2-P25 particles. The sensitizer RuL_3^{2+}, where (L = di-isopropyl 2,2'-bipyridine-4,4'-dicarboxylate)

L

122

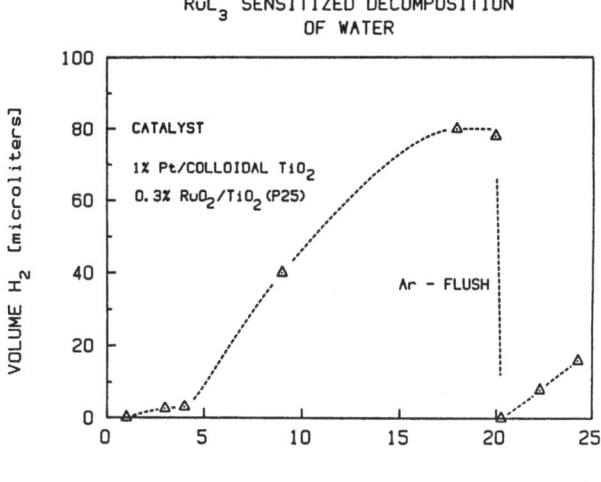

RuL$_3$ SENSITIZED DECOMPOSITION
OF WATER

CATALYST

1% Pt/COLLOIDAL TiO$_2$

0.3% RuO$_2$/TiO$_2$ (P25)

Ar - FLUSH

VOLUME H$_2$ [microliters]

IRRADIATION TIME [Hours]

Fig.5.7. Visible light induced H$_2$ generation from solutions containing RuL$_3^{2+}$ (10^{-4} M), PVS (10^{-2} M) 1g/l colloidal TiO$_2$ charged with 1% Pt and 1g/l TiO$_2$/0.3% RuO$_2$. pH 2, T = 60°C

was used together with the electron relay:

$$SO_3^- - (CH_2)_3 - N\bigcirc - \bigcirc N - (CH_2)_3 - SO_3^-$$

$$(PVS) \ .$$

The light-induced redox reaction

$$RuL_3^{2+} + PVS \longrightarrow RuL_3^{3+} + PVS^- \tag{5.17}$$

was coupled with a redox process, leading to water decomposition, i.e.,

$$2RuL_3^{3+} + H_2O \longrightarrow 2RuL_3^{2+} + \frac{1}{2}O_2 + 2H^+ \tag{5.18}$$

and

$$2PVS^- + 2H_2O \longrightarrow H_2 + 2OH^- + 2PVS \ . \tag{5.19}$$

Figure 5.7 shows generation of H$_2$ under visible ($\lambda > 420$ nm) irradiation of a solution containing 10^{-4}M RuL$_3^{2+}$, 10^{-2}M PVS and the two catalysts [5.90]. Sample volume is 10 ml and the pH employed is 2. The H$_2$ formation starts after an induction period of about 1 hour and hydrogen grows until a plateau is reached at about 80 µl. The system can be reactivated by degassing with Ar. The quantum yield achieved with this type of system is still poor but there is potential for significant improvement of the efficiency.

5.2.8 Photoluminescence of Colloidal CdS Particles

Recently, there have been extensive investigations of the photoluminescence from colloidal semiconductor sols. In our laboratory [5.91], colloidal particles of CdS have been produced by adding H_2S to an aqueous solution of $Cd(NO_3)_2$ (Method 1) or by rapid mixing of Na_2S and $Cd(NO_3)_2$ solutions (Method 2). Single crystals sized 50 and 20 Å have been obtained. While Type 1 particles are associated with dimers and trimers, Type 2 particles form large clusters of 2000-5000 Å size in aqueous solution. Luminescence studies carried out with Type 1 sol show the presence of a red emission (λ_{max} = 700 nm) arising from S vacancies and a very weak green fluorescence (λ_{max} = 515 nm) due to free carrier recombination, Fig.5.8. The red luminescence is extremely sensitive to the presence of acceptors such as MV^{2+}: 10^{-8}M MV^{2+} suffices to quench 50% of the emission. Kinetic analysis shows that only one MV^+ per CdS aggregate is required to quench the red luminescence, and this effect can be exploited to determine the aggregation number of CdS particles. Furthermore, MV^{2+} induces a green emission (λ_{max} = 530 nm) which is attributed to the formation of Cd vacancies. A green emission (λ_{max} = 520 nm) is also obtained by substitutional doping of Type 1 sols with chloride and this arises from interstitial S. Doping with Cu^{2+} produces the characteristic emission of this activator (λ_{max} = 820 nm) only if particles are prepared under aerobic conditions or illuminated in the presence of oxygen. The results have been interpreted in terms of a lattice defect model.

The extreme sensitivity/quenching of the luminescence by surface-adsorbed species has also been noted by *Rossetti* and *Brus* [5.92]. For example, PbS and p-benzoquinone along with several other compounds were found to cause 50% quenching of luminescence at 10^{-5}M, corresponding to far less than a monolayer surface coverage, while SH^--rich CdS colloids undergo a type of photochemical aging if left overnight in room light, during which the initial emission peak at 470 nm red shifts to about 505 nm with a tenfold increase in emission quantum yield. Resonance Raman [5.93] and electron microscopic studies have confirmed this aging process in which about 21 small crystallites (cubic CdS) dissolve and recrystalline onto one large 'seed' crystallite. The colloid remains transparent without CdS precipitation as it ages.

5.2.9 Luminescence and Photodegradation of Aerated Colloidal CdS Sols

In a series of papers, *Henglein* and colleagues [5.94-96] reported on the luminescence and photodegradation of aerated/oxygenated solutions of colloidal CdS solutions. The CdS sols were prepared by slow addition of $Cd(ClO_4)_2$

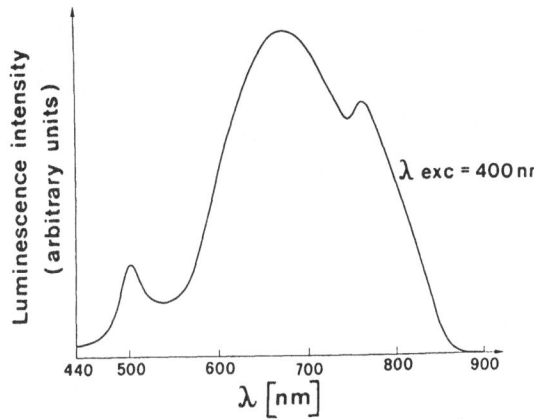

Fig.5.8. Photoluminescence from fresh Type 1 CdS particles prepared from aqueous $Cd(NO_3)_2$, H_2S and $(NaPO_3)_6$, measured with the MPF 44 spectrofluorimeter. Excitation wavelength = 400 nm

solution to a solution of Na_2S and the resulting CdS sols protected with commercial Ludox SiO_2. These sols show a weak fluorescence with a maximum at 620-660 nm (depending on the sample) and are quenched by anions (such as S^{2-}, Br^-, I^- at concentrations 10^{-3} M) and by cations (such as Tl^+, Ag^+, Pb^{2+}, Cu^{2+} at concentrations 10^{-4} M). Doping with 1% Cu^{2+} or Ag^+ enhanced the fluorescence intensity significantly.

Photolysis of aerated/oxygenated CdS sols leads to degradation of CdS. The quantum yields for photodegradation in the presence of oxygen alone is 0.04 and is enhanced in the presence of certain additives (0.24 with Tl^+, Pb^{2+} at concentrations 10^{-5} M, and for MV^{2+} at concentrations 10^{-4} M, 0.60 for MV^{2+} (10^{-4} M) and Na_2S (5×10^{-2} M). In aerated solutions, the well-established degradation mechanism occurs via photoanodic decomposition:

$$CdS + O_2 + 2H^+ \longrightarrow Cd^{2+} + S + H_2O_2 \ . \tag{5.20}$$

In the presence of hole scavengers such as S^{2-}, SO_3^{2-} or $S_2O_4^{2-}$ (dithionite) the enhanced degradation has been attributed to a photocathodic decomposition process:

$$2e^- + H^+ + CdS \longrightarrow Cd + SH^- \ . \tag{5.21}$$

The formation of Cd metal has been proven by optical absorption measurements and by reaction of Cd atoms with species such as H^+, N_2O or MV^{2+} to yield H_2, N_2 or MV^+, respectively [5.96]:

$$Cd + 2H^+ \longrightarrow Cd^{2+} + H_2 \tag{5.22}$$

$$2H^+ + Cd + N_2O \longrightarrow Cd^{2+} + N_2 + H_2O \tag{5.23}$$

$$Cd + 2MV^{2+} \longrightarrow Cd^{2+} + 2MV^+ \quad . \tag{5.24}$$

Flash photolysis studies with 347.2 nm light pulses have also shown that hydrated electrons are produced upon excitation of colloidal CdS and CdS–ZnS cocolloids. Intensity-dependence studies indicate the photoelectron production to be a monophotonic process. The yields of e_{aq}^- increase in the presence of S^{--} ions, reaching $\Phi = 0.15$ at $[Na_2S] = 2 \times 10^{-3}$ M.

5.3 Photoprocesses on 'Catalyst-Loaded' Semiconductor Dispersions

Studies of heterogeneous photocatalysis on 'naked' semiconductor dispersions have been undertaken already for a few decades, with early interests directed towards understanding the light sensitivity of the photographic emulsions and pigments (photochalking) in the dye industry. Elsewhere [5.12] we presented a comprehensive overview of recent studies in this area. A very recent development has been attempts to catalyze light-induced electron or hole transfer processes at the semiconductor-electrolyte interface with metal and metal-oxide deposits. The area is still in its infancy, a majority of the studies being exploratory in nature, but there is increasing evidence indicating that such catalysis is indeed possible.

In the majority of cases, band-gap excitation of direct band-gap semiconductor materials leads to generation of charge carriers (e^- and h^+) with rather high quantum efficiency. Yet there are several photoreactions, both of academic and industrial importance (e.g., photodecomposition of H_2O, HCl and H_2S into their constituents) which are grossly inefficient on naked semiconductors, because the majority of semiconductors are poor electrocatalysts for photocathodic/anodic reactions of interest. To facilitate H_2 evolution, one needs metallic deposits such as Pt to reduce the overvoltage requirements.

The success of 'electrocatalysis' experiments depends very much on the *electrical characteristics of the junction* between the metal or metal oxide and the semiconductor. Depending on the nature of the metal/metal oxide and the surface and bulk characteristics of the semiconductor, a metal (or metal oxide) semiconductor may give rise to a Schottky barrier (rectifying junction) or just be an ohmic contact. Since a Schottky barrier would drive electrons away from the metal, for metal deposits to act as reduction centers for the photogenerated electrons the contact must be ohmic. For example, Pt/TiO_2 appears to be so, as indicated by its efficient performance in water-reduction processes.

5.3.1 Characterization of Catalyst/Semiconductor Junctions

Only recently have there been a few electrochemical studies [5.98-101] aimed at the characterization of the catalyst junction with the semiconductor. *Hope* and *Bard* [5.98] recently made an illuminating electrochemical study of the Pt-TiO$_2$ interface. The nature of the electrical contact was investigated by current-voltage and impedence measurements as a function of surface preparation. Behavior characteristics of a Schottky barrier were found only for etched samples under certain conditions. The electrical properties of the contacts were strongly altered by thermal treatment of the sample, extended annealing producing low resistance ohmic junctions. Auger depth profiles of rectifying and ohmic samples show that interdiffusion of Pt and rutile is responsible for the formation of ohmic junctions with low contact resistance, while nondiffused samples can exhibit rectifying properties. It should be pointed out that simple theories of metal/semiconductor junctions predict that the Pt/TiO$_2$ contact would be such that irradiation of it would result in holes flowing to Pt and electrons accumulating in the TiO$_2$. The barrier height of such Schottky junctions is given by the difference in the Pt work function Φ (5.2 eV) and the TiO$_2$ electron affinity χ (4.0 eV). The existence of this barrier would prevent flow of electrons from TiO$_2$ to Pt until the Fermi level of electrons in TiO$_2$ was close to the top of the barrier. Thus, caution needs to be exercised before interpretations are made on simple models without regard to the nature of the treatment given to the metal/semiconductor junction.

Heller et al. [5.99,101] have also elaborated on the nature of the junction formed by platinum group metal (Pt, Ru, Rh, Pd) on n-, p-type semiconductor electrodes. Platinum forms an ohmic contact with p-InP while Rh and Ru are expected to form Schottky junctions with barrier height ψ up to 0.5 and 0.9 eV, respectively. Photoelectrochemical experiments with p-InP electrodes coated with a thin film of these noble metals evolve H$_2$ at very high efficiency. To interpret this apparent anomaly, it has been proposed that alloying of the metallic catalysts with H$_2$ at the hydrogen evolving sites reduces their work function considerably, thereby facilitating electron flow from the semiconductor. In support of this hypothesis, it has been shown that Pt, Rh, Ru and Pd deposits on n-CdS (which are Schottky junctions under normal conditions) form ohmic contacts upon hydrogen alloying. Thus H$_2$ diffusion modifies the junction properties up to 400 Å depth. Similar oxidation of catalysts at the oxygen-evolving sites is believed to occur, but due to the inefficient diffusion of oxygen onto metals, oxidation is restricted to less than two monolayers.

There have been a number of studies where catalytic effects have been obtained with RuO_2 deposits on n semiconductors and the results interpreted in terms of catalysis of hole transfer by RuO_2. Thus characterization of the RuO_2-semiconductor junction is of special interest, especially in the light of the fact that RuO_2 is a good electrocatalyst for H_2 evolution. *Gissler* and colleagues [5.102,103] made an electrochemical study on sputtered RuO_2 layers on n-CdS electrodes and observed formation of a Schottky barrier.

Characterizations of the *electronic interactions* between the metal and the semiconductor support are also of prime importance in thermal catalysis studies as well, to help understand the Strong-Metal-Support Interaction (SMSI) phenomenon. Noble-metal coated semiconductor materials are increasingly used and SMSI effects play a crucial role in controlling the activity and selectivity of these catalysts. For example, Pt/TiO_2 has received extensive scrutiny in this context by various physical methods in several laboratories [5.104-107]. When Pt is deposited onto TiO_2 and the catalyst reduced at high temperatures (> 773 K), significant SMSI effects have been observed (the metal shows reduced chemosorptive properties towards H_2, CO and markedly modified catalytic behavior for reasons such as Fischer-Tropsch synthesis, hydrogenolysis, etc.). *Herrmann* et al. [5.104,105] made electrical and photoconductivity studies of $Pt-TiO_2$ (Degussa P-25, anatase) catalysts to determine the electron transfer between the metal and the support. A migration of electrons from anatase to the Pt has been inferred and it has also been concluded that hydrogen atoms chemisorbed on the metal can migrate to the support (O^{2-} sites on TiO_2) with a simultaneous release of electrons in the semiconductor. Under SMSI conditions, excess electrons are believed to be present on Pt. *Chen* and *White* [5.106] have also reached similar conclusions. In the following sections we review briefly progress in selected areas of photoreactions on catalyst-loaded semiconductor dispersions.

5.3.2 Photodecomposition of Water on Catalyst-Loaded Semiconductors

One topic of wide interest is the photodecomposition of water into its constituents H_2 and O_2. By analogy with the 'in vivo' photosynthesis Z scheme, one can envisage cyclic water decomposition in two half-cycles: photogeneration of H_2 from water in the presence of external electron donors, and photogeneration of O_2 from water in the presence of external acceptors:

$$4 \, D^+ + 2 \, H_2O \longrightarrow 4 \, D + 4 \, H^+ + O_2 \qquad (5.25)$$

$$2 \, A^- + 2 \, H_2O \longrightarrow 2 \, A + 2 \, OH^- + H_2 \quad . \qquad (5.26)$$

An ideal cyclic water-splitting scheme is one where the two processes are achieved simultaneously with involvement of no external additives or at least the case where what is reduced in the O_2 half-cycle is used as external donor for the H_2 half-cycle. In systems which mimic either of these half-cycles, external additives are irreversibly consumed in oxidation/reduction steps, hence these additives have come to be known as 'sacrificial' agents. Owing to the large overvoltages and the nature of the multielectron transfer involved in the above reactions, in the absence of suitable redox catalysts these reactions either do not occur or are very inefficient.

Band-gap irradiation of metallized semiconductor dispersions in the presence of 'sacrificial' electron donors leads to very efficient production of H_2 from water [5.12]. Platinized CdS and TiO_2 have been the most studied systems, though there are reports using others, e.g., platinized Si, CdSe and phthalocyanine dispersions. A wide variety of substrates have been used, the common ones being EDTA, cysteine and alcohols such as methanol. *Taniguchi* et al. [5.108] have reported an interesting variation of the common system, whereby H_2 evolution from aqueous ethanol was observed for more than 150 hours (with $\phi = 0.021$ at 550 nm) upon illumination of n-Si powder photocatalysts whose anodic and cathodic surfaces were coated with polypyrrole and platinized Ag, respectively (photodeposition of polypyrrole and Ag was followed by subsequent photoplatinization). Several variations of these catalysts were also examined and the activity of different catalysts followed the order: Pt(Ag)polypyrrole > Ag(Si)polypyrrole > Pt-Si > bare Si. In the last two cases, the activities were lost within 100 hours of photolysis. In another educative study, *Memming* et al. [5.100] examined H_2 evolution with EDTA or S^{2-} on CdS-monograin membranes loaded on one side either with Pt or RuO_2 catalyst. Vigorous H_2 evolution was observed on the metal-metal-oxide surfaces of the solution side free of sulfide ions. *Mills* and *Porter* have also examined [5.48] many different semiconductors (TiO_2 rutile and anatase, CdS and $SrTiO_3$) platinized by various methods for water reduction with EDTA. Ultraviolet irradiation of RuO_2-TiO_2 in either the presence or absence of EDTA showed little ability to reduce water and $SrTiO_3$ was less active than anatase TiO_2. Photosensitized oxidation of water by WO_3 powder with Fe^{3+} as the electron acceptor occurs with a formal quantum efficiency of 3.1×10^{-3} M at 405 nm [5.75]. Deposition of RuO_2 onto the semiconductor catalyzed the rate of $\dot{O}_2$ production whereas deposits of Pt, Rh and Ru were only inhibitive.

The earliest report on cyclic water cleavage with catalyst-loaded semiconductor dispersion was by *Bulatov* and *Khidekel* [5.109], who reported O_2, H_2 production in uv photolysis of platinized TiO_2 in 1N H_2SO_4. This was followed by similar reports by *Wrighton* et al., who used platinized $SrTiO_3$ and $KTaO_3$ single crystals [5.110]. Since then, these systems have been examined in detail in several laboratories.

Pure TiO_2 has no activity in the photolysis of liquid or gas-phase water, but platinized TiO_2 does [5.49,51]. Severe photoadsorption of photoproduct O_2 often precludes observation of concomitant evolution of O_2 with H_2 during the photolysis. Even with $Pt-TiO_2$, continuous photodecomposition of gas-phase water does not take place, apparently because of the thermal back reaction between the products O_2 and H_2 on Pt. The photocatalytic activity of $Pt-TiO_2$ is much improved upon reduction of TiO_2 with H_2. There have been contradictory reports on attempts to displace the photoadsorbed oxygen by adding strongly adsorbing ions such as phosphate [5.47,48]. Photolysis of NaOH-coated platinized TiO_2 leads to hydrogen evolution with an initial quantum efficiency of 0.07 but declines thereafter significantly. *Somorjai* et al. have obtained similar results with NaOH-coated $SrTiO_3$ single crystals [5.111].

In photolysis studies of rhodium deposited $SrTiO_3$ dispersions under reduced pressure, high turnovers ($\simeq$ 12000 for Rh and $\simeq$ 50 for $SrTiO_3$) have been observed, with no significant decomposition of the semiconductor [5.53]. Positive results have also been obtained for deposits of Ru, Ir, Pd, Pt, Os and Re on $SrTiO_3$ with relative efficiencies (with respect to Rh) of the order of 24, 13, 11, 13, 10, 16 and 4%, respectively. Essentially similar results are obtained with $SrTiO_3$-$LaCrO_3$ dispersions [5.28]. Photocatalytic decomposition of water vapor and liquid over $NiO-SrTiO_3$ has also been subjected to detailed examination [5.73]. The activity for photodecomposition was increased by the pretreatment of reduction of the catalyst in hydrogen and reoxidation by oxygen before the reaction. Unlike the systems using platinized semiconductors, in these systems, the reverse reaction between H_2 and O_2 under illumination or in the dark is very slow, allowing the cyclic water photodecomposition to proceed steadily even in a closed system. For liquid water, activity is considerably enhanced at high alkali concentration ($\geq$ 1M).

In our laboratory, Pt [5.112] and Rh or Ru [5.113] loaded TiO_2 particles were prepared for water decomposition, emphasis being placed on achievement of very high dispersion. Cluster precursors $Rh_6(CO)_{16}$ and $Ru_3(CO)_{12}$ were used for preparation and the activity of these catalysts in mediating water decomposition through band-gap excitation was investigated. Activity increases in the order $Ru < RuO_2 < Rh \simeq Rh_2O_3 \simeq Pt$.

Bifunctional Rh/RuO_2 loaded TiO_2 exhibits optical performance, Fig.5.9, with overall light to chemical energy conversion efficiency of 0.13%. Lack of O_2 in the gas phase during photolysis observed with closed systems is due

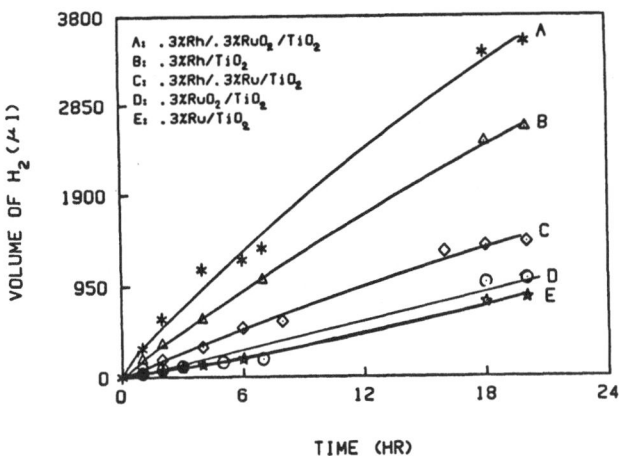

PHOTOACTIVITY OF VARIOUS Rh AND Ru CATALYSTS

A: .3%Rh/.3%RuO$_2$/TiO$_2$
B: .3%Rh/TiO$_2$
C: .3%Rh/.3%Ru/TiO$_2$
D: .3%RuO$_2$/TiO$_2$
E: .3%Ru/TiO$_2$

VOLUME OF H$_2$ (μl)

TIME (HR)

Fig.5.9. Hydrogen generation by photolysis of TiO$_2$-supported Rh and Ru catalysts; reaction temperature 30°C, pH = 14. Volume of H$_2$ (1 atm. T = 298 K) is
plotted as a function of irradiation time

to photo-uptake of O$_2$ by the TiO$_2$ particles. In alkaline solution the capa-
city for O$_2$ uptake is surprisingly high and the nature of the stored O$_2$ was
identified as a μ-peroxo bridged titanium species.

5. 3. 3 Photocleavage of Hydrogen Sulfide and Photosynthesis of Thiosulfate

Photocleavage of H$_2$S into H$_2$ and S is of wide industrial interest because of
its potential role in the hydrodesulfurization of crude petroleum products
(natural gas, oil and coal):

$$H_2S \xrightarrow{2h\nu} H_2 + S \quad (\Delta H = 9.4 \text{ kcal/mole}) \ . \tag{5.27}$$

Aqueous dispersions of CdS when illuminated by visible light cleave H$_2$S with
high efficiency [5.78-80]. While the reaction proceeds without the intervention of a noble metal catalyst, loading of CdS particles with RuO$_2$ enhances
markedly the efficiency of the process. The presence of oxygen has only a
small effect on the quantum yield of H$_2$S cleavage. Addition of ionic surfac-
tants greatly affects the efficiency of 'naked' semiconductor dispersions,
with a fourfold increase in quantum yield achieved by the addition of sodium-
laurylsulfate at concentrations below the CMC. Similar increases in efficiencies are observed upon addition of sulfite in studies with RuO$_2$-loaded CdS
dispersions [5.79,80]. The intervention of sulfite is rather indirect and involves reaction with S to yield S$_2$O$_3^{2-}$:

$$S + SO_3^{2-} \longrightarrow S_2O_3^{2-} \quad . \tag{5.28}$$

The reaction in the presence of sulfite is attractive for two reasons. First, thiosulfate is a more valuable product than sulfur since it is used in industrial processes such as photography. Secondly, sulfite favorably affects the rate of H_2S photocleavage and through removal of S makes it possible to sustain the reaction over long time periods. The overall reaction corresponds to the photogeneration of hydrogen and thiosulfate via

$$S^{2-} + SO_3^{2-} + 2H_2O \xrightarrow{2h\nu} S_2O_3^{2-} + 2OH^- + H_2 \quad . \tag{5.29}$$

While reaction (5.29) under standard conditions (pH 14, T 298 K) stores only 0.11 eV of free energy per absorbed photon, compared to 0.36 eV for the cleavage of H_2S, it has the advantage over the latter process in giving high yields of H_2 without formation of insoluble products. We were intrigued, therefore, by the idea of using it as the H_2 generating part in a hydrogen sulfide cleavage cycle and attempted to identify a photosystem capable of reducing $S_2O_3^{2-}$ back to S^{2-} and SO_3^{2-}. We found that conduction-band electrons produced by band-gap excitation of TiO_2 particles efficiently reduce thiosulfate to sulfide and sulfite [5.114]

$$2e_{cb}^-(TiO_2) + S_2O_3^{2-} \longrightarrow S^{2-} + SO_3^{2-} \quad . \tag{5.30}$$

This reaction was confirmed by electrochemical investigations with polycrystalline TiO_2 electrodes. The valence-band process in alkaline TiO_2 dispersions involve oxidation of $S_2O_3^{2-}$ to tetrathionate which quantitatively dismutates into sulfite and thiosulfate, the net reaction being

$$2h^+(TiO_2) + 0.5\ S_2O_3^{2-} + 1.5\ H_2O \longrightarrow SO_3^{2-} + 3H^+ \quad . \tag{5.31}$$

This photodriven disproportionation of thiosulfate into sulfide and sulfite

$$1.5\ H_2O + 1.5\ S_2O_3^{2-} \xrightarrow{h\nu} 2SO_3^{2-} + S^{2-} + 3H^+ \tag{5.32}$$

should be of great interest for systems that photochemically split hydrogen sulfide into hydrogen and sulfur, Fig.5.10.

5.3.4 Photoassisted Water-Gas Shift Reaction over Platinized Titania

Light-induced H_2 evolution from water using platinized TiO_2 dispersions and various carbon substrates is currently receiving intense interest and acti-

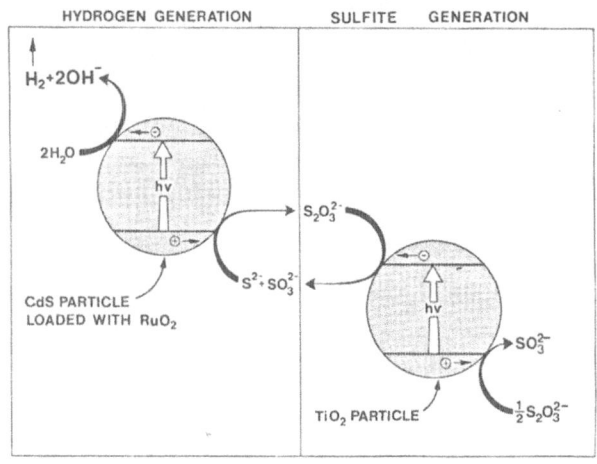

Fig.5.10. Scheme for light-induced H_2S decomposition with the $CdS/RuO_2/TiO_2$ system

vity. With CO as the hole reductant, the process corresponds to a photoassisted water-gas shift reaction

$$CO + H_2O \longrightarrow CO_2 + H_2 \quad . \tag{5.33}$$

Studies of the kinetics of this reaction over the temperature range 0-60°C by *White* et al. [5.70,71] have shown that on Pt-TiO_2, the reaction is zero order both in CO and water when $p_{CO} = 0.3$ Torr and $p_{H_2O} = 5$ Torr. The activation energy for the reaction is about 7.5 kcal/mole. The quantum efficiency was found to be 0.005. Later studies have shown [5.71] that the reaction rate does not depend on the method of Pt deposition, is first order in light intensity and is not dependent on the chemical state or Pt loading above 2 wt%, but depends on reduction of titania and is strongly dependent on the surface concentration of NaOH. The latter implies that the rate-determining step is the reaction of holes with surface hydroxyl ions. *Tsai* et al. [5.72] have also obtained similar results on the kinetics of the reaction. Based on platinum coverage studies, it is proposed that the periphery of deposited Pt islands is involved in the photogeneration of H_2.

Thewissen et al. [5.115] investigated aqueous suspensions of SiC coated with either Rh or RuO_2 catalysts for the photodecomposition of water, photoassisted water-gas shift reaction and the photoreduction of CO_2. Water-gas shift reaction occurs with light of sub-bandgap energy probably via surface states. A drastic increase in H_2 production occurs upon loading SiC with Rh deposits, whereas RuO_2 does not exhibit any catalytic activity.

5.3.5 Photoreactions of Organic Compounds

As with 'naked' semiconductor dispersions, in recent years there has been a growing number of studies of photoreactions on catalyst-loaded semiconductors involving various organic compounds as reactants. *Pichat* et al. [5.62] examined photocatalytic H_2 evolution from primary aliphatic alcohols (methanol, ethanol, ethanal and propan-1-01) as a function of Pt content on Pt/TiO_2. The optimal initial rate of H_2 production was found for Pt contents in the range 0.1-1.0 wt.% at an optimal temperature of 313 K. The maximum rate has been tentatively attributed to an optimum attraction of free electrons of titania by the Pt crystallites. At optimal temperature, the reaction of photoproduced holes with adsorbed alkoxide ions was rate determining, whereas the H_2 desorption rate played a part at lower temperatures.

Sakata and *Kawai* as well as *Sato* and *White* and several others [5.56-62] examined light-induced H_2 evolution from water using various carbonaceous materials (hydrocarbons, alcohols, active carbon, lignite, various biomass sources such as wood, cotton, protein, carbohydrates, to name a few): the list of possible organic substrates grows constantly. For ethanol-water mixtures using metallized TiO_2, the quantum efficiency at 380 nm for hydrogen production has been determined to be 6.5, 11.6, 19.0 and 38.0%, for loading with Ni, Pd, Rh and Pt, respectively. The gas-phase reaction using active carbon or lignite over illuminated, platinized TiO_2 leads to H_2, CO_2 and a small amount of O_2

$$H_2O(g) + \frac{1}{2} C(s) \longrightarrow H_2 + \frac{1}{2} CO_2 \quad . \tag{5.34}$$

The rate of the reaction declines with photolysis time owing to accumulation of H_2 and resulting loss of good contact between the catalyst and C. With active carbon as the substrate, the reaction is zero order with respect to water pressure, has an activation energy of about 5 kcal/mole and a quantum efficiency of 0.02 in the early stages of photolysis. In liquid phase, the oxidation of carbon is inhibited and water photodecomposition dominates.

There have also been exploratory studies on the feasibility of photoreduction of carbon dioxide to reduction products at different levels: HCHO, HCOOH, CH_3OH and CH_4 on catalyst-loaded semiconductors. Photoreduction as well as concurrent photooxidation of formaldehyde on uv-irradiated aqueous suspensions of $SrTiO_3$, TiO_2 powders loaded with transition metal oxides of Rh, Pt, Ru or Ir has been reported [5.61]. Both reactions are catalytic and the main oxidation products are H_2, HCOOH and CO_2. The reduction products are methanol, ethanol and small amounts of C_1-C_3 hydrocarbons. Formation of methanol is strongly dependent on the concentration of HCHO, pH and the amount

134

of the photocatalyst. On rhodiumoxide-loaded photocatalysts, the efficiency of methanol followed the order TiO_2(rutile) < $SrTiO_3$ < TiO_2 (anatase).

5.3.6 Photo-Kolbe Reaction

A reaction which clearly demonstrates the mechanistic features of photoprocesses on metallized semiconductor dispersions is the photo-Kolbe reaction studied extensively by *Bard* and co-workers [5.63-65]. Carried out in an electrochemical cell, at an illuminated n-TiO_2 electrode, decarboxylation of acetic acid-acetate yields CO_2 and ethane:

$$2 \ CH_3COO^- \xrightarrow[\text{electrode}]{h\nu, TiO_2} CH_3CH_3 + 2 \ CO_2 \ . \tag{5.35}$$

However, the same reaction carried out in aqueous media with platinized TiO_2 dispersions yields methane and CO_2 as major products

$$CH_3COOH \xrightarrow[\text{powder}]{h\nu, TiO_2} CH_4 + CO_2 \ . \tag{5.36}$$

Detailed investigations have shown that the photo-Kolbe reaction is very general in scope and has no specificity either to the semiconductor (occurs also on Pt-WO_3) or to acetic acid. On Pt-TiO_2, several aliphatic and aromatic carboxylic acids have been examined for their major photoproducts. The mechanism involves initial generation of alkyl (R.) radicals by the reaction of photogenerated holes with acetate

$$h^+ + RCO_2^- \longrightarrow R. + CO_2 \ . \tag{5.37}$$

At low light intensities, the large surface area of the catalyst results in low surface concentration of the radicals and this prevents second-order reactions such as dimerization and disproportionation. Also R. radicals produced near the reducing sites (Pt) on the powder facilitate reduction to RH as contrasted to R. radicals produced at illuminated electrodes. This behavior would account for the major production of methane with dispersions. However, carried out in the gas phase with water vapor, photolysis gives the normal electrolysis product, ethane, in high vields.

5.4 Conclusion

Colloidal semiconductors and semiconducting powders are being increasingly employed to mediate photosynthetic and photocatalytic processes. Through suitable choice of the material, desirable oxidation and reduction reactions with inorganic and organic substrates can be performed under illumination. An advantage with respect to conventional dark processes is specificity of product formation and this is achieved by fine tuning the valence- and conduction-band position to the chemical transformation envisaged. Surface phenomena such as adsorption of high-energy intermediates play a prime role in the catalytic effects exerted by these particles. The high surface area available in these systems is also advantageous in that it reduces undesirable side reactions such as photocorrosion frequently found with semiconductor electrodes.

Future work will comprise development of improved and highly specific catalysts. Molecular engineering through surface derivatization and/or adsorption is a promising strategy to achieve this goal. Furthermore, a large effort will be made to increase the efficiency of H_2O and H_2S cleavage catalysts which convert light into chemical energy and hence are attractive for solar energy conversion.

Acknowledgment. This work was supported by the Swiss National Science Foundation, a grant from the European Community, project D, and the United States Army European Research Office.

References

5.1 N. Sutin, C. Creutz: Pure Appl. Chem. **52**, 2717 (1980)
5.2 D.G. Whitten: Acc. Chem. Res. **13**, 83 (1980)
5.3 M. Grätzel: Acc. Chem. Res. **14**, 376 (1981)
5.4 J. Kiwi, K. Kalyanasundaram, M. Grätzel: Struct. Bonding **49**, 39 (1981)
5.5 M. Formenti, S.J. Teichner: Spec. Period. Rep. Catal. **2**, 87 (1978)
5.6 A.J. Bard: J. Photochem. **10**, 59 (1979)
5.7 R.I. Bickley: Spec. Period. Rep., Chem. Phys. Solids **7**, 118 (1978)
5.8 D. Lichtman, Y. Shapira: CRC Crit. Rev., Solid State Sci. **8**, 93 (1978)
5.9 F. Steinbach: Top. Curr. Chem. **25**, 117 (1972)
5.10 Th. Wolkenstein: Progr. Surf. Sci. **6**, 213 (1975)
5.11 E.P. Yesodharan, V. Ramakrishnan, J.C. Kuriokose: J. Sci. Ind. Res. **35**, 712 (1976)
5.12 K. Kalyanasundaram: In *Energy Resources by Photochemistry and Catalysis*, ed. by M. Grätzel (Academic, New York 1983)
5.13 J.-M. Herrmann, J. Disdier, M.N. Mozzanega, P. Pichat: J. Chem. Soc. Faraday Trans I, **77**, 2815 (1981)
5.14 G. Munuera, A. Navio, V. Rives-Arnau: J. Chem. Soc. Faraday Trans I, **77**, 2747 (1981)
5.15 N.V. Hieu, D. Lichtman: J. Catal. **73**, 329 (1982)
5.16 H. van Damme, W.K. Hall: J. Catal. **69**, 371 (1981)
5.17 J.P. Reymond, P. Vergnon, P.C. Gravelle, S.J. Teichner: Nouv. J. Chim. **1**, 197 (1977)
5.18 H. Mozzanega, J.-M. Herrmann, P. Pichat: J. Phys. Chem. **83**, 2251 (1979)
5.19 Y. Oosawa: J. Chem. Soc. Chem. Commun. 221 (1982)
5.20 H. Courbon, M. Formenti, P. Pichat: J. Phys. Chem. **81**, 550 (1977)

5.21 J. Cunningham, E.L. Goold, J.L.G. Fierro: J. Chem. Soc. Faraday Trans. I, **78**, 785 (1982)
5.22 J. Cunningham, E.L. Gold, E.M. Leahy: J. Chem. Soc. Faraday Trans. I, **75**, 305 (1979)
5.23 M.V. Rao, K. Rajeshwar, V.R. Pai Verneker, J. DuBow: J. Phys. Chem. **84**, 1987 (1980)
5.24 J.R. Harbour, M.L. Hair: NATO Adv. Study Inst., Ser. C**61**, 431 (1980)
5.25 J.R. Harbour, J. Tromp, M.L. Hair: J. Am. Chem. Soc. **102**, 1874 (1980)
5.26 S.N. Frank, A.J. Bard: J. Phys. Chem. **81**, 1484 (1977); J. Am. Chem. Soc. **99**, 4667 (1977)
5.27 H. Yoneyama, Y. Yamashita, H. Tamura: Nature **282**, 817 (1979)
5.28 D.H.M.W. Thewissen, K. Timmer, M. Eeuwhorst-Reinten, A.H.A. Tinnemans, A. Mackor: Isr. J. Chem. **22**, 173 (1982)
5.29 J.-M. Herrmann, P. Pichat: J. Chem. Soc. Faraday Trans. I, **76**, 1138 (1980)
5.30 B. Reiche, W.W. Dunn, A.J. Bard: J. Phys. Chem. **83**, 2248 (1979)
5.31 W.W. Dunn, A.J. Bard: Nouv. J. Chim. **5**, 651 (1981)
5.32 H. Hada, Y. Yonezawa, M. Saikawa: Bull. Chem. Soc. Japan **55**, 2010 (1982)
5.33 H. Yoneyama, N. Nishimura, H. Tamura: J. Phys. Chem. **85**, 268 (1981)
5.34 B. Aurien-Blajeni, M. Halmann, J. Manassen: Sol. Energy **25**, 165 (1980)
5.35 T. Inoue, A. Fujishima, S. Konishi, K. Honda: Nature **277**, 637 (1979)
5.36 G.N. Schrauzer, T.D. Guth: J. Am. Chem. Soc. **99**, 7189 (1977)
5.37 M.N. Mozzanega, J.-M. Herrmann, P. Pichat: Tetrahedron Lett. **34**, 2965 (1977)
5.38 P. Pichat, J.-M. Herrmann, J. Disdier, M.N. Mozzanega: J. Phys. Chem. **83**, 3122 (1979)
5.39 M.A. Fox, C.-C. Chen: Tetrahedron Lett. **24**, 547 (1983)
5.40 P.R. Harvey, R. Rudham, S. Ward: J. Chem. Soc. Faraday Trans. I, **79**, 1381 (1983)
5.41 J. Cunningham, B.K. Hodnett: J. Chem. Soc. Faraday Trans. I, **77**, 2777 (1981)
5.42 A. Walker, M. Formenti, P. Meriaudeau, S.J. Teichner: J. Catal. **50**, 237 (1977)
5.43 M.A. Fox, C.-C. Chen: J. Am. Chem. Soc. **105**, 4497 (1983)
5.44 M. Fujihara, Y. Satoh, T. Osa: Bull. Chem. Soc. Japan **55**, 666 (1982)
5.45 J. Cunningham, B. Doyle, E.M. Leahy: J. Chem. Soc. Faraday Trans. I, **75**, 2000 (1979)
5.46 J. Cunningham, D.J. Morrisey, E.L. Gold: J. Catal **53**, 68 (1978)
5.47 E. Borgarello, J. Kiwi, M. Grätzel, M. Visca: J. Am. Chem. Soc. **104**, 2996 (1982)
5.48 A. Mills, G. Porter: J. Chem. Soc. Faraday Trans. I, **78**, 3659 (1982)
5.49 T. Sakata, T. Kawai, K. Hashimoto: Chem. Phys. Lett. **88**, 50 (1982)
5.50 S. Sato, J.M. White: J. Catal. **69**, 128 (1981)
5.51 S. Sato, J.M. White: J. Phys. Chem. **85**, 592 (1981)
5.52 J.-M. Lehn, J.-P. Sauvage, R. Ziessel: Nouv. J. Chim. **4**, 623 (1980)
5.53 J.-M. Lehn, J.-P. Sauvage, R. Ziessel, L. Hilaire: Isr. J. Chem. **22**, 168 (1982)
5.54 B. Reichman, C.E. Byvik: J. Phys. Chem. **85**, 2255 (1981)
5.55 K. Kago, H. Yoneyama, H. Tamura: J. Phys. Chem. **84**, 1705 (1980)
5.56 T. Sakata, T. Kawai: Nouv. J. Chim. **5**, 279 (1981)
5.57 T. Sakata, T. Kawai: Chem. Phys. Lett. **80**, 341 (1981)
5.58 S. Sato, J.M. White: J. Phys. Chem. **85**, 336 (1981)
5.59 S. Sato, J.M. White: J. Phys. Chem. **86**, 3977 (1982)
5.60 I. Izumi, W.W. Dunn, K.P. Wilbourn, F.R.F. Fan, A.J. Bard: J. Phys. Chem. **84**, 3207 (1980)
5.61 A.H.A. Tinnemans, T.P.M. Koster, D.H.M.W. Thewissen, A. Mackor: Nouv. J. Chim. **6**, 373 (1982)

5.62 P. Pichat, M.-N. Mozzanega, J. Disdier, J.-M. Herrmann: Nouv. J. Chim. **6**, 559 (1982)
5.63 I. Izumi, F.R.F. Fan, A.J. Bard: J. Phys. Chem. **85**, 218 (1981)
5.64 H. Yoneyama, Y. Takao, H. Tamura, A.J. Bard: J. Phys. Chem. **87**, 1417 (1983)
5.65 S. Sato: J. Chem. Soc., Chem. Commun. 26 (1982)
5.66 W.W. Dunn, Y. Aikawa, A.J. Bard: J. Am. Chem. Soc. **103**, 6893 (1981)
5.67 H. Reiche, A.J. Bard: J. Am. Chem. Soc. **101**, 3127 (1979)
5.68 M.D. Ward, A.J. Bard: J. Phys. Chem. **86**, 3599 (1982)
5.69 M.D. Ward, J.R. White, A.J. Bard: J. Am. Chem. Soc. **105**, 27 (1983)
5.70 S. Sato, J.M. White: J. Catal. **69**, 128 (1981)
5.71 S.-M. Fang, B.-H. Chen, J.M. White: J. Phys. Chem. **86**, 3126 (1982)
5.72 S.-C. Tsai, C.-C. Kao, Y.-W. Chung: J. Catal. **79**, 451 (1983)
5.73 K. Domen, S. Naito, T. Onishi, K. Tamaur, M. Soma: J. Phys. Chem. **86**, 3657 (1982)
5.74 T. Kawai, T. Sakata: Chem. Phys. Lett. **72**, 87 (1980)
5.75 J.R. Darwent, A. Mills: J. Chem. Soc. Faraday Trans. II, **78**, 359 (1982)
5.76 K. Kalyanasundaram, E. Borgarello, M. Grätzel: Helv. Chim. Acta **64**, 362 (1981)
5.77 D. Meissner, R. Memming, B. Kastening: Chem. Phys. Lett. **96**, 34 (1983)
5.78 E. Borgarello, K. Kalyanasundaram, M. Grätzel, E. Pelizetti: Helv. Chim. Acta **65**, 243 (1982)
5.79 E. Borgarello, W. Erbs, M. Grätzel, E. Pelizetti: Nouv. J. Chim. **7**, 195 (1983)
5.80 D.H.M.W. Thewissen, A.H.A. Tinnemans, M. Eeuwhorst-Reinten, K. Timmer, A. Mackor: Nouv. J. Chim. **7**, 191 (1983)
5.81 M. Grätzel, A.J. Frank: J. Phys. Chem. **86**, 2964 (1982)
5.82 D. Duonghong, J. Ramsden, M. Grätzel: J. Am. Chem. Soc. **104**, 2977 (1982)
5.83 J. Moser, M. Grätzel: J. Am. Chem. Soc. **105**, 6547 (1983)
5.84 M. Grätzel, J. Moser: Proc. Nat'l. Acad. Sci. **80**, 3129 (1983)
5.85 J. Moser, M. Grätzel: Helv. Chim. Acta **65**, 1436 (1982)
5.86 A. Henglein: Ber. Bunsenges. Phys. Chem. **86**, 241 (1982)
5.87 R. Rossetti, L. Brus: J. Am. Chem. Soc. **104**, 7321 (1982)
5.88 M.A. Fox, B. Lindig, C.-C. Chen: J. Am. Chem. Soc. **104**, 5828 (1982)
5.89 K. Chandrasekaran, J.K. Thomas: Chem. Phys. Lett. **97**, 357 (1983); **99**, 7 (1983)
5.90 D. Duonghong, M. Grätzel: Helv. Chim. Acta **67** (1984) in press
5.91 J. Ramsden, M. Grätzel: J. Chem. Soc. Faraday Trans. (1984) submitted
5.92 R. Rossetti, L. Brus: J. Phys. Chem. **86**, 4470 (1982)
5.93 K. Metcalfe, R.E. Hester: J. Chem. Soc. Chem. Commun. 133 (1983)
5.94 A. Henglein: Ber. Bunsenges. Phys. Chem. **86**, 301 (1982); J. Phys. Chem. **86**, 2291 (1982)
5.95 Z. Alfassi, D. Bahnemann, A. Henglein: J. Phys. Chem. **86**, 4656 (1982)
5.96 M. Gutierrez, A. Henglein: Ber. Bunsenges. Phys. Chem. **87**, 474 (1983)
5.97 J. Kuczynski, J.K. Thomas: Chem. Phys. Lett. **88**, 445 (1982)
5.98 G.A. Hope, A.J. Bard: J. Phys. Chem. **87**, 1979 (1983)
5.99 A. Heller, E. Aharon-Shalom, W.A. Bonner, B. Miller: J. Am. Chem. Soc. **104**, 6942 (1982)
5.100 D. Meissner, R. Memming, B. Kastening: Chem. Phys. Lett. **96**, 34 (1983)
5.101 A. Heller, R. Memming, H. Tamura and their co-workers: Papers presented at the Electrochemical Society Spring Meeting, 1983, San Francisco, Extended Abstracts 505-507, 518
5.102 W. Gissler, A.J. McVoy: J. Appl. Phys. **53**, 1251 (1982)
5.103 W. Gissler, A.J. McVoy, M. Grätzel: J. Electrochem. Soc. **129**, 1733 (1982)
5.104 J.-M. Herrmann, J. Disdier, P. Pichat: In *Metal-Support and Metal-Additive Effects in Catalysis*, ed. by B. Imelik et al. (Elsevier, Amsterdam) p.27

5.105 J.-M. Herrmann, P. Pichat: J. Catal. **78**, 425 (1982)
5.106 B.-H. Chen, J.M. White: J. Phys. Chem. **86**, 3534 (1982)
5.107 W. Tsai, J.A. Schwarz, C.T. Driscoll: J. Phys. Chem. **87**, 1619 (1983)
5.108 Y. Taniguchi, H. Yoneyama, H. Tamura: Chem. Lett. 269 (1983)
5.109 A.V. Bulatov, M.L. Khidekel: Izv..Acad. Sci., USSR, Ser. Khim. 1902 (1976)
5.110 M.S. Wrighton, P.T. Wolczanski, A.B. Ellis: J. Solid State Chem. **22**, 17 (1977)
5.111 F.T. Wagner, G. Somorjai: J. Am. Chem. Soc. **102**, 5494 (1980)
5.112 J. Kiwi, M. Grätzel: J. Phys. Chem. **88** (1984) in press
5.113 E. Yesodharan, M. Grätzel: Helv. Chim. Acta **66**, 2145 (1983)
5.114 E. Borgarello, J. Desilvestro, M. Grätzel, E. Pelizzetti: Helv. Chim. Acta **66**, 1827 (1983)
5.115 D.H.M.W. Thewissen, A.H.A. Tinnemans, M. Eeuwhorst-Reinten, K. Timmer, A. Mackor: Nouv. J. Chim. **7**, 73 (1983)

6. Laser Studies of Surface Chemical Reactions

R. R. Cavanagh and D. S. King

With 11 Figures

6.1 Introduction

Currently lasers are being applied to study a very broad range of problems
in surface chemistry and surface physics. In many of these areas the introduction of laser probes complements or duplicates more conventional or traditional instrumentation. For instance, surface-enhanced [6.1,2] or unen-
hanced [6.3] Raman scattering provides the same type of vibrational structure information of surface adsorbates as obtained through reflection absorption infrared and electron energy loss spectroscopies [6.4]. The pulsed
laser atom probe is receiving attention as a potential technique for characterizing surface composition and depth profiling [6.5], but does not yet
significantly extend the temporal or spatial resolution available from the
more traditional atom probe techniques [6.6]. In several specialized research
areas laser-induced desorption, sputtering and multiphonon ionization are
replacing conventional mass spectrometer ionization sources, but result in
chemical speciation data comparable to conventional mass spectrometer [6.7]
sources. There are many specific and important areas of surface chemistry/
physics where the spatial resolution and high power densities available with
lasers present the potential for real scientific advancement [6.8,9]. One
obvious such area is semiconductor device fabrication [6.10]. Alternatively,
lasers provide the capability to distinguish between reactive (or product
species) of the same chemical identity but which occur with differing amounts
or types of internal excitation due to competing reaction pathways. This potential for detailed, quantum-state specific information about surface chemical processes opens the door to exploring and, hopefully, unraveling the
complex dynamical interplay in chemical processes occurring at the gas-solid
interface.

   This chapter will attempt to address recent experimental applications of
lasers aimed at characterizing the important steps in surface chemical reactions. Since in any collisional encounter there is the possibility for both

energy transfer and for chemical reaction, and since it is our ultimate goal to identify unique aspects of molecule-surface interactions, the development proceeds through analogy to advances in modern gas-phase chemical physics involving nonreactive (Sect.6.2) and reactive (Sect.6.4) scattering and also through a brief discussion of inelastic molecular-beam surface scattering (Sect.6.3). Section 6.5 deals in some detail with a series of experiments on a very simple chemical process — the thermal rupture of the NO-Ru{001} chemisorption bond, utilizing many of the same techniques successfully applied to problems mentioned in Sects.6.2-4. Finally, Sect.6.6 discusses laser-based experimental techniques suitable for determining the rates for energy transfer or energy flow between chemisorbed molecular species and surfaces.

6.2 Gas Phase Inelastic Scattering

Energy transfer in the gas phase has been the subject of many extensive studies, both theoretical and experimental [6.11,12]. The nonreactive, inelastic encounter of $A + BC$ $(v,J) \rightarrow A + BC$ (v',J') represents the simplest system which reflects the dynamics of molecular energy transfer. The interaction which governs the inelastic encounters for such a system will clearly have an orientational dependence. The interaction potential can be written as $V (R,\alpha) = V_0 (R) + V_a (R,\alpha)$, where the anisotropic part V_a distinguishes atom-molecule scattering from atom-atom scattering. Although V_a may not influence the hard sphere molecular cross section, the detailed nature of V_a will have pronounced and molecule specific effects on the *outcome* of the collision.

In the inelastic scattering of He off Na_2, for example, there is evidence for strong internal state dependences in the cross sections for rotational excitation of the final Na_2 species [6.13]. A statistical distribution of final rotational state population would be expected to be Boltzmann, that is for the probability to decrease monotonically and exponentially with final-state energy. In the He + Na_2 system, a local maximum in excited rotational state population is observed at high rotational excitation. This effect is known as a rotational rainbow or rainbow singularity. The physical source of this behaviour can be readily visualized. Simply stated, a collisional orientation of the atom and diatom exists which maximizes the rotational excitation in the diatom. If the angular anisotropy of the scattering potential gives rise to populations of final diatom rotational quantum states $P(J_f)$ as shown in Fig.6.1a, there will be a rainbow singularity in $P(J_f)$ when $\delta j_f(\alpha)/\delta\alpha$ tends to zero, shown in Fig.6.1b (which has been adapted from

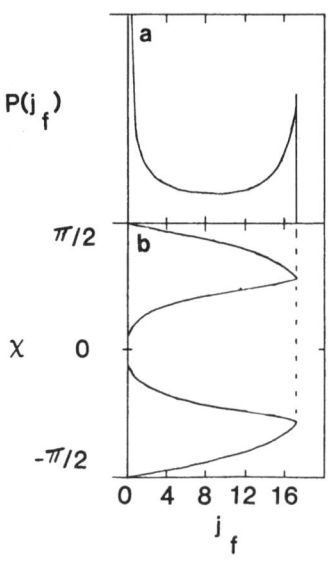

Fig.6.1a. Probability of rotational excitation ($P(j_f)$) as a function of final rotational state. b. Maximum rotational excitation as a function of collision angle χ

Fig.2a of Ref. [6.14]). In this model, the probability of a collision resulting in a highly rotationally excited final state is greatest when the incident collision angle $\alpha \sim 45°$. Figure 6.1 obviously represents a highly simplified interaction potential and ignores any initial rotational state distribution.

6.3 Molecule-Surface Inelastic Scattering

Replacing the atom with a surface, one obtains a more complex problem. The kinematic questions remain the same. One still has to address the changes in vibrational, rotational and translational energies of the colliding diatom. In analogy to atom-molecule scattering, one might expect a higher degree of rotational inelasticity in the diatom-surface collision because all collision trajectories result in backward scattering. However, a new consideration arising with the inclusion of the massive surface is that the surface, in essence, applies an external torque to the diatom, which removes the constraint of conservation of total angular momentum [6.14].

The basic question addressed in these nonreactive molecule-surface scattering experiments is what are the dominant interactions that control energy transfer between the diatom and the surface. Theoretical approaches to this problem have examined predictions based on rigid rotor-smooth rigid surface, rigid rotor-corrugated rigid surface, rigid rotor-nonrigid surface, and various other model systems. Much of the current work is focused on more re-

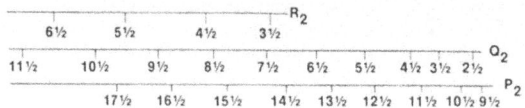

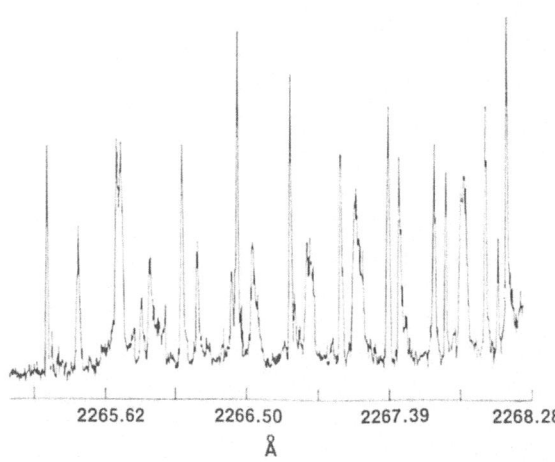

Fig.6.2. Laser-excited fluorescence spectrum of NO gas at thermal equilibrium (T = 300 K)

alistic, but more complex situations involving coupling of nonrigid rotors and nonrigid surfaces [6.14-17]. Other approaches have relied on semiclassical analysis coupled to trajectory calculations to follow empirically the results of various interactions [6.18]. In both instances the objective is to turn on or suppress certain physical interactions and observe the correlated changes in the scattering simulations. It is hoped that eventually a correct or adequate physical representation of the "surface" will emerge from comparison of calculations with detailed experimental results.

The requisite experimental results involve the measurement of quantum state to quantum state specific cross sections. Such detailed information is available through laser-excited fluorescence (LEF) studies of molecular-beam surface scattering. The LEF technique gives qualitatively identical information to absorption spectroscopy, but is in many cases considerably more sensitive. Figure 6.2 shows a laser-excited fluorescence excitation spectrum of the NO molecule in the origin region of its lowest energy symmetry allowed electronic transition, the $\tilde{A}\,^2\Sigma \leftarrow \tilde{X}\,^2\Pi$ or gamma band system. To obtain this spectrum the spontaneous fluorescence emission of NO in a bulb at room temperature is monitored by a photomultiplier tube while the wavelength of a tunable dye laser of ~0.2 cm^{-1} bandwidth is scanned across the absorption band, from 2265 to 2268 Å. The spectroscopy of NO in this spectral region is very well established [6.19]. The position of each individual line identifies the vibrational, rotational and spin-orbit quantum states of the mole-

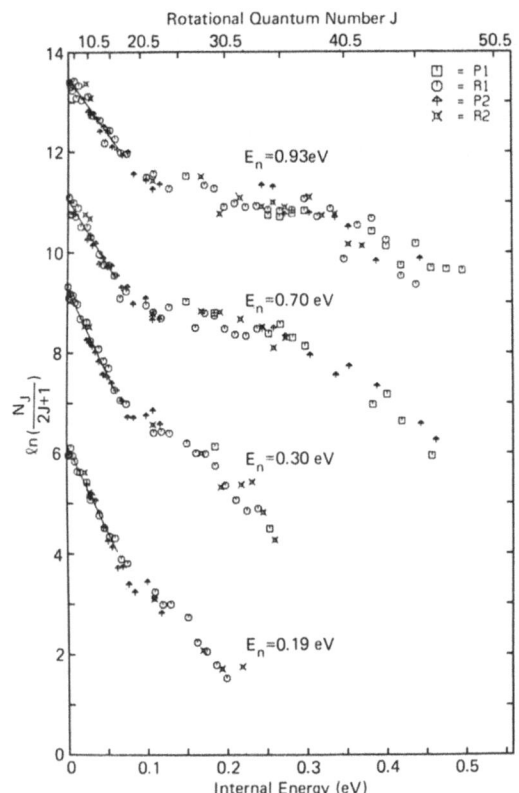

Fig.6.3. Rotational state distributions for scattered NO molecules as a function of internal energy $E_{int} = E_{rot} + E_{so}$. E_{so} is the spin-orbit energy, which is zero for P1 and R1 branch transitions, for which a J scale is given (top). For all data shown $\theta_i = 15°$, except $E_n = 0.19$ eV where $\theta_i = 40°$. $T_S = 650$ K. T_R as determined from the fit to the data (full line) at low J is from top to bottom: 550 ± 10 K, 475 ± 20 K, 375 ± 20 K and 340 ± 20 K

cules being probed, while the intensity obtained is directly related to the *density* of NO species in that state. In the scattering experiment then, by scanning the probe laser wavelength over a convenient (narrow) spectral region, one can map out initial and/or final vibrational and rotational state distributions. By using molecular-beam techniques the initial state distribution can be made exceedingly narrow, resulting in an accurate measure of state-to-state cross sections.

Perhaps the most thoroughly studied system to date is NO/Ag{111}[6.20-23]. Several interesting aspects of nonreactive scattering processes have become apparent in this work. The IBM group has demonstrated conclusively the appearance of rotational rainbows in the scattering of NO($J \approx 0$) off Ag{111}, which is a nominally flat crystal face. A portion of their data is reproduced in Fig.6.3, where the final rotational state population is plotted against rotational state energy content. A statistical or Boltzmann distribution would give a linear fit with a slope related to the effective rotational temperature. Although the low-energy portion of the final rotational state distribution might be described by a rotational temperature, the higher energy region is dominated by the rainbow effect (recall Fig.6.1).

To the extent that the low-energy portion (i.e., the majority of scattered species) can be described by a rotational temperature T_R, it was observed that T_R depended on the normal component of the incicent kinetic energy E_n and was approximately independent of surface temperature. The variation of T_R with E_n was approximately linear, following

$$T_R = a(E_n + \varepsilon) \quad ,$$

where $\varepsilon = 0.78 \pm 0.08$ eV and $a = 320 \pm 26$ K/eV over the range $0.1 < E_n < 1.0$ eV.

A degree of rotational polarization was observed for the scattered NO. The actual degree of polarization depended strongly on both the final rotational state being analyzed and the initial angle of incidence between the molecular beam and the surface normal. Experimentally, the distribution of orientation of final angular momentum vectors $\mathbf{J}$ (i.e., the axis of rotation) relative to the surface normal is measured by observing LEF intensities I probing a single rotational state while rotating the direction of polarization of the laser such that the angle between the polarization of the laser and the surface normal varies between 0 and π. The polarization anisotropy $\mathscr{P}$ is given by

$$\mathscr{P} = 5 \left[\frac{I(0) - I(\pi/2)}{I(0) + 2I(\pi/2)} \right]$$

and can vary from +5 for perfect alignment with $\mathbf{J}$ parallel to the surface normal to -2.5 for perpendicular alignment. In their experiments, *Luntz* et al. [6.22] observed values of $\mathscr{P}$ ranging from 0 (implying no preferential alignment) to -1.44 near $J = 30\frac{1}{2}$, implying a substantial anisotropy. Although significant, this is less than the value of -2.5 expected for the smooth surface limit. It is worth noting that the greatest degree of alignment is observed for those final states contributing to the rotational rainbow mentioned above.

6.4 Reactive Scattering

As complex and challenging as studies of inelastic scattering are, it is the question of reactive scattering which must be addressed in order to understand chemical processes at surfaces. That is, one must be assured that the effects are a result of chemical interactions and not only kinematics. Once again, a detailed picture of the dynamics of such a process will require both reliable theoretical input, and quantum state to quantum state cross sections measured for chemically reactive systems. The prototypical reaction $A + BC \rightarrow AB + C$ will be accompanied by both inelastic kinematic energy exchange, and will also reflect the exoergicity of the bond rupture and bond formation steps. For instance, the gas phase reaction $H + F_2 \rightarrow HF + F$ has been shown to have both a strong preference for a colinear approach geometry and to result in a non-Boltzmann population distribution in the vibrational levels of the nascent HF $(V = 4/V = 0 \sim 10^2)$ [6.24,25]. Simple model potential energy hypersurfaces indicate that such a population effect can be explained by an exoergicity of 29 Kcal/mol and the presence of a barrier in the entrance channel of 0.9 Kcal/mol for this reaction.

6.5 Thermal Desorption

As a starting point for the study of surface chemical reactions let us consider the simple thermal breaking of the chemisorption bond in a temperature-programmed desorption experiment. More complex reactions or reactive processes such as electron-stimulated desorption, atomic recombination, etc., can be studied using similar techniques while deriving the same type of detailed information.

Thermal desorption traces for NO from Ru{001} are shown in Fig.6.4 [6.26]. Both traces were obtained for a heating rate of 12 K s^{-1} and an initial NO coverage of ~0.8. The lower trace was recorded with a quadrupole mass spectrometer (QMS) set to monitor NO evolved from the surface into the ultrahigh vacuum chamber (base pressure ~10^{-10} torr) as the surface was resistively heated. The upper trace is the LEF signal recorded with the probe laser tuned to excite only these desorbing NO molecules in the J = 6 1/2 rotational quantum state of the v = 0 vibrationless level. Molecules in this state have 165 cm^{-1} = 0.02 eV of internal rotational energy. The QMS detection provides a higher signal-to-noise ratio resulting from high ion-collection efficiency, increased duty cycle (1 vs 10^{-5} for the LEF), and non-state-selective detection of all desorbing NO. However, the QMS does not provide any of the state-selective information available through the laser diagnostic techniques.

The quantum-state distribution of the desorbed NO is readily mapped out by repeating a large number of thermal desorption experiments and tuning the laser wavelength to probe product NO species in successive quantum states as the crystal is cleaned and redosed with fresh NO between each flash desorption. The resulting data are presented in Fig.6.5. This is a semilogarithmic plot of relative population in the final rotational state J versus the rota-

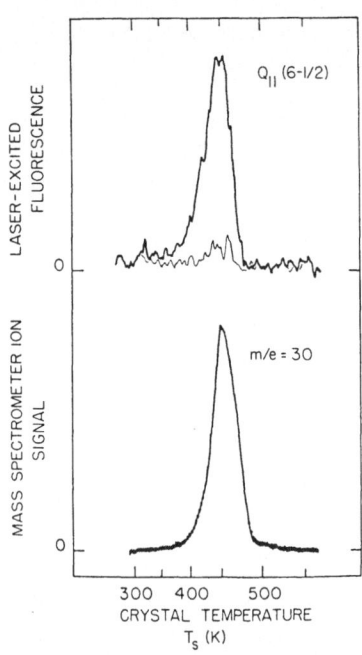

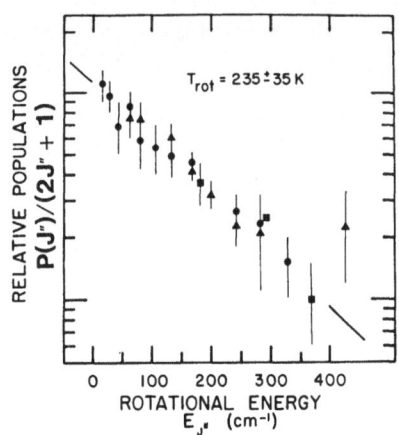

Fig.6.4. Laser-excited fluorescence and mass-spectrometer-detected thermal desorption spectra. Each trace is the result of a single temperature-programmed desorption. The base line LEF trace is the observed signal when the crystal is rotated to 3 cm below the laser beam

Fig.6.5. Rotational and spin-state population distribution for NO thermal desorption from Ru{001} at 455 ± 20 K. A single Boltzmann temperature, $T_{rot} = 235 \pm 35$ K, describes the total angular momentum distribution. The various symbols represent population densities for excitation on the $R_{11}(J")$ (circles), $Q_{11}(J")$ (trangles), and $Q_{22}(J")$ (squares) branch transitions

tional energy of that J^{th} state. A statistical distribution of final-state probabilities would result in a Boltzmann-like population distribution giving a linear fit. The fitting of the observed data to such a straight line is quite good, having a correlation factor of -0.98. Whether or not the actual population distribution is truly Boltzmann, its empirical exponential nature enables one to characterize the distribution by a single parameter — a rotational temperature T_R. The value of T_R derived from the NO/Ru{001} thermal desorption data is 235 ± 35 K, only one-half the temperature corresponding to the crystal temperature at the desorption maximum $T_s = 455 \pm 20$ K.

Several models have been specifically set forth to address the factor of two difference between the observed rotational temperature of the NO molecules desorbed from clean Ru{001}, and the temperature of the crystal at the desorption peak [6.27-29]. The basic physical processes for these models may provide useful limiting cases for idealized mechanisms, i.e., weakly interacting, strongly bound, etc. Since the models (i) differ significantly in their mechanisms, (ii) do not consider the specific molecular-surface interactions, and (iii) have only one experimental measurement to "explain",

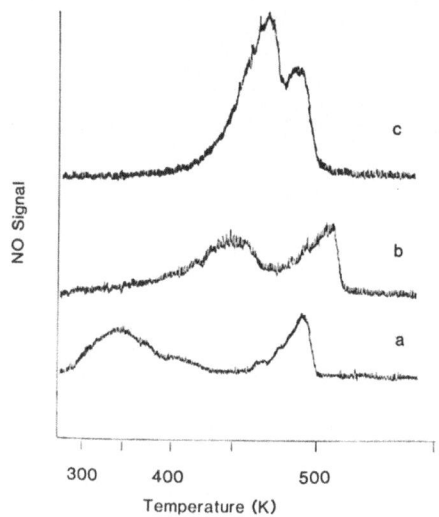

Fig.6.6. Mass-spectrometer-detected thermal desorption of NO from oxidized Ru. Temperature of annealing (a) 1110 K, (b) 1310 K, and (c) 1520 K. Heating rate 10 K/s

it is difficult to judge their general applicability to molecule-surface phenomena.

A more general appreciation of the quantum-state measurements of thermally desorbed NO can be gained by looking at other thermal desorption systems. In fact, the sensitivity of these quantum-state specific measurements to the precise details of the molecule-surface interaction is highlighted by parallel work on preoxidized Ru [6.30]. Conventional thermal desorption spectra of a series of cleaning flashes following Ru oxidation at 1465 K in $\sim 10^{-7}$ torr O_2 are shown in Fig.6.6. Note that while complex transformations are apparent in the NO desorption signals between 1110 and 1520 K, the desorption obtained using a 1110 K flash is dominated by features at 285 and 480 K. The high-temperature feature observed from the preoxidized Ru appears at approximately the same surface temperature (480 K) as observed for the clean Ru (455 K) — indicative of very similar desorption activation energies. The rotational state population distributions from the individual desorption features from the preoxidized Ru could be well described by Boltzmann distributions. The specific values of T_S and T_R are given in Table 6.1. From these data it is apparent that there is no simple scalar relationship between T_R and T_S. Any theoretical analysis of this type of data must consider dynamical aspects of the energy exchange between the adsorbed species and the surface, and the details of the interaction potential (including entrance/exit channel effects). One additional factor influencing the NO/Ru and NO/O/Ru systems is contributions from molecular dissociation which occurs during chemisorption and/or desorption. In thermal desorption experiments of NO from

Table 6.1. Thermally desorbed NO from clean and oxidized Ru

	Crystal temperature at desorption maximum (K)	Observed rotational temperature (K)	T_{ROT}/T_{surf}
Clean	455 ± 15	235 ± 30	0.5
Oxidized	475 ± 15	345 ± 30	0.7
	285 ± 15	255 ± 30	0.9

initially clean Ru{001} there is significant dissociation. Only about 5%-10% of a monolayer of NO is molecularly desorbed [6.31]. In the preoxidized system a large fraction of the initial NO coverage desorbs as molecular NO. It is certainly possible that the oxide overlayer closes off the dissociative channel that competes so favorably with molecular NO desorption in thermal desorption from clean Ru{001}, resulting in a significant shift in the dominant interactions. For the two desorption features observed near 460 K a dramatic difference in rotational final-state distributions is observed dependent on the history of the initial surface cleanliness prior to NO adsorption.

As implied in Sect.6.3, it is possible to obtain more detailed information than just the rotational population distribution using laser diagnostics. Ideally, one would like to know rotational, vibrational, translational, orientational, and spatial distributions. In addition, knowledge of each distribution is most useful if the measurements are made for specific (unaveraged) values of the remaining degrees of freedom. To date, there are no specific data on vibrational population distributions from thermal desorption experiments. There are some data for vibrational excitation observed following energetic molecular beam scattering off hot, i.e., $T_S > 650$ K, surfaces [6.32]. In their work on NO/Pt{111}, *Ascher* et al. [6.32] report that the ratio of the first vibrationally excited (v = 1) to ground vibrational (v = 0) states of the vibrationally inelastic scattered NO was somewhat lower than would be expected if described by a vibrational temperature equal to the surface temperature. From our signal-to-noise ratio we can say only that for the thermal desorption of NO from ruthenium, less than one in twenty desorbing NO species leaves the surface vibrationally excited. Given an improved signal-to-noise ratio it should be possible to characterize both the actual partitioning of vibrational energy and the dependence/independence of rotational and vibrational excitation in this chemisorption-bond breaking process.

One of the advantages of LEF detection is the potential to obtain velocity distributions for molecules with a well-defined degree and/or type of internal excitation by monitoring Doppler profiles. "Thermal" Doppler profiles of many light gas-phase molecules have full width at half-maxima of about 0.1 cm^{-1} ($\sim 10^{-5}$ eV), thus requiring the probe laser bandwidth to be on the order of 0.01 cm^{-1} or less in order to characterize the Doppler profile meaningfully. In more conventional surface experiments, velocity and angular flux distributions have been measured in molecular-beam scattering experiments. The determination of the angular flux distribution for several desorption-type experiments using mass spectrometers has also been reported. The flux distributions range from cos$^{16}\theta$ for CO_2 desorbed from platinum [6.33], to cos$^5\theta$ for atom-atom recombination on copper [6.34], to cos$^1\theta$ for a variety of "equilibrium" systems [6.35,36]. Even broader distributions have been theoretically predicted [6.37]. The large range of observed flux distributions suggests that this property may be very sensitive to the details of the specific surface chemical processes being studied. Equally important to the goal of understanding surface chemistry will be the ability to related and/or separate the internal excitation(s) of a desorbing product species to/from its external kinetic energy and angular flux distributions. This is not possible using QMS detection, but is readily achieved by Doppler-resolved LEF techniques.

When considering the contributions of molecular velocities to the shape of a spectral transition, one can easily express the observed Doppler shifted molecular transition frequency as $\nu = \nu_0[1 + (\mathbf{v} \cdot \hat{n}/c)]$, where ν_0 is the transition for a molecule at rest and $\mathbf{v} \cdot \hat{n}$ gives the projection of the molecular velocity on the Poynting vector of the probe radiation. If the molecular velocities are isotropically distributed in the probed sample, then the transition's spectral profile function is simply given by the Gaussian line shape

$$S(\nu) = S(\nu_0) \exp\left\{ - (\ln 2)\left[\frac{\nu - \nu_0}{(\Delta\nu)_d}\right]^2 \right\} \quad ,$$

where $2(\Delta\nu)_d$ is the Doppler width (FWHM) and is dependent only on the "translational temperature" $T_{K.E}$, as

$$\frac{(\Delta\nu)_d}{\nu_0} = \mathrm{const} \, (\frac{T}{M})^{\frac{1}{2}} \quad .$$

It is not reasonable, however, to assume that molecular desorption in ultra-high vacuum is isotropic. Indeed, the source crystal "shields" one hemisphere. Figure 6.7a displays an arbitrary flux distribution of cos$^5\theta$, with one iso-flux line drawn in, and a propagation direction for the laser. Under these conditions, a line shape symmetric about the rest frequency is anticipated

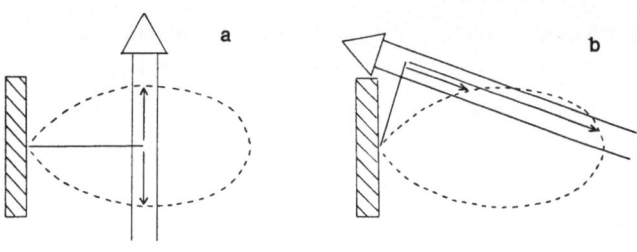

Fig.6.7

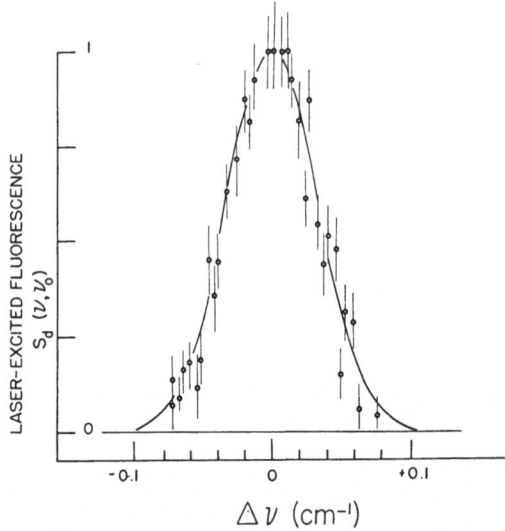

LASER-EXCITED FLUORESCENCE $S_d(\nu, \nu_0)$

$\Delta \nu$ (cm⁻¹)

Fig.6.7a,b. Isoflux contours for a point source with a $\cos^5$ flux distribution. a) Laser propagating parallel to the crystal face, giving a Doppler profile symmetric about the rest frequency. b) Laser propagating 70° away from parallel, giving a Doppler-shifted result

◄ Fig.6.8. Laser Doppler profile of NO (J"=9 1/2, E_{rot}=165 cm⁻¹) thermally desorbed from Ru {001} measured parallel to the crystal face. The crystal temperature at the thermal desorption maximum was 455 K. The laser bandwidth was 0.02 cm⁻¹ and the zero Doppler shift frequency was 44 252 cm⁻¹

due to geometrical considerations. A measurement of the tangential Doppler line shape was obtained in this geometry at an increased dye laser resolution of 0.01 cm⁻¹.

Data for desorbed NO molecules in the $v = 0$, $J = 6\ 1/2$ state are shown in Fig.6.8 [6.38]. Each datum corresponds to the average of 4-6 individual desorption experiments performed at the same probe laser wavelength. Clearly the observed Doppler profile is symmetric and centered about the rest frequency as compared to a thermal reference cell of NO. Further, unambiguous interpretation of the observed line shape is complicated. The actual line shape follows from two distinct factors: the distribution of molecular *speeds* combined with the *angular flux* distribution. Figure 6.9 depicts several computer-simulated "Doppler line shapes". All arise from the same translational temperature, i.e., same speed distribution, but correspond to angular flux distributions ranging from isotropic, through cosine to $\cos^5$. In fitting the experimental data in Fig.6.8, if we assume a cosine flux distribution we obtain a very good fit to a tangential velocity distribution characterized by a temperature $T_{\|} = T_S = 460$ K. Alternatively, assuming an isotropic

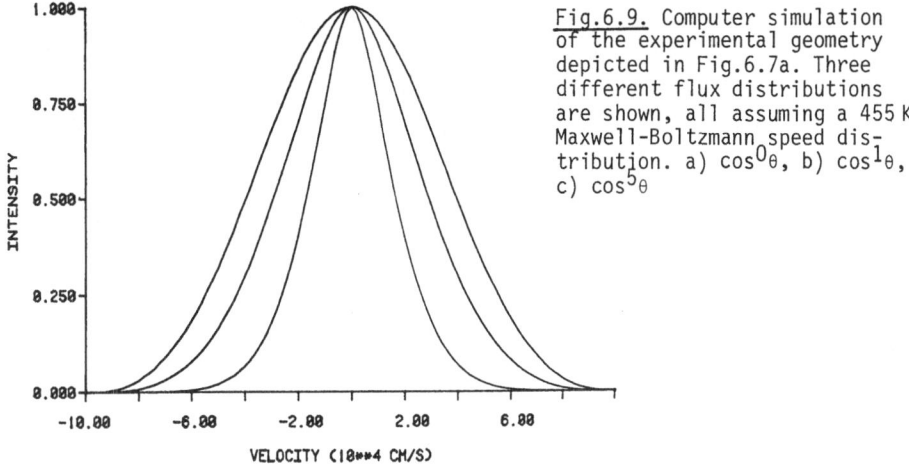

Fig.6.9. Computer simulation of the experimental geometry depicted in Fig.6.7a. Three different flux distributions are shown, all assuming a 455 K Maxwell-Boltzmann speed distribution. a) $\cos^0\theta$, b) $\cos^1\theta$, c) $\cos^5\theta$

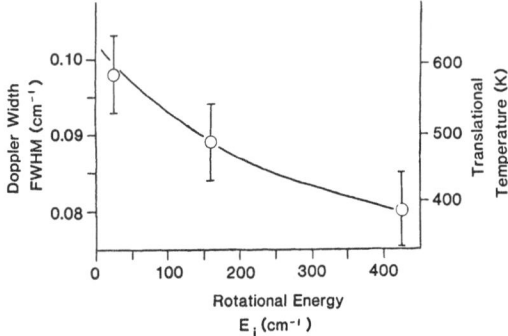

Fig.6.10. Rotational state dependence of the observed parallel Doppler profile. Observed full width at half-maximum is on the left-hand axis, and the corresponding temperature assuming a cos θ flux distribution is on the right-hand axis

flux distribution one obtains $T_{\parallel} \sim 245$ K. A highly forward peaked distribution would imply $T_{\parallel} \gg T_S$. Although the experimental geometry in Fig.6.8 gives no information on the polar angular flux distribution, from the observed symmetric line shape we know that there is symmetry associated with the flux relative to the surface normal. However, a stepped surface might be capable of exhibiting significant azimuthal asymmetry under similar experimental conditions.

Figure 6.10 shows the results of a series of measurements aimed at determining the correlation or independence between internal and external degrees of freedom. In this figure are plotted the Doppler widths (FWHM) observed for desorbed species in three different final rotational states. As indicated above, in addition to a polar angular flux distribution such Doppler profiles can be characterized by a single parameter, the translational temperature. Assuming a cosine flux distribution, the observed Doppler profiles are consistent with the temperature scale indicated on the right-hand side of the

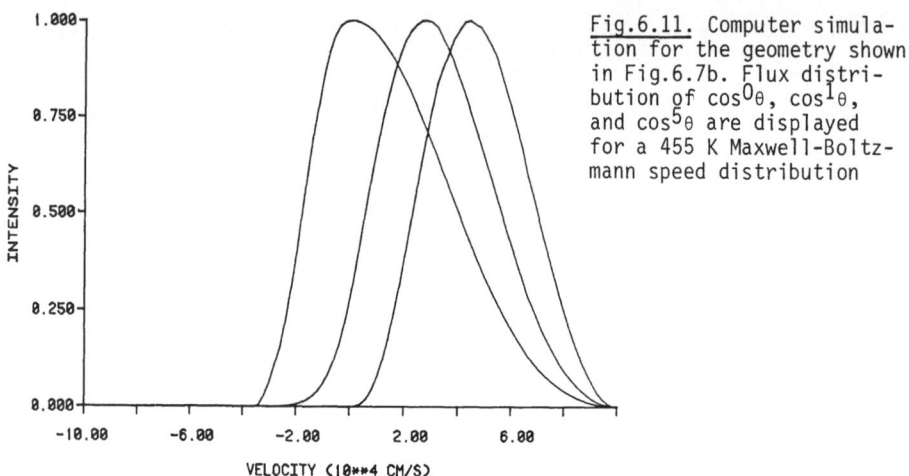

INTENSITY

VELOCITY (10**4 CM/S)

Fig.6.11. Computer simulation for the geometry shown in Fig.6.7b. Flux distribution of $\cos^0\theta$, $\cos^1\theta$, and $\cos^5\theta$ are displayed for a 455 K Maxwell-Boltzmann speed distribution

figure. The error bars, at the 95% confidence level, are appreciable, yet the data suggest a negative correlation between the rotational energy of a desorbing NO species and its kinetic energy.

At this point, experimentally, these results are consistent with a range of effects. The limiting cases are represented by the following two scenarios: (1) molecules with large degrees of rotational excitation have the same angular flux distributions as those with small degrees of rotational excitation but less kinetic energy, or (2) there is no dependence of kinetic energy on rotational state, but rather highly rotationally excited molecules are preferentially desorbed in a more highly foward peaked direction. The first rationalization suggests a conservation of energy argument while the second potentially illuminates subtle details in the potential energy hypersurface of the bond-breaking reaction process. If the latter is accurate, one might expect to see interesting alignment effects similar to those mentioned in Sect.6.3.

The flux versus kinetic energy ambiguity can be resolved by performing further Doppler profile measurements at other polar angles. Figure 6.7b illustrates an experimental geometry that appears to be very useful. An angle of ~20° is set between the laser and surface normal. Computer simulations for this geometry are shown in Fig.6.11, similar to those shown in Fig.6.9. The three curves correspond to isotropic, cosine and $\cos^5$ flux distributions for one speed distribution (T_{trans} = 455 K). Under these geometrical conditions the derived Doppler profile will no longer be either symmetrical nor necessarily maximized at the molecular rest frequency. The projection of molecular velocities on the propagation direction of the laser will result in predominantly red-shifted absorption. The resulting line shapes can be characterized by two parameters given the (known) experimental geometry (geometrically, the most important aspects are the angle between the laser and the surface normal, and the location of the LEF viewing region relative to the crystal face). These two parameters are the spectral shift from ν_0 in the peak position and the red-shifted edge's half-width at half-maximum. To a certain extent the desorbing flux may be likened to a beam of molecules which can be characterized by two properties: a most probable speed in the forward direction and a spread of speeds about the mean. In our thermal desorption experiments the Doppler profiles are narrow and peaked near the rest frequency. Assuming thermal speed distributions, there is a very strong correlation in all our modeling attempts between the shift in spectral peak position and the angular flux distribution, and between the red shifted half-width and the perpendicular translational temperature or speed distribution.

Doppler measurements in this near perpendicular geometry give independently both the angular flux and perpendicular speed distribution. Knowledge of the angular flux distribution then allows one to extract the parallel speed distribution from the tangential Doppler measurements. These three pieces of information then can be obtained for molecules as a function of extent or type of internal excitation.

Taken as a whole, the information available form these LEF experiments is quite detailed. Whether the molecular dynamics are relevant to the question of sticking (as in the beam/surface scattering example), or whether the dynamics are related to chemical processes or the desorption of molecules from surfaces (as in the thermal desorption experiment), the fundamental aspects of energy transfer at surfaces can be probed quantitatively. The ability to measure molecular speeds, flux distributions, and spatial orientation [6.38] as a function of rotational energy can provide the detailed experimental information necessary for definitive comparison to trajectory calculations of surface chemical processes. As these and complementary techniques are applied to larger and larger ranges of chemical phenomena at surfaces, the level of sophistication in dealing with potential energy hypersurfaces will expand. Already, laser-based analysis techniques are being applied to sputtering [6.39] and SIMS [6.40], and a number of groups anticipate application to ion-induced Auger, ESD, laser-induced desorption and second-order desorption processes.

6.6 Time Domain

One distinctly different area of lasers applied to surface chemistry should
be acknowledged. A critical question which still remains to be addressed is
which properties influence the transfer of energy between a surface and an
adsorbing (or desorbing) molecule. Consider a molecule bound to a surface.
If an internal mode of the adsorbed species becomes excited, on what time
scale will relaxation occur? If a molecule in the gas phase approaches a
surface, how is the heat of adsorption dissipated in order to form (or stabilize) the new surface chemical bond? These questions are the surface analog to energy transfer questions in solutions, in solids, and in the gas
phase [6.41]. A naive approach would be to extract lifetime information from
experimental linewidths by applying the Heisenberg uncertainty principle.
However, spectral lines can be influenced by both inhomogeneous and heterogeneous broadening, and thus the observed linewidth may have little relation
to the population lifetime of the levels of interest [6.42-44]. To extract
lifetime information and rates of energy transfer, real time measurements
using ultrashort laser pulses hold great promise.

   With the possible exception of the stimulated picosecond Raman gain
technique [6.45], no picosecond techniques have been demonstrated to have
the requisite sensitivity to conduct such real-time measurements at the face
of a single crystal surface. However, tunable infrared pump-probe techniques,
time-resolved fluorescence, and transient absorption all appear to offer
viable approaches to characterizing the dissipation and flow of energy in
chemisorbed layers on high surface area materials [6.46].

6.7 Summary

The use of lasers to probe the details of surface chemical reactions on the
molecular level provides new insight into the mechanisms and pathways of
energy transfer at surfaces. Internal energy distributions and real-time
population measurements represent two generic experimental approaches which
can address directly the issue of energy transfer at surfaces. Already,
laser-excited fluorescence has demonstrated a number of useful applications
to understanding chemical processes at surfaces. Such work, when coupled
with real-time measurements of energy transfer processes, represents a domain of great promise in surface chemical dynamics.

References

6.1 R. Chang, T.E. Furtak: *Surface Enhanced Raman Scattering* (Plenum, New York 1981)
6.2 R.P. Van Duyne: In *Chemical & Biochemical Applications of Lasers*, Vol.4, ed. by C.B. Moore (Academic, New York 1979) p.101
6.3 A. Champion, J.K. Brown, V.M. Grizzle: J. Vac. Sci. Technol. **20**, 893 (1982); Surf. Sci. **115**, L153 (1982)
6.4 A.T. Bell, M.L. Hair (eds.): *Vibrational Spectroscopies for Adsorbed Species*, ACS Symposium Series, 137 (ACS, Washington D.C. 1980)
6.5 W. Drachsel, Th. Jentsch, J.H. Block: In *Proc. 29th International Field Emission Symposium* (Almqviste and Wiksell International, Stockholm 1982) p.299
6.6 R. Wagner: *Field-Ion Microscopy*, Vol. 5 of Crystals Growth, Properties and Applications, ed. by G.J.M. Rocijmons (Springer, Berlin, Heidelberg, New York 1982)
6.7 G.M. Hieftje, J.C. Travis, F.E. Lytle: *Lasers in Chemical Analysis* (Humana Press, New Jersey 1981)
6.8 J.F. Ready: *Industrial Applications of Lasers* (Academic, New York 1978); *Effects of High-Power Laser Radiation* (Academic, New York 1971)
6.9 M. Bertolotti (ed.): *Physical Processes in Laser-Materials Interactions* (Plenum, New York 1983)
6.10 D.J. Erlich, R.M. Osgood, T.F. Deutsch: J. Vac. Sci. Technol. **21**, 23 (1982)
6.11 R. Schinke: J. Chem. Phys. **72**, 1120 (1980); **76**, 2352 (1982)
6.12 H. Loesch: Adv. Chem. Phys. **42**, 421 (1980)
6.13 K. Bergmann, U. Hefter, J. Witt: J. Chem. Phys. **72**, 4777 (1980)
6.14 J.C. Polanyi, R.J. Wolf: Ber. Bunsenges. Phys. Chem. **86**, 356 (1982)
6.15 R. Schinke: Surf. Sci. **127**, 283 (1983)
6.16 J.E. Hurst, G.D. Kubiak, R.N. Zare: Chem. Phys. Lett. **93**, 235 (1982)
6.17 J.A. Barker, A.W. Kleyn, D.J. Auerbach: Chem. Phys. Lett. **97**, 9 (1983)
6.18 J.C. Tully: Ann. Rev. Phys. Chem. **31**, 319 (1980)
6.19 R. Englemann, P.E. Rouse, H.M. Peek, V.D. Baiamonte: Los Alamos Scientific Lab. Report, LA-4364 (1970)
6.20 G.M. McClelland, G.D. Kubiak, H.G. Rennagel, R.N. Zare: Phys. Rev. Lett. **46**, 831 (1981)
6.21 A.W. Kleyn, A.C. Luntz, D.J. Auerbach: Phys. Rev. Lett. **47**, 1169 (1981)
6.22 A.C. Luntz, A.W. Kleyn, D.J. Auerbach: Phys. Rev. B**25**, 4273 (1982)
6.23 A.W. Kleyn, A.C. Luntz, D.J. Auerbach: Surf. Sci. **117**, 33 (1982)
6.24 J.C. Polanyi, J.L. Schreiber: Discuss. Faraday Soc. **62**, 267 (1977)
6.25 R.D. Levin, R.B. Bernstein: *Molecular Reaction Dynamics* (Oxford Univ. Press, New York 1974)
6.26 R.R. Cavanagh, D.S. King: Phys. Rev. Lett. **47**, 1829 (1981)
6.27 J.W. Gadzuk, U. Landman, E.J. Kuster, C.L. Cleveland, R.N. Barnett: Phys. Rev. Lett. **49**, 426 (1982); J. Electron Spectrosc. Relat. Phenom. **30**, 103 (1983)
6.28 J.M. Bowman, J.L. Gossage: Chem. Phys. Lett. **96**, 481 (1983)
6.29 S. Bialkowski: J. Chem. Phys. **78**, 600 (1983)
6.30 B.E. Hayden, K. Kretzschmar, A.M. Bradshaw: Surf. Sci. **125**, 366 (1983)
6.31 E. Umbach, S. Kulkarni, P. Feulner, D. Menzel: Surf. Sci. **88**, 65 (1979)
6.32 M. Asscher, W.L. Guthrie, T.-H. Lin, G.A. Somarjai: J. Chem. Phys. **78**, 6992 (1983)
6.33 T. Matsushima: Surf. Sci. **123**, L663 (1982)
6.34 G. Comsa, R. David: Surf. Sci. **117**, 77 (1982)
6.35 M. Balooch, M.J. Cardillo, D.R. Miller, R.E. Stickney: Surf. Sci. **46**, 358 (1974)
6.36 S.T. Ceyer, W.L. Guthrie, T.-H. Lin, G.A. Somarjai: J. Chem. Phys. **78**, 6982 (1983)

6.37 A.T. Modak, P.J. Pagni: J. Chem. Phys. **65**, 1327 (1976)
6.38 D.S. King, R.R. Cavanagh: J. Chem. Phys. **76**, 5634 (1982)
6.39 D. Grischkowsky, M.L. Yu, A.C. Balant: Surf. Sci. **127**, 315 (1983)
6.40 N. Winograd: This volume
6.41 A. Laubereau, W. Kaiser: Rev. of Mod. Phys. **50**, 607 (1978)
6.42 S.M. George, H. Auweter, C.B. Harris: J. Chem. Phys. **73**, 5573 (1980)
6.43 R.J. Abbott, D.W. Oxtoby: J. Chem. Phys. **70**, 4703 (1979)
6.44 K.E. Jones, A. Nichols, A.H. Zewail: J. Chem. Phys. **69**, 3350 (1978)
6.45 B.F. Levin, C.V. Shank, J.P. Heritage: IEEE J. Quantum Electron. QE-**15**, 1418 (1979)
6.46 K.B. Eisenthal, R.M. Hochstrasser, W. Kaiser, A. Laubereau (eds.): *Picosecond Phenomena III*, Springer Series in Chemical Physics, Vol.23 (Springer, Berlin, Heidelberg, New York 1982)

7. Surface Compositional Changes by Particle Bombardment

R. Kelly

With 14 Figures

7.1 Introduction

Bombardment-induced compositional change, also termed preferential sputtering, tends to be of very wide occurrence. The causes are not clear in all cases, namely, whether a preferentially lost species is lighter, less tightly bound, larger, subject to an electronic interaction, or subject to some aspect of diffusion. Nevertheless, the loss of C from TaC [7.1],

$$TaC + 1 \text{ keV He}^+ = TaC_{0.15} \quad ,$$

is probably correctly taken as loss occurring because C is lighter. Loss of Pb from Pb-In [7.2],

$$Pb_{0.29}In_{0.71} + 300 \text{ eV Ar}^+ = Pb_{0.00}In_{1.00} \quad ,$$

is a reasonable example where some aspect of chemical bonding is involved. Loss of Gd from Gd-Fe [7.3],

$$Gd_{0.24}Fe_{0.76} + 400 \text{ eV Ar}^+ = Gd_{0.01}Fe_{0.99} \quad ,$$

can be understood only if size (thence Gibbsian segregation) plays a role. The very common loss of halogen from ion-, electron- or photon-bombarded halides, e.g., the photographic process, is generally explained in terms of electronic interactions [7.4]:

$$AgBr + h\nu = Ag(s) + \tfrac{1}{2}Br_2(g) \quad .$$

Finally, superimposed on these four effects are three involving diffusion: diffusional relocation, diffusional deepening [7.5], and point-defect fluxes [7.6,7]. Electronic and diffusional effects will not, however, be discussed here except briefly in Sect.7.2, concerning examples of diffusional relocation. Also, only models and results relating to incident *ions* will be considered.

7.2 Changes Correlating with Mass

A first group of models for bombardment-induced compositional change is based on mass correlation. One model follows from recoil effects, i.e., recoil implantation and recoil sputtering. Another follows from a sputtering variant in which the escape depth is assigned a mass dependence. A third is based on near-threshold sputtering. It will be seen that only near-threshold effects have a broad basis in experiment.

The roles of recoil implantation and recoil sputtering can be inferred as follows. Consider a binary target made up mainly of mass 25 or 100 u, but with a limited amount of mass 10,25,50, or 100 u also present up to depth $\hat{x}$ in the case of recoil implantation and starting at $\overset{\vee}{x}$ in the case of recoil sputtering. The incident ions have mass 25 u and energy E_1. Then bombardment sets the surface atoms in motion (initially with energy T_2 and angle ψ) as schematized in Fig.7.1, and the quantities driven in (recoil implanted) or out (recoil sputtered) are governed ideally by simple relations of the type [7.8,9],

$$\text{number implanted} \approx N_2 \int_{x'=0}^{\hat{x}} \int_{T_2} d\sigma_{12}(E_1,T_2)dx'F_{23}(x-x',\psi) \qquad (7.1)$$

$$\text{number sputtered} \approx N_2 \int_{x'=\overset{\vee}{x}}^{\infty} \int_{T_2} d\sigma_{12}(E_1,T_2)dx'\{F_{23}(-\infty,\psi) - F_{23}(-x',\psi)\} . \qquad (7.2)$$

Here N_2 is the number density of the target, $d\sigma_{12}(E_1,T_2)$ is the differential scattering cross section (most simply of power-law form [7.10]), ψ is the recoil angle,

$$\cos\psi = (T_2/\gamma_{12}E_1)^{\frac{1}{2}} ,$$

γ_{12} is the energy-transfer factor $4M_1M_2(M_1+M_2)^{-2}$, M_i is atomic weight, and $F_{23}(x,\psi)$ is the usual Gaussian integral depth-distribution function which has, however, been generalized to allow for ψ other than zero [7.8]. Equations (7.1,2) can be expressed in closed form, numerical evaluations being shown in Table 7.1. Subscripts "1", "2", and "3" refer, both here and in what follows, to the incident ion, surface species, and bulk species.

Evidently the lighter component is subject to both implantation and sputtering to a significant extent, i.e., has a yield of order unity. By contrast, the heavier component shows much less movement, so that a major compositional change might be anticipated: a tendency for the target surface to become enriched in the heavier component while the lighter component is in part sputtered and in part driven in. A major qualification is in order,

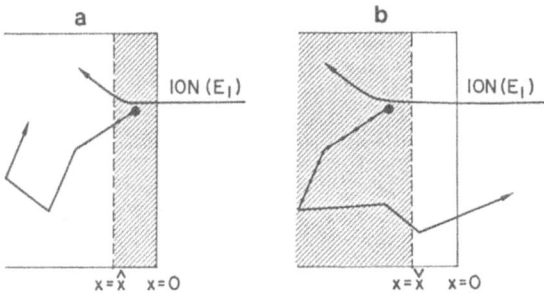

Fig.7.1a. Schematic trajectory for *recoil implantation* from a near-surface source extending from $x = 0$ to $x = \hat{X}$. The incident ion interacts directly with a source atom, the latter migrates randomly, and implantation is said to occur if it stops within the target, see (7.1). (b) Schematic trajectory for *recoil sputtering* from a near-surface source starting at $x = \check{X}$. The initial interaction is the same as for recoil implantation but only those events are considered where the source atom stops outside the target, see (7.2)

Table 7.1. Numerical examples of recoil-implantation and recoil-sputtering yields for thick sources according to (7.1,2). Example: $25 \rightarrow 10 \rightarrow 100$ means that an ion with $M_1 = 25$ u impacts on a near-surface atom with $M_2 = 10$ u, which in turn slows down in a target with $M_3 = 100$ u. The evaluations were made for $N_2 = 0.060$ Å^{-3} (as for Al), $m_{12} = m_{23} = 1/3$ (where m_{ij} is the power-law scattering parameter), $Z_i \propto M_i$ (where Z_i is the atomic number), and $x - \hat{X} = \check{X} = 0$

Masses [u]	Recoil-implantation yield, (7.1)	Recoil-sputtering yield, (7.2)
$25 \rightarrow 10 \rightarrow 100$	1.112	0.595
$25 \rightarrow 25 \rightarrow 100$	0.625	0.238
$25 \rightarrow 50 \rightarrow 100$	0.391	0.105
$25 \rightarrow 100 \rightarrow 100$	0.238	0.042
$25 \rightarrow 10 \rightarrow 25$	1.271	0.373
$25 \rightarrow 25 \rightarrow 25$	0.871	0.154
$25 \rightarrow 50 \rightarrow 25$	0.654	0.071
$25 \rightarrow 100 \rightarrow 25$	0.500	0.031

however. In real systems recoil effects are just a minor perturbation of the dominant slow collisional (cascade) sputtering, as can be seen in Fig.7.2 [7.11]. Recoil-related compositional change will therefore normally be unimportant.

A basically different treatment of the problem was undertaken by *Haff* [7.12]. He suggested that an incident ion sets up a cascade, the atoms in which can be approximated as having equal energy and undergoing a diffusion-like motion. Hence he wrote

$$amount\ lost \propto (Dt)^{\frac{1}{2}}$$

$$\propto v^{\frac{1}{2}} \propto M^{-\frac{1}{4}} \quad,$$

where $(Dt)^{\frac{1}{2}}$ is the diffusion length and v is velocity, and he finally concluded (in essence) that the loss ratio for a target consisting of A and B with

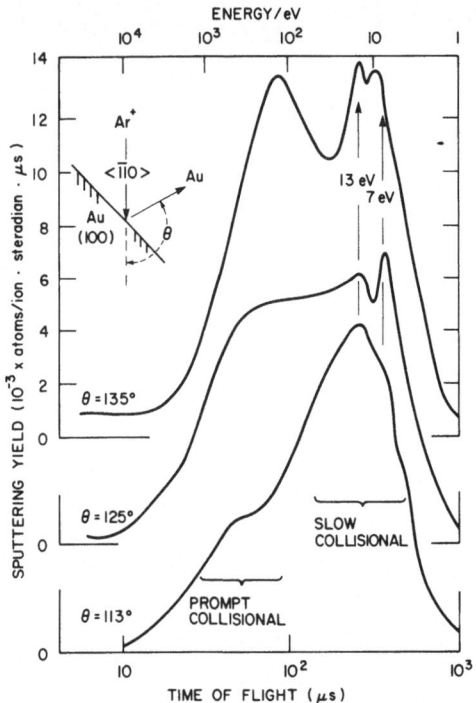

ENERGY/eV

SLOW
COLLISIONAL

PROMPT
COLLISIONAL

TIME OF FLIGHT (μs)

Fig.7.2. Time-of-flight spectra from the (100) surface of an Au single crystal during bombardment by 20 keV Ar+ ions. The inset shows the direction of incidence in relation to the direction of ejection and defines the angle θ. The high-energy component is termed here *prompt collisional sputtering* (recoil sputtering as in Fig.7.1b being a special case) and the low-energy component *slow collisional sputtering*. Bearing in mind that these are time-of-flight (rather than energy) spectra, it is clear that prompt collisional is much less important than slow collisional sputtering [7.11]

atom fractions x_A and x_B is

$$\frac{\text{loss of A}}{\text{loss of B}} = \frac{x_A M_B^{\frac{1}{4}}}{x_B M_A^{\frac{1}{4}}} \quad .$$

The result is that one might again predict that the surface of a binary target becomes enriched in the heavier component. It is unclear, however, if this treatment is acceptable. Sputtered atoms come dominantly from the outermost atomic layer [7.13,14] and not from distance $(Dt)^{\frac{1}{2}}$.

A third treatment of compositional change which shows a mass correlation concerns slow collisional (cascade) sputtering under *near-threshold conditions*. The most general expression for the slow collisional sputtering coefficient $S_{cascade}$ is [7.15]

$$S_{cascade} = \{\tfrac{1}{4}E_1 F(0)/U\}\{1 - (E_{th}/\gamma_{12}E_1)^{\frac{1}{2}}\} \quad , \tag{7.3}$$

which can be understood rather simply as the product of the outward-directed fraction of the collision cascade ($\tfrac{1}{4}$), the incident ion energy (E_1), the fraction of E_1 deposited in the outer atomic layer ($F(0)$), the reciprocal of the surface binding energy ($1/U$), and a somewhat more subtle term which in-

Table 7.2. Maximum energy transfers $\hat{T}_2 = \gamma_{12} E_1$ for low-energy ions incident on Ta_2O_5, Al_2O_3, or BeO

Incident energy, E_1, and ion	$\hat{T}_2$ for O [eV]	$\hat{T}_2$ for Ta [eV]	$\hat{T}_2$ for Al [eV]	$\hat{T}_2$ for Be [eV]
300 eV He^+	192	24	135	256
300 eV Ar^+	246	177	288	180

corporates a threshold energy (E_{th}) of order 10-80 eV [7.16]. [The first term of (7.3) is equivalent to the well-known *Sigmund* [7.17] expression.] Consider 300 eV He^+ or Ar^+ incident on Ta_2O_5, Al_2O_3, or BeO. Then, as seen in Table 7.2, the maximum energy transfers $\hat{T}_2 = \gamma_{12} E_1$ are such that surface enrichment of metal should occur for $He^+ \to Ta_2O_5$ but not for the other combinations.

Several experimental examples are thought-provoking though probably involve not mass-dependent change but rather *diffusional relocation*. For example, a surface deposit of 12 monolayers of Mo was found to be easily sputtered from W, somewhat less easily removed from Au or Cu, and removed only with difficulty from Al [7.18]. A mass correlation in which heavy components were retained would account for some of these observations, although they can also be explained [7.18] in terms of relocation on the surface by surface diffusion. The systems Cu on Si [7.19] and Au on Si [7.20] constitute further examples where heavy components were retained, but here the evidence was fairly clear that relocation by volume diffusion played a major role. The overlayers diffused into the bulk so were not subject to sputtering. While it was true that the heavy species were more difficult to remove, in fact in a study [7.21] of H, C, N, O, F, Ni, Pd, Pt, and Au overlayers on Si all species were removed up to two orders of magnitude more slowly than expected (Fig.7.3) [7.21]. Again relocation either by volume diffusion or cascade mixing is indicated. A series of very explicit examples involving relocation concerned metals such as Fe sputtered on to substrates such as Si [7.22]. While the object was to understand cone or pyramid nucleation, the important result in the present context is that the deposited metal atoms underwent surface diffusion with normal parameters. Lead on Si and U on Al_2O_3 are described [7.23] as inferring a reluctant sputtering of a heavy overlayer, but no further details are available.

A further group of examples concerns bombardment-induced oxygen loss from oxides [7.24] (e.g., Fig.7.4 [7.25]) as well as metal or halogen loss from halides [7.24]. A mass correlation has been proposed repeatedly but is actually very poor [7.26]: it fails, for example, to explain the differing

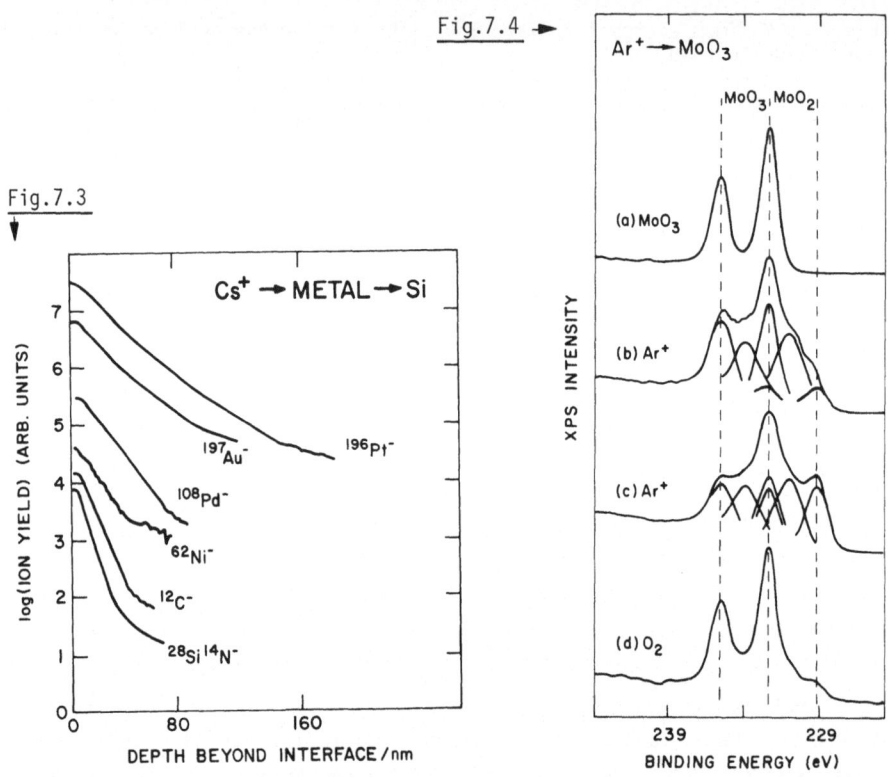

Fig.7.4 →

Fig.7.3

Fig.7.3. Secondary ion depth profiles of overlayer species on Si obtained by bombardment with 20 keV Cs^+. Data prior to the overlayer-silicon interfaces are not shown and the vertical position of each profile is arbitrary. All species were removed at up to two orders of magnitude more slowly than expected [7.21]

Fig.7.4. X-ray photoelectron spectra of Mo levels of MoO_3 exposed to Ar^+ ions or O_2 molecules consecutively, as follows: (a) commercial MoO_3 powder; (b) 400 eV Ar^+ (8×10^{15} ions/cm^2); (c) 400 eV Ar^+ (2×10^{16} ions / cm^2); (d) O_2 at 1.3×10^4 Pa for 5 min. Ar^+ bombardment causes preferential oxygen loss at the surface such that MoO_2 forms [7.25]

responses to ion impact of CdO, CuO, PbO, and PdO (which lose oxygen) and CoO, FeO, MnO, NbO, SnO, TiO, VO, and ZnO (which do not lose oxygen). Oxides are considered again in Sects.7.3,4 in relation to chemical bonds.

Until recently, binary alloys where the components had different masses were thought to constitute clear examples of systems showing mass-dependent compositional change. Thus Ag-Au, Al-Pd, Be-Cu, and Sn-Pt all lose the light component preferentially (e.g., Fig.7.5 [7.27]). A greater variety of systems has now been studied, including Au-Ni, Gd-Co, Pb-In, and Pd-Ni, all of which lose the heavier component preferentially, as well as Ag-Pd, Cu-Ni, and Mg-Al,

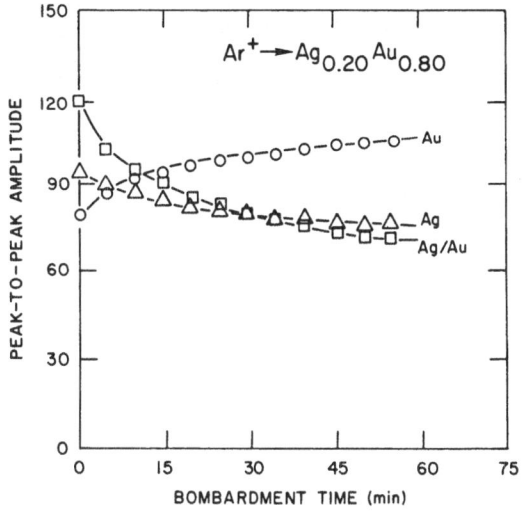

Fig.7.5. Change of Ag (350 eV) and Au (70 eV) Auger signals, as well as the ratio Ag/Au, for $Ag_{0.20}Au_{0.80}$ sputtered with 1 keV Ar^+. Ar^+ bombardment causes preferential Ag loss at the surface [7.27]

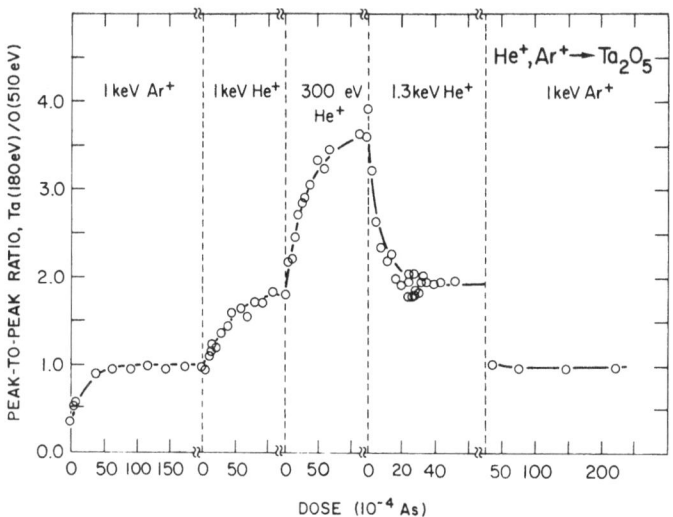

Fig.7.6. Development of the Ta (180 eV)/O (510 eV) Auger signal ratio during bombardment of an anodic Ta_2O_5 surface with Ar^+ or He^+ ions of various energies. The outer-layer changes are more pronounced for light ions and low energies, demonstrating the relevance of sputtering under near-threshold conditions to compositional change [7.28]

where preferential losses occur in spite of the masses being nearly equal. Binary alloys are considered again in Sects.7.3-5 in relation to chemical bonds and to Gibbsian segregation. It will emerge that the observed changes are *in all cases* understandable if Gibbsian segregation, either as observed or predicted, is taken into account.

Unlike the examples enumerated above, experiments at very low incident energies show clear evidence for mass-dependent compositional change, in all cases understandable in terms of near-threshold effects. *Taglauer* and *Heiland* [7.1,28] studied the impact of He$^+$ and Ar$^+$ on Ta$_2$O$_5$, TaC, WC, Al$_2$O$_3$, and BeO, and showed that the changes (loss of O or C) were more pronounced (i) for He$^+$ than for Ar$^+$ (Fig.7.6) [7.28], (ii) for lower incident energies (Fig.7.6), and (iii) for larger mass differences within the target. The changes were reversible. Such results are easily justified in terms of slow collisional sputtering under near-threshold conditions, (7.3) and Table 7.2. Further examples of near-threshold effects are given in [7.24,29]. Most work at higher energies is clearly not in a near-threshold regime, the main exceptions involving H$^+$, D$^+$, and He$^+$ bombardment of Ta$_2$O$_5$ [7.30], TiB$_2$ [7.31,32], and TiC [7.32].

7.3 The Role of Chemical Bonds in Slow Collisional Sputtering

Chemical bonding can contribute to bombardment-induced compositional change in at least three ways: in slow collisional (cascade) sputtering through the surface binding energy U, as in (7.3); in prompt thermal sputtering through the heat of atomization or vaporization; and in Gibbsian segregation, which is governed by both chemical bonds and strain energy. We treat slow collisional sputtering in this section, and recall, (7.3), that the yield scales as

$$S_{cascade} \propto 1/U \propto 1/\Delta H^a \quad , \tag{7.4}$$

where ΔH^a is the heat of atomization.

The slow collisional sputtering of alloys is conceptually straightforward, even if a bit lengthy, to treat provided preferential effects in the cascade [7.33] can be neglected. It has been shown [7.26] that $S_{cascade}$ can be expressed in a form, similar to that introduced in [7.34], which permits the subsurface and surface species to be designated. Specifically, sputtering due to a cascade involving the ith component of the subsurface (atom fraction $x_{i(2)}$, assuming that the subsurface shows the surface ("2") composition [7.35,36])[1] encountering the jth component of the surface (atom fraction $x_{j(2)}$ is described by

$$S_{cascade} \propto \sum_{i,j} x_{i(2)} x_{j(2)} \gamma_{ij}/U_j \quad . \tag{7.5}$$

---

1 It was assumed in [7.26], incorrectly as is now evident [7.35,36], that the subsurface shows the bulk composition.

Depending on the choice of indices, (7.5) gives either the total yield or
the yield for a particular component. If the target is taken to be a mixture
of A and B, then the sputtering coefficient for each component is readily
evaluated from (7.5), the final result being

$$\frac{\text{loss of A}}{\text{loss of B}} = \frac{S_A}{S_B} \equiv \frac{\gamma_A x_{A(2)}}{\gamma_B x_{B(2)}} = \left(\frac{x_{A(2)} + x_{B(2)}\gamma_{AB}}{x_{B(2)} + x_{A(2)}\gamma_{AB}}\right)\left(\frac{x_{A(2)}U_B}{x_{B(2)}U_A}\right) \quad , \tag{7.6}$$

where losses are seen to be governed mainly by chemical bonding (through U_i)
and, to a small extent, by mass (through γ_{AB}). Equation (7.6) distinguishes
between a yield defined in terms of ions incident on the entire target (S_i)
and a yield defined in terms of ions incident on one component of the target
(Y_i).

To explore the ratio U_B/U_A, it will be assumed [7.9,26] that alloys are
amenable to the *quasichemical* thermodynamic formalism, in which cohesion is
attributed solely to nearest-neighbor interactions [7.37]. Then U_{AA} is the
nearest-neighbor A-A bond strength, a negative quantity related to ΔH_A^a and
the bulk ("3") coordination number $Z_{A(3)}$ by

$$\Delta H_A^a = -\tfrac{1}{2} Z_{A(3)} U_{AA} \quad .$$

The term U_{BB} is defined similarly, while U_{AB}, the A-B bond strength, is related to the heat of mixing ΔH_m [7.38] by

$$\Delta H_m = x_{A(3)} x_{B(3)} Z_3 (U_{AB} - \tfrac{1}{2}[U_{AA} + U_{BB}]) \quad , \tag{7.7}$$

provided the alloy is random ("regular") and all relevant bulk coordination
numbers are equal. Here $x_{i(3)}$ is the bulk ("3") atom fraction. If it is again
assumed that the subsurface shows the *surface* composition [7.35,36], U_A follows as

$$U_A = -Z_2 [x_{A(2)} U_{AA} + x_{B(2)} U_{AB}] \quad ,$$

where Z_2, the surface coordination number, is assumed to have a single value.
The final result is

$$\frac{U_B}{U_A} \approx \frac{[2 - x_{A(2)}]\Delta H_B^a + x_{A(2)}\Delta H_A^a - x_{A(2)} h_m}{[1 + x_{A(2)}]\Delta H_A^a + [1 - x_{A(2)}]\Delta H_B^a - [1 - x_{A(2)}] h_m} \quad , \tag{7.8}$$

where h_m stands for $\Delta H_m/x_{A(3)} x_{B(3)}$. Equation (7.8) retains the restriction
that the system is random.

As bombardment proceeds, the surface composition will gradually alter until the following conservation relation is obeyed [7.39]:

$$\gamma_A x_{A(2)}^\infty / \gamma_B x_{B(2)}^\infty = x_{A(3)}/x_{B(3)} \quad , \tag{7.9}$$

where $x_{i(2)}^\infty$ is the steady-state ("∞") surface atom fraction.
The final composition follows by combining (7.6,8,9) but is not easily solved. We therefore note that by combining (7.6,9) alone, it is easily shown that the factor containing γ_{AB} in (7.6) *reduces* the extent of the predicted change. We therefore neglect this factor and obtain an upper limit to the change:

$$x_{A(2)}^\infty =$$

$$\frac{\Delta H_B^a - x_{A(3)} h_m - [x_{B(3)}(\Delta H_B^a)^2 + x_{A(3)}(\Delta H_A^a)^2 - 2x_{B(3)} x_{A(3)}(\Delta H_B^a + \Delta H_A^a) h_m + x_{B(3)} x_{A(3)} h_m^2]^{\frac{1}{2}}}{\Delta H_B^a - \Delta H_A^a + [x_{B(3)} - x_{A(3)}] h_m}$$

$$\tag{7.10}$$

Equation (7.10) suggests that that species which has the lower ΔH^a, i.e., the weaker chemical bonding, is lost preferentially from an alloy. Mass does not enter.

Table 7.3 gives a partial list of studied alloy systems, with the information on compositional change based in general on results as in Fig.7.5 [7.27] and the thermodynamic information taken where possible from [7.38]. All examples involve solid solutions or intermetallic compounds and therefore have an intimate mixing of the component atoms. Included in Table 7.3 are values of $x_{A(2)}^\infty$ for $x_{A(3)} = x_{B(3)} = 0.5$ deduced in all cases from (7.10) with the approximation $h_m = 0$ (as for an ideal solution) and, where possible, from (7.10) in its full form (as for a regular solution). Whether or not the term h_m is retained is seen to be unimportant, a result which shows that the regular solution approximation of (7.7,8,10) is better than required for the particular problem.

The $x_{A(2)}^\infty$ values of Table 7.3 reveal an important feature: while the trends tend to be correct in essentially all cases, the predicted values of $x_{A(2)}^\infty$ are not significantly different from the bulk value $x_{A(3)} = 0.5$. For example, with Pb-In, $x_{A(2)}^\infty$ is predicted to be 0.47, and would have been even closer to 0.5 had the factor containing γ_{AB} in (7.6) been retained, but is observed to be 0.00. Slow collisional sputtering, by itself, is thus quite incapable of causing the observed compositional changes with alloys, a detail which has not always been appreciated in the recent literature.

As already pointed out in Sect.7.2, all alloys will be seen in Sect.7.5 to be understandable if Gibbsian segregation is taken into account.

Table 7.3. The chemical state of alloy surfaces following high-dose ion impact (mostly from [7.9]). The component with the lower heat of atomization, ΔH^a, is listed first. We have excluded instances where alloy formation leads to a change of binding (e.g., Al-Si) or where the components tend to separate (e.g., Nb-U)

System	Metallurgical state[a]	Initial composition	Final composition	ΔH^a_A [eV]	ΔH^a_B [eV]	$x^\infty_A(2)$ from (7.10) with $x_{A(3)} = x_{B(3)} = 0.5$	$x^\infty_A(2)$ from (7.10) as before but with $h_m = 0$
Systems which lose the lighter component							
Ag-Au	ss	0.23-0.77	0.12-0.88	2.94	3.82	0.47	0.47
Al-Au	im	0.33-0.67	0.21-0.79	3.41	3.82	0.49	0.49
Al-Cu	ss (quenched)	0.95-0.05	0.90-0.10	3.41	3.49	0.50	0.50
Al-Pd	im	0.40-0.60	0.13-0.87	3.41	3.90	0.49	0.48
Be-Cu	ss or ss (quenched)	0.13-0.87	no Be	3.36	3.49	...	0.49
Cu-Pt	ss	0.25-0.75	0.15-0.85	3.49	5.85	0.44	0.44
Sn-Au	im or ss	various	loss of Sn	3.12	3.82	0.48	0.47
Sn-Pt	im	0.50-0.50	0.27-0.73	3.12	5.85	...	0.43
Systems which lose a heavier or equal-mass component							
Ag-Ni [7.40]	ss (mixed)	various	loss of Ag	2.94	4.46	...	0.45
Ag-Pd	ss	0.20-0.80	0.10-0.90	2.94	3.90	0.47	0.47
Au-Ni	ss (quenched)	0.74-0.26	0.53-0.47	3.82	4.46	0.48	0.48
Au-Pd [7.41]	ss	0.77-0.23	0.70-0.30	3.82	3.90	0.50	0.50
Cu-Ni	ss	0.38-0.62	0.24-0.76	3.49	4.46	0.47	0.47
Gd-Co	ss (amorphous)	0.45-0.55	0.24-0.76	4.12	4.44	...	0.49
Gd-Fe	ss (amorphous)	0.24-0.76	0.01-0.99	4.12	4.31	...	0.49
In-Ga [7.42]	liquid	0.17-0.83	loss of In	2.51	2.82	0.49	0.49
Mg-Al [7.70]	im	0.55-0.45	0.20-0.80	1.52	3.41	0.41	0.41
Pb-In	ss or im	0.29-0.71	0.00-1.00	2.02	2.51	0.47	0.47
Pb-Sn	ss	0.96-0.04	0.70-0.30	2.02	3.12	0.45	0.45
Pd-Ni	ss	0.80-0.20	0.71-0.29	3.90	4.46	0.48	0.48

[a]"ss" denotes solid solution; "im" denotes intermetallic.

Corresponding arguments for oxides can be made but run into the difficulty of how U will be defined. Reference [7.43] distinguished between an average *total* ΔH^a and an average *partial* ΔH^a. For example, for TiO_2 the total ΔH^a relates to

$$TiO_2(1) = Ti(g) + 2 \; O(g); \quad <U> \approx \text{total } \Delta H^a = 6.5 \text{ eV/gas atom} \quad , \quad (7.11)$$

Table 7.4. Total and partial heats of atomization (ΔH^a) in the sense of (7.11,12). Evaluated where possible from the JANAF tables (e.g., [7.44])

Substance[a]	Total ΔH^a [eV/gas atom]	Final state	Partial ΔH^a for conversion into final state [eV/gas atom]	Alternative final state	Partial ΔH^a for conversion into alternative final state [eV/gas atom]
Systems which lose O					
$CuO(1)$	3.6	$Cu_2O(s)$	2.9	$Cu(s)$	3.6
$Fe_2O_3(1)$	4.8	$Fe_3O_4(s)$	1.6	$FeO(s)$	4.4
$MoO_3(1)$	5.5	$MoO_2(s)$	3.8	$Mo(s)$	5.0
$Nb_2O_5(1)$	6.7	$NbO(s)$	6.0^b	$Nb(s)$	6.4
$TiO_2(1)$	6.5	$Ti_2O_3(s)$	5.4^b	$TiO(s)$	6.3
$U_3O_8(1)$	6.6	$UO_2(s)$	3.6	$U(s)$	7.1
$V_2O_5(1)$	5.6	$V_2O_3(s)$	4.0^b	$VO(s)$	4.8
Systems which sputter congruently					
$Cu_2O(s)$	3.8	$Cu(s)$	4.4		
$Fe_3O_4(s)$	5.0	$FeO(s)$	5.8	$Fe(s)$	5.5
$MoO_2(s)$	6.0	$Mo(s)$	5.7		
$NbO(s)$	7.3	$Nb(s)$	7.0		
$SnO(s)$	4.3	$Sn(s)$	5.6		
$SnO_2(1)$	4.6	$SnO(s)$	5.0	$Sn(s)$	5.3
$Ti_2O_3(s)$	6.7	$TiO(s)$	7.2	$Ti(s)$	7.9
$UO_2(s)$	7.3	$U(s)$	8.2		
$V_2O_3(s)$	6.2	$VO(s)$	6.3	$V(s)$	6.8

[a]"1", i.e., liquid, indicates that the substance amorphizes when bombarded. "s", i.e., solid, indicates that the substance remains crystalline.
[b]Formation of $NbO_2(1)$: 6.8 eV. Formation of $Ti_3O_5(1)$: 6.7 eV. Formation of $VO_2(1)$: 4.2 eV.

while one of several possible partial ΔH^a relates to

$$2TiO_2(1) = Ti_2O_3(s) + O(g); \quad <U> \approx \text{partial } \Delta H^a = 5.4 \text{ eV/gas atom} . \quad (7.12)$$

Here "s" stands for "solid" (i.e., crystalline), "1" for "liquid" (i.e., amorphous), and "g" for "gas" (i.e., sputtered). The reasoning is that, since $<U>$ of (7.12) is less than $<U>$ of (7.11), TiO_2 will lose O to form Ti_2O_3 when bombarded. Such reasoning works very well in practice [7.43] (Table 7.4), including with TiO_2. Thus the first 7 examples in Table 7.4 are predicted and observed to lose oxygen when bombarded, with formation of the first designated final states, whereas 7 of the next 9 examples are predicted and observed to sputter congruently, with only MoO_2 and NbO being problematical. Nevertheless, the difficulty encountered with alloys persists with oxides: the total and partial ΔH^a are sufficiently similar so that major changes such as those shown by the first 7 examples in Table 7.4 should not occur.

7.4 The Role of Chemical Bonds in Prompt Thermal Sputtering

Prompt thermal sputtering is the second aspect by which chemical bonds can govern bombardment-induced compositional changes. The relevant yield is given by the integrated vaporization flux in the region heated by an individual ion impact (Fig.7.7),

$$S_{thermal} = \iint p(T)(2\pi\dot{m}kT)^{-\frac{1}{2}} \cdot 2\pi y dy \cdot dt \quad , \tag{7.13}$$

where $p(T)$ is the temperature-dependent equilibrium vapor pressure and y is a dimension parallel to the target surface. Equation (7.13) has been evaluated in [7.45] with the result

$$S_{thermal} \propto \hat{p}(M\hat{T})^{-\frac{1}{2}} t_{eff} \propto M^{-\frac{1}{2}} \hat{T}^{3/2} (\Delta H^a)^{-2} \exp(-\Delta H^a/k\hat{T}) \quad , \tag{7.14}$$

where $\hat{T}$ is the "thermal-spike" temperature, or, more precisely, the maximum temperature increase at $x = y = 0$, t_{eff} is an effective time which scales as $(\hat{T}/\Delta H^a)^2$, and ΔH^a, the heat of atomization, is correctly used only when the vapor is monatomic. To estimate the effect of thermal sputtering on the surface composition of an alloy, it is useful to introduce the thermodynamic activity coefficient γ [7.37], $p_A(alloy) = p_A(pure)\gamma_A x_A$, and thus obtain

$$\frac{\text{loss of A}}{\text{loss of B}} = \frac{S_A}{S_B} \equiv \frac{\gamma_A x_{A(2)}}{\gamma_B x_{B(2)}} = \frac{\gamma_A x_{A(2)} M_B^{\frac{1}{2}} (\Delta H_B^a)^2}{\gamma_B x_{B(2)} M_A^{\frac{1}{2}} (\Delta H_A^a)^2} \exp\left\{\frac{\Delta H_B^a - \Delta H_A^a}{k\hat{T}}\right\} \quad . \tag{7.15}$$

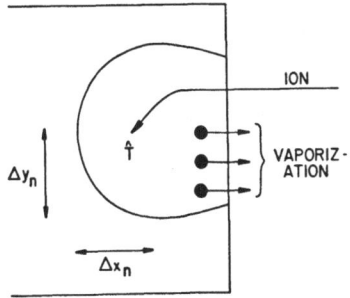

Fig.7.7. Prompt thermal sputtering. The incident ion is shown as depositing its energy in a truncated ellipsoid with dimensions Δx_n (the mean x-straggling of the elastic energy deposition) and Δy_n (the corresponding mean y-straggling). A transient high temperature $\hat{T}$ is assumed to result so that atoms vaporize from the surface

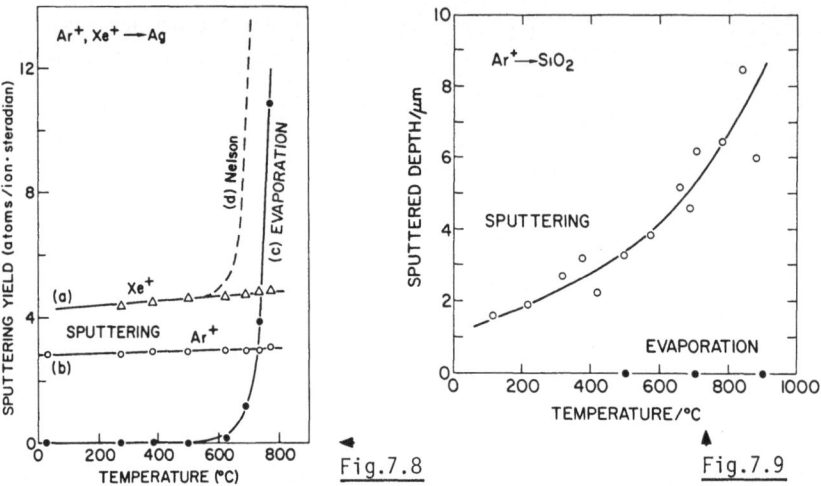

Fig.7.8
Fig.7.9

Fig.7.8. Sputtering yields of Ag vs the target temperature due to the impact of 8 keV Ar$^+$ and Xe$^+$ on polycrystalline Ag at normal incidence [curves (a) and (b)]. The experiments were done in ultrahigh vacuum and explicit corrections for vaporization were made with the beam off. Also shown is the vaporization component using the same detector as for the sputtering [curve (c)]. There is no up-turn at 625°C as in curve (d) such as was previously [7.49] taken as evidence for prompt thermal sputtering [7.48]

Fig.7.9. Sputtering yields of SiO$_2$ vs the target temperature due to the impact of 12 keV Ar$^+$ on amorphous SiO$_2$ at normal incidence. Also shown is the vaporization component calculated with the usual relation, $p(2\pi mkT)^{-\frac{1}{2}}$, assuming (i) that the duration of each experiment is 60 min and (ii) that the vaporization process proceeds as SiO$_2$(1) = SiO(g) + $\frac{1}{2}$O$_2$(g), "1" standing for liquid (i.e., amorphous) and "g" standing for gas [7.44]. Since the observed increase in S is not matched by an increase in vaporization, it follows that prompt thermal sputtering may be involved [7.50]

It would be possible at this point, by combining (7.9,15), to evaluate $x^\infty_{A(2)}$ for the systems listed in Table 7.3 and exactly the same trends as with the collisional relation, (7.10), would be found. It would also emerge that the magnitudes were interesting in that $Y_A/Y_B \gg 1$ [7.9], i.e., $x^\infty_{A(2)}$ was distinctly different from $x_{A(3)}$.

A new problem emerges, however: *with metals* thermal sputtering, like recoil effects, is just a minor perturbation of the dominant slow collisional sputtering [7.15]. For example, it appears to be absent with the highly volatile Mg [7.46], Zn [7.47], and Ag (Fig.7.8 [7.48]).

The situation *with oxides* is rather different because oxides can have a significantly higher volatility than metals. For example, Fig.7.9 [7.50] constitutes tentative evidence for thermal sputtering with SiO$_2$. By taking into account the full expression lying back of (7.14), it can be shown [7.45]

172

that thermal sputtering will be significant, i.e., $S_{thermal} \geq 1$, with any substance exhibiting a vapor pressure at $\hat{T}$ exceeding a critical value which was estimated to be $10^{2\pm1}$ atm. Figure 7.10 shows decomposition pressures for oxides. Not only are the trends correct, with oxides subject to 0 loss lying consistently above Cu_2O, but the magnitudes of the pressures are in accordance with the critical $10^{2\pm1}$ atm. The overall conclusion is that compositional changes with oxides can be described adequately in terms of thermal sputtering, with both trends and magnitudes being reasonable.

7.5 The Role of Gibbsian Segregation

In this final section we treat a framework in which both chemical bonds and strain energy can play a role in bombardment-induced compositional changes:

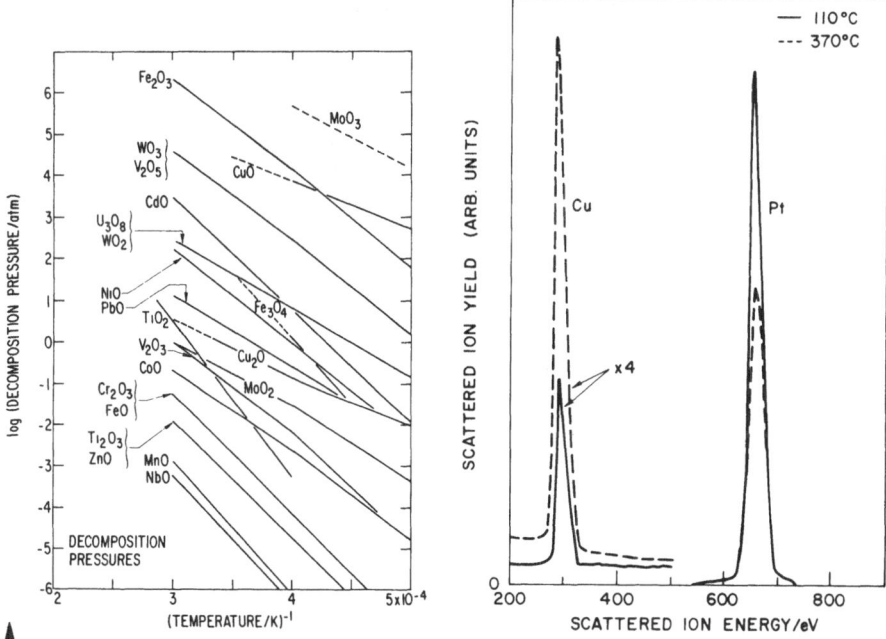

Fig.7.10. Decomposition pressure vs 1/T for oxides at very high temperatures. Those lying above Cu_2O show a bombardment-induced loss of 0, whereas the others do not [7.9,51]. The separation between the two groups occurs at about 10^2 atm for temperatures of 3000-4000 K. The curves are based where possible on the JANAF tables [7.44]

Fig.7.11. Energy distribution of 1000 eV Ne^+ ions scattered from polycrystalline $Cu_{0.20}Pt_{0.80}$ at 110°C and 370°C. Since low-energy ion scattering probes only exposed atoms, these results constitute a clear demonstration that Cu segregates to the surface to give $Cu_{0.55}Pt_{0.45}$. The compositional changes are substantial, in contrast to what is predicted for sputtering alone [7.52]

173

sputtering of any kind combined with Gibbsian segregation. Owing to the
paucity of information concerning oxides [7.24], only binary alloys will be
considered. It should be emphasized at the outset that segregation, in contrast to the predictions for collisional and thermal sputtering, is known to
lead to major compositional changes with alloys (e.g., Fig.7.11 [7.52]). The
changes are normally confined to the outermost atomic layer [7.53,54] but
this is precisely the layer from which about 80%-100% of sputtered atoms
originate [7.13,14].

Van Santen and *Boersma* [7.55] have presented a standard thermodynamic
argument which relates the steady-state surface concentration in a binary
system, $x_{A(2)}^{\infty}$, to the bulk concentration, $x_{A(3)}$. The following is based on
their treatment. It will be assumed that segregation alters only the outer
monolayer on the grounds that generalizations to several monolayers change
neither the trends nor the general magnitudes (Fig.7.12 [7.53]). It will be
assumed further that an alloy can be described as a random ("regular") solution, i.e., it has an arbitrary enthalpy of formation (7.7), but an ideal entropy of formation. The assumption regarding the entropy is equivalent to
proposing that the constituent atoms are located randomly, a situation which,
even if not necessarily true in equilibrium systems [7.52], should be a valid
description of a system subject to continuing bombardment. Thus, alloys normally approach a disordered rather than ordered state when bombarded at low
enough temperatures [7.56].

Let there be n_3 atoms in the bulk and n_2, with $n_2 \ll n_3$, in the outer monolayer. The whole system is initially described by a bulk atom fraction $x_{A(3)}$,
or simply x, but this evolves in the outer monolayer by virtue of segregation
to $x_{A(2)}^{\infty}$, i.e., $x + \delta$. The corresponding entropy change is

$$S = -kn_2[(1 - x - \delta)\ln(1 - x - \delta) + (x + \delta)\ln(x + \delta)]$$

$$- kn_3[(1 - x + n_2\delta/n_3)\ln(1 - x + n_2\delta/n_3)$$

$$+ (x - n_2\delta/n_3)\ln(x - n_2\delta/n_3)]$$

$$+ k(n_2 + n_3)[(1 - x)\ln(1 - x) + x\ln x]$$

$$= kn_2[(1 - x)\ln(1 - x) + x\ln x + \delta\ln(x/(1 - x))$$

$$- (1 - x - \delta)\ln(1 - x - \delta) - (x + \delta)\ln(x + \delta)] \quad .$$

The enthalpy change can be expressed in terms of what happens when $n_2\delta$
atoms of A are transferred from the bulk to the surface, and a similar num-

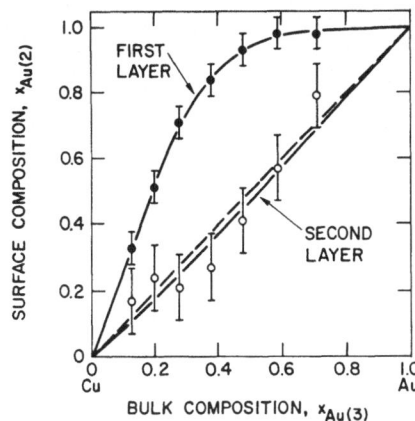

Fig.7.12. Steady-state surface composition at 300°C, expressed as atom fraction of Au, $x^\infty_{Au(2)}$, for the first (filled circles) and second (open circles) layers of a series of (111) Au-Cu alloys as a function of the bulk composition $x_{Au(3)}$. The surface compositions were derived from a knowledge of the Au Auger signals at two different energies. The solid lines represent the approximate trend of the calculated results; the dashed line would correspond to no surface enrichment. The important point is that the changes were found to be confined largely to the first layer [7.53]

ber of B atoms make the converse change. Let the energy of the system after the transfer, be it a bond energy or strain energy or otherwise, be represented by -Q, Q being the heat of segregation. Then the enthalpy change due to the transfer being carried out is

$$\Delta H = n_2 \int_0^\delta (d\delta')(-Q) \quad .$$

We form the free-energy change $\Delta G = \Delta H - T\Delta S$ and minimize by evaluating $\partial \Delta G / \partial \delta$. The result, with x replaced by $x_{A(3)}$ and $x + \delta$ by $x^\infty_{A(2)}$, is

$$\frac{x^\infty_{A(2)}}{1 - x^\infty_{A(2)}} = \frac{x_{A(3)}}{1 - x_{A(3)}} \exp\left\{\frac{Q}{kT}\right\} \quad . \tag{7.16}$$

This result is independent of the detailed description of Q.

The quantity Q can be evaluated in terms of either chemical bonds or strain energy and we first consider bonds. Let the bulk coordination number Z_3 (assumed equal for A and B) be composed of a lateral part Z_1 and a vertical part $2Z_v$:

$$Z_3 = Z_1 + 2Z_v \quad .$$

Also, let the quantities U_{AA}, U_{BB}, and U_{AB} be defined as in Sect.7.3 according to the quasichemical formalism. Then the energy of one atom of A in the bulk is

$$E_{A(3)} = Z_3[x_{A(3)}U_{AA} + x_{B(3)}U_{AB}]$$

and in the surface is

$$E_{A(2)} = Z_1[x^\infty_{A(2)}U_{AA} + x^\infty_{B(2)}U_{AB}] + Z_v[x_{A(3)}U_{AA} + x_{B(3)}U_{AB}] \quad .$$

175

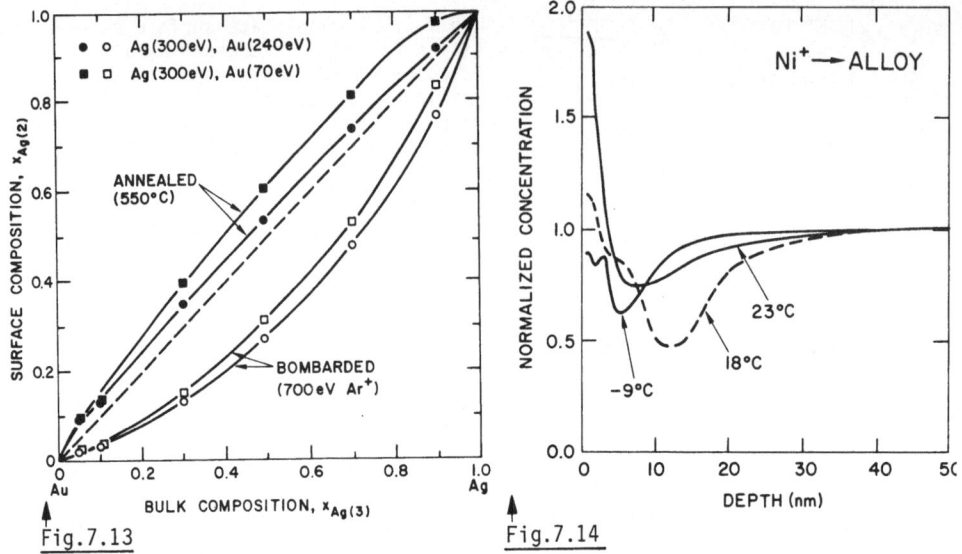

Fig.7.13.

Fig.7.14.

Fig.7.13. Steady-state surface composition, expressed as atom fraction of Ag, $x_{Ag(2)}^{\infty}$, of a series of annealed (550°C) and ion-bombarded (700 eV Ar$^+$) poly-crystalline Ag-Au alloys as a function of the bulk composition $x_{Ag(3)}$. The compositions were derived by Auger analysis. This example is interesting in showing explicitly that the species which segregates during annealing is the one which is lost preferentially in bombardment [7.57]

Fig.7.14. Measured depth profiles of Al in dilute Ni-based alloys which have been bombarded with 75 keV Ni$^+$ to doses of 1.6×10^{16} or 3.2×10^{16} ions/cm^2 at temperatures where only interstitials are mobile. Aluminum atoms have been pushed towards the surface by the flux of interstitial atoms. The signifi-cance of this example is that it demonstrates mass transport to occur at am-bient temperature instead of >500-700°C [7.59] when the point defects are generated extrinsically, in this case by bombardment. It should be recog-nized, however, that this example relates to bombardment-induced *redistri-bution* and not to Gibbsian segregation [7.58]

Provided the main contribution to the heat of segregation derives from bonds, we have

$$-Q = E_{A(2)} - E_{B(2)} + E_{B(3)} - E_{A(3)}$$

$$\approx \tfrac{1}{2} Z_v (U_{BB} - U_{AA}) \approx -\tfrac{1}{4}(\Delta H_B^a - \Delta H_A^a) \quad , \qquad (7.17)$$

where the approximations assume that the terms containing ΔH_m are unimportant and that $Z_v/Z_3 \approx 0.25$.

A result has thus been obtained similar in form to the ratio Y_A/Y_B for thermal sputtering as in (7.15). The energy term is reduced by a factor of 4 but, because the relevant temperature is ambient and not $\hat{T}$, the changes would be even more marked. And, unlike thermal sputtering, we are dealing

with an experimentally well-verified phenomenon. A surface showing segregation as discussed here would tend, when bombarded, to lose the segregated
component preferentially, thence the component with the weaker bonds, provided the necessary mass transport could occur. A *subsurface* depletion
would thus arise. An essentially explicit example of the correlation between
segregation and preferential sputtering is shown in Fig.7.13 [7.57], while
an example which suggests that under conditions of continuing bombardment
mass transport can take place at ambient temperature is shown in Fig.7.14
[7.58]. Further examples of low-temperature mass transport are given in
[7.40,60], while [7.36] shows that the resulting composition profiles are
similar to those of Fig.7.14.

The alternative is to evaluate Q in terms of strain energy. *McLean* [7.61]
argues that misfitted solute atoms will, in an equilibrium situation, concentrate at grain boundaries. If the atom is oversized it seeks an expanded
region of grain boundary and, if undersized, seeks a compressed region, so
that a misfit in either sense leads to grain-boundary segregation. This argument is commonly applied, at least in a qualitative sense, also to surface
segregation [7.62]. We would propose that both senses of misfit should not
be included, as a metal surface appears to involve a very slight expansion
[7.63]. It follows that oversized atoms are drawn to a surface but undersized
atoms are rejected. The numerical form of the heat of segregation, treated
as a problem in strain energy, has been shown [7.64] to involve the proportionality

$$-Q \propto (r_A - r_B)^2 \ , \tag{7.18}$$

where r is an appropriate atomic radius. Further quantification is difficult,
as it is not straightforward how to choose r_A and r_B. For convenience we shall
identify r_A and r_B with the pure-substance values [7.65] even though these
are considered [7.61] to be too large. A surface showing segregation based
on strain energy would again tend, when bombarded, to lose the segregated
component preferentially, thence the component with the larger size. Again,
mass transport at ambient temperature would be necessary and the composition
profiles would resemble those of Fig.7.14 [7.58].

Table 7.5 summarizes the correlation between surface segregation and
bombardment-induced compositional change for the same alloy systems as
listed in Table 7.3. Column 2 gives $\frac{1}{4}(\Delta H_B^a - \Delta H_A^a)$, which applies when (7.17)
is valid; column 3 gives $r_A - r_B$, which applies when (7.18) is valid; column
4 lists the observed or predicted (in brackets) segregation; and the final

Table 7.5. Correlation of Gibbsian segregation and compositional change with bond energies and atomic radii (mostly from [7.66]). An asterisk (column 2 or 3) denotes the assumed dominant driving force. Same systems as in Table 7.3

System	$\frac{1}{4}(\Delta H_B^a - \Delta H_A^a)$ [eV]	$r_A - r_B$ [7.65] [Å]	Species which segregates[a]	Species which is lost preferentially
Systems which lose the lighter component				
Ag-Au	0.22*	0.00	Ag	Ag
Al-Au	0.10*	-0.03	(Al)	Al
Al-Cu	0.02*	0.13*	Al	Al
Al-Pd	0.12*	0.04	(Al)	Al
Be-Cu	0.03	-0.15*	Be [7.67,7.68][b]	Be
Cu-Pt	0.59*	-0.10*	Cu	Cu
Sn-Au	0.18*	0.14*	Sn	Sn
Sn-Pt	0.68*	0.20*	Sn	Sn
Systems which lose a heavier or equal-mass component				
Ag-Ni	0.38*	0.19*	Ag [7.40]	Ag [7.40]
Ag-Pd	0.24*	0.07	Ag	Ag
Au-Ni	0.16*	0.19*	Au	Au
Au-Pd	0.02	0.07	Au [7.41]	Au [7.41]
Cu-Ni	0.24*	0.03	Cu	Cu
Gd-Co	0.08	0.55*	(Gd)	Gd
Gd-Fe	0.05	0.54*	(Gd)	Gd
In-Ga	0.08	0.22*	In [7.42]	In [7.42]
Mg-Al	0.47*	0.19*	Mg [7.69]	Mg [7.70]
Pb-In	0.12*	0.18*	Pb	Pb
Pb-Sn	0.28*	0.17*	Pb	Pb
Pd-Ni	0.14*	0.12*	Pd	Pd

[a]Observed segregation is indicated without brackets; predicted segregation is bracketed.
[b]Be is drawn to the surface of bombarded Be-Cu because of a strong coupling with interstitials [7.67,68].

column lists the sense of the compositional change. The correlation between the information in columns 2 and 3 and the observed segregation is very good, provided the following limits are taken:

bonds are unimportant if

$$\frac{1}{4}(\Delta H_B^a - \Delta H_A^a) < 0.10 \text{ eV} \quad ; \tag{7.19}$$

strain energy is unimportant if

$$r_A - r_B < 0.10 \text{ Å} \quad . \tag{7.20}$$

Equation (7.19) can be justified from (7.16,17), but (7.20) is wholly empirical. Only Be-Cu and Au-Pd present problems.

What is more important, however, is that we find a remarkable tendency for the species which is sputtered preferentially to be that which segregates, including Be-Cu and Au-Pd. Particularly important are the four instances where the compositional changes would have been difficult to understand in terms of bonds alone, but are straightforward when size is taken into account: Al-Cu, Gd-Co, Gd-Fe, and In-Ga. With Be-Cu, consideration of size suggests that Cu would segregate whereas it is Be which segregates in a point-defect flux [7.67,68] and which sputters preferentially. With Au-Pd, neither bonds nor size would have led to a prediction of segregation, yet segregation does occur and the preferential sputtering is that which would be expected. For other examples, see [7.66].

7.6 Conclusions

(a) A relation between mass and bombardment-induced compositional changes could arise in several ways. Recoil implantation and sputtering involve a direct interaction of an incident particle with a near-surface target atom such that the atom is driven deeper or expelled (Fig.7.1). They show a strong preference for low masses (Table 7.1), but should be of minor importance compared with slow collisional sputtering (Fig.7.2 [7.11]). The sputtering variant of [7.12] also leads to a mass correlation but is unacceptable for other reasons. Slow collisional sputtering at low incident energies with light particles, where the transferred energies are comparable to the sputtering threshold energy (Table 7.2), is a further process which shows a strong preference for low masses. The evidence that nearthreshold effects play a role is, unlike the other aspects of mass correlation, quite strong (Fig. 7.6 [7.28]).

(b) Slow collisional sputtering is governed mainly by chemical bonds rather than by mass. For example, for a binary alloy system A-B, the yield of A should be similar to

$$S_{A(cascade)} \propto [x_{A(2)} + x_{B(2)}\gamma_{AB}][x_{A(2)}/U_A] \quad , \tag{7.21}$$

where U_A is the surface binding energy. U_A can be evaluated from quasichemical theory as in (7.8) and, by combining (7.8,21) and the conservation relation of (7.9), values of the steady-state surface composition $x_{A(2)}^{\infty}$ can be derived. Interestingly, they do not differ significantly from the bulk values $x_{A(3)}$, showing that slow collisional sputtering by itself is incapable of causing the observed compositional changes with alloys (Table 7.3).

Cascade sputtering can also be applied to explain whether or not oxides show a preferential loss of oxygen. The main problem is that a somewhat artificial definition of the surface binding energy must be used, e.g., that the O binding energy in TiO_2 is given by the enthalpy change in the reaction

$$2TiO_2(1) = Ti_2O_3(s) + O(g); \quad <U> \approx \text{partial } \Delta H^a = 5.4 \text{ eV/gas atom} \quad .$$

Also, as with alloys, the problem persists that the magnitudes tend to be uninteresting (Table 7.4).

(c) Prompt thermal sputtering (Fig.7.7) is also governed by chemical bonds, but in a significantly stronger manner than cascade sputtering. Thus, for a binary alloy system A-B, one has

$$S_{A(thermal)} \equiv Y_A x_{A(2)} \propto Y_A x_{A(2)} M_A^{-\frac{1}{2}} \hat{T}^{3/2} (\Delta H_A^a)^{-2} \exp(-\Delta H_A^a/k\hat{T}) \quad .$$

Even though the ratio Y_A/Y_B normally differs markedly from unity, and does so in the right sense, it is difficult to see how thermal sputtering could be relevant to compositional changes with alloys. This follows from the inequality, inferred by recent results such as those for Ag (Fig.7.8 [7.48]), that

$$S_{A(cascade)} >> S_{A(thermal)} \quad .$$

Oxides differ from metals by virtue of having a considerably wider range of volatility (Fig.7.10). It is easily shown that, provided ion impact leads to localized temperatures high enough so that the following inequality is met,

$$p \geq p^* = 10^{2\pm1} \text{ atm} \quad ,$$

then volatilization should occur at a significant rate. Interestingly, the oxides lying above Cu_2O in Fig.7.10 all show a bombardment-induced loss of oxygen, though it remains true that the argument is more a trend analysis than an explicit proof.

(d) For similar sized atoms, Gibbsian segregation is governed by a relation contained in (7.16,17):

$$\frac{x_{A(2)}^{\infty}}{1 - x_{A(2)}^{\infty}} \approx \frac{x_{A(3)}}{1 - x_{A(3)}} \exp\left[\frac{\Delta H_B^a - \Delta H_A^a}{4kT}\right] \quad .$$

For atoms of dissimilar size, one must take strain energy into account. The usual interpretation is that both oversized and undersized atoms are drawn to a surface, but, given that metal surfaces are thought to be in a slightly expanded state [7.63], we prefer to believe that only oversized atoms are drawn. By taking into account both chemical bonds and strain energy, it turns out that most examples of segregation or the lack thereof are understandable (Table 7.5 and [7.66]). By postulating that bombardment leads to segregation even at ambient temperature by virtue of injecting point defects (Fig.7.14, [7.58]), it follows that the species which segregates will be lost preferentially so that a composition profile will be set up resembling those of Fig.7.14.

180

The correlation between segregation and preferential sputtering is found to be remarkably good (Table 7.5). Particularly important are the four instances where the compositional changes would have been difficult to understand in terms of bonds alone, but are straightforward when size is taken into account: Al-Cu, Gd-Co, Gd-Fe, and In-Ga. In one instance, Be-Cu, preferential sputtering agrees with segregation as is appropriate when a point-defect flux is present. In another instance, Au-Pd, segregation was not predicted yet did occur and was accompanied by preferential sputtering in the expected sense.

References

7.1 E. Taglauer, W. Heiland: In *Proc. Symp. on Sputtering*, ed. by P. Varga et al. (Inst. für Allgemeine Physik, Technische Univ., Wien, Austria (1980) p.423
7.2 S. Berglund, G.A. Somorjai: J. Chem. Phys. **59**, 5537 (1973)
7.3 N. Heiman, N. Kazama, D.F. Kyser, V.J. Minkiewicz: J. Appl. Phys. **49**, 336 (1978)
7.4 H. Kanzaki, T. Mori: Semicond. Insulators **5**, 401 (1983)
7.5 D.K. Murti, R. Kelly: Thin Sol. Films **33**, 149 (1976)
7.6 A.D. Marwick, R.C. Piller: Rad. Effects **33**, 245 (1977)
7.7 N.Q. Lam, G.K. Leaf, H. Wiedersich: J. Nucl. Mat. **85/86**, 1085 (1979)
7.8 S. Dzioba, R. Kelly: J. Nucl. Mater. **76**, 175 (1978)
7.9 R. Kelly: Surf. Sci. **100**, 85 (1980)
7.10 K.B. Winterbon, P. Sigmund, J.B. Sanders: Kgl. Danske Vid. Selsk., Mat. Fys.Medd. 37, No.14 (1970)
7.11 M.W. Thompson, I. Reid, B.W. Farmery: Philos. Mag. A**38**, 727 (1978)
7.12 P.K. Haff: Appl. Phys. Lett. **31**, 259 (1977)
7.13 D.E. Harrison, P.W. Kelly, B.J. Garrison, N. Winograd: Surf. Sci. **76**, 311 (1978)
7.14 M. Rosen, G.P. Mueller, W.A. Fraser: Nucl. Instrum. Methods **209/210**,63 (1983)
7.15 R. Kelly: Rad. Effects **80**, 273 (1984)
7.16 N. Matsunami, Y. Yamamura, Y. Itikawa, N. Itoh, Y. Kazumata, S. Miyagawa, K. Morita, R. Shimizu: Rad. Effects Lett.**57**, 15 (1980)
7.17 P. Sigmund: Phys. Rev. **184**, 383 (1969)
7.18 M.L. Tarng, G.K. Wehner: J. Appl. Phys. **43**, 2268 (1972)
7.19 R.R. Hart, H.L. Dunlap, O.J. Marsh: J. Appl. Phys. **46**, 1947 (1975)
7.20 P. Blank, K. Wittmaack: Rad.Effects Lett. **43**, 105 (1979)
7.21 P. Williams: Appl. Phys. Lett. **36**, 758 (1980)
7.22 R.S. Robinson, S.M. Rossnagel: J. Vac. Sci. Technol. **21**,790 (1982)
7.23 G. Carter: Private communication (Univ. of Salford, Salford, U.K. 1978)
7.24 R. Kelly: Nucl. Instr. Meth. **182/183**, 351 (1981)
7.25 K.S. Kim, W.E. Baitinger, J.W. Amy, N. Winograd: J. Electron Spectrosc. Relat. Phenom. **5**, 351 (1974)
7.26 R. Kelly: Nucl. Instr. Meth. **149**, 553 (1978)
7.27 W. Färber, P. Braun: Vakuum-Technik **23**, 239 (1974)
7.28 E. Taglauer, W. Heiland: Appl. Phys. Lett. **33**, 950 (1978)
7.29 J.E. Greene, R.E. Klinger, T.L. Barr, L.B. Welsh: Chem. Phys. Lett. **62**, 46 (1979)
7.30 H. von Seefeld, R. Behrisch, B.M.U. Scherzer, Ph. Staib, H. Schmidl: *Proc. Intern. Conf. on Atomic Collisions in Solids* (Moscow 1977)

7.31 G.C. Nelson, J.A. Borders: J. Nucl. Mater. **93/94**, 640 (1980)
7.32 M. Kaminsky, R. Nielsen, P. Zschack: J. Vac. Sci. Technol. **20**, 1304 (1982)
7.33 N. Andersen, P. Sigmund: In *Atomic Collisions in Solids*, ed. by S. Datz et al. (Plenum, New York 1975) p.115
7.34 H.F. Winters, P. Sigmund: J. Appl. Phys. **45**, 4760 (1974)
7.35 G. Betz: Surf. Sci. **92**, 283 (1980)
7.36 D.G. Swartzfager, S.B. Ziemecki, M.J. Kelley: J. Vac. Sci. Technol. **19**, 185 (1981)
7.37 R.A. Swalin: *Thermodynamics of Solids*, 2nd ed. (Wiley, New York 1972) p.141
7.38 R. Hultgren, P.D. Desai, D.T. Hawkins, M. Gleiser, K.K. Kelley: *Selected Values of the Thermodynamic Properties of Binary Alloys* (Am. Soc. for Metals, Metals Park, Ohio 1973)
7.39 H. Shimizu, M. Ono, K. Nakayama: Surf. Sci. **36**, 817 (1973)
7.40 J. Fine, T.D. Andreadis, F. Davarya: Nucl. Instrum. Methods **209/210**, 521 (1983)
7.41 A. Jablonski, S.H. Overbury, G.A. Somorjai: Surf. Sci. **65**, 578 (1977)
7.42 M.F. Dumke, T.A. Tombrello, R.A. Weller, R.M. Housley, E.H. Cirlin: Surf. Sci. **124**, 407 (1983)
7.43 H.M. Naguib, R. Kelly: Rad. Effects **25**, 1 (1975)
7.44 D.R. Stull, H. Prophet: *JANAF Thermochemical Tables*, 2nd ed. (Supt. of Documents, U.S. Govt. Printing Office, Washington DC 1971)
7.45 R. Kelly: Surf. Sci. **90**, 280 (1979)
7.46 M. Szymonski: Acta Phys. Polon. A**56**, 289 (1979)
7.47 M. Szymonski: Appl. Phys. **23**, 89 (1980)
7.48 K. Besocke, S. Berger, W.O. Hofer, U. Littmark: Rad. Effects **66**, 35 (1982)
7.49 R.S. Nelson: Philos. Mag. **11**, 291 (1965)
7.50 R.A. Dugdale, S.D. Ford: Trans. Brit. Ceram. Soc. **65**, 165 (1965)
7.51 J. Herion, G. Scharl, M. Tapiero: Appl. Surf. Sci. **14**, 233 (1982)
7.52 H.H. Brongersma, M.J. Sparnaay, T.M. Buck: Surf. Sci. **71**, 657 (1978)
7.53 J.M. McDavid, S.C. Fain: Surf. Sci. **52**, 161 (1975)
7.54 R. Bouwman, L.H. Toneman, M.A.M. Boersma, R.A. van Santen: Surf. Sci. **59**, 72 (1976)
7.55 R.A. van Santen, M.A.M. Boersma: J. Catal. **34**, 13 (1974)
7.56 E.M. Schulson: J. Nucl. Mater. **83**, 239 (1979)
7.57 M. Yabumoto, K. Watanabe, T. Yamashina: Surf. Sci. **77**, 615 (1978)
7.58 A.D. Marwick, R.C. Piller, P.M. Sivell: J. Nucl. Mater. **83**, 35 (1979)
7.59 J.J. Burton, C.R. Helms, R.S. Polizzotti: J. Chem. Phys. **65**, 1089 (1976)
7.60 H.H. Andersen, B. Stenum, T. Sørensen, H.J. Whitlow: Nucl. Instrum. Methods **209/210**, 487 (1983)
7.61 D.McLean: *Grain Boundaries in Metals* (Oxford U.P., Oxford 1957) p.116
7.62 R.S. Polizzotti, J.J. Burton: J. Vac. Sci. Technol. **14**, 347 (1977)
7.63 J.F. van der Veen, R.G. Smeenk, R.M. Tromp, F.W. Saris: Surf. Sci. **79**, 219 (1979)
7.64 B.J. Pines: J. Phys. Sov. Union **3**, 309 (1940)
7.65 H.M. Crosswhite: In *American Institute of Physics Handbook*, ed. by D.E. Gray (McGraw-Hill, New York 1972)
7.66 R. Kelly: In *Proc. Symp. on Sputtering*, ed. by P. Varga et al. (Inst. für Allgemeine Physik, Technische Univ., Wien, Austria 1980) p.390
7.67 H. Wollenberger: J. Nucl. Mat. **69/70**, 362 (1978)
7.68 R. Koch, R.P. Wahi, H. Wollenberger: J. Nucl. Mat. **103/104**, 1211 (1981)
7.69 C. Lea, C. Molinari: J. Mat. Sci. (in press)
7.70 P. Braun, M. Arias, H. Störi, F.P. Viehböck: Surf. Sci. **126**, 714 (1983)

8. Structure Determination of Small Metal Particles by Electron Microscopy

M. J. Yacaman

With 20 Figures

8.1 Introduction

The characterization of small metal particles is of direct importance for catalytic activity studies, since many of the most interesting catalysts are composed of metal particles (Rh, Pt, Re, etc.) supported on a carrier (γ-Al_2O_3, C, etc.). A systematic study of the crystal planes that a particle exposes to a gas during a reaction has direct impact on our knowledge of the physics and chemistry of surfaces.

In the last few years methods to characterize small particles have been greatly improved. Particularly Transmission Electron Microscopy (TEM) and Scanning Transmission Electron Microscopy (STEM) have produced a vast amount of new information about particle shape, structure and phase transformations.

In the present review it shall be discuss some of these methods and their application to catalysis and surface science. The application of the methods will be emphasized rather than the techniques.

8.2 Weak-Beam Dark Field

The basis of this method has been discussed at length in other publications [8.1-3] and will be discussed here only briefly. In this technique, *Yacamán* and *Ocaña* [8.1], images of a small particle are obtained by using a diffracted beam which is out of the Bragg condition. Figure 8.1 illustrates the principle of the method. When the beam is not in the Bragg condition its intensity shows periodic oscillations as a function of particle thickness. The oscillations cause tringes in the image, which are equal-thickness contours and can be used to produce a topographical map of the particle. Figure 8.2 shows an example of a bright field image (formed by using the transmitted electrons) and a weak-beam dark field image of the same particle. The particle with a square profile in the bright field shows a pyramidal, three-dimensional profile as indicated in Fig.8.2.

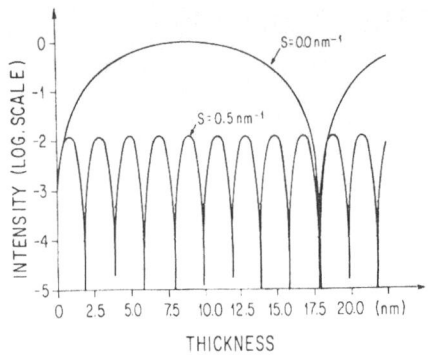

Fig.8.1. Plot of the intensity of a diffracted beam as a function of sample thickness for two different diffraction conditions (after M. Avalos [8.5])

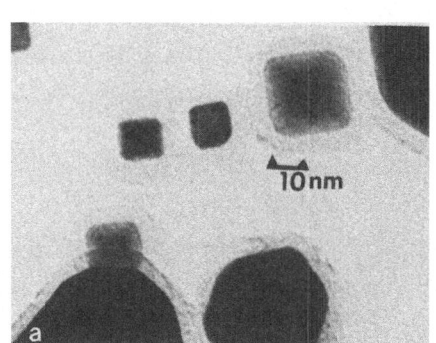

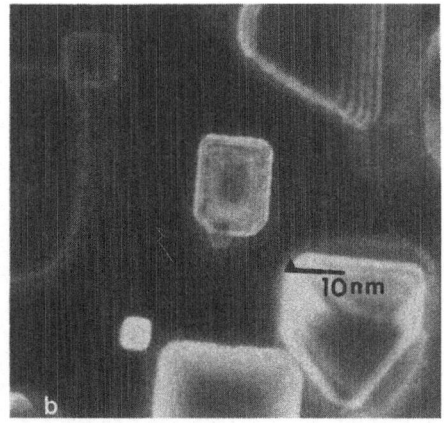

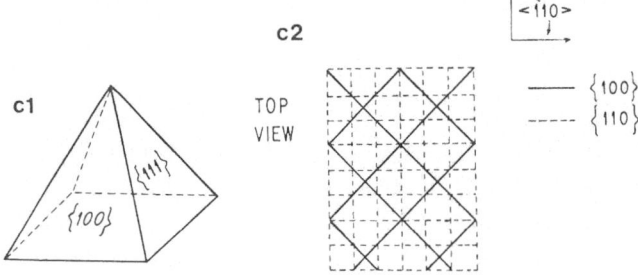

Fig.8.2a-c. Images of a gold particle; a) bright field image showing the square profile, b) weak-beam image using a {200} spot and c) reconstructed profile of the particle (after M. Avalos [8.5])

In principle, the character of scattering electrons by small particles requires application of the dynamical multibeam theory for electron diffraction [8.4]. However, a good approximation of the fringe periodicities can be obtained by using a kinematic theory as shown by *Avalos* [8.5] . The fringe spacing ξ_g is related to the deviation from the Bragg condition through the excitation error Sg [8.4] by the simple equation

$$\xi_g = \frac{1}{Sgm} \quad ,\tag{8.1}$$

where m = tanθ is the slope of the wedge producing the fringes. A typical value for Sg $\simeq 8 \times 10^{-2}$ Å^{-1} in weak-beam images; with θ $\simeq 54^\circ$ we obtain $\xi_g \simeq 10$ Å. However, multi-beam dynamical calculations predict a much lower value: $\xi_g \simeq 4$ Å. This means that this method can give reliable shape determinations of particles down to ~20 Å.

8.3 Electron Diffraction of Individual Particles

In recent years electron microscopy has been improved by the addition of scanning transmission electron microscopy (STEM) techniques. Using STEM electron optics it is possible to focus a fine electron beam of a diamter between 10-200 Å on an area of the sample and then obtain the corresponding diffraction pattern. Therefore, it is now possible to obtain patterns from individual small particles. This permits a straightforward determination of the crystal structure, avoiding complications in interpretation that arise when mixed patterns, corresponding to several particles, are obtained. On the other hand, the diffraction patterns contain fine structure which provides additional information about the particle structure and shape.

Figure 8.3 shows a microdiffraction pattern of a square particle. The overall structure can be easily recognized as fcc and corresponding to a <100> orientation. However, as can be seen, the spots split into several components.

It can be shown that this splitting is directly related to the presence of wedges on the particle as shown by Gómez et al. [8.6]. This phenomenon has several characteristics:

i)   The position and intensity of the split components depend on the **g** vector.

ii)   The amount of splitting depends on the deviation from the Bragg condition (magnitude of the S_g vector).

iii) The direction of the splitting is perpendicular to the edges of the particle.

To produce a full explanation of this effect, the dynamical theory for electron diffraction should be used. If we consider a wedge of slope m, the diffracted beams will have wave vectors $\mathbf{G}g^i$ given by

$$\mathbf{G}g^i = g + m\gamma^{(i)} \quad ,\tag{8.2}$$

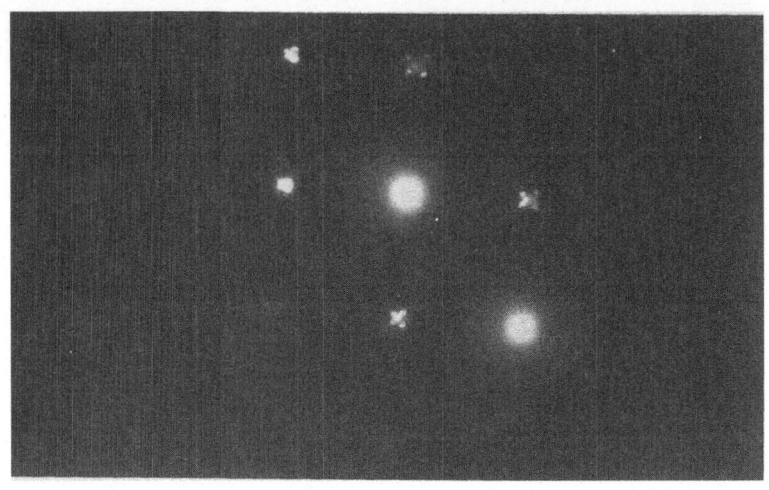

<u>Fig.8.3.</u> Microdiffraction pattern of a square particle as in Fig.8.2. Splitting of the diffraction spots is apparent

where γ^i are the eigenvalues of the scattering matrix, which can be obtained by standard calculation [8.7].

Equation (8.2) implies that in a many-beam case the spot corresponding to the $\mathbf{g}$ reciprocal vector will split into N components of different intensities. The kinematic theory for electron diffraction cannot provide a full explanation of the splitting, indicating the strong dynamic character of scattering in small particles. For a detailed discussion the reader is referred to reference [8.6]. This technique can be considered complementary to the weak-beam method described above. In fact, the splitting is the reciprocal space counterpart of the weak-beam fringes in real space. A splitting of Δg_i will produce in the image fringes with spacing (d),

$$d = \frac{1}{(\Delta g_i)m} \quad . \tag{8.3}$$

The splitting effect can be used also to give quantitative information about the wedges. For instance, for the pattern in Fig.8.3, the experimental split on the (0 2 0) spot was 0.0187 ± 0.005 Å^{-1}. A ten-beam dynamical calculation for $\gamma^{(i)}$ gave for the two strongest components values of: $\gamma^{(1)} = 2.66 \times 10^{-2}$ and $\gamma^{(2)} = 3.94 \times 10^{-2}$. Therefore, according to (8.2)

$$\Delta G_{(020)}^{(12)} = \left[\gamma^{(1)} - \gamma^{(2)} \right] \tan\theta \tag{8.4}$$

or

$$\tan\theta = \frac{0.0187}{0.0128} \quad \text{or} \quad \theta = 55.6^{\circ} \quad .$$

This value indicates that the faces of the pyramid are {111} planes and the base is a {100} plane, in agreement with the previous weak-beam determinations [8.8].

8.4 Single-Twinned Particles

In the following sections I shall describe a number of applications of these techniques. The first interesting case is a gold particle grown on a NaCl substrate marked by an arrow in Fig.8.4. This particle has a contrast that suggests a twinned structure, confirmed by the diffraction pattern of the particle. The pattern can be fully reproduced, assuming a single twin and that the particle has a <100> direction parallel to the <100> substrate direction.

This type of single-twinned particle has been previously reported by *Hayashi* et al. [8.9] for Al and by *Robinson* and *Gillet* [8.10] for Pd grown onto NaCl. The microdiffraction fully confirms the twin relationship. In many cases, several twins can be observed on a particle. This type of twinning is also frequently observed in catalytic systems, such as Pt/graphite, of which an example is shown in Fig.8.5.

8.5 Icosahedral and Decahedral Particles

A special type of particle observed in noble metals are icosahedral and decahedral particles. Figure 8.6 shows typical bright-field contrast of such particles. These shapes were first reported by *Ino* and *Owaga* [8.11] who described them as multiply twinned. The particles are composed of tetrahedral units packed together in a twin relationship. The icosahedron is composed of twenty tetrahedral units and the decahedron of fine units. *Gillet* [8.12] reviewed a number of cases in which these particles have been observed. However, there is a discrepancy in point group symmetries between the decahedron and icosahedron, with the polyhedron formed by piled up fcc tetrahedral units. This discrepancy implies that an "out-of-register gap" must exist between adjacent tetrahedral units. The strong interatomic repulsion created in an small particle by such a gap will create an energetically very unfavorable configuration. Some authors [8.8,13] have proposed that nonuniform strains exist on the particles in order to "close" the gap. An alternative model

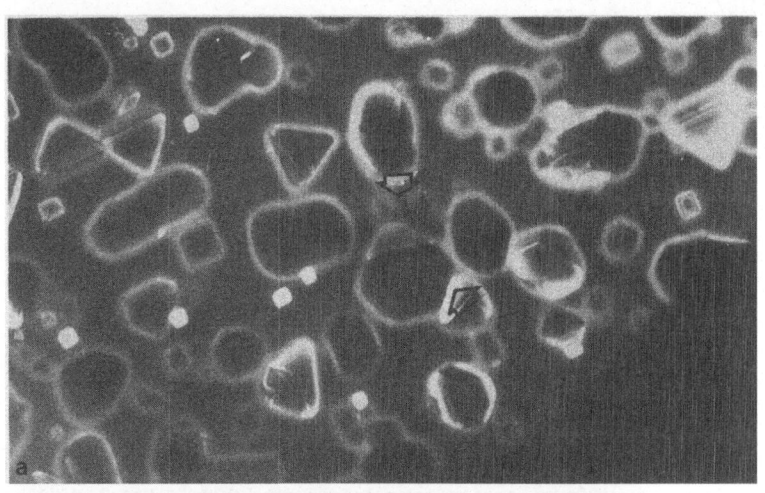

Fig.8.4a,b. Single-twinned gold particle produced by evaporated onto NaCl.
a) Weak-beam image, b) microdiffraction pattern

based on the ideas of *Bagley* [8.14] has been put forward by *Yang* [8.15],
whereby the tetrahedral units are no longer fcc. The icosahedron will have
a rhombohedral unit cell and the decahedron a body-centered orthorhombic
unit cell. This type of packing will satisfy the point group symmetrics and
no "gaps" or nonuniform strains will arise. Dark field contrast studies by
Yacamán et al. [8.16] supported this model. In addition *Roy* et al. [8.17]
and *Gómez* et al. [8.18] performed microdiffraction studies that have given
full confirmation to the non-fcc model. In Fig.8.7 microdiffraction patterns
of individual icosahedral particles are shown in three different orientations.

188

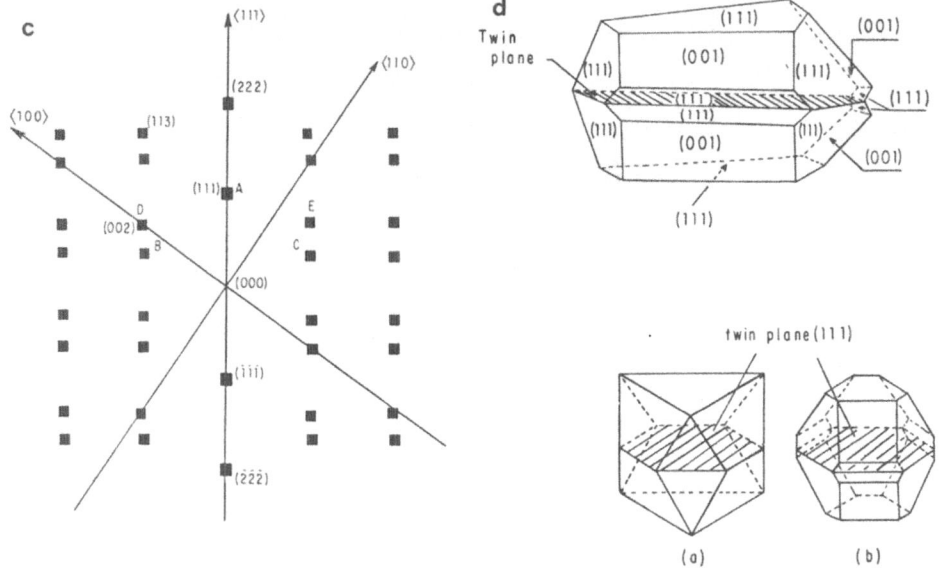

Fig.8.4c,d. Single-twinned gold particle produced by evaporation onto NaCl.
c) Calculated microdiffraction pattern of a <110> zone axis including diffraction spots by the two portions of the particle, d) reproduced shapes of
single-twinned particle indicating the twinning plane

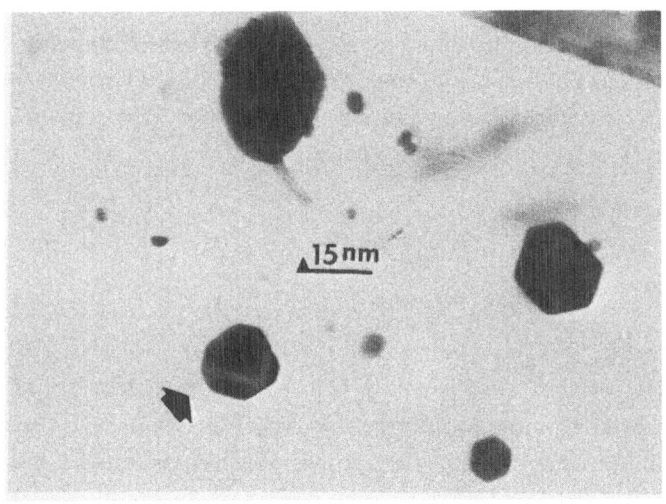

Fig.8.5. Single-twinned particle in a Pt/graphite catalyst

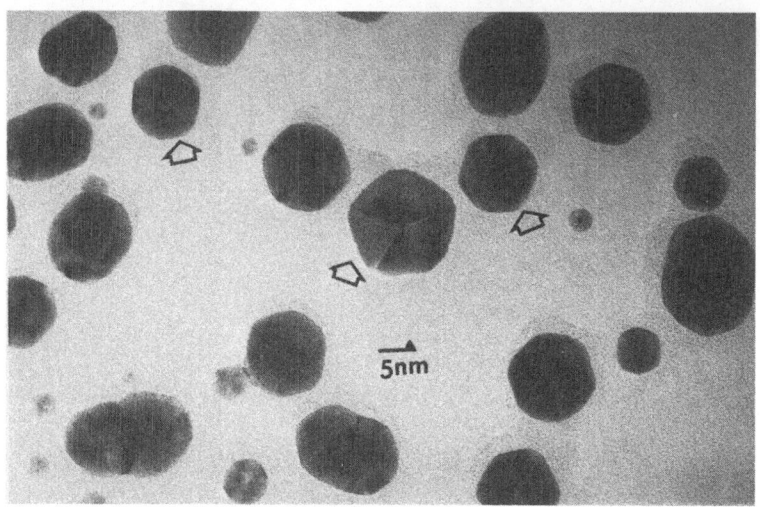

Fig.8.6. Bright-field contrast of decahedral and icosahedral particles

They correspond to the <111> zone axis or face orientation [8.18], the <112> or Edge orientation and <110> or fivefold orientation. This figure also shows the patterns calculated based on the Yang model, again with complete agreement between theory and experiment.

Fivefold particles are a very important example of departure from bulk symmetry in small particles. Recently *Fuentes* et al. [8.19] found icosahedral particles in Rh/SiO_2 catalysts. Generally speaking, these particles appear to be characteristic of cases in which the particle substrate interaction is very weak.

8.6 Regular fcc Shapes

Very important information for the study of catalytic systems concerns the shapes of fcc particles which do not have any twin boundaries and so can be considered as single crystals. The most frequently observed shapes are based on the octahedron and its truncations. These shapes obtained using weak-beam and microdiffraction techniques [8.20-22], are represented in Fig.8.8. A re-

Fig.8.7a-c. Diffraction patterns of an icosahedral particle in a) <111> orientation, b) <112> orientation, c) <110> orientation. The calculated diffraction pattern is shown in each case

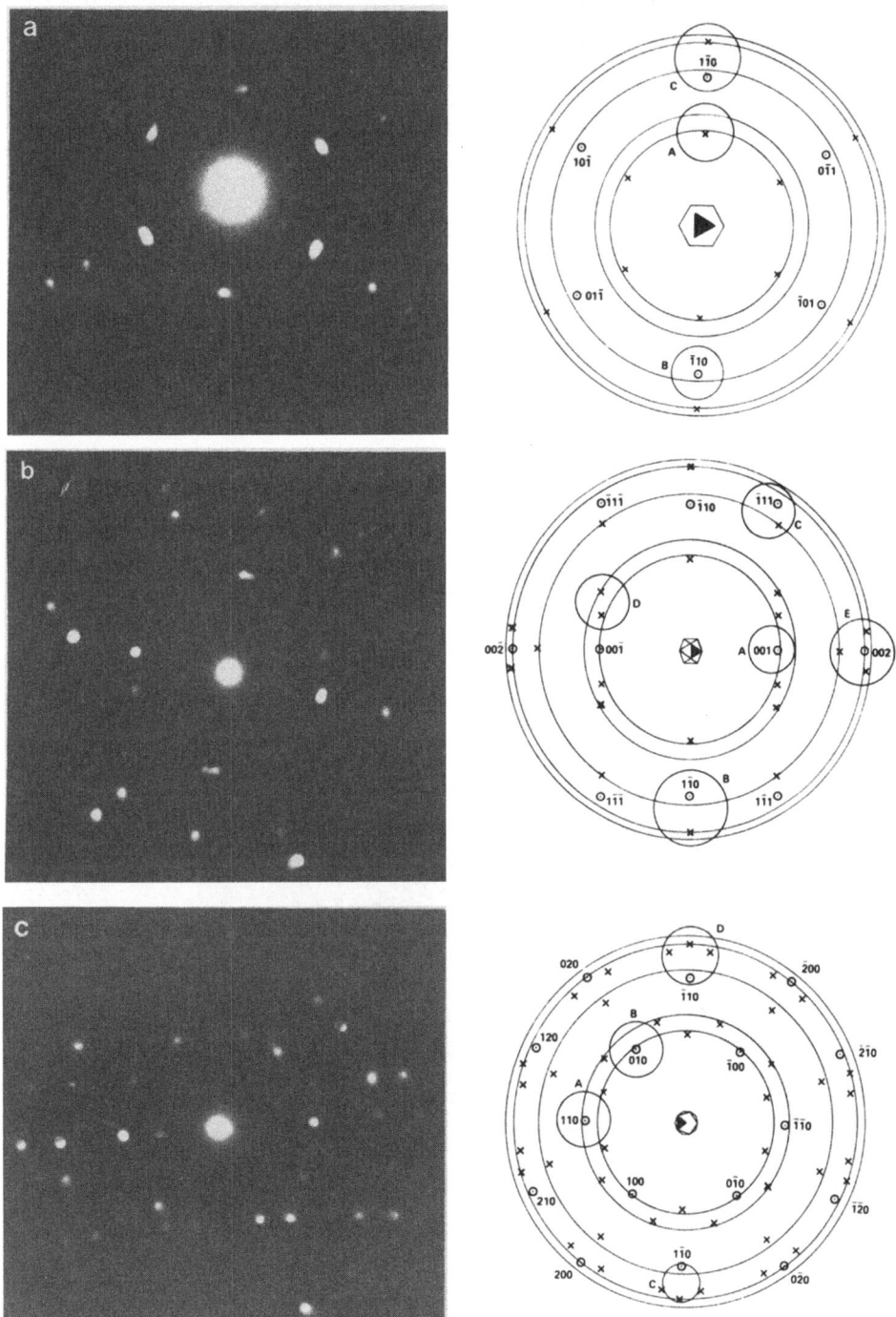

Fig.8.7a-c (caption see opposite page)

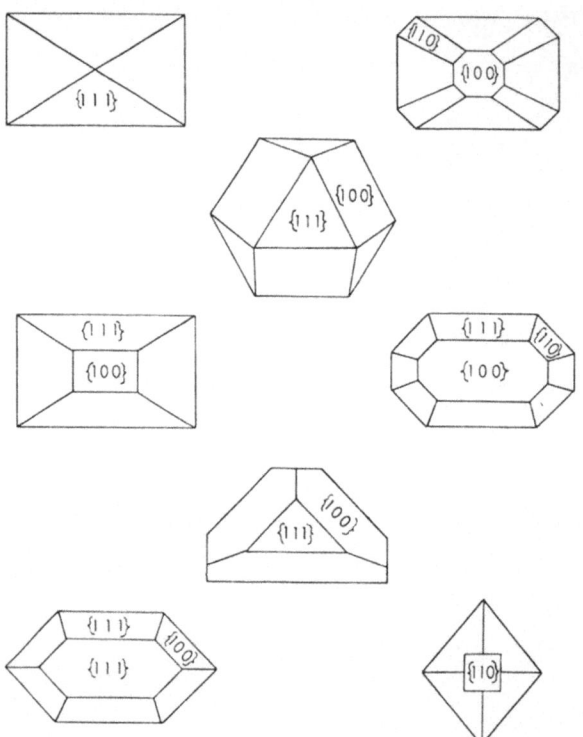

Fig.8.8. Typical shapes of fcc particles observed in evaporated films and supported metal catalysts

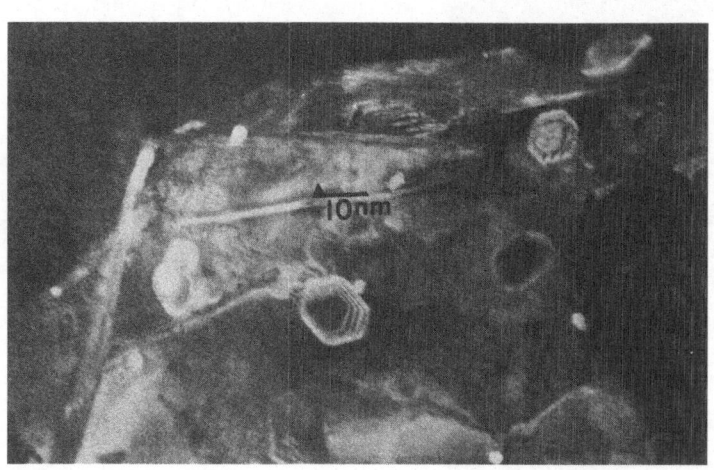

Fig.8.9. Weak-beam image of cubo-octahedral particles in a Pt/graphite catalyst

markable shape is the cubo-octahedron which contains {111} and {110} faces and is commonly observed in Pt, Ni and Rh based catalysts. Figure 8.9 shows an example of a Pt/graphite catalyst.

In some cases, the particles are truncated, producing a flat platelet structure (Fig.8.8). The shape strongly depends on the preparation conditions (temperature, reducing atmosphere, substrate, etc.) and each catalyst should be characterized individually. *Santiesteban* et al. [8.23] found significant changes of the particle shape after a methanation reaction at high temperatures.

8.7 Particle Surface Roughness

An important question that arises, is whether or not the particles have a smooth surface, as so far assumed in this chapter. Figure 8.10 shows a high-resolution weak-beam image of a square particle as in Fig.8.1. A close examination of the fringes shows that they are not straight but bent irregularly. Since the fringes follow the changes in thickness very accurately, these undulations imply that the surface of the particle is not smooth. The "lateral" resolution of a pattern (R) can be defined in the kinematic approximation as

$$R \simeq \frac{\xi_g}{2} \quad . \tag{8.5}$$

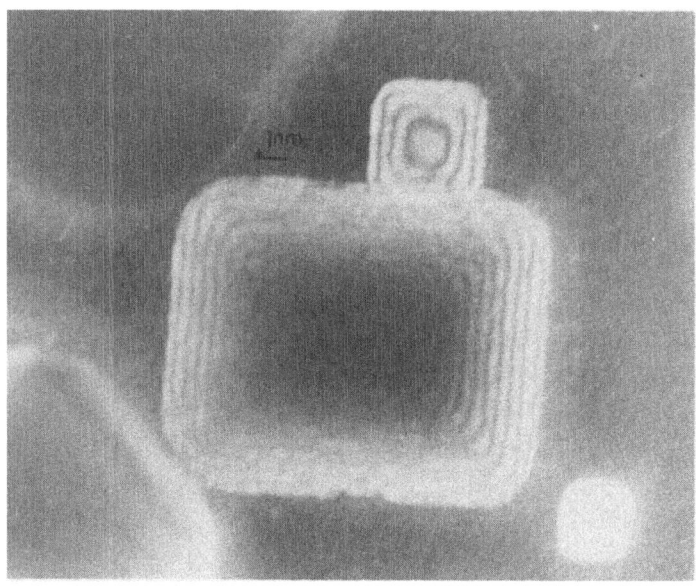

Fig.8.10. High-resolution weak-beam image of a square particle showing wavy thickness fringes

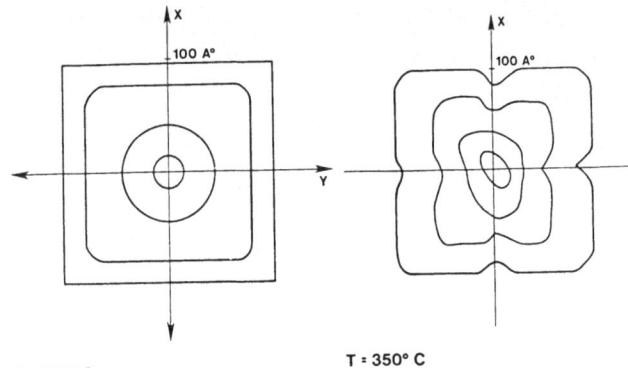

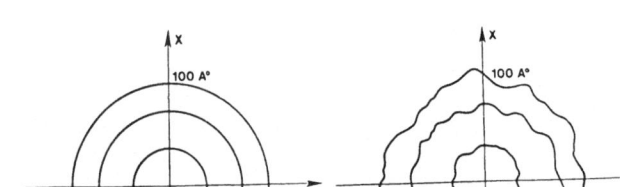

T = 250° C

T = 350° C

T = 550°

T = 650°

Fig.8.11. Topographic maps of a 100 Å square particle heated to different temperatures; 250°C, 350°C, 550°C, and 650°C

If a roughness feature has a size $r < R$ it will not be resolved. Then according to (8.5) with $\xi_g \simeq 5$ Å we obtain $R \simeq 2.5$ Å. This means that monolayer roughness can be detected by the weak-beam method in the high-resolution limit.

The general conclusion in our studies is that in most cases the particles have a rough surface structure. This statement is again valid for particles in catalysts [8.24]. Particle roughness can be studied systematically. Figure 8.11 shows the topographic maps obtained from square gold particles at different temperatures. Gold particles grown on a NaCl substrate at 10^{-9} Torr pressure and at $100°C$ were heated to different temperatures (without breaking the vacuum) and cooled down to room temperature for observation. The thickness contours in Fig.8.11, indicating at first profiles of a square pyramid, change to an almost spherical form at high temperatures and then to a very irregular shape at highest temperatures. These results have been discussed in terms of a roughening transition by *Cabrera* et al. [8.25].

8.8 Surface Sites on Rough Particles

To correlate structure with catalytic activity, more detailed information about the surface structure is required. On the rough surface of a particle, atom arrays might exist which may be active sites for catalytic reactions. To obtain a correlation with chemical activity more quantitative information about the surface roughness is necessary. This is an extremely difficult task. However, some information can be obtained through simple calculations of the structure, as shown by *Pérez* et al. [8.26]. These authors calculated the different stages of growth of cubo-octahedral particles. Their calculations assume that an arriving atom will occupy a local minimum energy site, producing a metastable configuration which contains incomplete layers. This situation corresponds to rapid growth of the particle in which overall equilibrium cannot be attained. Figure 8.12 shows a computer simulation sequence of growth of a cubo-octahedral particle, with the surface roughness reproduced. Figure 8.13 shows some of the different types of surface arrays of atoms that can be generated by surface roughness. The notation for each array is indicated in the figure. The distribution of the number of sites as a function of particle size can also be obtained, Fig.8.14. An important characteristic of this distribution is that there are oscillations in the number of sites, in contrast with the previous results of *van Hardeveld* and

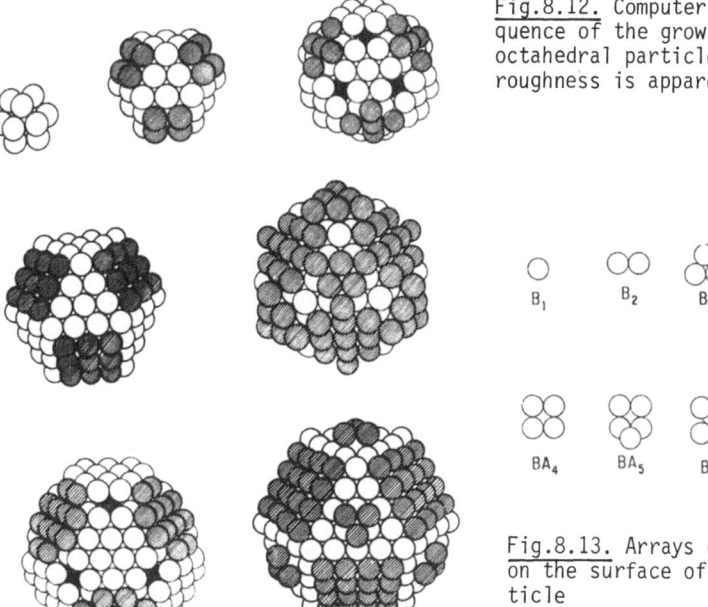

Fig.8.12. Computer-simulated sequence of the growth of a cubo-octahedral particle. Surface roughness is apparent [8.26]

B_1 B_2 B_3

BA_4 BA_5 BB_5

Fig.8.13. Arrays of atoms formed on the surface of a rough particle

195

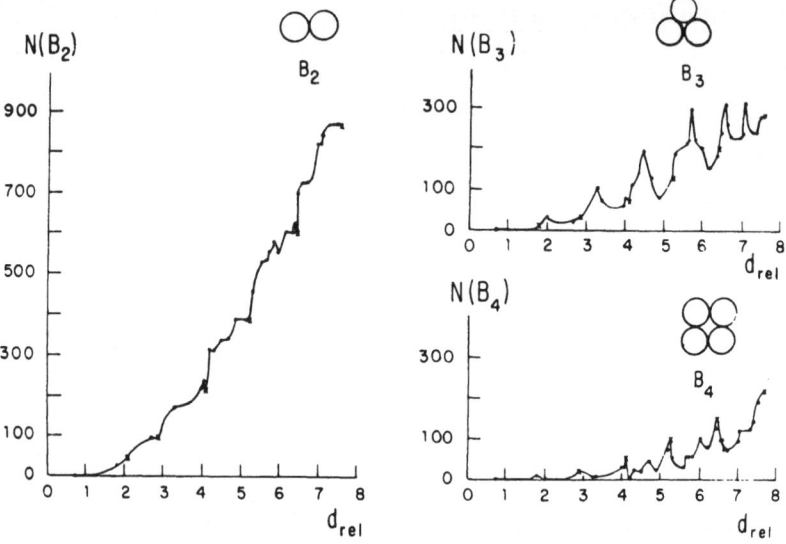

Fig.8.14. Curves for the number of sites on a particle as a function of the particle diameter normalized to the nearest-neighbor distance

Hartog [8.27]. When the site distributions in Fig.8.14 are convoluted with the experimental size distributions of the catalyst, an estimate of the total number of sites can be obtained.

This calculation is rather crude but can be used to interpret catalytic data, as shown in the following section. The reader is referred to the paper by *Pérez* et al. [8.26] for further details of the calculations.

8.9 Correlation of Catalytic Activity with Structure

Let us now discuss two structure-sensitive reactions that can be understood in terms of particle surface roughness: the hydrogenolysis of pentane by Rh particles [8.19] and the isomerization of 2-methyl pentane by Pt/γ-Al$_2$O$_3$ [8.28]. Pentane hydrogenolysis was studied at 150°C using Rh catalysts on different supports (SiO$_2$, Al$_2$O$_3$, TiO$_2$, carbon, etc.). Catalysts with different mean particle size were obtained. Figure 8.15 shows the experimental turnover number versus catalyst dispersion at the reaction temperature. Conventional gas chromatography was used to determine the reaction products: methane, ethane, butane and propane. An interesting result shown in Fig.8.15 is that catalysts with different crystal structure (cubo-octahedron or icosahedron) but with similar mean particle size yield similar activities. The

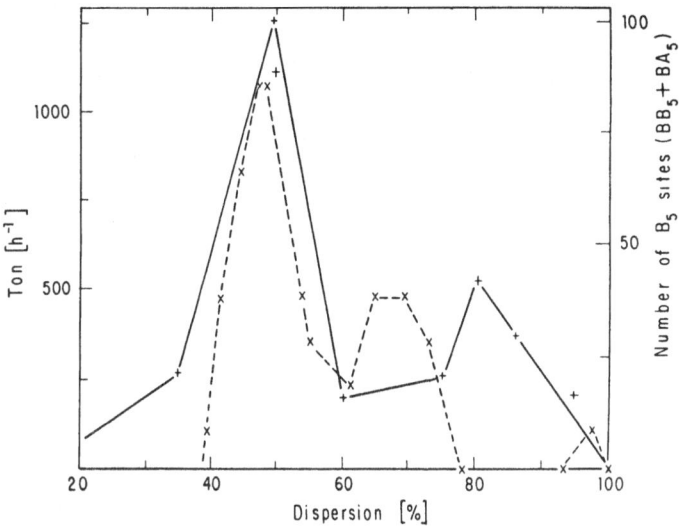

Fig.8.15. Turnover number for hydrogenolysis versus particle dispersion (experimental curve; ... theoretical curve)

particle size itself appears to be the most important factor in determining the activity. It was assumed that the most likely sites for hydrogenolysis were arrays of five atoms (BA_5 and BB_5), indicated in Fig.8.13, where the pentane molecule can be adsorbed. In Fig.8.15 the theoretical curve for the total number of B_5 sites is shown. This curve was obtained by convolution of the distributions by *Pérez* et al. [8.26] with the actual size distributions on the catalyst. The main experimental peaks of the activity (at 12-20 Å average size) are reproduced by the theoretical curve. The peak at higher dispersions is shifted, but this might be the result of errors in the particle size determination in the very small particle range. The agreement between experiment and theory is considered remarkable.

A second reaction is the isomerization of C-labeled hexanes which has been studied by *Gault* and co-workers [8.28]. In their experiments, they have come to the conclusion that there are two main isomerization mechanisms: the bond shift, which involves a simple bond displacement, and the cyclic mechanism which involes a cyclic intermediate. Their data for isomerization of 2-methylpentane by Pt/Al_2O_3 is shown in Fig.8.16, where the ratio between cyclic mechanism and bond shift is plotted as a function of catalyst dispersion. Using their published size distributions, we calculated the total number of sites of various kinds (the particles were assumed to be cubo-octahedrons, types of sites were tested by calculations). The best fit was obtained by assuming that the active sites for cyclic isomerization are single surface atoms

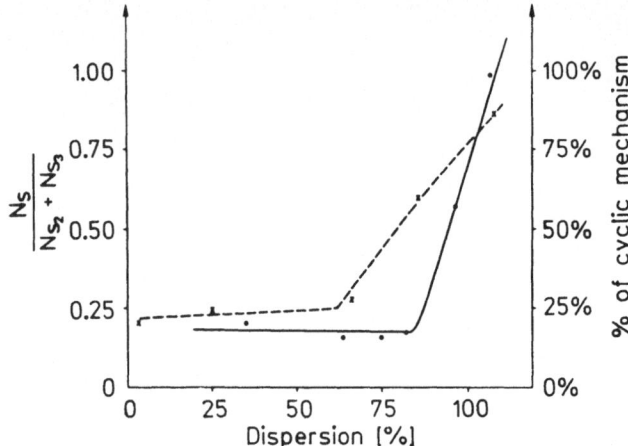

Fig.8.16. Percentage of cyclic mechanism to bond shift in isomerization of 2-methylpentane as a function of catalyst dispersion (experimental curve [8.22]; ... theoretical calculation)

(B_1 sites) and that the active sites for the bond shift mechanism are the arrays of two (B_2) and three (B_3) atoms shown in Fig.8.13. In Fig.8.16 we plotted the ratio

$$\left(\frac{\text{number } B_1}{\text{number } (B_2 + B_3)} \right)$$

as a function of particle dispersion. Again, the agreement between theory and experiment is very good. This shows that even with simple geometrical models interesting correlations between structure and activity can be obtained. This kind of work, although still preliminary, appears to be very promising.

8.10 Anomalous Structure in Electron Diffraction Patterns from Small Particles

So far we have considered the fine structure of diffraction patterns which is related to twinning or to shape features such a wedges. There is an important additional fine structure which is due to other factors. Figure 8.17 shows a diffraction pattern from gold particles with a nearly hexagonal shape. The particles do not show any twin contrast in dark field images. In addition to the normal spots of the <111> zone axis there are six anomalous spots corresponding to an interplanar distance of about 2.46 Å, with an intensity comparable to the regular spots. These spots have been observed before in con-

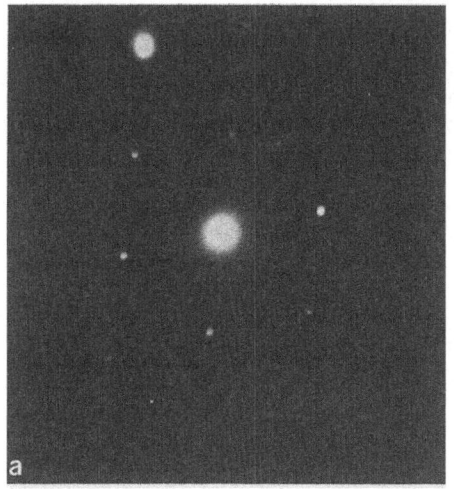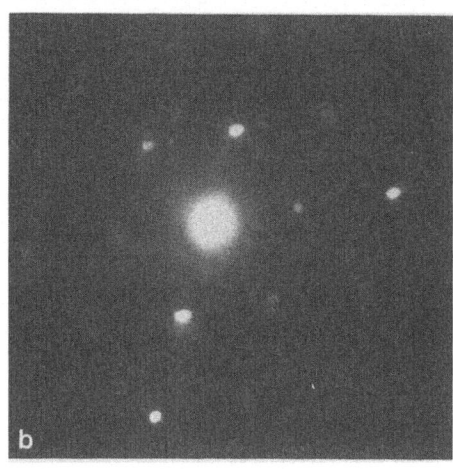

Fig.8.17. Microdiffraction pattern of an hexagonally shaped gold particle
at two orientations a) near the Laue condition, b) 1° off the Laue condition.
Six 2.46 Å spots are apparent. The pattern corresponds to a <00.1> hexagonal
zone axis

tinuous films of gold [8.29,30] and in gold particles [8.31,32]. Two main
recent interpretations have been offered as the origin of the extra spots.
The first, by *Cherns* [8.29], assumed that the spots were due to incomplete
stacking terminations of the fcc sequence in monoatomic steps. This will produce a weak finite intensity of hexagonal reflections. On the other hand,
Metois and *Heyeraud* [8.31] and *Tanishiro* et al. [8.32] associated the forbidden spots with the (23×1) surface superstructure found in {111} surfaces
of gold [8.33]. The extra spots are produced by the surface superstructure
and by double diffraction.

Neither of ESE models appears satisfactory to explain the results of
microdiffraction of individual particles. In both it is expected that the intensity of the forbidden spots is much less than that of the "bulk" spots.
As can be observed in Fig.8.17, the forbidden spots can have an intensity
close to that of the "bulk" spots. This is clearly noted in the two tilting
settings for the particle shown. On the other hand, the idea of surface steps
does not appear appropriate for small particles since steps imply long-range
crystalline order. The same problem arises for small particles (~ 100 Å) with
surface reconstruction. The full unit cell of the (23×1) superstructure
will be about 66 Å in size [8.33]. Therefore, reconstruction, at least in
the sense it is used for large crystals, does not seem likely on small particles. The surface roughness described in previous sections will produce

Table 8.1. Some hexagonal reflections of the fcc lattice

Hexagonal indices	d (Å)	fcc indices
(00.1)	7.05	irrational
(01.0) (10.0) (1$\bar{1}$.0)	2.47	irrational
(1$\bar{1}$.1) (01.1)	2.35	(111)
(1$\bar{1}$.3) (01.3)	1.71	irrational
(01.2)	2.$\overline{0}$3	(200)
(01.4)	1.45	(220)

certain effects in microdiffraction-patterns [8.34], such as streaks and weak extra spots, but will not provide an explanation for the strong forbidden intensities.

The anomalous diffraction features can be understood if one assumes that the fcc crystal diffracts like a hexagonal system. The hexagonal unit cell contained in the fcc cell will be defined by the following vectors:

$$\mathbf{a} = \frac{1}{2} a \, [1\bar{1}0] \quad \mathbf{b} = \frac{1}{2} a \, [01\bar{1}] \quad \mathbf{c} = a[111] \quad .$$

This will produce the reflections (in hexagonal indices) shown in Table 8.1. In terms of the hexagonal lattice, the pattern in Fig.8.17 corresponds to the <00.1> zone axis. Figure 8.18 shows another pattern corresponding to the <11.0> zone axis.

In a normal fcc crystal many of the reflections in Table 8.1 will be canceled out by scattering through successive layers, giving the normal fcc reflections, (111), (200), (220), etc.

Several mechanisms can explain breaking the diffraction rules for fcc crystals, e.g., repeated faulting in the stacking sequence along the particle. This, however, will not explain all the reflections which were observed [8.35] and it will also produce a characteristic contrast which could not be observed experimentally.

A possible mechanism to explain the results is that the atom positions in the particle are slightly displaced from those in the normal fcc structure. Lateral atom displacements will explain the intensities of the extra spots and the pattern geometry. The displacements might be the results of a minimum energy configuration in particles of small size. Shifts of this kind have been proposed in LEED intensity studies of reconstructed surfaces [8.36].

200

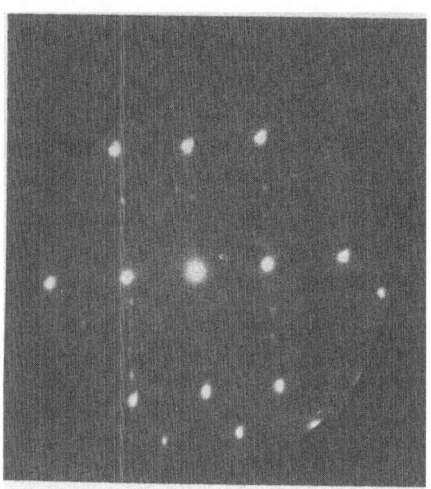

Fig.8.18. Diffraction pattern of a particle showing possible buckling of the layers

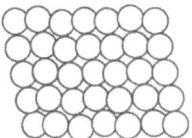 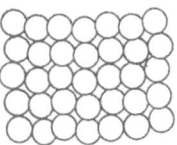

Fig.8.19. Possible models of an fcc particle showing buckling of the layers

The present phenomenon seems to be the equivalent of surface buckling in the case of small metal particles.

A detailed model of particle buckling will require a complete calculation of the minimum energy configuration, which is not available at present. However, from the information that has been obtained by LEED studies of Au, Pt and Ir surfaces, some models can be suggeste [8.37]. Figure 8.19 shows some models of a buckled particle. This type of layer displacement can be generated through internal strain in the particle due to size effects. The strain can be the result of partial dislocations produced during particle growth on a crystalline substrate.

An additional possibility to explain the displacements might be the formation of a charge density wave (CDW). In this model atoms are displaced from their ideal positions in a wave-line patterns by about 0.1 Å [8.34]. This shift will suffice to explain the presence of strong forbidden reflections. The pattern in Fig.8.18 has reflections which appears at positions 1/5, 2/5, 3/5, 4/5 of the distance between two strong spots. Therefore, this pattern can be described as a (1×5) supernet. Moreover some of the spots have a split corresponding to 1/20 of the distance between strong spots (in the perpendicular direction) and the supernet can be considered as a (5×20) structure. This type of supernet has been well established for Au {100} surfaces of single crystals [8.38].

The basic (1×5) structure can be generated through a CDW with a wavelength of $5a_s$ (a_s = nearest neighborg distance) and with a direction parallel to close rows of atoms. This is a very plausible model to explain the pattern in Fig. 8.18.

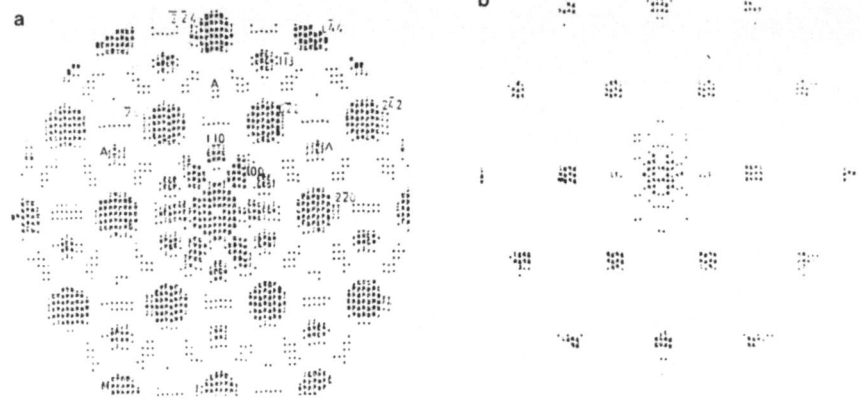

Fig.8.20. Simulated diffraction pattern of a cubo-octahedral particle corresponding to (a) 8 Å and (b) 35 Å thickness

Forbidden reflections might also be generated through other mechanisms.
In particular, size and shape effects might produce extra spots in the pattern [8.4] which are expected to be more conspicuous in the case of small
particles. The reciprocal lattice points of a faceted crystal are extended
in all directions perpendicular to the facets. The intensity of a spike is
roughly proportional to the area of the facet [8.39]: this effect has been
used by *Darby* et al. [8.40] to study faceting in thin gold films by electron
microscopy. On the other hand, the finite size of the crystal will produce
some additional spots. These reflections will always correspond to integer
indexes (in fcc notation); i.e., the spots will be forbidden only be zone
axis considerations. Therefore, this will not explain the irrational spots
in Table 8.1.

Vázquez-Polo [8.41] has carried out extensive calculations of the effect
of shape on diffraction patterns of cubo-octahedral particles. Figure 8.20
shows the calculated patterns for a cubo-octahedral particle at different
thickness. At 8 Å thickness (Fig.8.20a) two sets of extra spots can be observed. The first one can be fcc indexed as {110} corresponding to an interplanar distance of 2.85 Å. The second set corresponds to an interplanar
distance of 2.47 Å to 1/3 {422}, or {1$\bar{1}$.0} (in hexagonal notation). However,
when the thickness is increased to ~35 Å, the {110} reflections almost
vanish (Fig.8.20b). The remaining irrational reflections are very weak with
an intensity of about 10^{-2} Io (Io = intensity of the incident beam). For a
200 Å thick particle, the irrational reflections also disappear. Therefore,
it appears that the faceting alone is not sufficient to explain the observed
intensity of the extra spots. The strong dynamic character of the diffraction

of small noble metal particles might change the intensity values. However, the general behavior expected in a full dynamic calculation is not expected to be substantially modified [8.5].

Buckling of layers appears to be the most likely mechanism to explain the presence of strong hexagonal reflections in the microdiffraction patterns of small metal particles.

8.11 Conclusions

Modern electron microscopy offers new exciting possibilities for small-particle characterization.

We have shown that by combining weak-beam imaging methods with STEM microdiffraction, it is possible to obtain accurate information about the shape and crystal structure of small metal particles. The results are applicable to both evaporated particles and supported catalysts. A whole variety of particle structures has been observed: single or multiple twinning can exist in some cases, in others icosahedron and decahedral forms are present with non-fcc structures.

Single-crystal fcc particles are in most cases shaped as truncated octahedrons. They generate a number of different variations, such as the cubooctahedron, hexagonal or pentagonal platelets, etc.

In all cases a very general property of the small particles is their surface roghness, which is likely to generate active sites for catalytic reactions.

Finally, the study of anomalous diffraction intensities seems to indicate that the atom positions in the "bulk" of the particle are shifted with respect to the fcc ones. This might represent a very important property of small particles with strong implications for the electronic structure.

A more detailed understanding of the particle shape and properties opens up the possibility of systematic correlations between catalytic activity and structure which is one of the key goals of research in surface science.

Acknowledgement. The author would like to express his appreciation to all the members of the electron microscopy group of the Instituto de Fisica of the University of Mexico of their collaboration during the work described here, in particular to Drs. A. Gômez, D. Romeu, and G. Vázquez Polo for fruitful discussions.

The author is also indebted to Dr. M. Avalos for providing several samples used in this work, and to Mr. Francisco Ruîz and Ms. Rebeca Sosa for technical assistance.

References

8.1 M. José Yacamán, T. Ocaña: Phys. Stat. Sol. A **42**, 571 (1977)
8.2 M.J. Yacamán, K. Heinemann, H. Poppa: In *Surface Science: Recent Progress and Perspective*, CRC critical reviews in solid state and materials science, ed. by R. Vanselow (CRC, 1981)
8.3 M.J. Yacamán, A. Gómez, D. Romeu: Kinam **2**, 303 (1980)
8.4 P.B. Hirsh, A. Howie, R.B. Nicholson, Pashley, M.J. Whelan: *Electron Microscopy of Thin Crystals* (Butterworths, London 1967)
8.5 M. Avalos: Ph.D. Thesis, Stanford University (1982)
8.6 A. Gómez, P. Schabes, M.J. Yacamán, T. Ocaña: Philos. Mag. **47**, 169 (1983)
8.7 M. Goringe: In *Electron Microscopy in Materials Science*, ed. by U. Valdré, Z. Zichini (Academic, New York 1971)
8.8 T. Komoda: Jpn. J. Appl. Phys. **7**, 27 (1968)
8.9 T. Hayashi, T. Ohno, S. Yatsuya, R. Uyeda: Jpn. J. Appl. Phys. **16**, 705 (1977)
8.10 F. Robinson, M. Gillet: Thin Sol. Films **98**, 179 (1982)
8.11 S. Ino, S. Ogawa: J. Phy. Soc. Jpn. **27**, 1365 (1967)
8.12 M. Gillet: Surf. Sci. **67**, 139 (1977)
8.13 S. Ino: J. Phy. Soc. Jpn. **21**, 346 (1966)
8.14 B.F. Bagley: Nature **208**, 674 (1965)
8.15 C.Y. Yang: J. Cryst. Growth **47**, 274 (1979)
8.16 M.J. Yacamán, K. Heinemann, C.Y. Yang, H. Poppa: J. Cryst. Growth **47**, 187 (1979)
8.17 R.A. Roy, R. Messier, J.M. Cowley: Thin Solid Films **79**, 207 (1981)
8.18 A. Gómez, P.S. Schabes, M.J. Yacamán: Thin Solid Films **98**, 195 (1982)
8.19 S. Fuentes, F. Madera, M.J. Yacamán: J. Chim. Physique (in press)
8.20 J.M. Domínguez, M.J. Yacamán: J. Catal. **64**, 223 (1980)
8.21 M.J. Yacamán, J.M. Domínguez: J. Catal. **64**, 213 (1980)
8.22 J. Santiesteban, S. Fuentes, M.J. Yacamán: J. Mol. Catal. **20**, 213 (1983)
8.23 J. Santiesteban, S. Fuentes, M.J. Yacamán: J. Vac. Sci. Technol. A**1**, 1198 (1983)
8.24 O.L. Pérez, D. Romeu, M.J. Yacamán: J. Catal. **79**, 249 (1983)
8.25 M. Avalos, M.J. Yacamán: To be published
8.26 O.L. Pérez, D. Romeu, M.J. Yacamán: Appl. Surf. Sci. **13**, 402 (1982)
8.27 V. Van Hardeveld, F. Hartog: Surf. Sci. **15**, 189 (1969)
8.28 F.G. Gault, F. Garin, O. Maire: In *Growth and Properties of Metal Clusters*, ed. by J. Bourdon (Elsevier 1980)
8.29 D. Cherns: Philos. Mag. **30**, 549 (1974)
8.30 W. Krakow: Surf. Sci. 222, **111**, 503 (1981)
8.31 J.C. Hayeraud, J.J. Metois: Surf. Sci. **100**, 519 (1980)
8.32 Y. Tanishiro, H. Kanamori, K. Takayanagi, K. Yagi, G. Honjo: Surf. Sci. **111**, 395 (1981)
8.33 H. Nelle, E. Menzel: Z. Naturforsch. **32a**, 282 (1978)
8.34 C. Annett, W.R. Lawrence, R.E. Allen: Phys. Rev. B**10**, 4184 (1974)
8.35 J.E. Davey, R.H. Deiter: J. Appl. Phys. **36**, 284 (1965)
8.36 M.A. Van Hove, R.J. Hoestiner, P.C. Stair, J.P. Biberian, L.L. Hesmodel, I. Barios, E.A. Somorjai: Surf. Sci. **103**, 189 (1981)
8.37 J.J. Burton, G. Jura: In *Structure and Chemistry of Solid Surfaces*, ed. by G.A. Somorjai (Wiley, New York 1969)
8.38 D.G. Fedak, N.A. Gjostein: Surf. Sci. **8**, 77 (1967)
8.39 M. von Laue: Ann. Physik **26**, 55 (1936)
8.40 T.P. Darby, D.Y. Tuan, R.W. Balluffi: Surf. Sci. **72**, 357 (1978)
8.41 G. Vázquez-Polo: Ph.D. Thesis, National University of México, Facultad de Ciencias (1982)

9. Reconstruction of Metal Surfaces

P.J.Estrup

With 22 Figures

9.1 Introduction

Any fundamental understanding of the microscopic properties of a solid surface requires a reasonably detailed model of the atomic geometry. As a starting point, the simplest possible structural model is usually chosen: it is
assumed that the surface geometry is that which would result from a truncation of the bulk solid with no other changes. Thus, the two-dimensional (2D)
periodicity of the outermost layer is assumed to be identical to that of an
equivalent parallel layer in the bulk, and the spacing between adjacent
layers in the "selvedge" region is assumed to be the same as in the interior
of the crystal, Fig.9.1a.

However, the truncation model will not, in general, represent the most
stable structure. The coordination of the surface atoms and the electronic
structure in the selvedge region are not the same as in the bulk, and some
changes in the atomic geometry may therefore occur for a real surface, Fig.
9.1b. If the rearrangement leads to a change in the interlayer spacing perpendicular to the surface, it is referred to as a *"relaxation"*. If the rearrangement results in a different 2D geometry of one or more surface layers,
it is called a *"reconstruction"*.

It should be noted that this terminology is less precise than that used
in 3D crystallography. In Fig.9.1b a translation vector of the substrate surface is doubled due to the lateral displacements of the atoms; the same 2D
periodicity could be achieved in other ways, for example by raising every
other atom, producing a "buckled" surface, or by removing every other surface atom, producing microfacets. All three models would be examples of surface reconstruction. However, in 3D terminology only the last one would be
classified as a reconstructive rearrangement [9.1]. The change leading to a
structure such as that shown in Fig.9.1b would instead be termed *displacive*:
no interatomic bonds are broken and no new ones are formed; the topology of
the network is unaltered. For practical reasons this distinction between re-

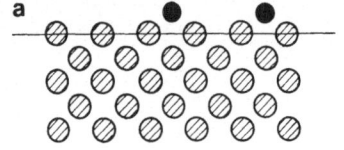 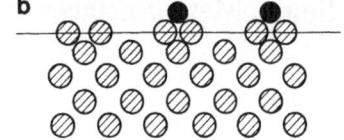

TRUNCATION   RECONSTRUCTION

Fig.9.1a,b. Surface models. a) Side view of truncation model. The shaded circles represent substrate atoms; the smaller filled circles represent adsorbed
atoms. b) Side view of a reconstructed surface. The 2D periodicity of the substrate surface has doubled. In addition the spacing between the first two
layers has been reduced

constructive and displacive changes is rarely made in discussions of surfaces;
in most cases where changes in the 2D periodicity have been observed it is not
yet known with certainty which type of substrate rearrangement occurs.

Surface reconstruction was first observed more than 25 years ago in LEED
(low-energy electron diffraction) studies of clean silicon and germanium crystals [9.2]. It is now recognized as a common phenomenon on semiconductors
[9.3], and the details of the resulting structures, as well as their relationship to the electronic and chemical properties of these surfaces continue to
be topics of intense research efforts. By contrast, for clean *metal* surfaces
it was thought that reconstruction was unimportant since, until fairly recently, the only known examples were some low-index faces of platinum [9.4], gold
[9.5,6], and iridium [9.7]. That view is no longer tenable. Surfaces of a
number of other metals, including molybdenum [9.8] and tungsten [9.8,9] have
been reported to reconstruct spontaneously; furthermore, it has become evident
that in many systems the presence of chemisorbed species can induce a rearrangement of the outermost metal atoms. Thus, the reconstruction phenomenon
is not only a fundamental interest in studies of the atomic geometry, the electronic structure and the phonon spectrum, but may also be essential for an
understanding of the nature of adatom-substrate and adatom-adatom interactions
in metals.

9.2 Surface Structures

9.2.1 Survey

Table 9.1 summarizes some of the available results for the reconstruction of
clean metal surfaces. The list includes surfaces of both face-centered (fcc)
and body-centered cubic (bcc) crystals. In the former group are the {110}

Table 9.1. Reconstructed clean metal surfaces

Surface	Superstructure	References
Al {110}	(5×1)	9.10
Ir {110}	(1×2)	9.11-17
Pt {110}	(1×2)	9.17-21
Au {110}	(1×2)	9.17,22-30
Al {100}	$(\sqrt{2} \times \sqrt{2})$	9.10
Ir {100}	(1×5)	9.7,31-33
Pt {100}	(20×5), "hex"	9.4,31-36
Au {100}	(20×5); c(26×68)	9.5,6,32,33,37-39
Au {111}	$(\sqrt{3} \times 22)$	9.32,40,41
V {100}	(5×1)	9.42,43
Cr {100}	$(\sqrt{2} \times \sqrt{2})$	9.44,45
Mo {100}	I - $(\sqrt{2} \times \sqrt{2})$	9.8,46
W {100}	$(\sqrt{2} \times \sqrt{2})$	9.8,9,47-49

and {100} faces of Al, Ir, Pt and Au and the {111} face of Au; in the latter group are the {100} faces of Mo and W, and possibly V and Cr.

The experimental evidence for the surface structure has come primarily from investigations by LEED [9.50-52] and, more recently, by HEIS or RBS [9.54] (i.e., high-energy ion scattering, also known as Rutherford backscattering), by FIM (field ion microscopy) [9.16,66-57], by helium atom diffraction [9.58], by electron microscopy [9.30,59], and by scanning tunneling microscopy [9.25,60]. Additional, but less direct, information has been obtained by correlation of structural data with measurements concerning the surface electronic structure, e.g., photoemission [9.61] and work function [9.62], and with chemisorption studies involving kinetics of adsorption, diffusion and desorption, and surface phase diagrams. Some examples are given in the following sections.

There has been considerable controversy over the cleanliness of the surfaces used experimentally. It is difficult, in general, to rule out completely that impurities play a role in stabilizing the reconstruction, but at least for Al, Ir, Pt, Au, Mo and W the impurity content is below the detection limit of Auger electron spectroscopy (AES). There is some doubt, however, that superstructures observed on V{100} [9.42] and Cr{100} [9.44] indeed are intrinsic to clean surfaces. Thus, the V{100}(5×1) structure has recently been ascribed to an oxygen impurity [9.43], and the formation of the $(\sqrt{2} \times \sqrt{2})$ structure on clean Cr{100} has not been reproduced [9.45].

A thorough search for reconstruction, with negative results, has been made for some surfaces, notably Nb{100} [9.63] and Ta{100} [9.64], but the absence of a particular surface from the list (Table 9.1) does not necessarily mean that it does not reconstruct. Many surface orientations have not yet been studied, and each requires careful investigations over a wide temperature range. The difficulties are exemplified by the Mo{100} reconstruction [9.8]; the formation of the modulated (incommensurate) I-($\sqrt{2} \times \sqrt{2}$) structure is observed only below room temperature and is inhibited by both structural imperfections and chemical impurities.

There are indications that metals which do not ordinarily exhibit surface reconstruction may do so if deposited as a thin film on another substrate. Thus, Co on Cu{100} shows a ($\sqrt{2} \times \sqrt{2}$) reconstruction already in the first layer, and the reconstruction persists as several additional epitaxial layers are deposited [9.65]. On the other hand, in studies of Au{110} films on Ag{110} it was found that the reconstruction does not set in until the Au film is 4-5 layers thick [9.66]. Clearly, composite systems of this type are promising candidates for studies of the reconstruction phenomenon.

Reconstruction may also be induced by chemisorption. Since new chemical bonds are formed, a perturbation of the substrate geometry should always be expected. However, in an increasing number of chemisorption systems it is found that the adsorbate causes unexpectedly large (0.1 - 0.2 Å) lateral displacements of substrate atoms. These structural changes can explain the long-standing puzzle of how a fractional monolayer of weakly scattering adatoms, such as hydrogen, can produce strong extra beams in the LEED pattern. Among the known examples are H/Ni{110} [9.67-69], H/Mo{100} [9.46], H/Pd{110} [9.70], and H/W{100} [9.71-77]. Other adsorbate-substrate combinations for which reconstruction has been demonstrated, or at least made highly plausible, are S/Fe{110} [9.78], CO/Mo{100} [9.79], N/W{100} [9.80], and O/W{110} [9.81]. Of course, if the adsorbing gas is highly reactive and if its pressure is sufficiently high, 3D compounds (e.g., oxides, sulfides, halides) may form, causing major disruption of the metal lattice. Such processes will not be considered as surface reconstructions in the present discussion and they are beyond the scope of this survey.

Several reviews of metal surface relaxation, with or without simultaneous reconstruction, are available [9.50-52]. Clean surfaces with dense packing, e.g., fcc{111} and bcc{110}, usually show a very small contraction of d_{12}, the distance between the first and second layers, of magnitude 1% or less. More "open" surfaces, e.g., bcc{100} and fcc{110}, can have larger contractions, typical values being 5%-10%, but adsorbates tend to restore d_{12} to the bulk value. The experimental data, mainly from LEED and HEIS, have most often been fitted with d_{12} as the only adjustable parameter. Recent theoretical studies [9.82] suggest, however, that *multilayer* relaxation may be important.

Experimental results for copper surfaces [9.83-85] confirm that prediction; for the Cu{110} surface d_{12} is found to be contracted by 5%-8%, and d_{23} to be expanded by 2%-3%. Similar data are not yet available for reconstructed surfaces.

Below are given some additional structural details for reconstructed metal surfaces, the possible driving mechanism for the transformations are considered, and chemical consequences of the reconstruction are discussed.

9.2.2 fcc{110} Surfaces

The {110} surfaces of iridium, platinum and gold all reconstruct in the clean state to give a (1×2) structure, i.e., to have double periodicity in the {001} direction. Figure 9.2 shows the resulting LEED pattern. Many attempts have been made to find a single model that would agree with all the experimental data obtained for these three surfaces, but so far none has been entirely successful. Figures 9.3b,c and d show three models most often considered. The structure in Fig.9.3b is the "paired-row" model which is formed from the truncated solid (Fig.9.3a) by lateral displacements of the close-packed rows in the first layer so that the (1×2) geometry is obtained. Figure 9.3c shows the "buckled-surface" model, in which alternate rows in the first layer are raised and lowered with respect to the second layer. The "missing-row" model is shown in Fig.9.3d; in this model every other close-packed row in the first layer has been removed.

On the basis of LEED intensity analyses, the preferred model is the missing-row structure, even though the R factor (which measures the agreement

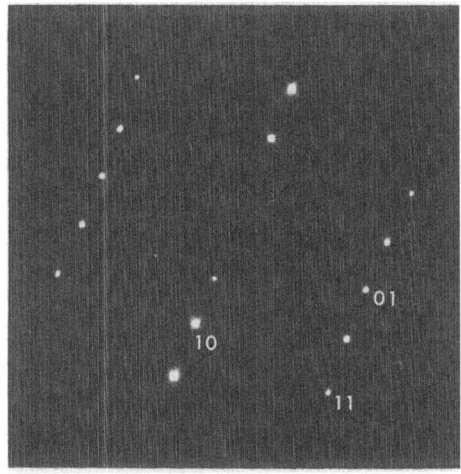

Fig.9.2. LEED pattern from a clean Au{110} surface, showing the (1×2) structure

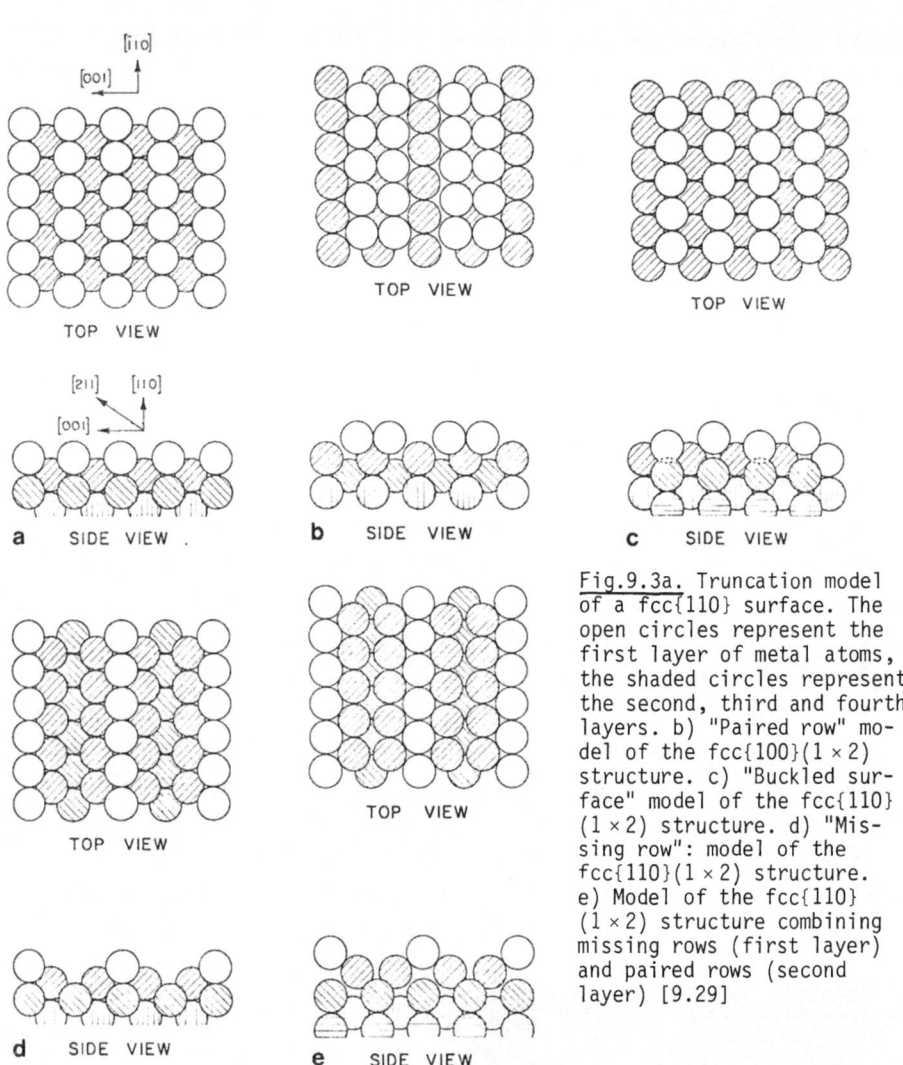

TOP VIEW

TOP VIEW

TOP VIEW

a SIDE VIEW

b SIDE VIEW

c SIDE VIEW

TOP VIEW

TOP VIEW

d SIDE VIEW

e SIDE VIEW

Fig.9.3a. Truncation model of a fcc{110} surface. The open circles represent the first layer of metal atoms, the shaded circles represent the second, third and fourth layers. b) "Paired row" model of the fcc{100}(1 × 2) structure. c) "Buckled surface" model of the fcc{110} (1 × 2) structure. d) "Missing row": model of the fcc{110}(1 × 2) structure. e) Model of the fcc{110} (1 × 2) structure combining missing rows (first layer) and paired rows (second layer) [9.29]

between model and experiment [9.86,87]) is only in the marginally satisfactory range for Ir{110} [9.15] and Pt{110} [9.19] and seems too high for Au{110} [9.23,24]. It is possible that improved agreement could be achieved by changing the positions of the atoms in the second, and perhaps even the third layer. A model with such modifications is shown in Fig.9.3e, combining missing rows in the first layer with paired rows in the second. The model was proposed to explain X-ray diffraction data for the Au{110} surface [9.29]. This technique, which has only recently been applied to surface structural problems, has the advantage that the diffraction data can be directly inverted to yield interatomic distances. It should therefore be free from the main

difficulties inherent in a LEED intensity analysis. Nevertheless, the model
(Fig.9.3e) appears to have unrealistic features; in particular the distance
between the first and second layer has been expanded by $\simeq 40\%$ relative to the
bulk value.

The most convincing evidence for a missing-row model is provided by several direct imaging techniques. First, FIM observations of Ir{110} favor this
model possibly with a lateral shift of the second layer [9.16,88]. Second,
the surface reconstruction of small gold particles has been observed with
transmission electron microscopy (TEM) and the atomic-scale images confirm
the (1×2) periodicity with missing rows [9.30]. Third, the Au{110} surface
has been investigated by scanning tunneling microscopy (STM) [9.25] and the
results demonstrate the large corrugation associated with the missing-row
model. The STM is done by scanning a very thin metal tip over the surface
and measuring the electron tunnel current [9.89]. In the simplest picture
this current varies monotonically with the distance between the surface and
the tip and a 3D picture of the surface topography can therefore be constructed. Figure 9.4 shows an STM image of a partially disordered Au{110} surface
[9.90]. In the area to the left of point A the double periodicity in the
<001> direction is clearly seen. Between A and B there are regions with
(1×3) and (1×4) corrugation. All steps are made of {111} facets, suggesting that the formation of these close-packed planes provides the driving
force for the reconstruction.

The principal shortcoming of the missing-row model is its apparent inability to explain the rapid phase change induced by certain adsorbates [9.18,
21,91,92]. When CO, NO or Cl_2 adsorbs on Ir{110} or Pt{110}, the transformation $(1 \times 2) \rightarrow (1 \times 1)$ is observed in LEED, which would indicate that the missing rows somehow are being restored. To accomplish this, half a monolayer
of metal atoms must, in a minute or so, diffuse over distances of perhaps
50 Å or more. Estimates of the diffusion rate for Pt suggests, however, that
at room temperature a metal atom would on the average make less than a single
diffusion jump in the available time interval [9.17], and the phase change
therefore seems impossible.

To resolve this problem yet another model for the fcc{110}(1×2) surface
has been proposed [9.17]. It is the "sawtooth" structure shown in Fig.9.5.
Every other close-packed row has been rolled sideways and upwards, but no
atom has had to move more than one lattice spacing. The structure has {100}
and {111} facets and the unit cell therefore has lower symmetry than in the
missing-row model (Fig.9.3d). Measurements by spin-polarized LEED have indeed indicated that this should be the case [9.26,27]. Furthermore, the very

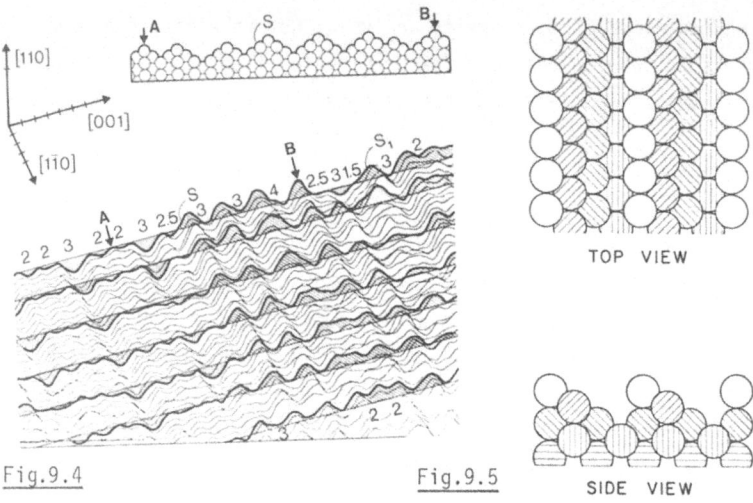

$[\bar{1}10]$

$[001]$

$[1\bar{1}0]$

TOP VIEW

SIDE VIEW

Fig.9.4.   Fig.9.5.

Fig.9.4. STM [9.90] picture of a partially disordered Au{110}(1 × 2) surface. The reconstructed surface consists of ribbons of narrow {111} facets along the <110> direction. Divisions on the crystal axes are 5 Å. The straight lines help visualize the terraced structure with monolayer steps (e.g., at S). Below each line the missing rows, and above each line the remaining rows, are enhanced. The numbers on the top scan give distances between maxima in units of the bulk lattice spacing. The inset shows the proposed structural model for the observed corrugation between A and B

Fig.9.5. "Sawtooth model" of the fcc{110}(1 × 2) structure [9.17]. It can be formed from the truncated solid by displacing every other close-packed row (in the first layer) upwards and laterally along the [22$\bar{1}$] direction

large corrugation tends to be supported by helium diffraction experiments [9.58]. Unfortunately, the sawtooth model does not fit the X-ray diffraction data [9.29] and it is inconsistent with the STM picture (Fig.9.4) since half-integer lattice spacings should then be seen between the maxima from a (1 × 2) and an adjacent (1 × 3) channel [9.90].

At a temperature of about 700 K the clean Au{110}(1 × 2) structure disorders in a reversible phase transition. A study [9.22] of the (0½) LEED beam profiles indicates that one-dimensional disordering occurs consisting of a break-up of long <110> channels. There should, therefore, be little change in the local configuration of an individual surface atom. This is consistent with HEIS results [9.93] which show no abrupt change in the area of the surface peak (a measure of the number of displaced surface atoms) across the phase transition. It should be possible to obtain information about the mechanism and energetics of the reconstruction by statistical-mechanical modeling of the transition [9.94] but such efforts are hampered by the lack of a definitive structural model and by indications that the thermal stability of the (1 × 2) structure may vary from sample to sample [9.95,96].

Table 9.1 also lists Al{110} as an fcc metal surface undergoing spontaneous reconstruction. Aluminum is of particular interest in this context because it shows that reconstruction is not restricted to transition, or d-band, metals. The samples were prepared by epitaxial growth on GaAs(100) and the reconstruction was independent of the Al film thicknesses which ranged from ~0.1 to 1 μm [9.10]. It should be noted that the observed superstructure had (5 × 1) periodicity, i.e., quite different from that on the other metals. No structural models have been proposed for this surface.

9.2.3 fcc{100} Surfaces

Iridium{100}, Pt{100} and Au{100} can be prepared with a (1 × 1) periodicity, by appropriate surface treatments [9.97-99]. However, all three are metastable and at elevated temperature reconstruct spontaneously and irreversibly to produce similar—but not identical—superstructures. The simplest case is Ir{100} which in LEED shows a (1 × 5) periodicity [9.7,31]. Two models proposed for this surface are sketched in Fig.9.6 [9.32,33]. In both models the first layer of Ir atoms forms a hcp arrangement which periodically comes into registry with the square array of the second layer. The (1 × 5) periodicity requires 4% compression relative to a bulk {111} plane, but this strain is assumed to be relieved by buckling of the first layer. In the model of Fig. 9.6a the atoms in registry occupy twofold bridge sites; in Fig.9.6b, they are positioned directly on top of atoms in the second layer. If it is assumed that bond lengths are the same as in the bulk, the buckling in a is about 0.5 Å, (i.e., ±0.25 Å relative to the middle plane); in b the buckling is somewhat larger, ~ 0.8 Å [9.32]. A LEED intensity analysis [9.33] gives best agreement with model a, but with somewhat less buckling, i.e., with reduced bond length of the surface atoms. Evidence in support of the hexagonal model has come from photoelectron spectroscopy [9.62] which shows that the electronic structure of the reconstructed {100} surface is very similar to that of the ordinary {111} surface. As in the case of the {110} faces, the tendency to form a close-packed structure provides a possible rationale for the reconstruction.

A problem with the models shown in Fig.9.6 is that the density in the first layer is 20% higher than in the (metastable) {100} structure and, as for {110}, it is difficult to explain the kinetics of the structural transformations. To overcome this difficulty it has been proposed [9.32,100] that the models in Fig.9.6 should be modified to contain rows of vacancies in the top layers, possibly combined with some lateral shifts of the atoms. The rows must of

213

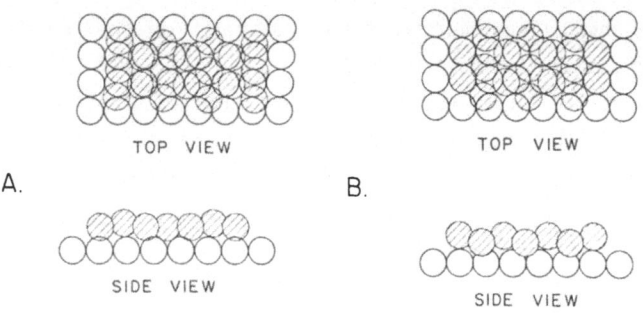

TOP VIEW   TOP VIEW

A.   B.

SIDE VIEW   SIDE VIEW

Fig.9.6A,B. Models of the Ir{100}(1 × 5) surface. The atoms in the first layer, shaded circles, form a hexagonal arrangement, which comes into registry with the second every fifth lattice spacing. In A the atoms in registry use "two-bridge" sites; in B "top/center" registry is used [9.32]

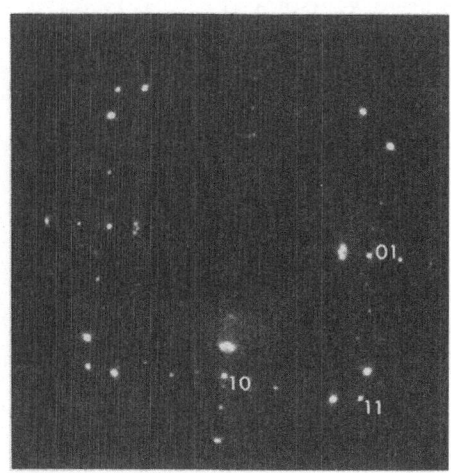

Fig.9.7. LEED pattern from a clean Au{100} surface. The structure has been labeled (20 × 5) or c(26 × 68)

course be regularly spaced to maintain the (1 × 5) periodicity. Such models could be acceptable according to LEED analyses [9.33] but would indicate a considerably more complex driving mechanism for the reconstruction.

A LEED pattern from the stable surface of clean Au{100} is shown in Fig. 9.7 [9.38]. Although the extra beams are near the 1/5 order positions, the structure clearly does not have a simple (1 × 5) periodicity. A (20 × 5) structure has been proposed for this surface [9.101]; it gives rise to most but not all of the observed spot splitting, and the best current estimate is that the surface has a c(26 × 68) unit mesh [9.32] but an incommensurate structure is obviously another possibility. This periodicity can be produced from the hexagonal (1 × 5) arrangement (Fig.9.6) by an additional periodic modulation, for example, in the form of regularly spaced dislocations. However, the size of the unit mesh prevents a detailed LEED intensity analysis.

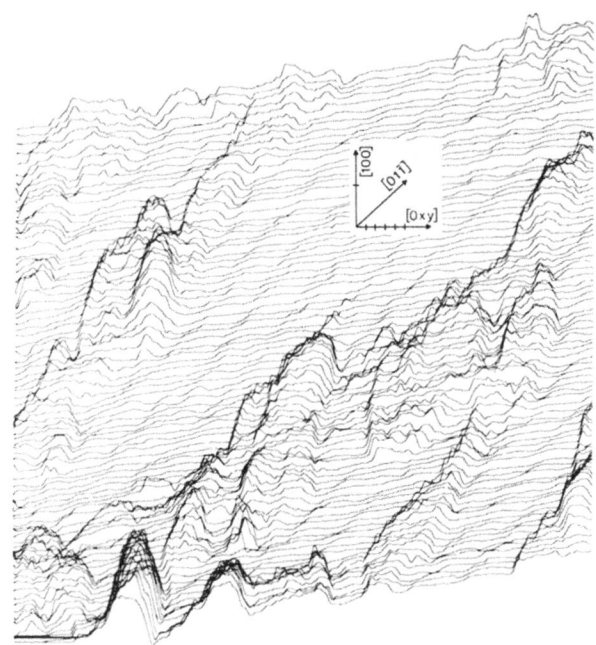

Fig.9.8. STM picture of a Au{100}(1 × 5) surface. The smooth areas are the clean surface. The areas with pronounced peaks are attributed to islands containing C. The divisions on the crystal axes are 5 Å apart. The scale along [011] and the [0XY] direction (roughly 60° from [011]) are uncertain due to experimental difficulties [9.101]

Both HEIS [9.37] and He atom diffraction [9.39] data confirm that the top layer of reconstructed Au{100} closely resembles a hcp layer. The only significant corrugation observed in He diffraction is that corresponding to the (1 × 5) periodicity. A series of models with this unit cell has been tested but only moderate agreement with the measurements was obtained [9.39]. It seems likely that a successful model must include a rearrangement also of the second and perhaps deeper layers.

A direct image of the Au{100} surface has been obtained by STM [9.101], (Fig.9.8). It exhibits smooth areas with apparent (1 × 5) periodicity but in this case the resolution is not sufficient to resolve the detailed atomic arrangement. (The clusters of large peaks in the picture are ascribed to localized surface states, due to "dangling bonds" associated with carbon impurities, for which an enhanced resolution is predicted [9.102].)

The LEED pattern from a reconstructed Pt{100} surface also shows splitting of the 1/5 order beams [9.4,103]. As for Au{100}, the structure is often referred to as (20 × 5) [9.36,104], but apparently any one of a series of closely related structures may be produced, possibly depending on steps and other defects [9.33]. These structures may be derived from the hexagonal arrangement by a small periodic modulation. A detailed LEED intensity analysis is impossible also in this case due to the very large unit mesh.

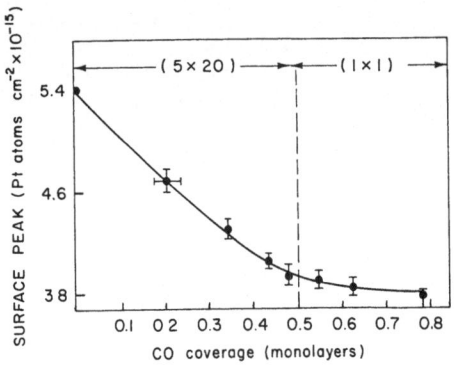

Fig.9.9. HEIS results for CO adsorption on Pt{100}. The ordinate is the <100> surface backscattering yield of 1 MeV $^4\mathrm{He}^+$ ions, measured at 198 K [9.36]

Several studies have been made of the transformation of Pt{100}(20×5) to (1×1) which can be induced by adsorption of H_2,CO, and other gases [9.35,36, 98,104-106]. The transformation is reversed when the gas is desorbed. The effect of CO adsorption at 198 K is shown by the HEIS data in Fig.9.9 [9.36]. The yield of backscattered ions decreases monotonically with increasing CO coverage and then levels off at about 0.5 monolayers, which is also the coverage at which the 1/5 order beams can no longer be seen in the LEED pattern. As seen from the change in the magnitude of the surface peak, about 1.65×10^{15} Pt atoms cm^{-2} move back into registry with the bulk during the $(20 \times 5) \rightarrow (1 \times 1)$ transition. Since the density for a {100} plane is only 1.28×10^{15} Pt atoms cm^{-2}, more than the first layer must be involved in the reconstruction. Thus, according to this measurement, the structure must contain displacements (larger than 0.1 Å) also in deeper lying layers. However, it is conceivable that a different interpretation of the data can be made if the induced transition does not lead to an ordered (1×1) surface but to a disordered arrangement, as suggested by recent LEED observations [9.107].

The Al{100} reconstruction does not follow the pattern of the other metals in this group. In RHEED studies of Al{100} films epitaxially grown on GaAs, it was found that the surface reconstructs at low temperature ($\sim$ 200 K) to give a $c(2 \times 2)$ periodicity [9.10]. If the temperature is raised, the ½ order beams weaken and disappear near 420 K. However, this transition can be reversed by again lowering the temperature. A weak $c(2 \times 2)$ reconstruction was also observed on a {100} surface of a bulk Al crystal [9.10]. The rearrangement is believed to be displacive, but no structural models have been proposed.

9.2.4 fcc{111} Surfaces

The only member of this group known to reconstruct is Au{111} [9.32,40,41]. In the LEED pattern [9.106] from the clean surface each integral order spot is surrounded by hexagonal arrays of extra spots. The interpretation again involves a distorted hexagonal array. Thus, the pattern appears to arise from a superposition of three equivalent domains, rotated $120°$ from each other, each having $(\sqrt{3} \times 22)$ unit mesh [9.32]. The position of the extra spots in the pattern indicates a uniaxial contraction of the first layer by 4.5%. It should be noted that an almost identical interpretation has been given of images of crystallite surfaces observed by electron microscopy (TEM) [9.59, 108].

9.2.5 bcc{100} Surfaces

As mentioned in Sect.9.2.1, there is reason to doubt that the superstructures observed on V{100} and Cr{100} are characteristic of the clean surfaces. Only W{100} and Mo{100} will, therefore, be considered here.

When a clean W{100} surface is cooled below room temperature, new $(\tfrac{1}{2}\tfrac{1}{2})$ beams appear in the LEED pattern (Fig.9.10), showing the formation of a $(\sqrt{2} \times \sqrt{2})$ — also called c(2 × 2) — structure [9.8,9,109]. The $(\tfrac{1}{2}\tfrac{1}{2})$ beam intensity varies gradually with temperature (Fig.9.11), and the change is very rapid and completely reversible, indicative of a displacive phase transition. The behavior of Mo{100} is very similar [9.8] except that the low-temperature phase is a modulated $(\sqrt{2} \times \sqrt{2})$ structure which gives a quartet of spots surrounding the $(\tfrac{1}{2}\tfrac{1}{2})$ position in the LEED patern (Fig.9.12). The separation of the extra spots is between 1/8 and 1/9 of the reciprocal lattice parameter, and possibly the superstructure is incommensurate with the bulk lattice.

A model of the clean W{100}$(\sqrt{2} \times \sqrt{2})$ structure is sketched in Fig.9.13a. The atoms in the top layer (shaded circles) have been displaced along <11> to form zigzag chains; the resulting arrangement has p2mg symmetry [9.47]. The direction of **d**, the surface component of the displacement, can be inferred from the symmetry of the LEED pattern. In most experiments the pattern has fourfold symmetry about the origin, as in Fig.9.10, because the pattern is a sum of intensities from equivalent domains rotated $90°$ with respect to each other. Surface steps can lift this rotational degeneracy, and the extra spots along a <11> line through the origin are then found to be weak or absent [9.47,110]. This is symptomatic of a glide line along the corresponding direction for the real structure and it unambiguously determines the symmetry elements of the unit mesh, and hence the direction of **d**. The systematic extinctions in the LEED pattern also prove that all the W atoms in the top layer lie in the same plane [9.49]. By LEED intensity analyses [9.72,111] and HEIS measurements [9.73], the magnitude of **d** has been determined to be ∼0.2 Å (0.16-0.25 Å). (The displacements have been exaggerrated in Fig.9.13 for clari-

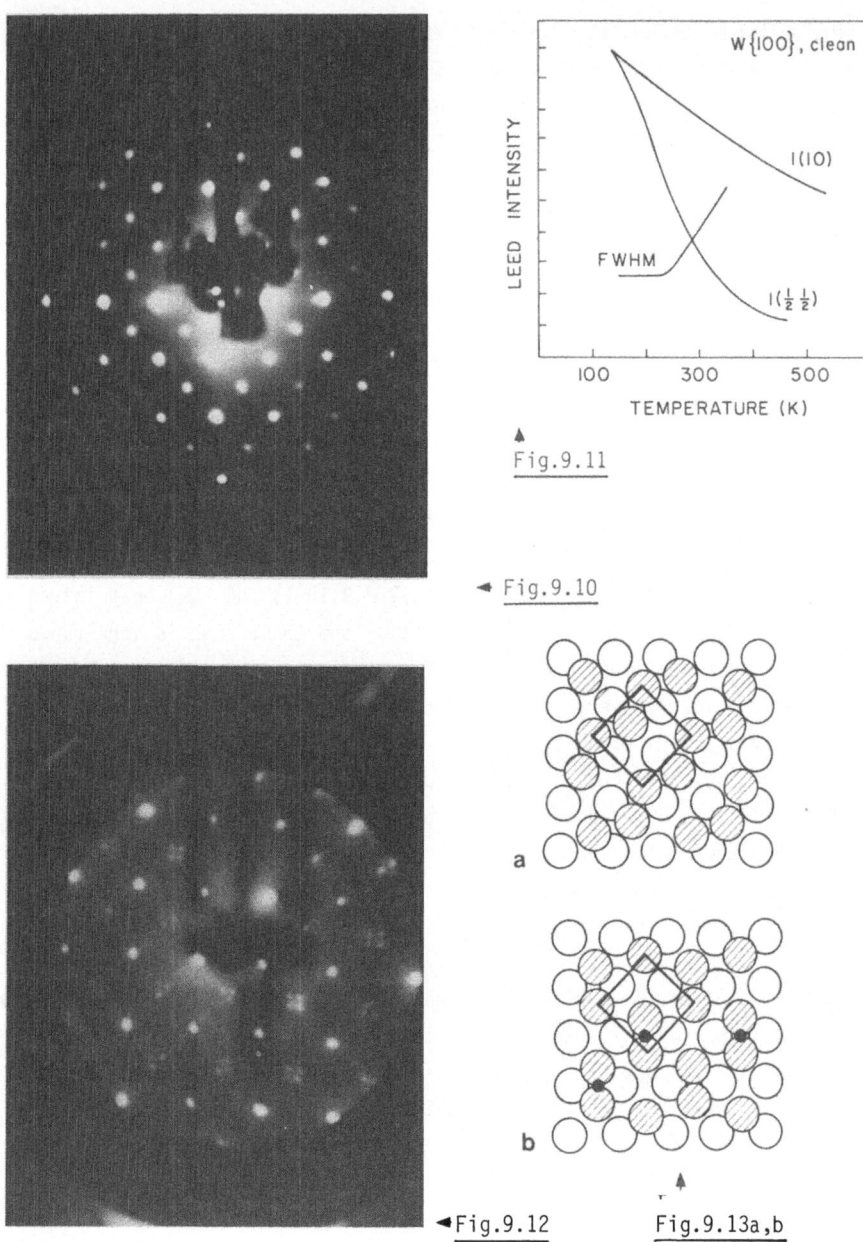

Fig.9.11

Fig.9.10

a

b

Fig.9.12   Fig.9.13a,b

Figs.9.10-13 (captions see opposite page)

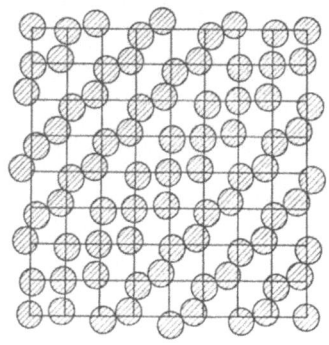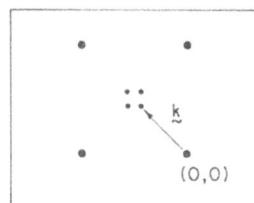

Fig.9.14. Model of the clean reconstructed Mo{100} surface, corresponding to the LEED pattern sketched in the inset below (cf. Fig.9.12). The wavelength of the displacement wave is $8/7(\sqrt{2}\,a_0)$ which is a more realistic choice than that used in a previous illustration [9.46]

ty.) The interlayer spacing d_{12} is found to be 1.49 Å, corresponding to a 6% contraction [9.111].

Fewer studies have been made of the reconstructed Mo{100} surface; the large unit mesh will in any case make a LEED intensity analysis difficult. To construct a possible model of the 2D arrangement it is convenient to start with the (1 × 1) structure and then impose a periodic lattice distortion (PLD) with wave vector **k** determined from the diffraction pattern [9.112]. The inset in Fig.9.14 shows the appropriate **k** vector for Mo{100} and for purposes of illustration its magnitude is taken to be $k = \frac{1}{2}\,(\frac{7}{8}\,\sqrt{2}\,a^{*})$, where a^{*} is the reciprocal lattice parameter. The displacement wave, assumed to be simple harmonic, is then given by $\mathbf{d}(r) = \mathbf{d}\,\sin(kr)$, where r denotes the lattice positions. For the nth atom in a row along <11> (parallel to **k**) the displacement is $\mathbf{d} = \mathbf{d}\,\sin(n\,\frac{7\pi}{8})$. The direction (but not the magnitude) of **d** can also be found from the LEED pattern in this case. The (kinematical) diffraction

Fig.9.10. LEED pattern from a clean W{100} surface at ~150 K. The reconstructed surface has $(\sqrt{2} \times \sqrt{2})$ —also called c(2 × 2) —periodicity

Fig.9.11. Temperature dependence of LEED beams from W{100}. I(10) and I($\frac{1}{2}\frac{1}{2}$) denote the (integrated) intensity of the (10) and ($\frac{1}{2}\frac{1}{2}$) beam, respectively. FWHM is the angular width of the ($\frac{1}{2}\frac{1}{2}$) beam

Fig.9.12. LEED pattern from a clean Mo{100} surface at ~150 K. The periodicity is that of a modulated, possibly incommensurate, $(\sqrt{2} \times \sqrt{2})$ structure

Fig.9.13a. Model of the reconstructed W{100} surface (top view). The $(\sqrt{2} \times \sqrt{2})$ unit mesh is shown; it has p2mg symmetry. b) Model of the hydrogen-induced reconstruction of W{100}. The unit mesh has c2mm symmetry. The small filled circles represent H atoms occupying bridge sites on "dimers" or W atoms

amplitude is $A \propto \sum_j \exp(i\mathbf{k} \cdot \mathbf{r}_j)$, and the leading term in the expression for the diffracted intensity of the extra beams, at normal incidence, becomes [9.75,112]

$$I(\mathbf{k}) = A(\mathbf{k})A^*(\mathbf{k}) \propto (\mathbf{k} \cdot \mathbf{d})^2 \quad .$$

When d is perpendicular to $\mathbf{k}$, or nearly so, the diffracted beams are predicted to be absent or weak. Application of this argument to the Mo{100} case shows that the distortion wave must be longitudinal [9.46], i.e., as for W{100} the displacements are along <11>. The model in Fig.9.14 has been drawn accordingly; it shows zigzag chains similar to those in Fig.9.13a, but with a modulated lateral separation of the atoms.

Attempts have been made to image directly the reconstructed surfaces by FIM. However, if ordinary procedures are used no lateral displacements can be seen in the micrographs [9.55]. Presumably the reason is that the energy difference between the nonreconstructed and reconstructed surfaces (or between reconstructions with different displacement directions) is small [9.113] and that the large electric field which is applied in FIM significantly perturbs the surface [9.57,114,115]. In particular, on clean W{100} and Mo{100} the electric field may cause a change in the direction of the atomic displacements from in plane (M_5 phonon mode) to out of plane (M_1 phonon mode) [9.113,116]. The occurrence of the latter under FIM conditions is indicated by the results illustrated in Fig.9.15,16 [9.56]. It is found that field evaporation from the metal surface does not remove atoms at random; instead vacancies are produced such that the remaining atoms form first a (2×2) and then a $c(2 \times 2)$ structure. On W{100} this effect can be observed at temperatures up to 500-600 K.

The in-plane distortions, present when no field is applied, also persist at quite high temperatures. Figure 9.11 shows the temperature dependence of several features in the LEED pattern from clean W{100}[9.8,117,118]. The intensity of the $(\frac{1}{2}\frac{1}{2})$ beam decreases much faster than that of an integral order beam, and the width of the $(\frac{1}{2}\frac{1}{2})$ beam increases noticeably for $T \gtrsim 250$ K. This temperature is taken to be the critical temperature T_c, above which the long-range order of $(\sqrt{2} \times \sqrt{2})$ vanishes. The long "tail", of the $(\frac{1}{2}\frac{1}{2})$ intensity curve above T_c is therefore ascribed to distortions possessing only short-range order.

The modification of reconstruction by adsorbates, in particular by hydrogen, has been studied for both W{100}[9.71,75-77,119] and Mo{100}[9.46,112]. In the presence of hydrogen a variety of new surface structures is observed, depending on the coverage and temperature. A portion of the phase diagram for

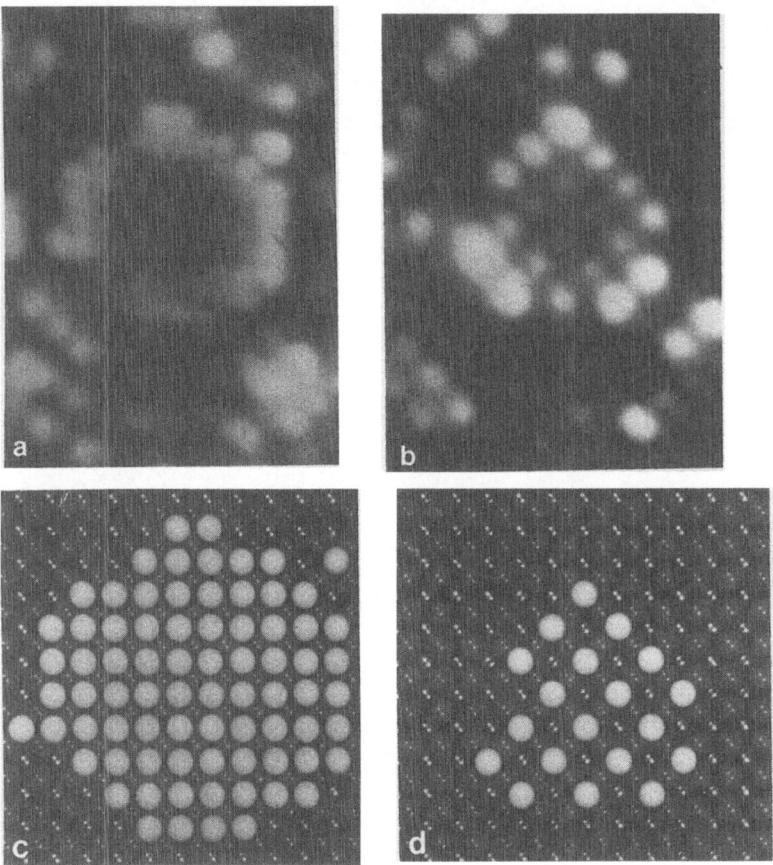

Fig.9.15a-d. Field ion micrographs of W{100} at low temperature. a) An initial apparent (1 × 1) structure obtained by low-temperature field evaporation. b) A final ($\sqrt{2} \times \sqrt{2}$) arrangement obtained by field evaporation at 430 K. c,d) Hard-sphere models reproducing (a) and (b) respectively [9.56]

H/W{100} is shown in Fig.9.17 [9.120]. In addition to an H-induced ($\sqrt{2} \times \sqrt{2}$) structure the phase diagram contains a prominent region where the surface is incommensurate with the bulk lattice. At coverages $\theta \gtrsim 0.5$ a phase with long-range order in only one direction is produced. At still higher coverages the adsorbate-induced features in the LEED patterns are weak. This agrees with HEIS measurements [9.73] which indicate that W atom displacements diminish in this coverage range (Fig.9.18). Not included in the phase diagram is a transition which occurs at $\theta \lesssim 0.1$ between the ($\sqrt{2} \times \sqrt{2}$) structure of clean W{100} (Fig.9.13a) and the hydrogen-induced structure (Fig.9.13b). In this transition the direction of the W atom displacements switches from <11> to <10> [9.75,77]. The adsorbed H atoms, which scatter too weakly to be located

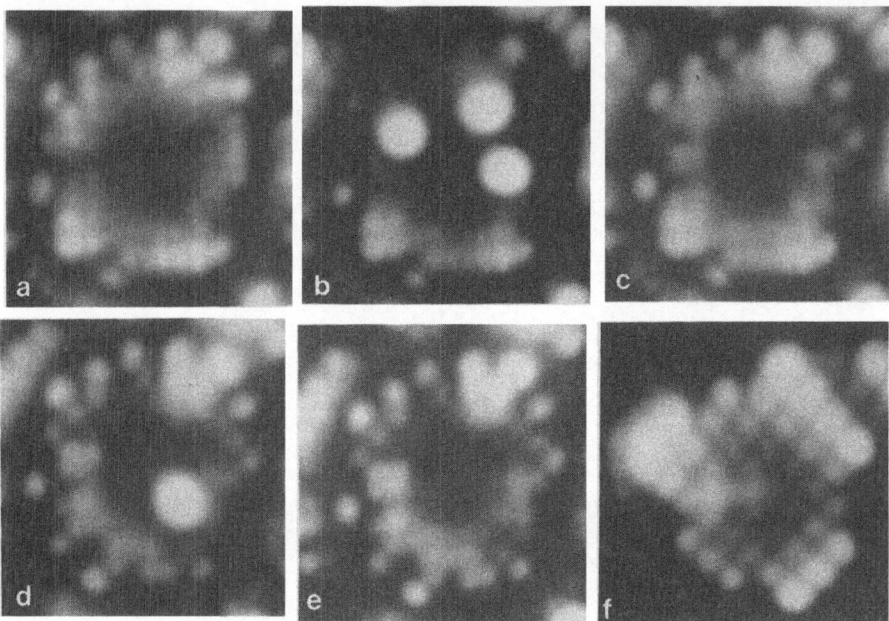

Fig.9.16a-f. Low-temperature field ion micrographs of Mo{100}, showing vacancy creation mechanism in field evaporation at elevated temperature, leading to the formation of apparent $(\sqrt{2} \times \sqrt{2})$ configuration from an initial apparent (1×1) structure. a) Initial plane formed by field evaporation at low
temperature. b) Same plane after specimen heated to 450 K with the field held
at just below evapoaration. c) Same plane after removal of adatoms, showing
a few vacancies. d,e) Same plane after repeating procedures in (b) and (c).
f) Final configuration of the same plane after further field evaporation at
450 K. The plane contains 42 atoms which, with the exception of only two
atoms, conforms to a $(\sqrt{2} \times \sqrt{2})$ structure [9.56]

by LEED, have been shown by EELS [9.74] to occupy bridge sites on the tungsten
dimers, Fig.9.13b. Structural models for the other phases seen in Fig.9.17
are still under investigation. Among the open questions is the precise geometry of the high-temperature (1×1) phase [9.76,77,114,120,121] and of the incommensurate phase [9.76,77,120]; and the effect of steps and other defects
on the reconstruction [9.76,77,122,123].

There have been fewer studies of H/Mo{100} than of H/W{100} but the available data show strong similarities between the two systems. The phase diagram
for H/Mo{100} contains a series of commensurate and incommensurate phases
[9.112] and an adsorbate-induced switching of direction of the displacements
occurs also for this substrate [9.46].

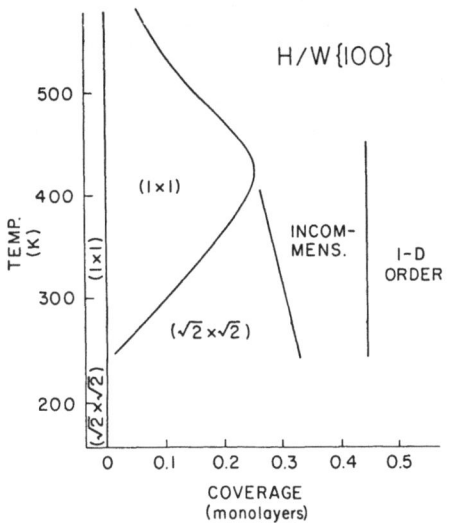

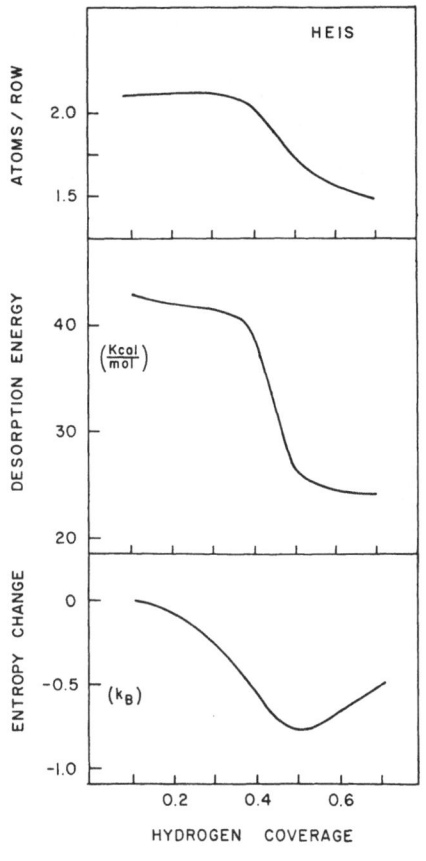

Fig.9.17. Phase diagram for the H/W{100} chemisorption system in the temperature-coverage plane. The narrow strip to the left of zero coverage represents the clean surface [9.120]

Fig.9.18.. Effects of hydrogen adsorption on W{100}. The panel on top shows the decrease in the W atom displacements as measured by HEIS. The middle panels shows the desorption energy for H₂. The bottom panel shows the entropy change per surface W atom (in units of Boltzmann's constant) [9.120]

9.3 Origin of Reconstruction

It seems clear that surface reconstruction results from a tendency towards minimum surface free energy. However, an accurate calculation of the free energy is very difficult and theoretical predictions of the equilibrium geometry of metal surfaces are almost completely lacking.

The experimental evidence suggests that reconstructive rearrangements are favored on fcc metals whereas displacive rearrangements are favored on bcc metals. As discussed in Sects.9.2.2,3, the preferred models for Ir, Au and Pt can be rationalized on the assumption that the surface atoms tend to form a close-packed {111} configuration. The reconstructed surface then has fewer "dangling bonds" per surface atom; apparently, for {110} faces, the resulting reduction in energy more than compensates for the increase in surface area

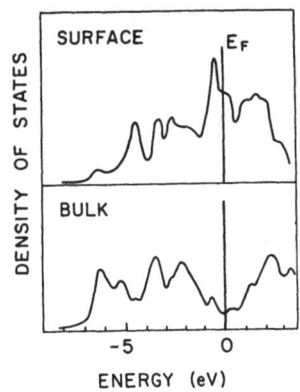

DENSITY OF STATES

SURFACE E_F

BULK

-5   0

ENERGY (eV)

Fig.9.19. Calculated electronic density of states for W{100}. The upper panel shows the DOS for the topmost surface layer. The lower panel shows the corresponding DOS in the bulk crystal [9.126]

which accompanies the microfacetting. Obviously, this cannot be true in general, since no reconstruction is seen on clean Ag and Pd surfaces, for example. Thus, at least a semiquantitative calculation of the electronic energy may be required in each case.

The explanation of the displacive rearrangement on clean W{100} and Mo{100} also depends on a detailed understanding of the surface electronic structure [9.119,124-128]. The density of states (DOS) for these two substrates has been calculated and Fig.9.19 shows representative results for W{100} [9.126]. The DOS of the unreconstructured surface layer (upper curve) is quite different from that of a parallel layer in the bulk (lower curve). In particular, the DOS contains a large peak just below the Fermi energy E_F. A change in atomic geometry may split these states into lower-lying (filled) and higher-lying (empty) states, thereby reducing the total electronic energy. This mechanism is analogous to a Jahn-Teller transition in which a partially filled degenerate level is split into bonding and antibonding states (and it deemphasizes somewhat the role previously assigned to the electrons at the Fermi level [9.8, 124,126,129,130]. On Nb{100} and Ta{100} E_F falls below the peak in the DOS; no reconstruction is therefore expected and none is observed experimentally [9.63,64].

The displacive transformations, driven by lowering the electronic energy, raise the elastic energy by distorting the lattice. To determine the final (equilibrium) geometry, a detailed lattice-dynamical calculation is necessary. Model calculations for W{100} indicate that the $(\sqrt{2} \times \sqrt{2})$ structure indeed is favored over a wide range of values of the force constants [9.113]. This distortion corresponds to the "freezing out" of either M_5 (in-plane displacements) or M_1 (out-of-plane displacements) phonon modes [9.113,116]. Incommensurate structures also appear in such model calculations, made possibly by the difference between the force constants in the top and the lower layers.

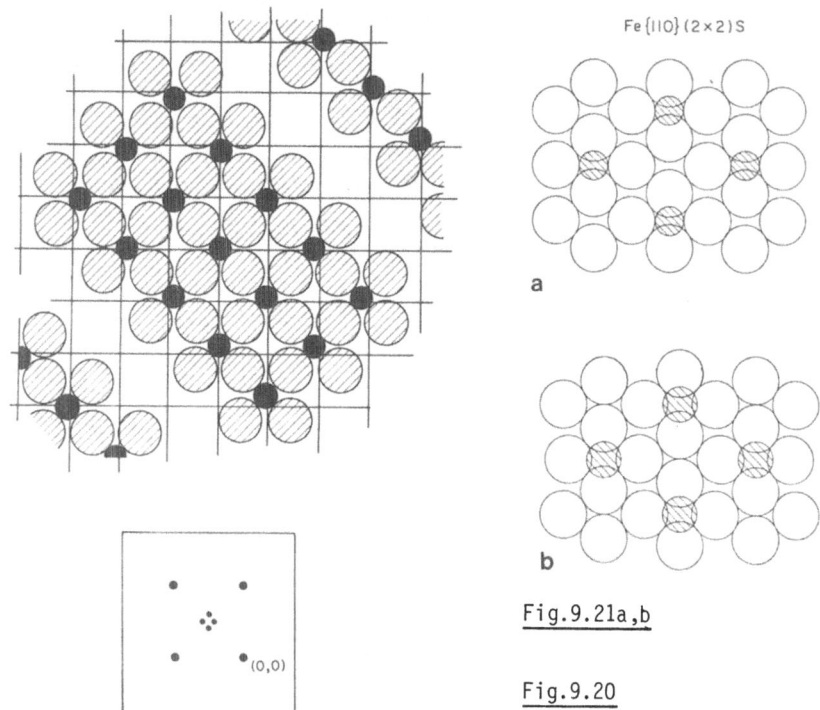

Fe{110}(2×2)S

a

b

Fig.9.21a,b

Fig.9.20

Fig.9.20. Model of the N/W{100} surface at a fractional nitrogen coverage of $\lesssim 0.4$. The filled circles represent N atoms. The large shadded circles represent W atoms in the top layer which for an unreconstructed surface would be located at the center of the squares. The inset below shows the corresponding LEED pattern [9.80]

Fig.9.21a,b. Model of the Fe{110} p(2×2)-S surface. The shaded circles represent S atoms, the large open circles iron atoms in the first substrate layer. a) Unreconstructed surface; b) reconstructed surface [9.78]

Several theoretical studies have been made of the effect of adsorbates on the displacive transformations [9.128,131-133]. For a mobile adsorbate, like hydrogen, on W{100} the adsorbate-substrate interactions can be shown to lead to the experimentally observed switching in the displacement direction, to the increase in T_c for the $(\sqrt{2} \times \sqrt{2})$ structure, and — at higher coverages — to the formation of an incommensurate phase. In this system the adsorbate only modifies the substrate phase transition. However, in cases where the adatom-substrate interaction is very strong it may dominate the equilibrium structure. The N/W{100} surface, a model of which is shown in Fig.9.20, appears to be such a case [9.80]. The top layer of W atoms forms contracted domains, so producing the most favorable N-W coordination. Thus, long-range cooperative effects are probably of minor significance in this reconstruction. Another example

of a reconstruction driven primarily by adsorbate bonding forces is shown
in Fig.9.21b. Sulfur adsorbs on the Fe{110} surface to give a p(2 × 2) periodicity. On the unreconstructed surface (Fig.9.21a) the S-Fe bonds are either
quite long or quite short; the reconstruction positions the substrate atoms
so that the four S-Fe bonds are nearly equivalent and have the most favorable length [9.78]. It should be noted that a similar driving mechanism cannot explain the H/W{100} structure sketched in Fig.9.13b. On that surface the
H atoms form a dilute layer [9.134] and a large fraction of the W dimers do
not have an adatom on top.

9.4 Reconstruction Effects on Chemisorption

As discussed above, an adsorbate may strongly affect the properties of the
substrate. By the same token, substrate reconstruction can profoundly influence the behavior of an adsorbed layer. The existence of this effect was
realized only recently, but a few illustrations will suffice to show its
importance.

Figure 9.22 give some results for the H/Mo{100} surface. The upper curve
shows the coverage dependence at 140 K of the $\frac{1}{2}\frac{1}{2}$ LEED beam intensity; it increases at first as an H-induced reconstruction sets in, and then decreases
as an incommensurate structure is formed and the substrate distortions diminish [9.112]. The curves below (Fig.9.22) show that the H^+ current observed
in ESD (electron-stimulated desorption) behaves in a very similar manner. It
is evident that at 140 K the ESD cross section is a sensitive function of
the substrate structure. The cross section is also seen to decrease with temperature, as a result of the thermally induced phase transitions in the
H/Mo{100} surface [9.112].

Another example is provided by the rate of H_2 desorption from W{100}[9.120].
The desorption activation energy, determined from adsorption isobars, is
found to drop to about half of its initial value at a hydrogen coverage of
$\theta \lesssim 0.5$, as shown in Fig.9.18 (middle panel). This drop coincides with the
reduction in W-atom displacements observed by HEIS [9.73] (upper panel). Thus,
the substrate reconstruction is a dominating factor in the desorption kinetics
of this system. The participation of the substrate degrees of freedom in the
desorption process is demonstrated also by the large entropy change per surface W atom, shown at the bottom of Fig.9.18 [9.120,134,135].

Large reconstruction effects have also been observed in chemisorption
studies on fcc metals. A dramatic example is the kinetic oscillations which

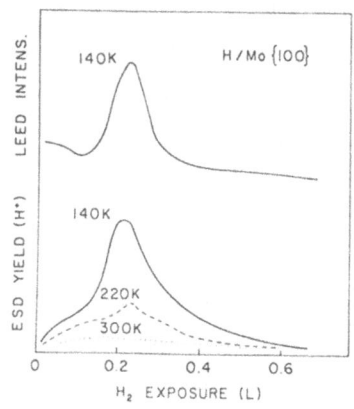

Fig.9.22. Correlation of LEED and ESD data for Mo{100} with adsorbed hydrogen. The abscissa is the hydrogen exposure in L (1 L = 10^{-6} torr s). The upper curve shows the intensity of the $(\frac{1}{2}\frac{1}{2})$ LEED beam at 140 K. The curves in the lower panel show the electron-stimulated desorption yield of H^+ at 140 K, 220 K and 300 K

occur in Pt-catalyzed oxidation of CO, as the {100} substrate changes back and forth between the (1×1) and the (20×5) structures [9.136].

Acknowledgments. The author is indebted to G. Ehrlich, W. Graham, H. Krakauer, H. Rohrer, I. Robinson, and D. Zehner for help in the form of figures, reprints and comments. This work was supported by NSF grant DMR 8305802 and by facilities made available by the Brown University Materials Research Laboratory.

References

9.1   See, for example, H.D. Megaw: *Crystal Structures, A Working Approach* (W.B. Saunders, Philadelphia 1973)
9.2   R.E. Schlier, H.E. Farnsworth: In *Semiconductor Surface Physics* (Univ. of Pennsylvania Press, Philadelphia 1957) p.3; J. Chem. Phys. **30**, 917 (1959)
9.3   For a review, see C.B. Duke: CRC Crit. Rev. Solid State Mater. Sci. **8**, 69 (1978)
9.4   S. Hagstrum, H.B. Lyon, G.A. Somorjai: Phys. Rev. Lett. **15**, 491 (1965)
9.5   D.G. Fedak, N.A. Gjostein: Surf. Sci. **8**, 77 (1967); Acta Metall. **15**, 827 (1967)
9.6   P.W. Palmberg, T.N. Rhodin: Phys. Rev. **161**, 568 (1967)
9.7   J.T. Grant: Surf. Sci. **18**, 228 (1969)
9.8   T.E. Felter, R.A. Barker, P.J. Estrup: Phys. Rev. Lett. **38**, 1138 (1977)
9.9   M.K. Debe, D.A. King: J. Phys. C**10**, L303 (1977)
9.10   R. Ludeke, G. Landgren: Phys. Rev. Lett. **47**, 875 (1981)
9.11   K. Christman, G. Ertl: Z. Naturforsch. **28a**, 1144 (1973)
9.12   C.M. Chan, S.L. Cunningham, K.L. Luke, W.H. Weinberg, S.P. Withrow: Surf. Sci. **78**, 15 (1978)
9.13   C.M. Chan, M.A. Van Hove, W.H. Weinberg, E.D. Williams: J. Vac. Sci. Technol. **16**, 642 (1979)
9.14   C.M. Chan, M.A. Van Hove, W.H. Weinberg, E.D. Williams: Solid State Commun. **30**, 47 (1979)
9.15   C.M. Chan, M.A. Van Hove, W.H. Weinberg, E.D. Williams: Surf. Sci. **91**, 440 (1980)
9.16   J.D. Wrigley, G. Ehrlich: Phys. Rev. Lett. **44**, 661 (1980)
9.17   H.P. Bonzel, S. Ferrer: Surf. Sci. **118**, L263 (1982)

9.18 H.P. Bonzel, R. Ku: Surf. Sci. **33**, 91 (1972)
9.19 D.L. Adams, H.B. Nielsen, M.A. Van Hove, A. Ignatiev: Surf. Sci. **104**, 47 (1981)
9.20 A.M. Lakee, W. Allison, R.F. Willis, K.H. Rieder: Surf. Sci. **126**, 654 (1983)
9.21 W.N. Unertl, T.E. Jackman, P.R. Norton, D.P. Jackson, J.A. Davies: To be published
9.22 D. Wolf, H. Jagodzinsky, W. Moritz: Surf. Sci. **77**, 265 (1978)
9.23 J.R. Noonan, H.L. Davis: J. Vac. Sci. Technol. **16**, 587 (1979)
9.24 W. Moritz, D. Wolf: Surf. Sci. **88**, L29 (1979)
9.25 G. Binning, H. Rohrer, C. Gerber, E. Weibel: Phys. Rev. Lett. **49**, 57 (1982)
9.26 B. Reihl, B.I. Dunlap: Appl. Phys. Lett. **37**, 941 (1980)
9.27 N. Müller, M. Erbudak, D. Wolf: Solid State Commun. **39**, 1247 (1981)
9.28 S.H. Overbury, W. Heiland, D.M. Zehner, S. Datz, R.S. Thoe: Surf. Sci. **109**, 239 (1981)
9.29 I.K. Robinson: Phys. Rev. Lett. **50**, 1145 (1983)
9.30 L.D. Marks, D.J. Smith: Nature **303**, 316 (1983)
9.31 P. Heilmann, K. Heinz, K. Mueller: Surf. Sci. **83**, 487 (1979)
9.32 M.A. Van Hove, R.J. Koestner, P.C. Stair, J.P. Biberian, L.L. Kesmodel, I. Bartos, G.A. Somorjai: Surf. Sci. **103**, 189 (1981)
9.33 M.A. Van Hove, R.J. Koestner, P.C. Stair, J.P. Biberian, L.L. Kesmodel, I. Bartos, G.A. Somorjai: Surf. Sci. **103**, 218 (1981)
9.34 P.R. Norton, J.A. Davies, D.P. Jackson, N. Matsunami, J.U. Andersen: J. Vac. Sci. Technol. **15**, 650 (1978)
9.35 P.R. Norton, J.A. Davies, D.P. Jackson, N. Matsunami: Surf. Sci. **85**, 269 (1979)
9.36 P.R. Norton, J.A. Davies, D.K. Creber, C.W. Sitter, T.E. Jackman: Surf. Sci. **108**, 205 (1981)
9.37 B.R. Appleton, D.M. Zehner, T.S. Noggle, J.W. Miller, O.E. Schow III, L.H. Jenkins, J.H. Barret: In *Ion Beam Surface Layer Analysis*, ed. by O. Meyer, G. Linkes, F. Kappler (Plenum, New York 1976) Vol.2, p.607
9.38 J.F. Wendelken, D.M. Zehner: Surf. Sci. **71**, 178 (1978)
9.39 K.H. Rieder, T. Engle, R.H. Swendsen, M. Manninen: Surf. Sci. **127**, 223 (1983)
9.40 J. Perdereau, J.P. Biberian, G.E. Rhead: J. Phys. F**4**, 798 (1978)
9.41 D.M. Zehner, J.F. Wendelken: Proc. 7th Internat. Vac. Congr. and 3rd Internat. Conf. of Solid Surfaces (Vienna 1977) p.517
9.42 P.W. Davies, R.M. Lambert: Surf. Sci. **95**, 571 (1980)
9.43 V. Jensen, J.N. Andersen, H.B. Nielson, D.L. Adams: Surf. Sci. **116**, 66 (1982)
9.44 G. Gewinner, J.C. Peruchetti, A. Jaegle, R. Riedinger: Phys. Rev. Lett. **43**, 935 (1979)
9.45 J.S. Foord, A.P.C. Reed, R.M. Lambert: Surf. Sci. **129**, 79 (1983)
9.46 R.A. Barker, S. Semancik, P.J. Estrup: Surf. Sci. **94**, L 162 (1980)
9.47 M.K. Debe, D.A. King: Phys. Rev. Lett. **39**, 708 (1977)
9.48 J.A. Walker, M.K. Debe, D.A. King: Surf. Sci. **104**, 504 (1981)
9.49 D.P. Woodruff: Surf. Sci. **122**, L 653 (1982)
9.50 F. Jona: J. Phys. C. Solid State Phys. **11**, 4271 (1978)
9.51 M.A. VanHove, S.Y. Tong: *Surface Crystallography by LEED*, Springer Ser. Chem. Phys. **2** (Springer, Berlin, Heidelberg, New York 1979)
9.52 M.A. Van Hove: In *The Nature of the Surface Chemical Bond*, ed. by T.N. Rhodin, G. Ertl (North-Holland, Amsterdam 1979) p.275
9.53 K. Heinz, K. Müller: In *Structural Studies of Surfaces*, Springer Tracts in Mod. Phys. **91** (Springer, Berlin, Heidelberg, New York 1982) p.1
9.54 L.C. Feldman, J.W. Mayer, S.T. Picraux: *Materials Analysis by Ion Channeling* (Academic, New York 1982)
9.55 T.T. Tsong, J. Sweeney: Solid State Commun. **30**, 767 (1979)

9.56 R.T. Tung, W.R. Graham, A.J. Melmed: Surf. Sci. **115**, 576 (1982)
9.57 Chi-fong Ai, T.T. Tsong: Surf. Sci. **127**, L 165 (1983)
9.58 T. Engel, K.H. Rieder: In *Structural Studies of Surfaces*, Springer
     Tracts in Mod. Phys. **91** (Springer, Berlin, Heidelberg, New York 1982)
     p.55
9.59 K. Takayanagi: In *Electron Microscopy 1982* (Deutsche Gesellschaft f.
     Elektronenmikroskopie, Frankfurt) Vol.1, 43 (1982)
9.60 G. Binnig, H. Rohrer: Surf. Sci. **126**, 236 (1983)
9.61 P. Heimann, J.F. Van der Veen, D.E. Eastman: Solid State Commun. **38**,
     595 (1981)
9.62 J. Küppers, H. Michel: Appl. Surf. Sci. **3**, 179 (1979)
9.63 A.J. Melmed, S.T. Ceyer, R.T. Tung, W.R. Graham: Surf. Sci. **III**, L 701
     (1981)
9.64 A. Titov, W. Moritz: Surf. Sci. **123**, L 709 (1982)
9.65 L. Gonzalez, R. Miranda, M. Salmeron, J.A. Verges, F. Yndurain: Phys.
     Rev. B**24**, 3245 (1981)
9.66 Y. Kuk, L.C. Feldman: Private communication
9.67 T.N. Taylor, P.J. Estrup: J. Vac. Sci. Technol. **11**, 244 (1974)
9.68 J. Demuth: J. Colloid Interface Sci. **58**, 184 (1977)
9.69 T. Engel, K.H. Rieder: Surf. Sci. **109**, 140 (1981)
9.70 M.G. Cattania, V. Penka, R.J. Behm, K. Christmann, G. Ertl: Surf. Sci.
     126, 382 (1983)
9.71 R.A. Barker, P.J. Estrup: Phys. Rev. Lett. **41**, 130) (1978)
9.72 R.A. Barker, P.J. Estrup, F. Jona, P.M. Marcus: Solid State Commun.
     25, 375 (1978)
9.73 I. Stensgaard, L.C. Feldman, P.J. Silverman: Phys. Rev. Lett. **42**, 247
     (1979)
9.74 R.F. Willis: Surf. Sci. **89**, 457 (1979)
9.75 P.J. Estrup, R.A. Barker: In *Ordering in Two Dimensions*, ed. by S.K.
     Sinha (North-Holland, Amsterdam 1980) p.39
9.76 D.A. King, G. Thomas: Surf. Sci. **92**, 201 (1980)
9.77 R.A. Barker, P.J. Estrup: J. Chem. Phys. **74**, 1442 (1982)
9.78 H.D. Shih, F. Jona, D.W. Jepsen, P.M. Marcus: Phys. Rev. Lett. **46**, 731
     (1981)
9.79 S. Semancik, P.J. Estrup: J. Vac. Sci. Technol. **18**, 541 (1981)
9.80 K. Griffiths, C. Kendon, D.A. King, J.B. Pendry: Phys. Rev. Lett. **46**,
     1584 (1981)
9.81 R.J. Smith, S. Fine, N. Holland, M.W. Kim: Bull. Am. Phys. Soc. **28**
     (3), 470 (1983)
9.82 U. Landman, R.N. Hill, M. Mostoller: Phys. Rev. B**21**, 448 (1980);
     R.N. Barnett, U. Landman, C.L. Cleveland: To be published
9.83 H.L. Davis, J.R. Noonan: Surf. Sci. **126**, 245 (1983)
9.84 I. Stensgaard, R. Feidenhansl, J.E. Sörensen: Surf. Sci. **128**, 281 (1983)
9.85 D.L. Adams, H.B. Nielsen, J.N. Andersen: Surf. Sci. **128**, 294 (1983)
9.86 E. Zanazzi, F. Jona: Surf. Sci. **62**, 61 (1977);
     F. Jona, H.D. Shih: J. Vac. Sci. Technol. **16**, 1248 (1979)
9.87 D.L. Adams, H.B. Nielsen, M.A. Van Hove: Phys. Rev. **820**, 4789 (1979)
9.88 J.D. Wrigley: Ph.D. Thesis, University of Illinois 1982)
9.89 G. Binnig, H. Rohrer: Surf. Sci. **126**, 236 (1983); Helv. Phys. Acta **55**,
     726 (1982)
9.90 G. Binnig, H. Rohrer, C. Gerber, E. Weibel: Surf. Sci. **131**, L379 (1983)
9.91 J.L. Taylor, D.E. Ibbotson, W.H. Weinberg: J. Chem. Phys. **69**, 4298
     (1978)
9.92 W. Erley: Surf. Sci. **114**, 47 (1982)
9.93 D.P. Jackson, T.E. Jackman: J.A. Davies, W.N. Unertl, P.R. Norton:
     Surf. Sci. **126**, 226 (1983)
9.94 D. Wolf, H. Jagodzinski, W. Moritz: Surf. Sci. **77**, 283 (1978)
9.95 J.R. Noonan: Private communication

9.96 I.K. Robinson: Private communication
9.97 T.N. Rhodin, G. Broden: Surf. Sci. **60**, 466 (1976)
9.98 H.P. Bonzel, C.R. Helms, S. Keleman: Phys. Rev. Lett. **35**, 1237 (1935)
9.99 J.F. Wendelken, D.M. Zehner: Surf. Sci. **71**, 178 (1978)
9.100 J.J. Burton, G. Jura: In *Structure and Chemistry of Solid Surfaces*, ed. by G.A. Somorjai (Wiley, New York 1969) pp.21-1 ???
9.101 G. Binnig, H. Rohrer: To be published
9.102 A. Baratoff: To be published
9.103 H.B. Lyon, G.A. Somorjai: J. Chem. Phys. **46**, 2539 (1967)
9.104 M.A. Barteau, E.I. Ko, R.J. Madix: Surf. Sci. **102**, 99 (1981)
9.105 G. Broden, G. Pirug, H.P. Bonzel: Surf. Sci. **72**, 45 (1978)
9.106 P.A. Thiel, R.J. Behm, P.R. Norton, G. Ertl: Surf. Sci. **121**, L553 (1982)
9.107 H.B. Nielsen, D.L. Adams: Surf. Sci. **97**, L351 (1980)
9.108 J.C. Heyraud, J.J. Metois: Surf. Sci. **100**, 519 (1980)
9.109 K. Yonehara, L.D. Schmidt: Surf. Sci. **25**, 238 (1971)
9.110 G.-C. Wang, T.-M. Lu: Surf. Sci. **122**, L635 (1982)
9.111 J.A. Walker, M.K. Debe, D.A. King: Surf. Sci. **104**, 405 (1981)
9.112 P.J. Estrup: J. Vac. Sci. Technol. **16**, 635 (1979)
9.113 A. Fasolino, G. Santaro, E. Tosatti: Phys. Rev. Lett. **44**, 1684 (1980)
9.114 P.J. Estrup, L.D. Roelofs, S.C. Ying: Surf. Sci. **123**, L703 (1982)
9.115 A.J. Melmed, W.R. Graham: Surf. Sci. **123**, L706 (1982)
9.116 A. Fasolino, G. Santaro, E. Tosatti: In Proc. Internat. Conf. on Phonon Physics, Bloomington, Indiana (to be published)
9.117 M.K. Debe, D.A. King: Surf. Sci. **81**, 193 (1979)
9.118 P. Heilman, K. Heinz, K. Müller: Surf. Sci. **89**, 84 (1974); Proc. Conf. on Determination of Surface Structure by LEED, 1980, ed. by P. Marcus (to be published)
9.119 L.D. Roelofs, P.J. Estrup: Surf. Sci. **125**, 51 (1983)
9.120 A. Horlacher Smith, R.A. Barker, P.J. Estrup: Surf. Sci. **136**, 327 (1984)
9.121 S.C. Ying, L.D. Roelofs: Surf. Sci. **125**, 218 (1983)
9.122 T.M. Gardiner, E. Bauer: Surf. Sci. **119**, L353 (1982)
9.123 G.C. Wang, T.M. Lu: Surf. Sci. **122**, L635 (1982)
9.124 E. Tosatti: Solid State Commun. **25**, 637 (1978)
9.125 J.E. Inglesfield: J. Phys. C**12**, 149 (1979)
9.126 H. Krakauer, M. Posternak, A.J. Freeman: In *Ordering in Two Dimensions*, ed. by S. Sinha (Nort-Holland, Amsterdam 1980) p.47
9.127 I. Terakura, K. Terakura, N. Hamada: Surf. Sci. **111**, 479 (1981)
9.128 A. Fasolino, G. Santaro, E. Tosatti: Surf. Sci. **125**, 317 (1983)
9.129 J.C. Campuzano, D.A. King, C. Somerton, J.E. Inglesfield: Phys. Rev. Lett. **45**, 1649 (1980)
9.130 M.I. Holmes, T. Gustafsson: Phys. Rev. Lett. **47**, 443 (1981)
9.131 K.H. Lau, S.C. Ying: Phys. Rev. Lett. **44**, 1222 (1980)
9.132 T. Inaoka, A. Yoshimori: Surf. Sci. **115**, 301 (1982)
9.133 S.C. Ying, L.D. Roelofs: Surf. Sci. **125**, 218 (1983)
9.134 A. Horlacher Smith, J.W. Chung, P.J. Estrup: J. Vac. Sci. Technol. **A2**, 877 (1984)
9.135 P.J. Estrup: (to be published)
9.136 G. Ertl, P.R. Norton, J. Rüstig: Phys. Rev. Lett. **49**, 177 (1982)

10. Surface Crystallography by Means of SEXAFS and NEXAFS

J. Stöhr

With 13 Figures

10.1 Introduction

The availability of high brightness (flux/unit area) monochromatized synchrotron radiation [10.1] has allowed the development of new experimental techniques which tune into specific surface atoms and probe their structural environment. Two such techniques, the subject of this chapter, are the surface extended X-ray absorption fine structure (SEXAFS) and near-edge X-ray absorption fine structure (NEXAFS) techniques. Both measure the X-ray absorption by a specific atom on the surface which is distinguished from other atoms by one of its main absorption edges (usually K or L edge).

SEXAFS [10.2,3] is simply based on the application of the well-established extended X-ray absorption fine structure (EXAFS) [10.4,5] technique to surfaces, and it is therefore governed by the same physical processes, theory and data analysis procedures as bulk EXAFS. For this reason this chapter will not review the concepts of EXAFS spectroscopy, but rather the reader is referred to extensive reviews elsewhere [10.5]. Here we address only the specific concepts and experimental details which distinguish SEXAFS from EXAFS. Examples will be given with emphasis on exploiting concepts specific to surface crystallographic determinations.

NEXAFS refers to the detailed structures near the absorption edge of an EXAFS or SEXAFS spectrum. Previously, some researchers referred to NEXAFS also as the X-ray absorption near-edge structure (XANES) [10.6-8]. The fine structure of the edge is, in general, determined by complex scattering processes of the photoelectron created by X-ray absorption. At present it appears that the NEXAFS structures are best explained by a multiple-scattering theory [10.7-9] which takes into account not only the backscattering from the atomic cores of the neighbors (like EXAFS) but also the detailed charge distribution of the valence electrons. Such calculations have been successfully applied to explain the NEXAFS of diatomic molecules [10.9,10] molecular complexes [10.11,12] and crystalline [10.13] and amorphous [10.14] solids.

It has been proposed [10.15] that a conceptually simpler single-scattering formalism, similar to that used in EXAFS, may in fact be adequate to describe the NEXAFS spectra. At present the suitability of the single-scattering formalism is still an open question. The most pronounced and yet most easily understood near-edge structures are observed for chemisorbed molecules [10.16, 17] and it is NEXAFS studies of such sytems which will be discussed here. Examples will be given of chemisorbed di- and polyatomic molecules on different metal surfaces to illustrate the unique structural and bonding information contained in the NEXAFS spectra.

The structure of this chapter is as follows. Section 10.2 outlines the physical principles and processes which determine the NEXAFS and SEXAFS spectra. The measurement technique and experimental details are discussed in Sect.10.3. Section 10.4 will illustrate the application of the techniques to elucidate the structure of surface complexes formed by gas-solid and solid-solid interactions. Conclusions are drawn in Sect.10.5

10.2 Principles of SEXAFS and NEXAFS

10.2.1 Physical Processes which Determine SEXAFS and NEXAFS

When the energy of the incident X-ray radiation exceeds the excitation threshold of a core electron of an atom, a photoelectron is created. This photoelectron leaves behind a core hole, and we shall define the probability of exciting the photoelectron or equivalently of creating the core hole as the *absorption coefficient* of a particular shell (e.g., K, L or M). NEXAFS and SEXAFS are based on the measurement of the absorption coefficient of a surface atom near or above its characteristic excitation threshold ("absorption edge").

For an isolated atom, the absorption coefficient above the absorption edge is a smoothly varying function mostly determined by the overlap of the initial state wave function of the core electron and the final state wave function of the photoelectron. If the absorbing atom is bonded to any neighbors, the final state wave function of the escaping photoelectron may be modified by scattering due to the neighbor atoms, Fig.10.1. The detailed scattering processes depend on the kinetic energy of the excited photoelectron. The NEXAFS and SEXAFS regimes of the absorption spectrum shown in Fig.10.1 are characterized by different excitation energies (kinetic energies) above the absorption threshold and therefore correspond to different scattering processes.

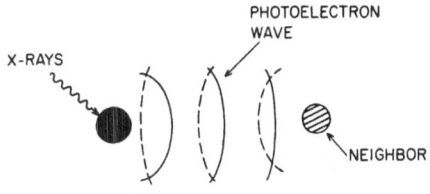

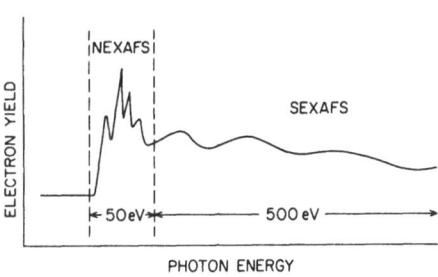

Fig.10.1. Absorption of an X-ray by an atom results in emission of a photoelectron. The interference of the outgoing and backscattered photoelectron wave by a neighbor atom leads to modulation of the X-ray absorption coefficient, called NEXAFS, in the region just above a core absorption threshold (edge), and (S)EXAFS, well above the edge

In the NEXAFS regime, the photoelectron kinetic energy is comparable to valence electron binding energies, and the scattering processes of the photoelectron are thus influenced by the potential due to the detailed valence electron charge distribution. The NEXAFS regime is characterized by complicated strong multiple-scattering processes. Because photoelectron mean free paths are long at low kinetic energies ($\lesssim$ 50 eV), the NEXAFS is, in general, also influenced by scattering contributions from neighbors which are more than ~5 Å from the absorbing atom. For general surface complexes, the calculation of the NEXAFS spectrum is a nontrivial task [10.7,8] and in its complexity, it is similar to a full dynamical calculation of low-energy electron diffraction (LEED) intensities [10.18] or a calculation of normal photoelectron diffraction intensities [10.19]. As discussed below, the case of a chemisorbed molecule is a particularly simple exception, because the NEXAFS is dominated by *intramolecular* scattering processes, and in first order, the effect of the substrate is merely to hold the molecule and orient it. Thus, the theory developed for gas-phase molecules [10.9,10] can be applied to interpret NEXAFS spectra of chemisorbed molecules.

In contrast, the EXAFS or SEXAFS regime is characterized by single-scattering processes of the photoelectron with high kinetic energy off the atomic cores of neighbor atoms. Because the electron mean free paths are at a minimum over most of the SEXAFS range (50-300 eV), usually only the closest (<4 Å) neighbors contribute.

To illustrate typical excitation processes which determine the NEXAFS and SEXAFS spectrum, Fig.10.2 shows a diatomic molecule with internuclear spacing

233

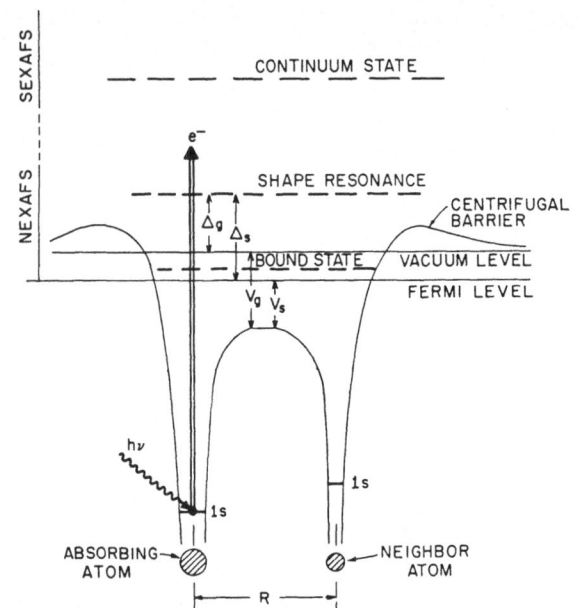

Fig.10.2. Potential wells in a diatomic molecule. Energies are referenced to the vacuum level (E_V) in gas phase and to the Fermi level (E_F) in chemisorbed molecules. NEXAFS spectra are usually dominated by two resonances. One corresponds to transitions to an empty or partially empty molecular state with a transition energy less than the core ionization energy (bound state transition), the other to an empty molecular state in the continuum (shape resonance). The latter is best described as a multiple-scattering resonance. The (S)EXAFS region corresponds to higher energy transitions into continuum states which can be described by plane waves and a single-scattering theory. The molecular potential in the interstitial region between the two atoms is nearly constant. Its separation from E_V or E_F is called the "inner potential" and defines the zero of kinetic energy of the photoelectron

R. The lowest energy absorption structure usually corresponds to a transition of a 1s electron into an empty or partially filled molecular orbital [10.20]. Since the excitation energy for this transition is less than the 1s ionization energy (binding energy relative to the vacuum level), the observed absorption structure is called a bound state resonance. The bound final state of the excited core electron has a long lifetime on the molecule, and the excitation energy is lowered below the core ionization energy by the Coulomb interaction with the core hole. For gas-phase molecules, narrow bound state transition often with giant intensities may be found up to ~10 eV below the core ionization threshold [10.20]. Both the initial and final states involved in the bound state transition have a well-defined symmetry. Thus, for chemisorbed molecules, dipole selection rules can be applied to deduce the molecular orientation on the surface. This is done by measuring the bound state resonance intensity as a function of the electric field vector (**E**) orientation of the linearly polarized synchrotron radiation. Furthermore, the presence or absence of final states of a particular symmetry may be used in some cases to gain information on the hybridization of the molecular bond. For example, for molecules with π bonding the K-edge NEXAFS spectrum will exhibit a strong bound state resonance which corresponds to a transition of a 1s electron into the antibonding π^* orbital. The π resonance is maximized if the **E** vector is perpendicular to the molecular symmetry axis and it vanishes for **E** parallel to it. The presence or absence of the π^* resonance is thus a direct indication of the hybridization of the bond.

For excitation energies exceeding the ionization energy, other pronounced structures are typically found. These structures in the continuum were first associated by *Nefedov* [10.21] and *Dehmer* [10.22] with final states trapped on the molecule by an effective potential barrier. Such quasi-bound states can decay away from the molecule with a lifetime determined by the tunneling probability through the barrier (Fig.10.2). A more general and more quantitative description of the resonances was first suggested by *Dill* and *Dehmer* [10.23] who used multiple-scattering theory. They also introduced the name *shape resonances* in analogy to those observed in electron scattering by gas-phase molecules. Currently, the shape resonances are best thought of as arising from resonant back-and-forth scattering of the photoelectron wave between the absorbing atom and its neighbor(s). This simple scattering picture suggests that the resonance position should be correlated with the intramolecular distance(s). As discussed in more detail below, this correlation indeed exists, and it leads to a powerful, simple approach to determine intramolecular bond length changes induced by the chemisorption bond. Shape resonances exist for all molecules and are strongest when the **E** vector lies along the internuclear axis between the neighboring atoms. The polarization dependence of the shape resonance intensity can therefore be used to determine the molecular orientation of chemisorbed molecules.

The discussion above depicts the traditional picture where bound state resonances are associated with transitions to at least partially empty molecular orbitals and shape resonances with continuum states which, through resonance scattering, have an enhanced amplitude on the molecule. Closer inspection of experimental results for gas-phase molecules show that this separation of bound state resonances and shape resonances is somewhat artificial, since in some cases (e.g., O_2 molecule), the K-edge shape resonance falls below the 1s ionization threshold, i.e., is a bound state resonance [10.24]. In addition, for chemisorbed molecules, core electron binding energies are usually referenced to the Fermi level and not to the vacuum level, and the separation into bound state and continuum resonances is less meaningful. In a more generalized picture, the near-edge resonances in molecules can be viewed as scattering resonances characterized by different symmetries and lifetimes of the final photoelectron state.

The multiple-scattering theory describing NEXAFS is the most general formulation of the scattering processes which determine the X-ray absorption spectrum. For excitation energies well above ($\gtrsim 50$ eV) the edge, this general theory can be simplified, because the scattering becomes insensitive to the valence electron charge distribution, and the dominant scattering process is that in which the photoelectron is backscattered by the core potential of the neighbor atoms. The spherical photoelectron wave created on the absorbing atom can be approximated by a plane wave, and the interference of the outgoing and backscattered wave gives rise to a periodic modulation of the absorption coefficient (i.e., EXAFS).

10.2.2 The Basic Equations

The multiple-scattering theory used to describe the NEXAFS of molecules extends the concepts of electron scattering by a single center with spherical symmetry to many scattering centers characterized by their phase shifts. The photoelectron motion in the interstitial region between the centers is approximated by free propagation with wave vector k, which is related to the kinetic energy E_k of the photoelectron by

$$k = 0.5123\sqrt{E_k} \quad , \tag{10.1}$$

where k is in Å^{-1} and E_k is in eV units. The potential in the interstitial region (Fig.10.2) is given by the constant muffin-tin *inner potential* $V_g (<0)$ referenced to the vacuum level for gas-phase molecules [10.8]. For chemisorbed molecules, the inner potential V_s ($V_s = V_g - \phi$, where ϕ is the work function) is referenced to the Fermi level. The inner potential determines the zero of kinetic energy and wave vector of the photoelectron. Recently, *Natoli* [10.8] derived a particularly simple correlation between the intramolecular bond length R in a diatomic molecule and the photoelectron wave vector k_r at the σ shape resonance excitation. He pointed out that within the framework of multiple-scattering theory, the maximum in absorption or scattering cross section corresponds to a pole in the scattering matrix. This leads to an equation which links k_r, R and the atomic scattering phase shifts $\phi_i(k)$ of the absorbing atom and its neighbor. If the k dependence of $\phi(k)$ is smooth, a particularly simple expression is obtained

$$k_r R = \text{const.} \quad , \tag{10.2}$$

where the constant depends on the phase shifts $\phi_i(k)$. If, as shown in Fig. 10.2, we define Δ_g (Δ_s) as the energy difference between the σ shape resonance excitation energy E_σ and the 1s binding energy E_B^{1s} relative to the vacuum level (Fermi level), we can rewrite (10.2) as

$$(\Delta_g - V_g)R^2 = C_0 = \text{const.} \tag{10.3a}$$

for gas-phase molecules or

$$(\Delta_s - V_s)R^2 = C_0 = \text{const.} \tag{10.3b}$$

for chemisorbed molecules. The value of E_B^{1s} can be obtained from XPS [10.25]. For chemisorbed molecules, the onset of the K absorption corresponds to transitions to states just above the Fermi level, and it can therefore be used as a measure of E_B^{1s}. Equation (10.3) can be used to determine the intramolecular bond length R from the measured σ shape resonance position Δ_g (Δ_s) relative to the vacuum level (Fermi level) provided the constants V_g (V_s) and C_0 are known. Here it should be pointed out that from a practical point of view, the shape resonances due to scattering processes involving H atoms are weak and can be neglected. Therefore, in the following we consider only shape resonances arising from scattering processes between heavier atoms (i.e., C, N and O).

The inner potential V_g for *gas-phase* molecules unfortunately depends not only on the atomic pair involved in the scattering process (e.g., C-C versus C-O) but also on the details of the intramolecular bond between these atoms (e.g., triple bond in C_2H_2 versus single bond in C_2H_6). By suitable calibration procedures, empirical rules for V_g can be derived which allow bond lengths in molecules to be determined with an accuracy of <0.05 Å [10.24]. For *chemisorbed* molecules, it has been found empirically [10.26] that V_s is less dependent on the atomic pair involved in the scattering and on the bond order (as long as we are dealing with atoms with similar atomic number, e.g., C, N or O). It appears that V_s for chemisorbed molecules is affected by the metallic charge which tends to equilibrate the inner potential of the free molecule with that of the metal ($V_s \sim -9$ eV) [10.26,27].

The constant C_0 in (10.3) depends on the *atomic* scattering phase shifts $\phi_i(k)$ and hence is fixed for a given atomic pair. Because the scattering phase shifts for C, N and O are very similar, the constant C_0 will be nearly the same for different low-Z molecules. Furthermore, because of the small energy range in which the shape resonance occurs (<15 eV), it will be a good approximation to neglect the k dependence of $\phi(k)$. We have found empirically [10.26,27] that the value of $C_0 = 36$ eV $Å^2$ describes all studied cases of C, N and O containing molecules.

The basic equations describing the EXAFS signal are well established. For s initial states (K or L_1 edges), the EXAFS signal as a function of the photoelectron wave vector k is given by [10.5]

$$\chi(k) = - \sum_i A_i(k) \sin(2kR_i + \tau_i(k)) \quad . \tag{10.4}$$

The summation extends over all neighbor shells i separated from the absorbing atom by a distance R_i, and $\tau_i(k)$ is the total phase shift which the photoelectron wave experiences from the absorbing and backscattering atoms.

Equation (10.4) shows that the EXAFS signal from a given neighbor shell i is characterized by an amplitude function $A_i(k)$ (discussed below) and a phase function $2kR_i + \tau_i(k)$. To determine R_i, it is necessary to know $\tau_i(k)$ or vice versa. Since the EXAFS signal is determined by scattering processes off the atomic cores, the phase shifts $\tau_i(k)$ are transferable for a given absorber and backscatterer pair, independent of the bonding between the atoms [10.5]. Thus, $\tau_i(k)$ can be derived from a model compound of known structures (i.e., R) and used for the EXAFS analysis of a structurally unknown system. This important phase shift transferability concept allows a theory-independent determination of neighbor distances by (S)EXAFS.

The amplitude $A_i(k)$ in (10.4) is given by

$$A_i(k) = (N_i^*/kR_i^2)F_i(k) \, e^{-2\sigma_i^2 k^2} \, e^{-2R_i/\lambda(k)} \quad . \tag{10.5}$$

Here $F_i(k)$ is the backscattering amplitude of the neighbor atoms and the exponential terms in (10.5) are the Debye-Waller-like term and the damping term due to inelastic scattering [mean free path $\lambda(k)$] of the photoelectrons.

For SEXAFS the parameter N_i^* in (10.5) is of particular importance: it is the effective coordination number of the absorbing atom at a distance R_i (i^{th} neighbor shell), given by

$$N_i^* = 3 \sum_{j=1}^{N_i} \cos^2 \alpha_j \quad . \tag{10.6}$$

Here, the sum extends over all neighbor atoms j (total number N_i) in the ith shell and α_j is the angle between the electric field vector **E** of the X-rays at the central atom site and the vector R_{ij} from the central atom to the jth atom in the ith shell. The E vector can be envisioned as a "search light" revealing all neighbors in a given direction.

For single-crystal materials, or in the case of oriented molecules or atoms on surfaces, (10.6) has to be evaluated for an assumed model geometry. For higher than twofold symmetry N^* does not depend on the azimuthal substrate

orientation with respect to **E**, but only on the polar orientation θ. If θ is
the angle between **E** and the surface normal and β is the angle between the
adsorbate-substrate internuclear axis and the surface normal, the following
expression arises [10.3]

$$N_i^* = 3N_i \cos^2\theta\cos^2\beta + (3N_i/2) \sin^2\theta\sin^2\beta \quad . \tag{10.7}$$

Equation (10.5) shows that the coordination N_i (i.e., the chemisorption site)
can be determined from analysis of the amplitude of the SEXAFS oscillations.
In practice, one measures the SEXAFS amplitude for two different **E** vector
orientations with respect to the surface normal (i.e., normal incidence
$\theta = 90°$, and grazing incidence $\theta \leq 20°$). Comparison of the measured amplitude
ratio and that calculated (10.5,6) for different chemisorption geometries is
often sufficient to determine unambiguously the chemisorption site. In addition, the measured amplitudes can be compared to that measured for a model
compound. This allows one to determine the absolute coordination number on
the surface [10.3].

10.3 Measurement Technique

For surface crystallographic studies, a detection technique is needed which
enables one to distinguish and optimize the signal from surface atoms relative to the unwanted background signal from the bulk. Since electrons have
a shorter scattering length in solids than X-rays, it is advantageous to use
the electron-yield signal from the sample as a monitor of the absorption coefficient.

The X-ray absorption process by a core electron in shell A of an atom is
depicted in Fig.10.3. The hole left behind by the excited photoelectron can
be filled by a radiative fluorescent X-ray process or by a nonradiative Auger
deexcitation process. Since the statistical average of both processes is proportional to the annihilation of the core hole, the number of Auger electrons
or fluorescent X-rays is a direct measure of the absorption coefficient
(i.e., the probability that the core hole exists). SEXAFS and NEXAFS may
therefore be carried out by monitoring the Auger electron intensity of a surface atom as a function of photon energy. Other detection variants [10.3]
such as partial or total electron-yield collection also measure the surface
absorption coefficient, because their dominant contribution originates from
the Auger deexcitation channel [10.28].

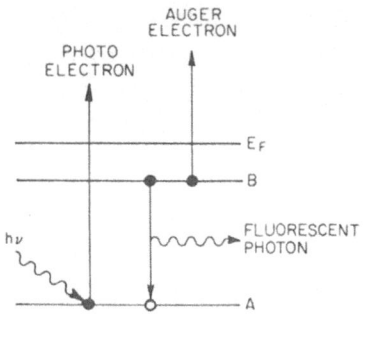

AUGER ELECTRON

PHOTO ELECTRON

E_F

B

FLUORESCENT PHOTON

$h\nu$

A

◄ Fig.10.3. Schema of a photon absorption process by a core electron and the annihilation processes of the created core hole

Fig.10.4. Experimental arrangement for electron yield SEXAFS and NEXAFS studies. Electrons are detected with a CMA for elastic Auger yield measurements and with a two-grid retarding detector for total yield and partial Auger yield studies. Ions may be detected and mass analyzed with the time-of-flight detector opposite the CMA ▼

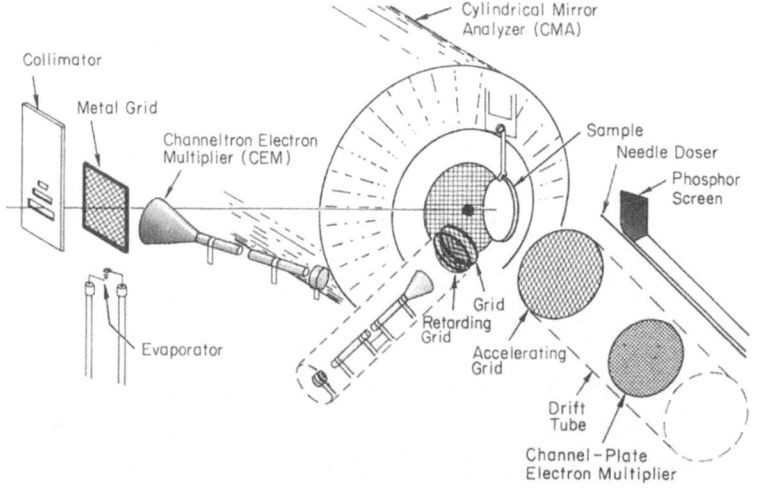

Cylindrical Mirror Analyzer (CMA)

Collimator

Metal Grid

Channeltron Electron Multiplier (CEM)

Sample
Needle Doser

Phosphor Screen

Evaporator

Grid
Retarding Grid

Accelerating Grid

Drift Tube

Channel–Plate Electron Multiplier

Because of the finite sampling depth L (typically 5-100 Å, depending on the detection mode), the measured electron-yield signal will often contain only a small fraction (1%-50%) from the outermost surface layer. Therefore, electron-yield NEXAFS and SEXAFS are not well suited for the study of *clean* surfaces. The power of the techniques lies in the ability to study the environment of atoms which are present only in the surface complex and not in the bulk substrate. In this case, the surface atoms can be selected by their atom-specific absorption edge.

The electron-yield measurements reported below were carried out using the experimental arrangement shown in Fig.10.4. (Ion-yield SEXAFS studies [10.29] are not discussed here.) The monochromatic X-ray beam coming from the monochromator is first trimmed to reduce scattered light. The transmitted beam impinges on a high transmission (80%) metal grid which can be coated in situ. The electron yield from this grid amplified by a channeltron electron multiplier serves as a dynamic intensity monitor [10.30]. The coating material is chosen so as not to exhibit any absorption edges in the photon energy range of interest. For studies of the C, N and O K edges, Cu is a good coating material. The X-ray beam is then incident on the sample. The total or partial electron-yield signal from the sample is measured with a simple retarding grid detector consisting of two hemispherical grids and a high gain spiraltron electron multiplier. For total yield detection, the grids are operated at a small, positive voltage. For partial yield detection, the first grid is kept at ground and the second grid at the chosen retarding voltage (typically -400 V). In general, the output signal of the electron multiplier is high enough to employ current measurement techniques using a floating battery box and a current amplifier [10.31]. Typical output currents in our measurements were of the order of 10^{-8} - 10^{-7} A. Auger yield SEXAFS measurements are carried out using an electron energy analyzer, e.g., a cylindrical mirror analyzer (CMA). The analyzer window is kept fixed at the kinetic energy of the elastic Auger electrons from the adsorbate.

10.4 Applications

In this section I present some selected examples of surface studies by NEXAFS and SEXAFS. The NEXAFS studies deal with both chemisorbed low-Z atoms and molecules on metal surfaces but focus on the latter. Such systems can be regarded as models for basic surface complexes present in heterogeneous catalysis. The molecular NEXAFS spectra are quite easily understood, and we believe that in the future the most important and powerful application of NEXAFS will lie in the monitoring of molecular reactions on surfaces.

In particular, results will be presented for chemisorbed atomic (O on Ni{100}), [10.32] di- and pseudodiatomic (CO and CH_3O on Cu{100}) [10.27] and polyatomic (C_6H_6 and C_5H_5N on Pt{111}) [10.33] species.

Unfortunately, at present almost no SEXAFS studies of chemisorbed low-Z molecules have been carried out. Such experiments have been impeded by the low photon flux (1×10^9 photons/s) available in the spectral range 250-1000 eV where the K edges of C (285 eV), N (400 eV) and O (530 eV) are located. Because the SEXAFS structures are about an order of magnitude smaller and stretched out over a ten times larger energy range, significantly higher photon flux is needed than for NEXAFS. SEXAFS studies for CH_3O/Cu{100} [10.34] and O_2/Pt{111} [10.35] yielded the oxygen-metal distance but did not reveal the intramolecular bond length. Because, at present, such studies and those of the chemisorbed low-Z atoms C, N and O [10.3,36] are still plagued by

signal-to-noise problems, we shall not discuss them here. Rather, we present
SEXAFS results obtained at higher X-ray energies (hν > 3000 eV) where the
available high photon flux (>1 × 10^{11} photons/s) allows one to collect data
with good signal-to-noise ratios. Thus, the potential of SEXAFS for surface
crystallography is best revealed. It is clear that in the near future similar structure determinations will be possible for low-Z atoms and molecules.
The presented SEXAFS studies involve different types of surface complexes
such as clusters with no long-range order (Cu on amorphous graphite) [10.37],
chemisorption systems with well-defined registry with respect to the substrate
(Cl on Si{111})[10.38] and nonreactive (Ag) and reactive (Pd) interface formation of one solid with another (Si) [10.39-41].

10.4.1 NEXAFS Studies

10.4.1.1 Atomic Chemisorption

The first example is of a periodic atomic adsorbate (oxygen) layer on a
single-crystal surface (Ni{100}). For such a system, the NEXAFS spectrum is
determined by single- and multiple-scattering processes of the excited O 1s
photoelectron by the Ni substrate atoms and to a smaller degree by the other
O neighbors on the surface. The oxidation of Ni{100} is characterized by the
consecutive formation of a p(2 × 2), a c(2 × 2) and a NiO 1 × 1 low-energy electron diffraction (LEED) pattern [10.42].

Experimental NEXAFS spectra above the O K edge for the progressive oxidation of Ni{100} are shown in Fig.10.5 [10.32]. Spectra recorded at exposures
of 1.5, 10, 20 and 40 L (1 L = 10^{-6} Torr s) were found to be identical within
statistical error and to exhibit a pronounced polarization dependence. The
p(2 × 2) LEED pattern exhibited maximum contrast around 1.5 L exposure. The
c(2 × 2) pattern was first visible around 5 L and became clearest around 20-30
L. The NEXAFS features change above 50 L due to oxide formation, as seen in
Fig.10.5 for the spectra recorded at 80 L. At 160 L exposure, the polarization dependence has almost vanished. For 280 L (not shown), the NEXAFS spectrum looks identical to that in Fig.10.5(h). It has become completely isotropic
and is identical to that for fcc bulk NiO. Since the edge fine structure is
more easily recorded (collection time ~15 min) than a complete SEXAFS spectrum (collection time >2 h), Fig.10.5 represents a good example of how an
unknown system can be characterized first by NEXAFS before addressing certain
structural issues with SEXAFS.

The NEXAFS spectrum itself can yield detailed structural information when
compared to an appropriate multiple-scattering calculation as shown in Fig.
10.6 for the c(2 × 2) O on Ni{100} phase. The calculations were performed

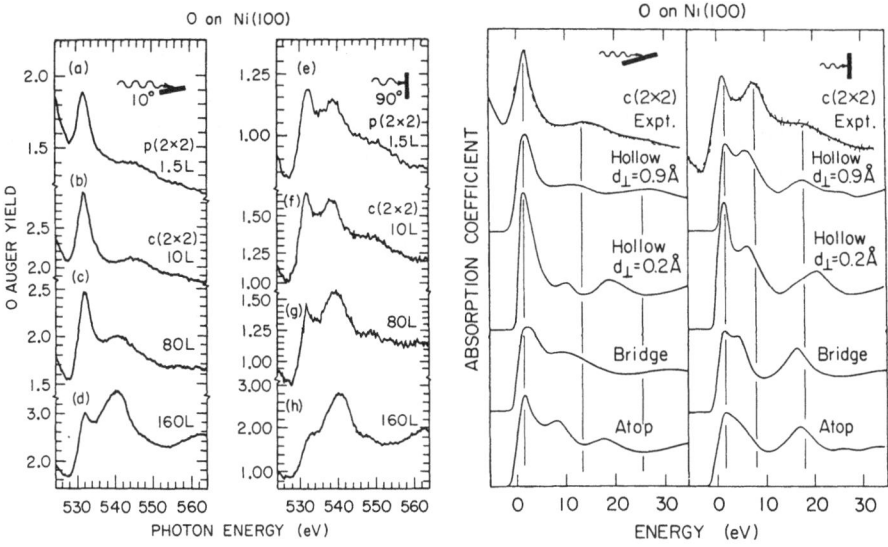

O on Ni(100)

O on Ni(100)

Fig.10.5

Fig.10.6

Fig.10.5. Near-edge fine structure spectra around the O K edge for increasing oxygen coverage on Ni{100} and two different X-ray incidence angles. Note that spectra show strong polarization dependence at low coverage which vanishes at high coverage due to cubic NiO formation

Fig.10.6. Experimental and calculated NEXAFS spectra for a c(2 × 2) O on Ni{100} overlayer. The calculated spectra included full multiple scattering and assumed different chemisorption sites as indicated

using a computational scheme based on a cluster method [10.7]. The calculation proceeds by first dividing the cluster into shells of atoms around a central (absorbing) atom. The scattering properties of each shell are described by a set of scattering phase shifts, and the multiple-scattering equations are solved consecutively within each shell. The final step is to calculate multiple scattering between shells and the assembly of the whole cluster. When combined with an atomic matrix element linking the core and excited electron states, the reflection matrix so obtained gives the NEXAFS cross section exactly in one-electron theory. Many-body processes which limit the lifetime of the core hole and photoelectron are included as a complex (absorptive) potential whose effect is to broaden spectral features on the appropriate scale.

The calculations shown in Fig.10.6 show best agreement with experiment if the O atom is placed in the fourfold hollow Ni site at a vertical spacing of $d_{\perp} = 0.9$ Å [10.32,36]. These results show that polarization-dependent NEXAFS spectra provide a sensitive tool to monitor structural changes at surfaces.

Furthermore, it is possible to model the experimental data by a multiple-scattering cluster approach and discriminate between different structural models. The use of NEXAFS to obtain structural information for *atomic* adsorption on solid surfaces thus appears to be possible. There are two advantages with respect to SEXAFS. NEXAFS measurements can be performed for lower adsorbate coverages (~1/100 monolayer), and they can be carried out even if absorption edges occur too close together to allow a SEXAFS analysis. From a practical point of view, it is a major disappointment, however, that the NEXAFS spectra of *atomic* adsorbates are not dominated by just the NN substrate atoms but rather that the scattering from as many as 30 neighbor atoms has to be included in the calculations before convergence is achieved. This is in contrast to the NEXAFS of *molecular* adsorbates, discussed below, where the spectra are dominated by intramolecular scattering with only small or negligible scattering contributions from the surface substrate atoms.

10.4.1.2 Chemisorption of Diatomic Molecules

NEXAFS spectra of chemisorbed diatomic or pseudodiatomic (i.e., ignoring the hydrogen atoms) molecules are dominated by resonance structures which closely resemble those observed in the equivalent gas-phase spectra [10.16,17]. This is surprising, because we have seen in the previous section that for chemisorbed atoms scattering processes involving surface substrate atoms give sizable and detailed contributions. These scattering contributions clearly have to be present for chemisorbed molecules, too, but are overshadowed by the larger intramolecular scattering resonances.

This is illustrated by Fig.10.7 where NEXAFS spectra are shown above the O K edge for two molecules with carbon-oxygen bonds chemisorbed on Cu{100} [10.27]. The two molecules differ in two important aspects. Methoxy (CH_3O) is bonded by a single C-O bond, and the bond length is therefore relatively large (R = 1.43 Å). Carbon monoxide, on the other hand, is triple bonded with two orthogonal π orbitals perpendicular to the C-O axis. The triple bond results in a 0.3 Å shorter bond length (R = 1.13 Å).

The spectrum of CH_3O on Cu{100} at grazing incidence is dominated by a large resonance at ~538 eV which vanishes at normal incidence. This structure is the σ shape resonance, and its dependence on X-ray incidence (**E** vector) establishes that methoxy stands up on the surface. In contrast, CO/Cu{100} exhibits two resonance structures, of which the one at ~536 eV is a π bound state resonance and that at ~552 eV a σ shape resonance. The orientational dependence of the resonances proves that CO also stands up on the surface.

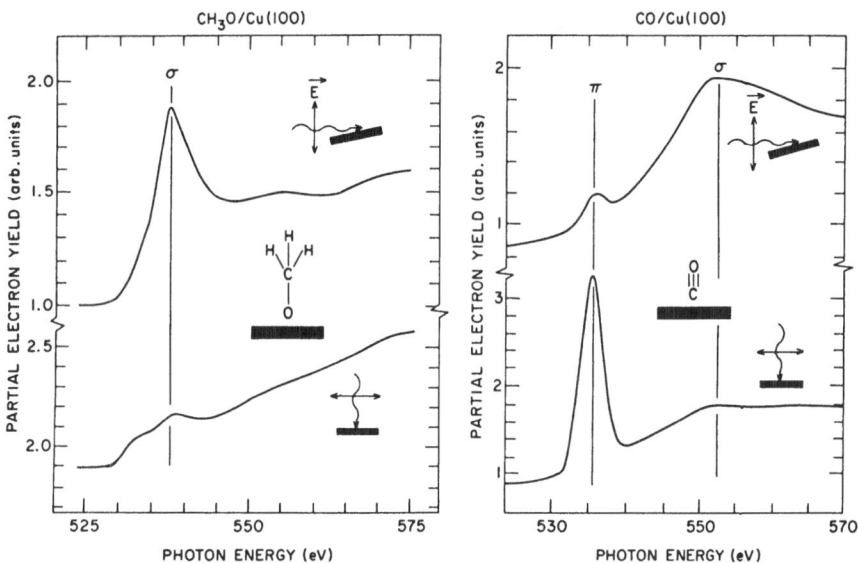

Fig.10.7. NEXAFS spectra for CH$_3$O and CO on Cu{100}. CH$_3$O exhibits only a σ shape resonance because of its C-O single-bond nature (no π orbital). CO is triple bonded (two orthogonal π orbitals) and therefore exhibits a π as well as σ resonance. The polarization dependence determines the molecular orientation as shown and the σ resonance position is characteristic of the C-O bond length

Note that from the NEXAFS spectra we cannot determine whether the molecules bond via the C or O end.

The position of the σ shape resonance in Fig.10.7 exhibits a dramatic shift of ~14 eV caused by the change (0.30 Å) in the carbon-oxygen bond length between CO and CH$_3$O. As discussed quantitatively in detail elsewhere [10.27], this correspondence between changes in σ shape resonance position and intramolecular bond length is in good accord with (10.3b).

From the examples shown in Fig.10.7, we find that NEXAFS spectra of chemisorbed diatomic molecules provide us with at least three types of information. First, the presence or absence of the π resonance is a direct indication of the hybridization of the intramolecular bond. Although our example is for carbon-oxygen bonds, the same conclusion is also valid for other atomic pairs (e.g., carbon-carbon bond) [10.26]. Secondly, the dependence of the spectra on E vector orientation establishes the molecular orientation on the surface. As discussed previously [10.16,17], a more detailed orientational study and fit of the measured intensity variation to theory (cos^2 or sin^2 functions) allows the determination of the molecular orientation to ~10°. Finally, the

245

σ shape resonance position is a direct measure of the intramolecular bond length. Quantitative bond length determinations by means of (10.3) can be carried out with an accuracy of <0.05 Å [10.26,27].

10.4.1.3 Chemisorption of Polyatomic Molecules

The resonances observed for diatomic molecules arose from a bound state transition to antibonding π orbitals and from resonance scattering along the bond direction of the absorbing atom and its neighbor. Both resonances are also expected to be present for poylatomic molecules. Again, the π resonance will be observable only if the molecule contains π bonds and the σ shape resonance should be present for all molecules if E lies along an internuclear axis. In addition, one might expect resonances which arise from scattering processes involving second or third nearest-neighbor atoms.

Results for deuterated benzene (C_6D_6) and pyridine (C_5H_5N) on Pt{111}, chemisorbed at room temperature and saturation coverage, show that these expectations are well founded [10.33]. As seen in Fig.10.8, C_6D_6 exhibits a strong, broadened π resonance (peak A) and two resonances (peaks B and C) which behave like σ resonances as a function of E vector orientation. Because the π orbitals are perpendicular to the molecular plane and shape resonances arise from in-plane scattering, the observed spectral dependence on X-ray incidence unambiguously determines that the C_6D_6 ring lies flat on the surface. This is in marked contrast to C_5H_5N, which from the NEXAFS spectra in Fig. 10.8 is clearly found to stand up with the ring plane parallel to the surface normal. Note that for this ring orientation the changes of X-ray incidence are not as clear-cut as for the lying-down geometry because some of the π orbitals and internuclear axes have a residual projection on the E vector even for the minimum-intensity geometry (grazing incidence for π, normal for σ resonance). The π resonance is significantly narrower than for benzene because the π orbitals are not involved in the chemisorption bond. Studies above the N K edge completely support the molecular orientation deduced from the C K edge data in Fig.10.8 [10.33].

Peak B arises from C-C NN scattering and is the "normal" σ shape resonance as for diatomic molecules. The resonance position indicates a slight (0.03 Å) symmetric stretch of all C-C bonds upon chemisorption for C_6D_6 and no change for C_5H_5N. Peak C is a novel resonance which is absent for diatomic molecules. It may therefore be thought to arise from different scattering processes between carbon-carbon nearest neighbors (NN), second NN and third NN [10.26]. For a symmetric hexagon the second and third NN distances are larger by fac-

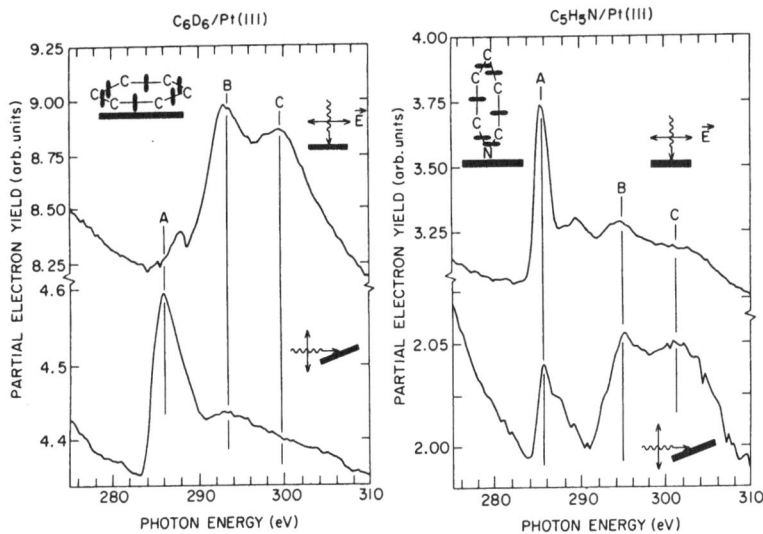

Fig.10.8. NEXAFS spectra of C_6D_6 and C_5H_5N on Pt{111}. Peak A is a π reson-ance and peaks B and C are associated with σ shape resonance excitations as discussed in the text. The polarization dependence proves C_6D_6 to lie down and C_5H_5N to stand up, as shown. Note peak A is broadened for C_6D_6 because of π bonding to the surface

tors $\sqrt{3}$ and 2, respectively, than the first NN distance. However, the posi-tion of peak C cannot be explained by (10.3b). Rather, for the longer second NN distance (~2.4 Å) the normal shape resonance would fall below the K-edge threshold (1s binding energy relative to the Fermi level) and is therefore not observed. At the time of this writing the origin of peak C has not been established unambiguously. Preliminary calculations indicate that it may arise from a shake-up excitation associated with the NN σ shape resonance (peak B).

In summary, for polyatomic molecules the NEXAFS spectra provide similar information on orientation, intramolecular bonding and the first NN bond length as for diatomic molecules

10.4.2 SEXAFS Studies

10.4.2.1 Distance Determination in Clusters

The structural parameter most easily and reliably obtained from an EXAFS spectrum is the nearest-neighbor distance. In particular, the analyis of the EXAFS data is independent of crystal structure and is applicable to materials with no long-range order. The first application of EXAFS which we shall discuss involves one of the fundamental problems in solid-state physics, namely the formation of a solid. By studying clusters of atoms ranging from dimers to continuous solid films, we can monitor changes in the average nearest-neighbor distance.

As our model system, we choose Cu clusters evaporated onto amorphous graphite [10.37]. Depending on the Cu coverage, clusters are formed on the surface with an average size, determined by electron microscopy [10.43]. Copper is known to interact very weakly with amorphous graphite such that the clusters are merely held in place by it. In addition, X-ray absorption of the graphite substrate is relatively small compared to that of the heavier Cu atoms and the signal-to-background ratio is favorable. The EXAFS measurements were carried out above the Cu K edge by monitoring the X-ray fluorescence [10.37]. The samples consisted of about 50 layers of Cu clusters sandwiched between carbon layers ~50-100 Å thick. Although the reported measurements were carried out in air we nevertheless discuss them here because similar measurements will undoubtedly be done in the future in ultrahigh vacuum by electron-yield SEXAFS. In fact, this latter approach offers significant advantages, such as in situ sample preparation, and it enables one to measure a single layer of metal clusters or less deposited onto a bulk substrate (e.g., graphite or different oxides).

Results for the derived Cu-Cu nearest-neighbor distance as a function of Cu coverage are shown in Fig.10.9. The distance remains nearly the same down to very low coverages ($\sim 1 \times 10^{16}$ atoms/cm^2), corresponding to a cluster size of ~40-45 Å in diameter. For lower coverages a contraction in the average bond length is observed, up to 9% at the lowest measured coverage of 8×10^{14} atoms/cm^2, a particle size which is below the limit of detection by standard transmission electron microscopy. Note that for most single-crystal surfaces

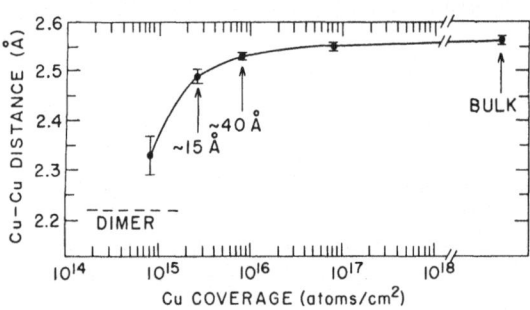

Fig.10.9. Average Cu-Cu nearest-neighbor distance in Cu clusters supported on graphite vs bulk Cu metal determined by EXAFS. For the smallest measured clusters (8×10^{14} Cu atoms/cm^2 coverage) the distance approaches that for the gas-phase Cu$_2$ dimer

this coverage would correspond to about 1 monolayer, which is easily measured by electron-yield SEXAFS. At the lowest coverage the average Cu-Cu distance is 2.33 ± 0.04 Å as compared to the bulk value of 2.56 Å and approaches that of gas-phase Cu_2 (2.22 Å).

The observed contraction of the average Cu-Cu bond length in small clusters can be explained by a reduction in repulsive interactions between atoms located in surface sites of the cluster. The contraction scales as the surface-to-volume ratio, which increases as the reciprocal diameter of the cluster [10.37].

10.4.2.2 Chemisorption Geometry on Single-Crystal Surfaces

Because of the intrinsic anisotropy of single-crystal surfaces SEXAFS studies of adsorbates can utilize a powerful concept involving the polarized nature of synchrotron radiation. The so-called *search-light effect* discussed in Sect.10.2.2 allows one to search preferentially for neighbor atoms along the **E** direction. This is beautifully demonstrated in the work by *Citrin* et al. [10.38] shown in Fig.10.10 for atomic Cl chemisorbed on both Si{111} 7 × 7 and Ge{111} 2 × 8.

Figure 10.10 displays SEXAFS data recorded by Auger electron detection above the Cl K edge (2825 eV) for two sample orientations. For $\theta = 10°$, the **E** vector lies approximately along the surface normal and the SEXAFS spectra for both substrates exhibit sizable oscillations above the edge, which are shown again in the lower half of the figure after background subtraction. In contrast, for $\theta = 90°$ the **E** vector lies in the surface plane and the SEXAFS oscillations disappear almost completely. Therefore, the first nearest Si or Ge neighbor of a Cl atom on the surface has to be located directly underneath, i.e., Cl chemisorbs atop a surface substrate atom. The SEXAFS frequency allows one to determine the Cl-Si and Cl-Ge NN distances to be 2.03 ± 0.03 Å and 2.07 ± 0.03 Å, respectively, by using $SiCl_4$ and $GeCl_4$ model compounds.

The atop chemisorption site revealed by the polarization dependence of the NN SEXAFS signal is independently confirmed by determining the second NN Cl-Si and Cl-Ge distance and by directly deriving the NN coordination number N (10.6) from comparison of the SEXAFS amplitudes of the surface complexes to those recorded for the models $SiCl_4$ and $GeCl_4$. These independent ways of deriving structural parameters and checking for their mutual consistency render SEXAFS such a definitive and reliable technique for surface crystallography.

Finally, we mention that the chemisorption geometry derived from SEXAFS is also supported by the strong polarization dependence of the NEXAFS struc-

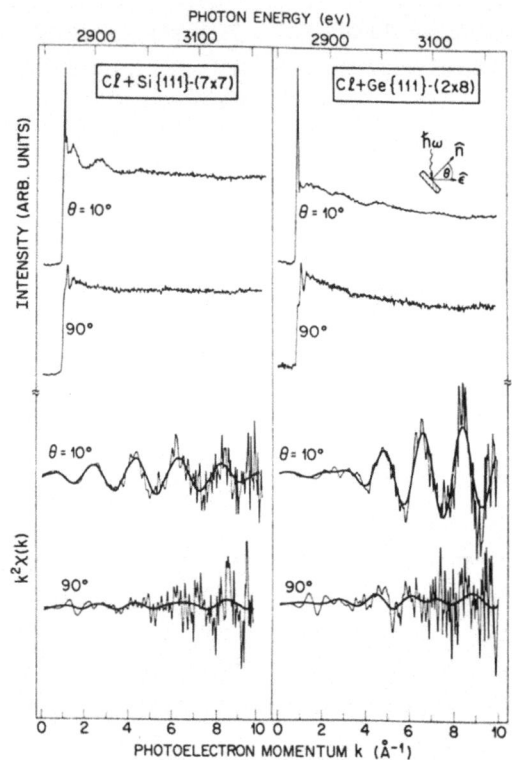

Fig.10.10. *Top*: Raw SEXAFS data from the Cl K edge taken at different polarizations. *Bottom*: Background-substracted raw SEXAFS data and filtered data (*smooth line*) corresponding to first neighbor Cl-substrate distance. Deviations between the raw and filtered data are due primarily to SEXAFS from the second neighbor Cl-substrate distance. Note strong polarization dependence of SEXAFS amplitudes in both the Cl-Si and Cl-Ge systems [10.38]

ture seen in Fig.10.10. For $\theta = 10°$ a strong threshold resonance is observed which almost vanishes for $\theta = 90°$. As pointed out by *Citrine* et al. [10.38] this resonance corresponds to a transition to an unoccupied antibonding σ state. The polar nature of the Cl-Si or Cl-Ge bond and the bonding geometry account for the large intensity and polarization dependence of the threshold structure. In general, one would expect the NEXAFS spectra of chemisorbed atoms on covalently bonded semiconductors to be richer in structure than for metal substrates because of the directionality and localization of the bonding.

10.4.2.3 The Structure of Solid-Solid Interfaces

The physics and chemistry of solid-solid interfaces play the key role in many technologies. They determine the characteristics and quality of devices in the electronics industry or the activity of supported catalysts in the oil industry. In the following, we shall discuss structural investigations by SEXAFS of the interface formed between a metal and a covalent semiconductor. We choose the Pd and Ag metal interfaces with a Si{111} 7×7 surface [10.39-41] because the two systems are textbook examples of the different interfaces which can be formed during the early stages of metal deposition.

Figure 10.11 compares the L_2 SEXAFS spectrum recorded by Auger yield detection of Ag metal to that of 2.5 monolayer (ML) of Ag evaporated onto a clean Si{111} 7×7 surface. The spectra are almost identical, indicating that silver must form metallic clusters or crystallites on the surface at 2.5 ML coverage. The same result is obtained at even lower coverage of 1 ML [10.41], corresponding to 7.8×10^{14} atoms/cm^2. Both the Ag-Ag nearest-neighbor distance and average coordination are almost unchanged relative to Ag metal. From our results for Cu clusters on graphite we know that the Ag clusters or crystallites on the Si{111} surface must be rather large ($\gtrsim 50$ Å) to account for the bulk-like SEXAFS spectra. It is this clustering tendency of Ag which accounts for its use in the photographic industry [10.43].

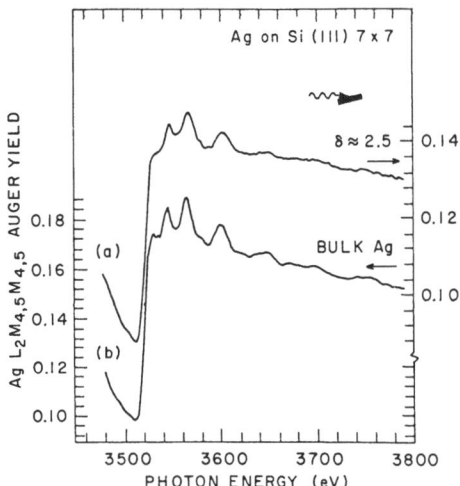

Fig.10.11. SEXAFS spectra above the Ag L_2 edge for 2.5 monolayer of Ag on Si{111} 7×7 and for bulk Ag metal

The findings for Ag are in contrast to those for its neighbor in the periodic table, Pd, as seen from Fig.10.12. Here we compare the L_2 SEXAFS spectra for a 1.5 ML Pd layer on Si{111} 7×7 to that for a thermally grown Pd$_2$Si film and for Pd metal. Without analysis of the spectra it is apparent that the 1.5 ML spectrum is almost identical to that for bulk Pd$_2$Si and differs from that of Pd metal. Note especially the resonance at threshold [10.40] and the SEXAFS frequency at higher energies. These results prove that even at the

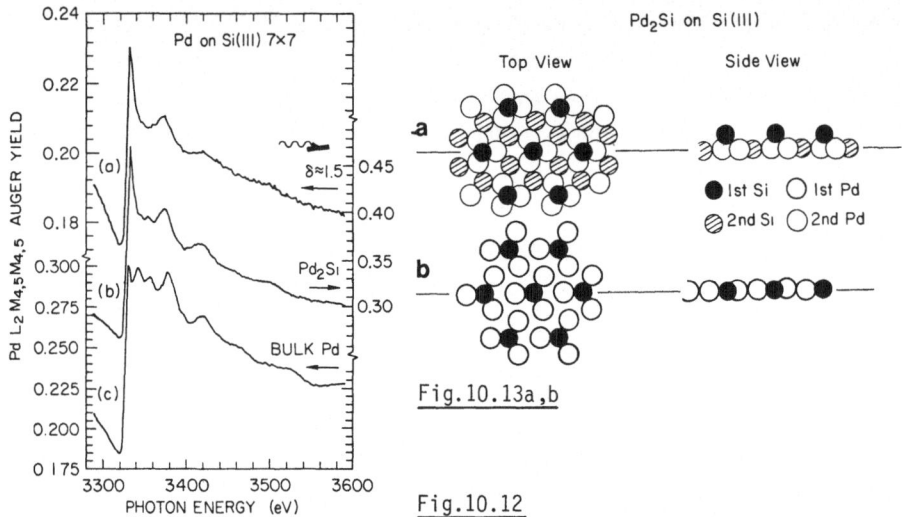

Fig.10.12

Fig.10.13a,b

Fig.10.12. SEXAFS spectra above the Pd L$_2$ edge for Pd on Si{111} 7 × 7; a) 1.5 monolayers of Pd deposited at RT on Si{111} 7 × 7, b) a thick layer of Pd$_2$Si grown at 500°C after deposition of 300 Å of Pd, c) a 300 Å layer deposited at RT

Fig.10.13a,b. The two alternating atomic planes normal to the C$_0$ axis in Pd$_2$Si. The large clear circles are Pd atoms and the smaller darker circles represent Si. a) is the base plane of the hexagonal unit cell with the black Si atoms raised half way along the C$_0$ axis; b) is the plane of the black Si atoms [10.44]

early stages of Pd deposition a compound is formed which strongly resembles Pd$_2$Si. This can proceed locally by replacement of a Si surface atom by three Pd atoms as shown in Fig.10.13a [10.44]. The replaced Si atom, shown black in Fig.10.13, is pushed up and rides on top of the three Pd atoms. It is co-ordinated by three new in-plane Pd atoms as in Fig.10.13b. The lattice spacings in the Si{111} plane and in the Pd$_2$Si plane perpendicular to the hexagonal C$_0$ axis match almost perfectly. These results clearly demonstrate the potential of SEXAFS to reveal the structure of complex metal-semiconductor surfaces.

10.5 Conclusions

The results presented demonstrate that SEXAFS is a powerful structural technique applicable to a large variety of surface structural problems. Its theory is well understood and its limitations well established because of its direct correspondence to the bulk EXAFS technique [10.5]. Because of the intrinsic anisotropy of surfaces, SEXAFS can often take advantage of the search-light effect which is much less useful in typical bulk EXAFS studies. Otherwise SEXAFS offers the same virtues and suffers from the same limitations as EXAFS. No detailed discussion of these points is required here since they have been extensively covered elsewhere [10.5].

The NEXAFS region of the spectrum is in general less understood than the EXAFS one. It is clear that NEXAFS offers additional and complementary information, including details on the valence electron charge distribution (bonding orbitals). NEXAFS is the part of the absorption spectrum which exhibits the largest and most detailed structures. The spectra can be recorded in cases where it is impossible to obtain EXAFS data because of signal-to-background problems arising from high dilution or interference of other absorption edges or structures [10.3]. A thorough theoretical understanding of NEXAFS is therefore highly desirable.

We have seen that the case of chemisorbed molecules is exceptional in that the NEXAFS is dominated by local intramolecular scattering processes, similar to gas-phase molecules. The observed resonances are easily interpreted and their presence, position, and polarization dependence provides unique information on the intramolecular bonding, bond length and the molecular orientation. It appears that the condition for resonance scattering between two atoms (σ shape resonance) is well fulfilled for typical bond lengths between two low-Z atoms (1.1 - 1.5 Å) but not for the longer bond lengths between a low-Z atom and a typical substrate atom (1.8 - 2.2 Å). This may explain why the simple intramolecular scattering processes dominate over the complicated long-range ones which involve surface substrate atoms.

Acknowledgements. The experiments reported here obviously could not have been performed without the help of skillful and competent collaborators. The work reflects an experimental program carried out over the last five years in numerous collaborations indicated by the co-authors of the listed publications. I should especially like to thank F. Sette for helping me understand the rules which govern the NEXAFS spectra of molecules. All work reported here was done at SSRL which is supported by the office of Basic Energy Sciences of DOE and the Division of Materials Research of NSF.

References

10.1 For a review see: *Synchrotron Radiation Research*, ed. by H. Winick, S. Doniach (Plenum, New York 1980)

10.2   The first SEXAFS measurements were published in 1978:
        P.H. Citrin, P. Eisenberger, R.C. Hewitt: Phys. Rev. Lett. **41**, 309
        (1978);
        J. Stöhr, D. Denley, P. Perfetti: Phys. Rev. B**18**, 4132 (1978);
        J. Stöhr: Jpn. J. Appl. Phys. **17**, Suppl. **17-2**, 217 (1978)
10.3   For reviews of SEXAFS spectroscopy see:
        J. Stöhr: In *Emission and Scattering Techniques*, ed. by P. Day (Reidel,
        Dordrecht 1981) and SSRL Report 80/07;
        J. Stöhr, R. Jaeger, S. Brennan: Surf. Sci. **117**, 503 (1982);
        J. Stöhr: In *Principles, Techniques and Applications of EXAFS, SEXAFS
        and XANES*, ed. by R. Prins, D. Koningsberger (Wiley, New York 1984)
10.4   D.E. Sayers, F.W. Lytle, E.A. Stern: Phys. Rev. Lett. **27**, 1204 (1971)
10.5   For reviews of EXAFS spectroscopy see:
        a) E.A. Stern: Contemp. Phys. **19**, 289 (1978);
        b) P. Eisenberger, B.M. Kincaid: Science **200**, 1441 (1978);
        c) D.R. Sandstrom, F.W. Lytle: Ann. Rev. Phys. Chem. **30**, 215 (1979);
        d) P. Rabe, R. Haensel: In *Festkörperprobleme 20* (Pergamon-Vieweg,
        Stuttgart, 1980) p.43;
        e) P.A. Lee, P.H. Citrin, P. Eisenberger, B.M. Kincaid: Rev. Modern
        Physics **53**, 769 (1981);
        f) B.K. Teo, D.C. Joy (eds.): *EXAFS Spectroscopy, Techniques and Appli-
        cations* (Plenum, New York 1981)
10.6   A. Bianconi: Appl. Surf. Sci. **6**, 392 (1980)
10.7   P.J. Durham, J.B. Pendry, C.H. Hodges: Comput. Phys. Comm. **25**, 193
        (1982)
10.8   C.R. Natoli: Proceedings of the First Internat. Conf. on EXAFS and
        XANES, Frascati, Italy, 1982. Springer Ser. Chem. Phys., Vol.27
        (Springer, Berlin, Heidelberg, New York 1982) p.43
10.9   J.L. Dehmer, D. Dill: In *Electron-Molecule and Photon-Molecule Colli-
        sions*, ed. by T. Rescigno, V. McKoy, B. Schneider (Plenum, New York
        1979) p.225
10.10   J.L. Dehmer, D. Dill: J. Chem. Phys. **65**, 5327 (1976)
10.11   C.R. Natoli, D.K. Misemer, S. Doniach, F.W. Kutzler: Phys. Rev. A**22**,
        1104 (1980)
10.12   A. Bianconi, M. Del'Ariccia, P.J. Durham, J.B. Pendry: Phys. Rev. B**26**,
        6502 (1982)
10.13   G.N. Greaves, P.J. Durham, G. Diakun, P. Quinn: Nature **294**, 139 (1981)
10.14   P.H. Gaskell, D.M. Glover, A.K. Livesey, P.J. Durham, G.N. Greaves:
        J. Phys. C: Solid State Phys. **15**, L597 (1982)
10.15   J.E. Müller, W.L. Schaich: Phys. Rev. B**27**, 6489 (1983)
10.16   J. Stöhr, K. Baberschke, R. Jaeger, T. Treichler, S. Brennan: Phys.
        Rev. Lett. **47**, 381 (1981)
10.17   J. Stöhr, R. Jaeger: Phys. Rev. B**26**, 4111 (1982)
10.18   J.B. Pendry: *Low Energy Electron Diffraction* (Academic, New York 1974);
        C.B. Duke: In *Surface Effects in Crystal Plasticity*, ed. by R.M. La
        Tanision and J.F. Fourie (Nordhoff, Leyden 1977);
        F. Jona: J. Phys. C**11**, 4271 (1978);
        M.A. Van Hove, S.Y. Tong: *Surface Crystallography by LEED*, Springer
        Ser. Chem. Phys. Vol.2 (Springer, Berlin, Heidelberg, New York 1979)
10.19   S.D. Kevan, D.H. Rosenblatt, D. Denley, B.-C. Lu, D.A. Shirley: Phys.
        Rev. Lett. **41**, 1565 (1978);
        C.H. Li, S.Y. Tong: Phys. Rev. Lett. **43**, 526 (1979)
10.20   For a review see:
        A.P. Hitchcock: *Bibliography of Atomic and Molecular Inner Shell Exci-
        tation Studies*, J. Electron Spectrosc. **25**, 245 (1982)
10.21   V.I. Nefedov: J. Struct. Chem. **11**, 277 (1970)
10.22   J.L. Dehmer: J. Chem. Phys. **56**, 4496 (1972)
10.23   D. Dill, J.L. Dehmer: J. Chem. Phys. **61**, 692 (1974)

10.24 F. Sette, J. Stöhr, A.P. Hitchcock: J. Chem. Phys. (to be published)
10.25 For gas-phase molecules see:
      K. Bomben, C.J. Eyermann, W.L. Jolly: Jan. 18, 1983 update of A.A.
      Bakke, H.W. Chen, W.L. Jolly: J. Electron Spectrosc. **20**, 333 (1980)
10.26 J. Stöhr, F. Sette: To be published
10.27 J. Stöhr, J.L. Gland, W. Eberhardt, D. Outka, R.J. Madix, F. Sette,
      R.J. Koestner, U. Döbler: Phys. Rev. Lett. **51**, 2414 (1983)
10.28 J. Stöhr, C. Noguera, T. Kendelewicz: Phys. Rev. B (to be published)
10.29 R. Jaeger, J. Feldhaus, J. Haase, J. Stöhr, Z. Hussain, D. Menzel,
      D. Norman: Phys. Rev. Lett. **45**, 1870 (1980)
10.30 J. Stöhr, R. Jaeger, J. Feldhaus, S. Brennan, D. Norman, G. Apai:
      Appl. Opt. **19**, 3911 (1980)
10.31 J. Stöhr, D. Denley: In *Proceedings, International Workshop on X-ray
      Instrumentation for Synchrotron Radiation*, ed. by H. Winick, G.S.
      Brown. Stanford, April 1978, SSRL Report 78/04
10.32 D. Norman, J. Stöhr, R. Jaeger, P.J. Durham, J.B. Pendry: Phys. Rev.
      Lett. **51**, 2052 (1983)
10.33 A.L. Johnson, E.L. Muetterties, J. Stöhr: J. Am. Chem. Soc. **105**, 7183
      (1983)
10.34 J. Stöhr, D. Outka, R.J. Madix, U. Doebler: Unpublished results
10.35 J. Stöhr, J.L. Gland, R.J. Koestner: Unpublished results
10.36 See, e.g., J. Stöhr, R. Jaeger, T. Kendelewicz: Phys. Rev. Lett. **49**,
      142 (1982)
10.37 G. Apai, J.F. Hamilton, J. Stöhr, A. Thompson: Phys. Rev. Lett. **43**,
      165 (1979)
10.38 P.H. Citrin, J.E. Rowe, P. Eisenberger: Phys. Rev. B**28**, 2299 (1983)
10.39 J. Stöhr, R. Jaeger: J. Vac. Sci. Technol. **21**, 619 (1982)
10.40 G. Rossi, R. Jaeger, J. Stöhr, T. Kendelewicz, I. Lindau: Phys. Rev.
      B**27**, 5154 (1983)
10.41 J. Stöhr, R. Jaeger, G. Rossi, T. Kendelewicz, I. Lindau: Surf. Sci.
      134, 813 (1983)
10.42 For a review see:
      C.R. Brundle: In *Aspects of the Kinetics and Dynamics of Surface Reac-
      tions - 1979*, ed. by U. Landman, AIP Conference Proceedings No.61
      (American Institute of Physics, New York 1980)
10.43 J.F. Hamilton, R.C. Baetzold: Science **205**, 1213 (1979)
10.44 W.D. Buckley, S.C. Moss: Solid State Electron. **15**, 1331 (1972)

11. Determination of Surface Structure Using Atomic Diffraction

T. Engel

With 17 Figures

11.1 Introduction

Atom diffraction from surfaces is both a very old and a rather new area of research in surface science. The first study of this phenomenon [11.1] verified the wave nature of light particles by scattering He from Li F{100}. Subsequent work was almost entirely confined to scattering from Li F{100} and graphite {0001} for many years [11.2-4] largely due to fact that only inert surfaces could be studied because of the poor vacuum conditions available to most of these investigators. Although this restricted the types of surfaces which could be studied, much was learned about the atom-surface potential and a theoretical framework was developed to calculate diffraction intensities from surfaces. The first UHV study of atom diffraction [11.5] rekindled interest in this area and was followed by a variety of studies on close packed metal surfaces. Thes original studies are discussed in recent comprehensive reviews of atom diffraction from surfaces [11.6,7]. These articles also refer to earlier excellent reviews of the subject which cannot be discussed in the present article for reasons of brevity.

Atom diffraction from surfaces became a technique which could be more easily compared to other techniques in surface science due to introduction of UHV techniques coupled with the usual surface spectroscopies, and noteworthy advances were the investigation of semiconductor surfaces [11.8-10] and adsorbate-covered surfaces [11.11,12]. At present it can be concluded that atom diffraction can be observed from all reasonably well-ordered single-crystal surfaces. The question remains whether the technique can produce quantitative structural information such as that available through analysis of LEED intensities. The present review will attempt to discuss progress toward this goal which is as yet not realized. It is also useful to place atom diffraction in the matrix of surface-science techniques. Its primary use as discussed in this article is structural analysis. However, diffraction experiments also give detailed information on the atom-surface potential [11.13]. Addition of time-of-flight capabilities allow inelastic scattering studies which have provided detailed information on surface phonons [11.14]. It should also be mentioned that the same apparatus, with reactive gases substituted for those suitable for diffraction, allow the most detailed kinetic studies which can currently be carried out in the field of surface science [11.15, 16]. Atom scattering therefore has a broader range of applications than most surface techniques, but this must be balanced against the experimental complexity of the technique, discussed in [11.6], as well as in the considerable cost associated with such an apparatus.

11.2 The Atom-Surface Potential

The key to interpreting experimental data and extracting structural information from them is a knowledge of the gas-surface potential. A schematic illustration describing the interaction of an incoming particle with a surface is shown in Fig.11.1. Not a great deal is known about the detailed nature of this potential for a realistic surface although efforts ranging from summation of pairwise interaction potentials to attempts to take many-body effects into account have been made.

For distances far from the surface the potential is knwon to have the form [11.18]

$$V_{attr}(u) = C_3 Z^{-3} \quad , \tag{11.1}$$

whereas there is no general way to predict the repulsive part of the potential. *Nørskov* and co-workers [11.19,20] proposed that the repulsive potential is given by a linear function of the surface electron density sampled by the diffracting atom at its classical turning point. This potential is surface independent and has the form

$$V_{rep}(\mathbf{r}) = 305 \, \rho(\mathbf{r}) \tag{11.2}$$

with $V(\mathbf{r})$ in eV for ρ in atomic units. Although the value of the constant is disputed and other values ranging from 176 eV(au)$^{-3}$ to 520 eV(au)$^{-3}$ have been obtained [11.21-24], the dependence on density is not disputed.

A possible way to generate a realistic atom surface potential is to calculate $\rho(\mathbf{r})$ for the surface in question and to use (11.2) to generate $V_{rep}(\mathbf{r})$. This can be combined with (11.1) to generate the total potential if C_3 can be evaluated. Note that the periodic part of this potential which is due to the modulation in $\rho(\mathbf{r})$ parallel to the surface is solely contained in the repulsive portion, since $V_{attr}(Z)$ depends only on the distance normal

258

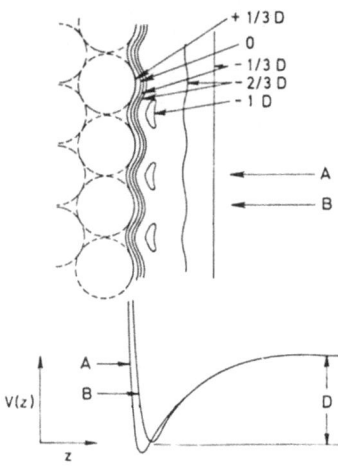

Fig.11.1. Equipotential lines for the interaction of an atom with a single-crystal surface. The (negative) potential energies are
given in terms of the depth of the potential
well, D. The bottom part of the diagram shows
the potential as a function of z, the distance normal to the surface for two different incoming trajectories, A and B [11.17]

to the surface. This is a simplification which must be examined more closely
in future studies.

Calculations of $\rho(\mathbf{r})$ for a surface are nontrivial and two approaches have
been developed recently. *Hamann* [11.25] has carried out calculations in a
self-consistent local density limit, which requires extensive computational
effort. *Haneman* and *Haydock* [11.26], reasoning from (11.2) that the diffracting atom is sampling electron densities of 10^{-4} (au)$^{-3}$, have suggested calculating $\rho(\mathbf{r})$ by a simple superposition of atomic charge densities. The jus-
tification for this approach is that the true change density in molecules
differs significantly from a simple superposition of the atomic charge densities only in the bonding region. Since the classical turning point of the
diffracting atoms is far from the bonding region between the surface atoms,
a simple superposition ignoring the interactions between atoms should give
reasonable results.

Comparison of the two methods of calculating $\rho(\mathbf{r})$ at a distance of 6 atomic units (1 atomic unit = 0.529 Å) above the surface shows that variations
of up to a factor of 3 are observed [11.26]. More significant than the absolute charge density is the periodic variation parallel to the surface.
For the Ni{110} surface for which the absolute charge densities varied by a
factor of 3, the corrugation amplitudes agreed within 10%. This suggests that
the superposition of atomic charge densities may be a sufficiently accurate
method to calculate $\rho(\mathbf{r})$ although a more stringent test would be a detailed
comparison on a covalently bonded substance such as Ga As, for which *Hamann*
[11.25] has carried out calculations.

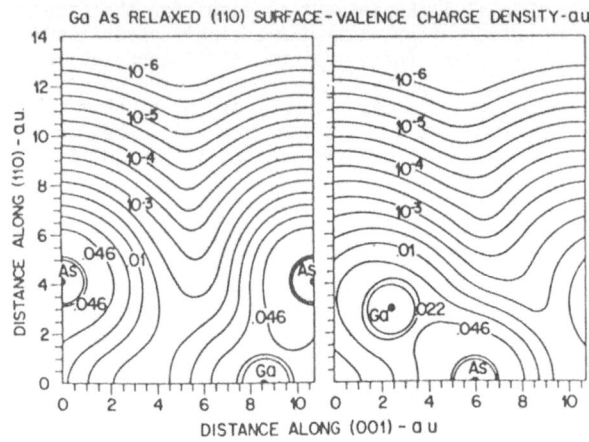

GaAs RELAXED (110) SURFACE - VALENCE CHARGE DENSITY - au

Fig.11.2. Charge densities for the relaxed (1×1) surface of GaAs {110}. Note the exponential falloff of the density along the [110] direction which is perpendicular to the surface [11.25]

Figure 11.2 shows electron density contours for the GaAs{110} surface and it can be seen that the falloff in $\rho(\mathbf{r})$ perpendicular to the surface is exponential. Therefore a reasonable ansatz for the atom-surface potential is

$$V_j(x,y,z) = C \exp(-\kappa[z - \zeta_j(x,y)]) - \frac{c_0}{(z + z_0)^3} \quad , \qquad (11.3)$$

where c_0 and z_0 are to be regarded as fitting parameters and C, K and the corrugation function $\zeta_j(x,y)$ are obtained from the calculation of $\rho(\mathbf{r})$. The subscript j on $\zeta(x,y)$ and $V(x,y,z)$ indicates the energy dependence of the potential which here enters through (11.2). The value of the charge density appropriate to the experiment is obtained from (11.2) and as is seen from Fig.11.2, ζ_{max}, which is the peak-to-peak amplitude of $\zeta_j(x,y)$, increases with ρ.

To implement a model potential such as that of (11.3) practically the constants C_0 and Z_0 must be evaluated. They can be obtained using the phenomenon of resonant scattering or selective adsorption [11.6,13]. Figure 11.3 shows the variation of the specular diffraction beam corresponding to $\mathbf{G} = 0$ (Sect. 11.3.1) with the polar angle of incidence θ_i which is measured from the surface normal. Sharp minima and maxima are observed from which the bound states in the atom-surface potential can be calculated without any assumptions about the atom-surface potential. By fitting to model potentials, the constants C_0 and Z_0 can be obtained [11.13,27].

As discussed above, a model potential like that given as (11.3) contains assumptions and simplifications, the most important of which are the relationships between the repulsive potential and the charge density, those involved in the calculation of the electron density, and the lack of a lateral

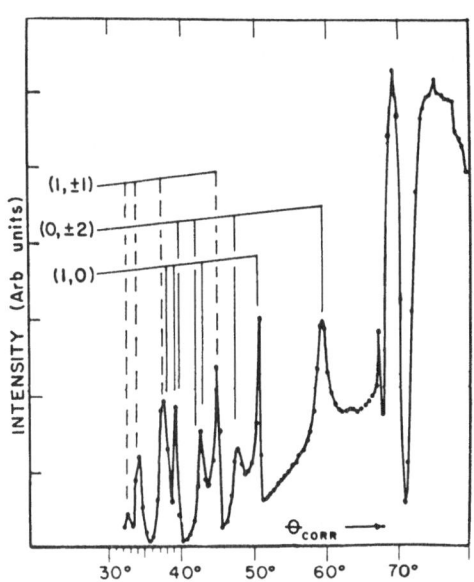

INTENSITY (Arb units)

(1,±1)
(0,±2)
(1,0)

Θ_{CORR} ⟶

30°   40°   50°   60°   70°

Fig.11.3. Variation of specular intensity with the polar angle θ_i for He scattering from Li F{100}. The number pair in parenthesis indicates the G vector involved in the transition and the energy level quantum numbers 0,1,2,3 increase from right to left [11.4]

variation of the attractive part of the potential. Since resonant scattering experiments measure only the laterally averaged potential V_0 (Sect.11.3) diffraction experiments are needed to measure $\zeta_j(x,y)$ if the total potential is to be verified experimentally. In the following section the quantum theory of particle diffraction will be presented and formalism for calculating the intensities for given potentials will be discussed.

11.3 Quantum Theory of Particle Diffraction

11.3.1 General Considerations on Diffraction

Diffraction beams are characterized by their intensities and the angle at which they occur. The latter information together with the wavelength of the incident atomic beam give the unit cell vectors a_1 and a_2. The structure within the unit cell, which is contained in $\zeta_j(x,y)$, must be determined from the relative intensities of the diffracted beams [11.6].

In developing a formalism for atom diffraction, the relationship between the unit cell vectors of the direct lattice (a_1 and a_2) and those of the reciprocal lattice (b_1 and b_2), which are given below, will be required:

$$|b_1| = \frac{2\pi}{|a_1|\sin\beta} \qquad |b_2| = \frac{2\pi}{|a_2|\sin\beta} \qquad (11.4)$$

and

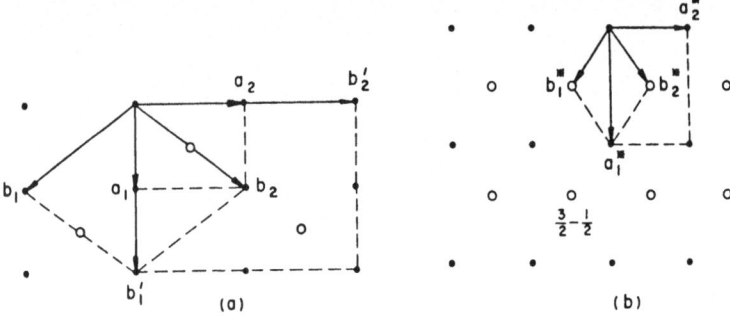

Fig.11.4a. Three choices for units cell's direct lattice; that indicated by the vectors $\mathbf{b}_1'$ and $\mathbf{b}_2'$ is nonprimitive. b) Reciprocal lattice corresponding to the direct lattice of (a) with the reciprocal lattice vectors $\mathbf{a}_i^*$ and $\mathbf{b}_i^*$, related to $\mathbf{a}_i$ and $\mathbf{b}_i$ by (11.5) [11.28]

$$a_p b_q = 2\pi\delta_{pq} \quad .$$

$$(11.5)$$

To illustrate these relationships, a direct lattice and its corresponding reciprocal lattice are shown in Fig.11.4.

Consider a particle of energy E_i incident on a single crystal surface at a polar angle θ_i. The wavelength of the particle is given by

$$\lambda_i = \frac{h}{\sqrt{2mE_i}}$$

$$(11.6)$$

which corresponds to 0.57 Å for a helium atom of 63 meV incident energy. This is a convenient reference energy since it corresponds to the mean energy of a fully expanded nozzle beam of a monatomic gas with a nozzle temperature of 300 K.

The wave vector of the incident beam which also describes the angle of incidence is related to the wavelength by

$$k_i = |\mathbf{k}_i| = \frac{2\pi}{\lambda_i} \quad .$$

$$(11.7)$$

It is convenient to separate $\mathbf{k}_i$ into its component parallel and perpendicular to the surface

$$\mathbf{k}_i = (K, k_{iz}) = (k_i \sin\theta_i, -k_i \cos\theta_i) \quad .$$

$$(11.8)$$

The condition for which diffraction can occur is that the wave vector describing the outgoing beam

$$\mathbf{k}_G = (K_G, k_{Gz}) = (k_i \sin\theta_G, k_i \cos\theta_G)$$

$$(11.9)$$

is related to $\mathbf{k}_i$ by

$$K_G = K + G \quad , \tag{11.10}$$

where

$$G = jb_1 + \ell b_2 \tag{11.11}$$

are the reciprocal lattice vectors of the surface. Since the scattering is elastic, the total energy of the scattered particle is unchanged and

$$k_i^2 = k_G^2 \quad . \tag{11.12}$$

This restricts the number of diffracted beams to that set for which

$$k_{Gz}^2 = k_i^2 - (K + G)^2 \geq 0 \quad . \tag{11.13}$$

However, evanescent waves for which $k_{Gz}^2 < 0$, although not observable, play an important part in atom diffraction.

11.3.2 The Corrugated Hard Wall Model

In the corrugated hard wall model, the atom-surface potential discussed in Sect.11.2 is simplified to

$$V(z) = 0 \quad \text{for} \quad Z > \zeta(x,y)$$

$$V(z) = \infty \quad \text{for} \quad Z \leq \zeta(x,y) \quad . \tag{11.14}$$

In this model, two significant features of a realistic potential, namely the attractive well and the finite steepness (softness) of the repulsive part are neglected. The modulation of the potential parallel to the surface remains and a physical connection between the electron charge density at the surface and $\zeta(x,y)$ can be made by using (11.1), equating E_i with $V_{rep}(r)$ to obtain an appropriate value of $\rho(r)$, and identifying $\zeta(x,y)$ with the periodic lateral variation of this charge density contour if calculations of $\rho(r)$ are available.

The far field solution of the scattering problem is given by [11.6]

$$\psi(r) = \exp[i(KR + k_{iz}z)] + \sum_G A_G \exp[i(K + G)R]\exp[ik_{Gz}z] \quad . \tag{11.15}$$

The complex scattering amplitudes A_G are related to the experimentally observed intensities P_G by

$$P_G = \left| \frac{k_{Gz}}{k_{iz}} \right| |A_G|^2 \quad . \tag{11.16}$$

The factor in front of the $|A_G|^2$ ensures that the total particle flux crossing any plane parallel to the solid surface is zero.

Since all the scattering from the corrugated hard wall is elastic, the sum of the P_G is given by

$$\sum_G P_G = 1 \quad . \tag{11.17}$$

Calculating A_G can be significantly simplified if it is assumed that the far field solution given by (11.15) holds up to the surface. Since the surface presents an infinite barrier,

$$\psi[\mathbf{R}, Z = \zeta(x,y)] = 0 \tag{11.18}$$

and (11.15) with this boundary condition becomes

$$\exp\{i[\mathbf{K}\mathbf{R} + k_{iz}\zeta(x,y)]\} + \sum_G A_G \exp\{i[\mathbf{K}\mathbf{R} + \mathbf{G}\mathbf{R} + k_{Gz}\zeta(x,y)]\} = 0 \quad , \tag{11.19}$$

or in its more usual form

$$\sum_G A_G \exp[ik_{Gz}\zeta(x,y)] \exp[i\mathbf{G}\mathbf{R}] = -\exp[ik_{iz}\zeta(x,y)] \quad . \tag{11.20}$$

Since these equations must be satisfied for each point within the unit cell, a set of equations can be set up to solve for A_G.

The assumption that the far field solution is also valid at the surface, which is called the Rayleigh hypothesis, is valid only for small corrugation amplitudes and for special cases such as a one-dimensional sinusoidal corrugation

$$\zeta(x) = \frac{1}{2} \zeta_1 \cos \frac{2\pi x}{a} \quad , \tag{11.21}$$

and a two-dimensional corrugation with a square unit cell

$$\zeta(x,y) = \frac{1}{4} \zeta_1 \left(\cos \frac{2\pi x}{a} + \cos \frac{2\pi y}{a} \right) , \tag{11.22}$$

the Rayleigh limits are $\zeta_1 = 0.143a$ and $\zeta_1 = 0.188a$ respectively [11.29-31]. The addition of higher terms to the expansion

$$\zeta(x,y) = \sum_G \zeta_G \exp[i(\mathbf{G}\mathbf{R})] \tag{11.23}$$

reduces the range of validity of the Rayleigh hypothesis. Corrugations for many clean and adsorbate-covered surfaces are within the range of validity of the Rayleigh hypothesis although corrugation functions describing surface reconstructions involving extensive rearrangement or large adsorbates on most surfaces will not satisfy the criteria.

However, this is not a limitation since a variety of methods have been developed to solve for the A_G given $\zeta(x,y)$ without invoking the Rayleigh hypothesis. The interested reader is referred to [11.6] for a discussion of these. A notable recent development is the special points method of *Salanon* and *Armand* [11.32] which allows exact calculations to be carried out without increasing the computational effort beyond that required for those methods utilizing the Rayleigh hypothesis.

Despite these methods, the application of the hard corrugated wall model to surfaces of appreciable corrugation amplitude is questionable as recent attempts to fit experimental data have shown [11.33]. The reason for the inability of the model to fit the data is probably due to the neglect of the softness of the repulsive wall rather than to the absence of the attractive well. However, as has been noted elsewhere [11.25,34], the inclusion of the attractive well influences the repulsive part of the potential and has the effect of making it steeper at lower energies. Without question the main contribution of the hard corrugated wall model has been the success with which diffraction data from relatively smooth surfaces could be interpreted with a minimum of computational effort. The drawbacks to the model are mainly due to the lack of a microscopic prescription to calculate $\zeta(x,y)$ for a given surface geometry. Thus although diffraction data taken from a real surface with a realistic atom-surface potential of the type given by (11.3) can be mapped into a hard corrugated wall problem, there is no obvious way to go back to a realistic surface potential having found the best fit to $\zeta(x,y)$ in the hard wall limit. The identification of $\zeta(x,y)$ with the $\zeta_j(x,y)$ of (11.3) using electron density calculations is a step in this direction. However, it does not seem probable that structural determinations in which atomic positions are specified to within 0.1 Å are possible with the simplied surface potential of the hard corrugated wall model.

11.3.3 The Inversion Problem in the Hard Corrugated Wall Limit

Experimentalists usually encounter the problem of structural determination in the inverse form to that described above, since one begins with diffraction intensities P_G and wishes to determine $\zeta(x,y)$. Recently, two procedures have been developed to calculate $\zeta(x,y)$, given the P_G [11.35,36], of which that in [11.36] has the advantage that no assumptions need be made about $\zeta(x,y)$, and so is described briefly below. Although both methods are at present restricted to one-dimensional corrugation functions, this is not a limitation in principle but rather in developing an efficient algorithm for the two-dimensional case.

The inversion procedure [11.36] utilizes the observation that for trial values of $\zeta(R)$ suitably close to their true values, rapid convergence to the true values can be attained using an iterative method. Beginning with a set of A_G corresponding to the true corrugation $\zeta(R)$, a set A_G^{cal} is calculated from (11.19) using $\zeta_1 = 10^{-8}$ and $\zeta_n = 0$, $n > 1$ from (11.21). There A_G^{cal} are used to calculate a new set of A_G' given by

$$A_G' = m \, A_G^{cal} |A_G^{exp}| / |A_G^{cal}|, \quad G \neq 0 \quad , \tag{11.24}$$

where initially $2^{-8} < m < 2^{-5}$. In generating the A_G', the calculated phases of the complex amplitudes A_G^{cal} are retained and the intensities of all beams for which $G \neq 0$ are reduced by the factor m to mimic a corrugation of much smaller amplitude. A renormalized set of amplitudes A_G'' given by

$$A_G'' = A_G' \frac{\sum\limits_{G} |A_G^{exp}|}{\sum\limits_{G} |A_G'|} \tag{11.25}$$

are then used to solve for $\zeta(R_i)$, where the R_i are a discrete set of points in the unit cell. The imaginary part of $\zeta(R_i)$ is discarded and a least-squares procedure is used to calculate new coefficients ζ_i. The ζ_i are used to repeat the procedure outlined above until no further change in their values is observed. At this point m is increased to a somewhat higher value. The entire procedure is repeated until $m = 1$, when the corrugation has slowly been brought up to the true value and the ζ_i are the best-fit parameters to the true corrugation $\zeta(R)$.

Such a procedure has the advantage that experimental data can be quickly examined with a laboratory sized computer (the method was developed on the LSI 11/23) and the main structural features can be extracted. More detailed modeling can be carried out at a later data, whereas it can be established immediately if experiments under other conditions of parameters such as E_i or θ_i would be advantageous. In view of the complexity of the measurements involved, the immediate feedback is of great utility.

11.3.4 The Coupled Channel Method for Calculating Diffraction Intensities

The above discussion has emphasized the need for calculating diffraction intensities using a potential which is more realistic than the hard corrugated wall. The coupled channel method is well suited to do this and in the following we follow the arguments of *Liebsch* and *Harris* [11.37] and *Wolken* [11.38].

Since the atom-surface potential will show the two-dimensional periodicity characteristic of the surface, it can be written as

$$V(\mathbf{r}) = \sum_{\mathbf{G}} V_{\mathbf{G}}(z) \exp(i\mathbf{G}\mathbf{R}) \quad . \tag{11.26}$$

If $V(\mathbf{r})$ is known, $V_{\mathbf{G}}(z)$ can be calculated from

$$V_{\mathbf{G}}(z) = A^{-1} \int V(\mathbf{r}) \exp(-i\mathbf{G}\mathbf{R}) \, d^2\mathbf{R} \tag{11.27}$$

where the integration is over the unit cell of area A.

The wave functions describing the scattered particles must satisfy the Schrödinger equation

$$\left[-\left(\frac{h^2}{2m}\right)\nabla_r^2 + V(\mathbf{r}) - \left(\frac{h^2k^2}{2m}\right) \right] \psi_\alpha(\mathbf{r}) = 0 \tag{11.28}$$

and because of the two-dimensional surface periodicity, $\psi_\alpha(\mathbf{R})$ can be written as

$$\psi_\alpha(\mathbf{r}) = \sum_{\mathbf{G}} \psi_{\mathbf{G}\alpha}(z) \exp[i(\mathbf{K} + \mathbf{G})\mathbf{R}] \quad , \tag{11.29}$$

where the $\psi_\alpha(z)$ are unknown functions of z. Substituting (11.29) in (11.28), multiplying on the left by $\exp[-i(\mathbf{K} + \mathbf{G})\mathbf{R}]$ and integrating over the two-dimensional unit cell, (11.28) becomes

$$\sum_{\mathbf{G}} \delta_{\mathbf{G}\mathbf{G}'} \left[\frac{h^2}{2m} (\mathbf{K} + \mathbf{G})^2 - \frac{h^2k^2}{2m} - \frac{d^2}{dz^2} \right] \psi_{\mathbf{G}\alpha}(z) + \sum_{\mathbf{G}} \psi_{\mathbf{G}}(z) \left\{ A^{-1} \int \exp(-i\mathbf{G}'\mathbf{R})V(\mathbf{r}) \right.$$
$$\left. \exp(i\mathbf{G}\mathbf{R}) \, d^2\mathbf{R} \right\} = 0 \quad . \tag{11.30}$$

Substituting (11.26) into the above equation we obtain

$$\frac{h^2}{2m} \left\{ \left[k^2 - (\mathbf{K} + \mathbf{G})^2 \right] + \frac{d^2}{dz^2} \right\} \psi_{\mathbf{G}\alpha}(z) = \sum_{\mathbf{G}'} V_{\mathbf{G}-\mathbf{G}'}(z) \psi_{\mathbf{G}'\alpha}(z) \quad . \tag{11.31}$$

This coupled set of linear second-order differential equations is to be solved for the particular solutions $\psi_{\mathbf{G}\alpha}(z)$. The solution to the scattering problem must satisfy the boundary conditions

$$\psi(\mathbf{r}) \longrightarrow 0 \quad \text{as} \quad Z \longrightarrow -\infty$$
$$\psi(\mathbf{r}) \longrightarrow 0 \quad \text{as} \quad Z \longrightarrow +\infty$$

for all closed channels which satisfy the conditions $k_{\mathbf{G}z}^2 < 0$ and

$$\psi(\mathbf{r}) \longrightarrow \exp[i(\mathbf{K}\mathbf{R} + k_{iz}z)] + \sum_{\mathbf{G}} A_{\mathbf{G}} \exp[i(\mathbf{K} + \mathbf{G})\mathbf{R}] \exp[ik_{\mathbf{G}z}z] \tag{11.32}$$

as $z \rightarrow +\infty$, and for all open channels which satisfy the condition $k_{\mathbf{G}z}^2 \geq 0$. Here $\psi(\mathbf{r})$ can be expressed as a linear combination of $\psi_\alpha(\mathbf{R})$ and the coefficients define the complex scattering amplitudes $A_{\mathbf{G}}$. Methods of solving these equations have been discussed elsewhere [11.37,38]. In principle the expansions for $V(\mathbf{r})$, (11.26), and $\Psi(\mathbf{r})$, (11.29), have an infinite number of terms but only a relatively small number of components need be considered. The number of coupled diffraction channels depends on λ_i, θ_i and the strength of the coupling parameters $V_{G-G'}(z)$ but lies in the range 20-80. For the relatively smooth surfaces which have to date been examined with the method only 1-3 coefficients $V_{G-G'}(z)$ have been considered.

The importance of the coupled channel approach to calculating diffraction intensities is that it is in principle exact and can accommodate realistic potentials. The coupling of the incident and diffracted waves to the bound states is fully included and resonant scattering effects such as those shown in Fig.11.3 can be calculated. However, the inclusion of inelastic effects is necessary to describe scattering intensities accurately in the vicinity of resonances [11.39].

11.4 Experimental Tests of the Atom-Surface Potential

Resonant scattering experiments have been used to map out the atom-surface potential, most extensively for the helium-graphite interaction [11.40]. However, the experiments are most sensitive to the attractive part of the potential and give little information on the repulsive part of the potential. Since the opposite is true for diffraction experiments, a suitable test for the total potential should be its compatibility with both resonant scattering and diffraction experiments.

Diffraction intensities have been recently measured for Cu{110} [11.41] and Ni{110} [11.42] for a number of incident energies E_i and have been analyzed using the close coupling methods discussed in Sect.11.3.

The copper results [11.41] show the expected increase in $A_\mathbf{G}$ with E_i consistent with an increase in the amplitude of $\zeta(x,y)$ as the incoming particle penetrates the electron charge distribution more deeply. In interpreting these results, *Garcia* et al. [11.43] have proceeded from the potential given in (11.3) using $\rho(\mathbf{r})$ calculated from the superposition of atomic charge densities. The Nørskov potential (11.2) was used including an appropriate averaging of $\rho(\mathbf{r})$ over the spatial extent of the He atom.

The corrugation function $\zeta(x,y)$ was expressed as

$$\zeta(x,y) = \frac{1}{2}\left[\zeta_{10}\cos\frac{2\pi x}{a_1} + \zeta_{01}\cos\frac{2\pi y}{a_2}\right] + \frac{1}{2}\zeta_{11}\cos\left(\frac{2\pi x}{a_1}\right)\cos\left(\frac{2\pi y}{a_2}\right) \quad,(11.33)$$

Table 11.1. Comparison of corrugation parameters

	E_z(mev)	ζ_{10}(Å)	ζ_{01}(Å)	α(Å^{-1})
Garcia et al. [11.43]	3.5	0.004	0.057	2.02
(surface layer relaxed by 10%)	46.0	0.015	0.107	2.02
Liebsch et al. [11.41]	3.5	-	0.052	2.63
(unrelaxed surface)	46.0	-	0.079	2.63

where $a_1 = 2.54$ Å and $a_2 = 3.60$ Å are the unit cell lengths in the [1$\bar{1}$0] and [110] azimuths respectively. The charge density calculations gave the results shown below in Table 11.1 for an inward relaxation by 10% of the topmost layer as has been found with LEED [11.44]. In these calculations ζ_{11} was found to be negligible. The same data was analyzed by *Liebsch* et al. [11.41] using a potential

$$V(\mathbf{r}) = \int_{-\infty}^{\varepsilon} F \, d\varepsilon g(\varepsilon)\rho(\varepsilon,\mathbf{r}) \quad , \tag{11.34}$$

in which ε_F is the Fermi energy of the metal, $g(\varepsilon)$ is a known smooth function of energy which depends only on the properties of an isolated He atom, and $\rho(\varepsilon,\mathbf{r})$ is the local density of states of the surface.

The resultant potential can be written as

$$V(\mathbf{r}) = V \exp\{-\alpha[z - h(x,y,z)]\} - \frac{C_0}{(z - z_0)^3} \tag{11.35}$$

with

$$h(x,y,z) = \left(\frac{1}{2} \zeta_{10} \cos \frac{2\pi x}{a_1}\right) \exp[-\beta_{10}(z - z_0)] \tag{11.36}$$

and the authors have chosen to neglect the corrugation along the [1$\bar{1}$0] azimuth. This potential is very similar to that of (11.3), the only difference being the unequal falloff lengths α and β_{10} for the $\mathbf{G} = 0$ and $\mathbf{G} \neq 0$ components of $V(\mathbf{r})$.

In both these studies, the parameters V, α, C_0, z_0 and β were obtained from charge density calculations, previously determined values from the van der Waals potential, and in part they have been used as fitting parameters. Table 11.1 compares the results of the two studies at two values of the normal component of energy, E_z. There is some uncertainty in the values of ζ_{01} obtained by *Liebsch* et al. [11.41] due to our inability to extract accurate data from their Fig.5.

Although these potentials differ somewhat, both give good fits to the diffraction intensities as is seen in Figs.11.5,6. Figure 11.7 shows the resultant potential and the dependence of ζ_G on the normal component of energy using the results of *Garcia* et al. [11.43] and Fig.11.8 shows the corresponding results obtained by *Liebsch* et al. [11.41].

A further interesting comparison between these two studies is the electron charge distribution at the surface which for *Garcia* et al. [11.43] is an input parameter in calculating the diffraction intensities, while for *Liebsch* et al. [11.41] it is derived from a best fit to the diffraction data. Figure 11.9 shows the laterally averaged and corrugated parts of the surface electron density for the charge density $n(\mathbf{r})$ calculated from the local density of states. Also shown in this figure is $n^A(\mathbf{r})$, calculated from a superposition of atomic charge densities. Absolute charge densities calculated in this manner are not very accurate. However, $n_1(z)$ and $n_1^A(z)$, which are the corru-

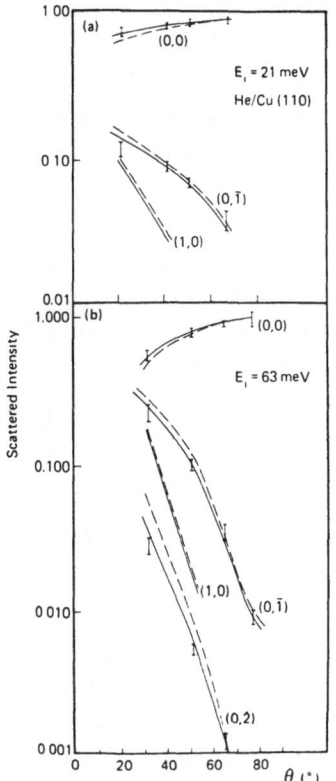

Fig.11.5a. Experimental intensities for E_i = 21 meV as a function of the polar angle of incidence θ_i for He diffraction from Cu{110} are indicated by vertical bars. Intensity calculations for a relaxed and unrelaxed surface layer are shown by solid and dashed lines, respectively; b) as above for E_i = 63 meV [11.43]

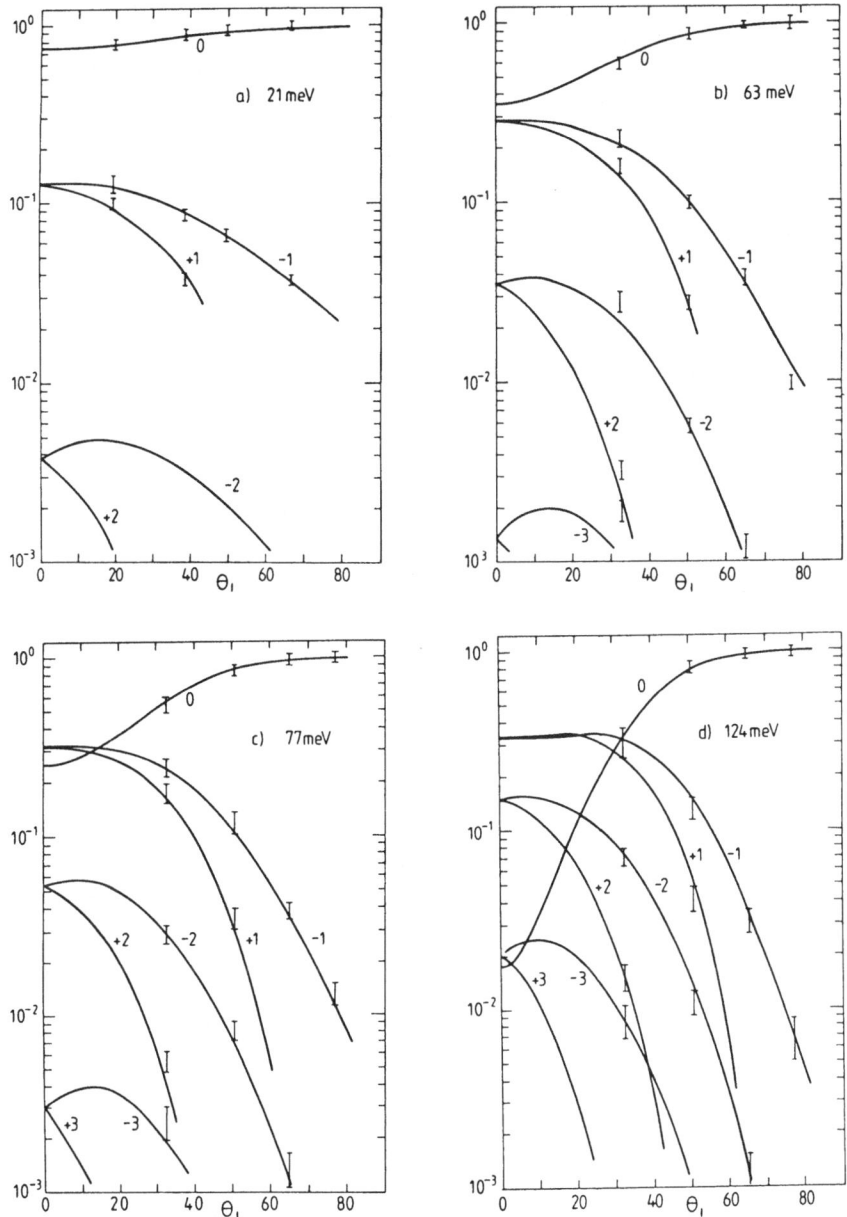

Fig.11.6a. Comparison of calculated (solid lines) and measured (vertical
bars) He diffraction intensities from Cu{110} as a function of θ_i for an incident energy $E_i = 21$ meV; b) as above for $E_i = 63$ meV; c) as above for 77 meV;
d) as above for 124 meV [11.41]

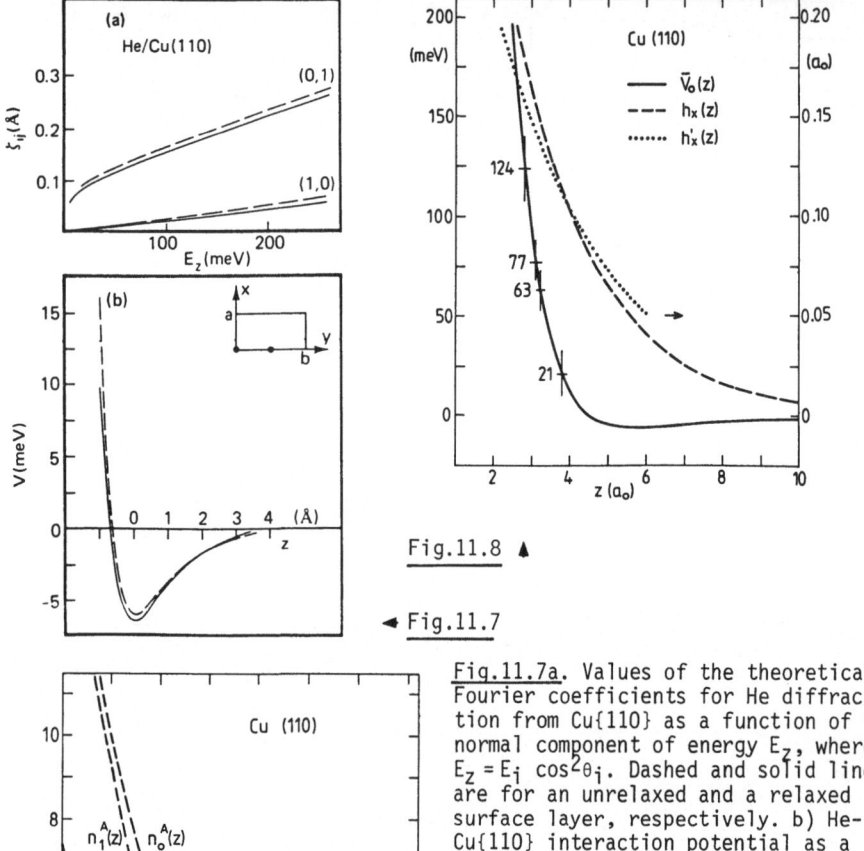

Fig.11.7a. Values of the theoretical Fourier coefficients for He diffraction from Cu{110} as a function of the normal component of energy E_z, where $E_z = E_i \cos^2\theta_i$. Dashed and solid lines are for an unrelaxed and a relaxed surface layer, respectively. b) He-Cu{110} interaction potential as a function of Z for x = y = 0 (dashed line) and x = 0, y = b/2 (solid line). Only the relative position of the two potential curves along the Z axis is relevant [11.43]

Fig.11.8. Laterally averaged potential $\overline{V_0}(z)$ and the corrugation parameter ζ_{01} [denoted by $h_x(z)$ and $h_x'(z)$] as a function of z, where the equilibrium He-surface distance is 3.0 Å. The numbered vertical bars indicate the z value of the classical turning point for various energies E_i [11.41]

Fig.11.9. Variation of the surface electron charge with the distance Z for a Cu{110} surface. The solid lines represent the densities extracted from a fit to the diffraction data and the dashed lines represent the density calculated from an asymptotic form of overlapping atomic charge densities. The subscripts 0 indicate the laterally averaged and the subscripts 1 indicate the corrugated parts of the electron charge density [11.36]

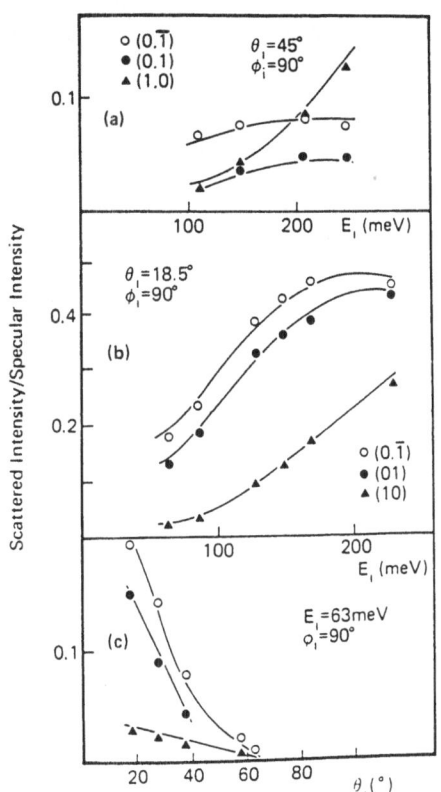

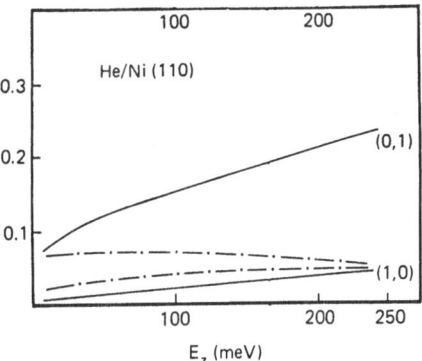

Fig.11.10. Helium diffraction intensities from Ni{110} as a function of θ_i for various incident energies E_i. The symbols are the experimental intensities, the solid lines the calculated intensities [11.45]

Fig.11.11. Values of the Fourier coefficients of the corrugation function as a function of E_z, the Z component of the incident energy, for He diffraction from Ni{110} [11.45]

gated parts, taken at the same value of the laterally averaged change density, allow a more meaningful comparison. For the relevant densities, the corrugation amplitude from atomic charge superposition is a factor of 4 larger than that derived from the diffraction intensities. The authors attribute this to the itinerant nature of the metallic electrons, which is not taken into account by superposing charge densities.

Therefore although both groups can fit the diffraction intensities well, as shown in Figs.11.5,6, there is a lack of quantitative agreement on the parameters of the potential and even as to whether the surface corrugation can be accurately modeled using the superposition of atomic charge densities.

Similarly, calculations of the electron charge density for the Ni{110} surface using both methods discussed in Sect.11.2 agree on the maximum amplitude of 0.15 Å [11.25,26]. However, this cannot be reconciled with the experimentally determined value of 0.05 Å at $E_i = 63$ meV [11.11]. *García* et al. [11.45] have also tried unsuccessfully to fit diffraction data on Ni{110} [11.42] in which the corrugation parameters along the [1$\bar{1}$0] and [001] directions do not vary with E_i as would be predicted. Figure 11.10 shows a comparison of the calculated and experimentally determined diffraction intensities which agree well. The energy dependence of the corrugation parameters for this best fit are shown in Fig.11.11 along with those calculated from a superposition of atomic charge intensities and it is seen that the ζ_{01} coefficient corresponding to the [001] direction remains constant with E_z, whereas the calculated value shows the expected dependence. The reason for this discrepancy is at present not understood.

Although the above results show that diffraction from relatively smooth surfaces can be modeled using realistic potentials and coupled channel methods, less success has been reported on the more strongly corrugated semiconductor surfaces where similar approaches have not been successful in generating structural models [11.10]. It is apparent from these efforts that it is more critical to approximate the true potential closely by a trial potential for strongly corrugated surfaces than it is for relatively flat surfaces.

In summarizing this section, it can be said that rapid progress in the development of a realistic atom-surface potential has been made. Corrugation surfaces based on atomic models can readily be constructed and as shown in the following section, are often able to decide between alternative models of surface structure. However, a clearer understanding of the limits of validity of various methods for calculating $\rho(\mathbf{r})$ coupled with progress in formulating a microscopic atom-surface potential are necessary before atom diffraction can become truly quantitative.

11.5 Current Applications of Atom Diffraction in Surface Science

11.5.1 Atom Diffraction as a Surface Technique

Due to the extremely low incident energies ($E_i = 10 - 200$ meV) and large size of the incident particles, atom diffraction is a highly surface-sensitive technique with no penetration of the topmost layer for reasonably close packed surfaces. It is also the only surface structural technique where the scattering is from the electron charge distribution rather than from the ion cores. Consequently, incident atoms are scattered strongly from surface consisting of atoms of low atomic number. Primarily·for these reasons, atom diffraction is particularly useful in the following areas: hydrogen chemisorption, surface reconstructions and incommensurate layers, and insulator surfaces. Research in these areas is discussed individually below. Although atom diffraction experiments can be carried out on any well-ordered surfaces, only areas will be emphasized in which atom diffraction has particular advantages over other techniques. The reader is referred to [11.6] for a more comprehensive discussion.

11.5.2 Chemisorption of Hydrogen on Surfaces

Despite its importance in a wide range of chemical processes at solid surfaces ranging from catalysis to energy storage, little is known about the structure of hydrogen adlayers. Only one LEED study in which I-V curves have been analyzed has involved hydrogen [11.46], primarily because of its weak scattering power for electrons in the energy range suitable for LEED. Figure 11.12 shows a corrugation surface $\zeta(x,y)$ for hydrogen dissociatively adsorbed on Ni{110} at a coverage of $\theta_H = 0.8$ [11.11] together with the corresponding structural model. The model is derived from $\zeta(x,y)$ by associating the maxima in the electron charge distribution with underlying hydrogen atoms. Since the nickel substrate is very weakly corrugated, it is not possible from the diffraction data alone to establish the registry between the adlayer and the substrate. The model chosen, however, has sites of highest symmetry and is eminently plausible from chemical arguments.

The most striking feature of this structure is the preference of a majority of the hydrogen atoms for asymmetric sites. This can be understood on the basis of repulsive interactions in the adlayer and upon closer examination it is seen that the hydrogen adatoms form a distorted hexagonal array which results from an energy balance between adsorption at sites of high symmetry and a hexagonal array in which the repulsive forces are minimized. Further adsorption of hydrogen up to a coverage of $\theta_H = 1$ leads to the structure shown in Fig.11.13 in which all the hydrogens are at sites of low symmetry, again showing the dominance of repulsive interactions in determining the adlayer structure.

These examples, which resulted from an analysis of diffraction data using the hard corrugated wall model (Sect.11.3), illustrate that considerable insight can be gained about the surface geometry without a realistic atom-surface potential. It would be difficult to obtain the same structural information with LEED due to the low scattering cross section of hydrogen and to the size of the unit cell for the (2×6) structure shown in Fig.11.11 which exceeds that routinely dealt with in LEED calculations. On the basis of these examples, it can be concluded that atom diffraction is currently able to reveal more structural details of hydrogen adlayers than alternative surface-science techniques. However, until an accurate atom-surface potential is

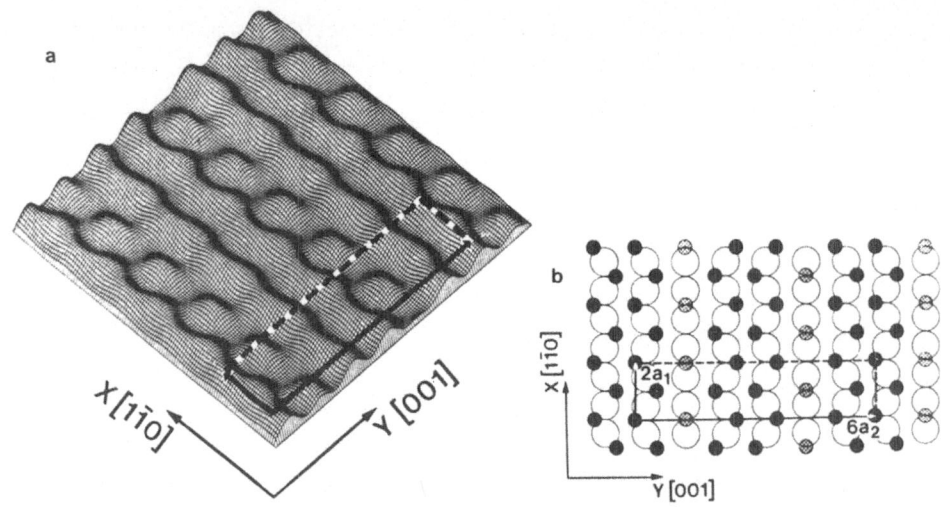

Fig.11.12a. Best fit corrugation function for the (2×6) phase of hydrogen adsorbed on Ni{110}. The surface unit cell is indicated: the smaller dimension is 4.98 Å. The vertical scale has been expanded and is 0.26 Å peak to peak. b) Hard sphere model for the (2×6) phase. The small filled and shaded circles represent H atoms and the large circles represent the outermost layer of the Ni{110} substrate. The surface unit cell is indicated [11.11]

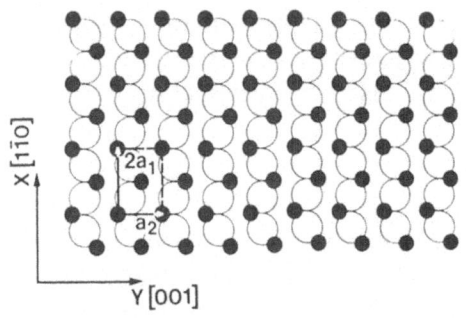

Fig.11.13. Hard sphere model for the (2×1) adsorption phase. The small filled circles represent H atoms and the large circles represent the outermost layer of the Ni{110} substrate. The surface unit cell is indicated [11.11]

available, the positions of the hydrogen atoms vertical to the surface cannot be established in other than a qualitative sense.

11.5.3 Reconstruction of Adlayer Surfaces

As seen in the previous section, adsorption on top of the substrate gives rise to a two-dimensional corrugation with well-defined maxima corresponding to the adsorbed atoms. It is also known [11.47,48] that adsorption can lead to a reconstruction of the surface in which the adsorbate lies in or below the substrate plane. This phenomenon is particularly important in corrosion

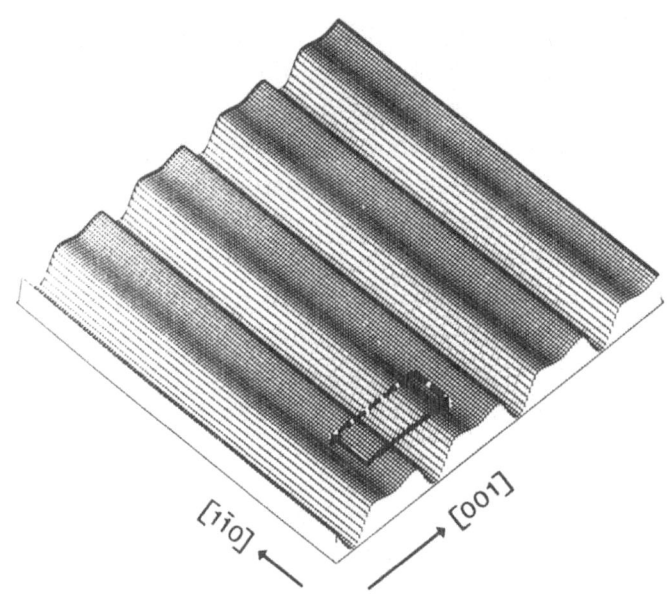

Fig.11.14. Best fit corrugation function of the (1 × 2) phase of hydrogen adsorbed on Ni{110}. The surface unit cell is indicated. Note the one-dimensional character of the surface when compared with Fig.11.12a [11.1]

since it precedes the formation of 3-dimensional oxide nuclei and material loss. However, it is difficult to establish the occurrence of such a reconstruction and unless quantitative LEED calculations or Rutherford backscattering [11.49] data are available, indirect evidence such as work function changes [11.50] or ion-scattering intensities [11.51] have been used to verify reconstruction.

Surface reconstruction can lead to characteristic changes in $\zeta(x,y)$ which allow one to distinguish it clearly from adsorption on the surface as is seen in Fig.11.14. Upon adsorption of more than one monolayer of hydrogen, the Ni{110} surface corrugation shows a highly anisotropic form. This cannot be due to adsorption on the surface since those electrons localized on the adsorbate atoms will show an approximately clyindrical symmetry about the bond axis. The itinerant electrons in the metal can show such a large anisotropy and this suggests that the hydrogen atoms are located in the plane of the first layer nickel atom. The occurrence of a surface reconstruction has been confirmed with LEED [11.52] and can be easily established by atom diffraction simply from the form of the corrugation function. Similarly, dissociative adsorption of oxygen has been observed on Ni{110} [11.53] and Cu{110} [11.54] to lead to surface reconstruction using atom diffraction. However, in these cases, the surface is reconstructed in the [1$\bar{1}$0] azimuth rather than in the [001] azimuth as is the case for hydrogen adsorption.

In cases such as those described above, the structural information obtained using atom diffraction is inherently qualitative in that the adsorbate atoms cannot be located even if the atom-surface potential is accurately known, since they make no appreciable contribution to the electron density sampled by the diffracting atoms. However, the substrate positions can be ascertained from experimentally determined corrugation functions if scattering potentials are available through modeling studies.

11.5.4 Diffraction from Incommensurate Layers or Layers with Large Unit Cells

Complex LEED patterns have been frequently observed for clean [11.55] and adsorbate covered surfaces [11.48] and have been attributed to multiple diffraction involving the adlayer and the substrate [11.56]. Although these patterns contain the complete structural information of the selvedge region, it is often too complex to allow a unique interpretation, and it would be of great advantage to investigate the topmost layer separately. Since low-energy He atoms do not penetrate the surface, this separation is possible for reasonably close packed layers using atom diffraction. A recent example is the investigation of the reconstructed Au{100} surface [11.57]. LEED investigations show a C(26 × 68) periodicity [11.55], which is not meant to imply a commensurate layer, whereas $\zeta(x,y)$ as determined from atom diffraction experiments [11.57] is one-dimensional and is consistent with the topmost layer having a (1 × 5) periodicity. The corrugation function is shown in Fig.11.15 and two possible models consistent with this result are shown in Figs.11.16, 17. It is not uniquely possible to establish the structure since the flat portions in both unit cells would have a vanishingly small corrugation and therefore not allow a distinction between the two symmetries shown.

However, the atom diffraction experiment shows clearly that the major reconstruction has a (1 × 5) unit cell which may be due to dislocation lines originating from the high surface packing density. It can be concluded that the C(26 × 68) LEED symmetry is due to secondary atomic shifts rather than to major structural features.

The unique surface sensitivity of atom diffraction is well suited for investigating similar cases and has been able to show that the Si{111}-(7 × 7) reconstruction is characteristic of the surface layer rather than being a multiple diffraction effect [11.9].

278

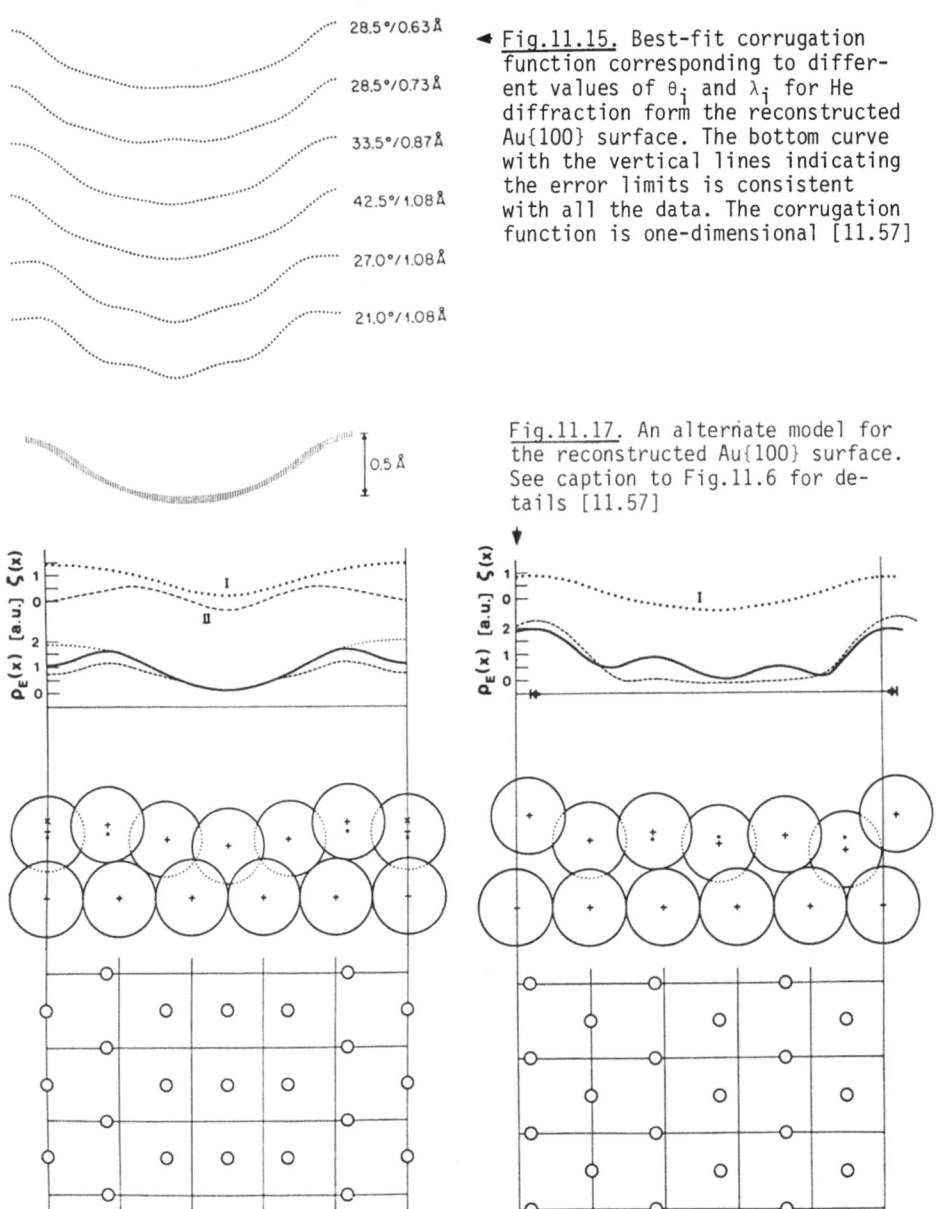

28.5°/0.63Å

28.5°/0.73Å

33.5°/0.87Å

42.5°/1.08Å

27.0°/1.08Å

21.0°/1.08Å

◄ Fig.11.15. Best-fit corrugation function corresponding to different values of θ_i and λ_i for He diffraction form the reconstructed Au{100} surface. The bottom curve with the vertical lines indicating the error limits is consistent with all the data. The corrugation function is one-dimensional [11.57]

0.5 Å

Fig.11.17. An alternate model for the reconstructed Au{100} surface. See caption to Fig.11.6 for details [11.57]

Fig.11.16. *Bottom*: view onto the Au{100} surface. Continuation of the bulk structure up to the surface would have atoms at the intersection of the solid lines. The open circles indicate the position of the Au atoms in the topmost layer. *Middle*: side view of the atom portions. *Top*: corrugation functioning $\zeta(x)$ and surface charge densities $\rho(x)$ for different vertical positions of the atoms in the topmost layer [11.57]

279

11.5.5 Insulator Surfaces

Insulators are a class of materials which are difficult to investigate structurally due to electrostatic charging which arises when using charged particle beams. Since atoms are electrically neutral, no charging occurs when using atomic beams. Of particular interest in the study of insulator surfaces is the possibility of charge redistribution at the surface between ionic species which may become unstable when not surrounded by their full coordination shell.

One of the first surfaces to be investigated was LiF{100} [11.58] and it is reasonably certain that no appreciable charge redistribution or relative movement of neighboring surface atoms takes place on this surface. However, the oxide surfaces which have been examined, NiO{100} [11.59] and MgO{100} [11.60], both show corrugation amplitudes which are appreciably less than what would expected from the ideal surface geometry and the known ionic radii.

For NiO{100}, the maximum corrugation amplitude is 0.28 Å [11.59], which is much smaller than the difference of the ionic radi of 0.68 Å [11.61]. LEED calculations indicate that the positions of the ion cores do not deviate significantly from the ideal configuration [11.62], indicating that a charge redistribution at the surface has taken place. Similarly, for MgO{100} the maximum corrugation amplitude is 0.36 Å, whereas the difference in ionic radii would predict 0.68 Å [11.60].

In these cases, the complementary nature of LEED and atom diffraction is well illustrated. LEED measurements indicate that NiO{100} is unreconstructed since the ion core positions are unchanged, but atom diffraction experiments show that charge redistribution has taken place at the surface, reflecting the charge in chemical bonding.

11.6 Summary

This review of atom diffraction from solid surfaces has been intended to supplement [11.5,6], and particular emphasis has been placed on recent work which has made substantial progress in deriving realistic atom-surface potentials from microscopic considerations. Theories relating the repulsive potential to the electron charge density at the surface together with the application of coupled channel methods for calculating diffraction intensities have been applied to reasonably smooth metal surfaces with considerable success. However, the lack of quantitative agreement of these surfaces and the present inability to generate satisfactory scattering potentials for more strongly corrugated surfaces show that further work is necessary to derive and extend existing potentials.

Even at the present stage of development of atom diffraction in which only qualitative information regarding displacements of surface atoms normal to the surface can be made, the technique has made significant contributions, discussed in some detail in Sect.11.5. Since nearly all our current understanding of the atom-surface potential has evolved in the last three years, further rapid development is to be expected in the coming years. It does not seem unduly optimistic to predict that atom diffraction will provide highly quantitative structural information in the immediate future. In view of the versatility of the beam surface-scattering technique, which without modifications in apparatus is capable of inelastic scattering and reaction kinetic studies, significant applications of the technique in surface science can be expected soon.

References

11.1 J. Estermann, O. Stern: Z. Phys. **61**, 95 (1930)
11.2 G. Boato, P. Cantini, L. Mattera: Surf. Sci. **55**, 191 (1976)
11.3 H.U. Finzel, H. Frank, H. Hoinkes, M. Luschka, H. Nahr, H. Wilsch, V. Wonka: Surf. Sci. **49**, 577 (1975)
11.4 D.R. Frankl, D. Wesner, S.V. Krishmaswamy, G. Derry, T. O'Gorman: Phys. Rev. Lett. **41**, 60 (1978)
11.5 D.V. Tendulkar, R.E. Stickney: Surf. Sci. **27**, 516 (1971)
11.6 T. Engel, K.H. Rieder: "Structural Studies of Surfaces", in Springer Tracts in Modern Physics, Vol. 91 (Springer, Berlin, Heidelberg, New York 1982) p.55
11.7 K.H. Rieder: In *Dynamics of Gas-Surface Interaction*, Springer Series in Chemical Physics, Vol.21 (Springer, Berlin, Heidelberg, New York 1982) p.61
11.8 M.J. Cardillo, G.E. Becker: Phys. Rev. B**21**, 1497 (1980)
11.9 M.J. Cardillo, G.E. Becker: Phys. Rev. Lett. **42**, 508 (1979)
11.10 M.J. Cardillo, G.E. Becker, S.J. Sibener, D.R. Miller: Surf. Sci. **107**, 409 (1981)
11.11 T. Engel, K.H. Rieder: Surf. Sci. **109**, 140 (1981)
11.12 J. Lapujoulade, Y. LeGruer, M. Lefort, Y. Lejay, E. Maurel: Surf. Sci. **118**, 103 (1982)
11.13 H. Hoinkes: Rev. Mod. Phys. **52**, 933 (1980)
11.14 G. Brusdeylins, R.B. Doak, J.P. Toennies: Phys. Rev. Lett. **46**, 437 (1982)
11.15 J.A. Schwarz, R.J. Madix: Surf. Sci. **46**, 317 (1974)
11.16 T. Engel, G. Ertl: J. Chem. Phys. **69**, 1276 (1978)
11.17 R.E. Stickney: Adv. At. Mol. Phys. **3**, 143 (1967)
11.18 E. Zaremba, W. Kohn: Phys. Rev. B**13**, 2270 (1976)
11.19 N. Esbjerg, J.K. Nørskov, C. Umrigar: J. Phys. L**7** (1982)
11.20 M. Manninen, J.K. Nørskov, C. Umrigar: J. Phys. F**12** (1982)
11.21 M.J. Stott, E. Zaremba: Phys. Rev. B**22**, 1564 (1980)
11.22 M.J. Ruska, R.M. Nieminen, M. Manninen: Phys. Rev. B**24**, 3037 (1981)
11.23 J. Harris, A. Liebsch: J. Phys. C**15**, 2275 (1982)
11.24 N.D. Lang, J.K. Nørskov: Phys. Rev. B**21**, 2131 (1980)

11.25 D.R. Hamann: Phys. Rev. Lett. **46**, 1227 (1981)
11.26 D. Haneman, R. Haydock: J. Vac. Sci. Technol. **21**, 330 (1982)
11.27 L. Mattera, C. Salvo, S. Terrjni, F. Tommasini: Surf. Sci. **97**, 158 (1980)
11.28 P.J. Estrup, E.G. McRae: Surf. Sci. **25**, 1 (1971)
11.29 R. Petit, M. Cadilhac: C.R. Acad. Sci. Ser. B**262**, 468 (1966)
11.30 R.F. Millar: Proc. Cambridge Philos. Soc. **69**, 217 (1971)
11.31 P.M. Van den Berg, J.T. Fokkema: J. Opt. Soc. Am. **69**, 27 (1979)
11.32 B. Salanon, G. Armand: Surf. Sci. **112**, 78 (1981)
11.33 M. Manninen, J.K. Nørskov, C. Umrigar: Surf. Sci. **119**, 393 (1982)
11.34 E. Zaremba, W. Kohn: Phys. Rev. B**15**, 1769 (1977)
11.35 K.H. Rieder, N. Garcia, V. Celli: Surf. Sci. **108**, 169 (1981)
11.36 R. James, D.S. Kaufman, T. Engel: Surf. Sci., **133**, 305 (1983)
11.37 A. Liebsch, J. Harris: Surf. Sci. **123**, 355 (1982)
11.38 G. Wolken, Jr.: J. Chem. Phys. **58**, 3047 (1973)
11.39 J.M. Soler, V. Celli, N. Garcia, K.H. Rieder, T. Engel: Surf. Sci. **108**, 1 (1981)
11.40 W.E. Carlos, M. Cole: Phys. Rev. B**21**, 3713 (1980)
11.41 A. Liebsch, J. Harris, B. Salanon, J. Lapujoulade: Surf. Sci. **123**, 338 (1982)
11.42 K.H. Rieder, N. Garcia: Phys. Rev. Lett. **49**, 43 (1982)
11.43 N. Garcia, J.A. Barker, I.P. Batra: J. Electron Spectrosc. **30**, 137 (1983)
11.44 J.R. Noonan, H.L. Davis: Surf. Sci. **99**, L 424 (1980)
11.45 N. Garcia, J.A. Barker, and K.H. Rieder: Solid State Comm. **45**, 567 (1983)
11.46 K. Christmann, R.J. Behm, G. Ertl, M.A. Van Hove, W.H. Weinberg: J. Chem. Phys. **70**, 4168 (1979)
11.47 R.G. Smeenk, R.M. Tromp, F.W. Saris: Surf. Sci. **107**, 429 (1981)
11.48 H.M. Kramer, E. Bauer: Surf. Sci. **93**, 407 (1980)
11.49 L.C. Feldman, J.W. Mayer, S.T. Picraux: *Materials Analysis by Ion Channeling* (Academic, New York 1982)
11.50 H.M. Kramer, E. Bauer: Surf. Sci. **92**, 53 (1980)
11.51 J.A. Van der Berg, L.K. Verheij, D.G. Armour: Surf. Sci. **91**, 218 (1980)
11.52 T.N. Taylor, P.J. Estrup: J. Vac. Sci. Technol. **11**, 244 (1974)
11.53 T. Engel, K.H. Rieder: To be published
11.54 E. Bauer, T. Engel: Surf. Sci. **71**, 695 (1978)
11.55 M.A. Van Hove, R.J. Koestner, P.C. Stain, J.B. Biberian, L. Kesmodel, I. Bartos, G.A. Somorjai: Surf. Sci. **103**, 187 (1981); **103**, 218 (1981)
11.56 E. Bauer: Surf. Sci. **7**, 351 (1967)
11.57 K.H. Rieder, T. Engel, R.H. Swendsen, M. Manninen: Surf. Sci. **127**, 233 (1983)
11.58 N. Garcia: J. Chem. Phys. **67**, 897 (1977)
11.59 P. Cantini, R. Tatarek, G.P. Felcher: Phys. Rev. B**19**, 1161 (1979)
11.60 K.H. Rieder: Surf. Sci. **118**, 57 (1982)
11.61 H. Megaw: *Crystal Structures —A Working Approach* (Saunders, Philadelphia 1973)
11.62 J.A. Walker, C.G. Kinniburgh, J.A.D. Matthew: Surf. Sci. **78**, 221 (1977)

12. An Atomic View of Crystal Growth[*]

G. Ehrlich

With 12 Figures

12.1 Introduction

Studies of crystal growth have been of scientific interest at least since
the time of Gibbs, but an atomic view of the subject really began to emerge
in the 1920s, through the work of Volmer, Stranski, and Kossel [12.1]. From
these early investigations, largely based on visual observations or on studies with the light microscope, it became evident that growth from the vapor
occurs in a series of steps, shown in Fig.12.1. Atoms from the vapor are captured at a crystal surface on colliding with it. To become part of the crystal they must still reach a growth site, that is a kink in a lattice step.
At low supersaturations of the vapor, incorporation can occur only if the
lifetime of the adatom on the flat terraces is long enough for the atom to
diffuse to a step prior to evaporation. At higher vapor pressures the concentration of adatoms on the flat may be large enough to allow the creation
of new nuclei, at which growth can then proceed [12.2]. All of these processes have been nicely illustrated in the Monte Carlo simulations reviewed
by *Gilmer* [12.3]. The creation of epitaxial layers of a foreign material on
a crystalline substrate is, however, more complicated [12.4]. As is stressed
in *van der Merwe's* review [12.5], epitaxy is significantly affected by the
strength of the interactions between adatoms as well as between adatoms and
substrate. Nevertheless, the atomic events in both crystal and overlayer
growth involve the same steps:

condensation,
migration of adatoms,
atomic incorporation,
and formation of surface clusters.

---

*Supported by the National Science Foundation under Grant NSF DMR 82-01884.

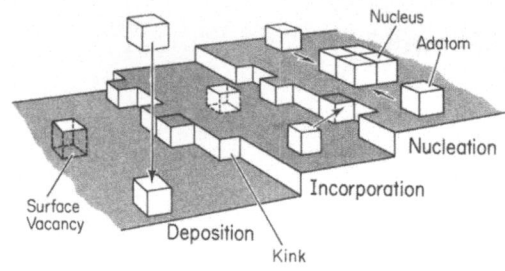

Fig.12.1. Volmer-Stranski picture of crystal growth from the vapor phase

This brief review [12.6] will concentrate on actual observations on the atomic level of these individual steps.

12.2 Observation of Single Adatoms

Only two experimental techniques have so far been demonstrated to provide information about surface events on the atomic level: field ion microscopy [12.7] and electron microscopy [12.8,9]. In the field ion microscope a surface shaped in the form of a fine needle is imaged by applying a high field. The latter causes the preferential ionization of rare-gas atoms, introduced into the microscope to a pressure of $<10^{-3}$ Torr, at atomic protrusions on the surface. The ions so created are accelerated by the high field toward a suitable detector, where an image such as in Fig.12.2 is formed with atomic resolution. Although operation of the microscope requires high fields, this has not limited the application of the field ion microscope to metals ranging from aluminum to tungsten, as well as to semiconductors.

Scanning electron microscopy has also been successful in visualizing individual atoms at a surface; so far, observations have been limited to heavy atoms on an amorphous carbon substrate to optimize contrast [12.10]. No investigations of atomic behavior on crystals have yet been reported. However, transmission electron microscopy [12.11] has begun to yield interesting information about the growth of monatomic metal layers on metal substrates, and about the properties of very small particles [12.12].

A new technique, scanning tunneling microscopy, has recently been reported by IBM Zurich as having the capability of revealing surface information on the angstrom level [12.13]. This method appears to be an adaptation of the Topografiner developed at the National Bureau of Standards in 1972 [12.14]. The IBM reports, though preliminary, are exciting and it will be interesting to see how far this promising method can be pushed. At the moment, however, field ion microscopy is still unique in providing data about the atomic pro-

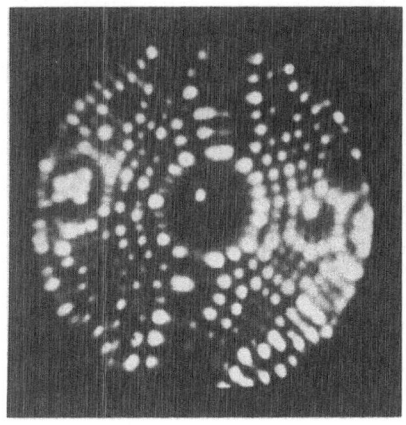

Fig.12.2. Field ion image of a <211> oriented W emitter, with a single atom on the central {211} plane. Courtesy of K. Stolt

cesses in layer growth, and all the information presented here has been obtained using this rather simple technique.

12.3 The Atomic Steps in Layer Growth

12.3.1 Capture and Condensation of Atoms

The capture of metal atoms, on metal surfaces at least, is known to occur with high efficiency [12.15]; rather less is known about the rate at which energy is transferred from the colliding atom to the lattice. This process is important for understanding crystal growth. If deexcitation occurs instantly, on the time scale of a vibrational period, then subsequent movement of adatoms toward growth sites must occur as an ordinary diffusion process. On the other hand, if deexcitation after capture is slow, the possibility exists that excited adatoms may contribute significantly toward transport.

The latter possibility has been tested in the field ion microscope [12.16]. In the arrangement customary for studies of metal adatoms, deposition on the sample occurs from the side. If thermalization of captured atoms is indeed rapid, then their location on the surface provides a faithful indication of where they first land and the deposit should end at the geometrical shadow line. However, if excited atoms contribute significantly to diffusion, then adatoms could cover even those portions of the sample not in direct line of sight to the source. Experiments to test thermalization have been done with tungsten atoms from a source at ≈ 3000 K condensing on a tungsten surface at ≈ 20 K. As is apparent from Fig.12.3, there is little spreading beyond the boundaries set by geometry. We conclude that thermalization after capture is

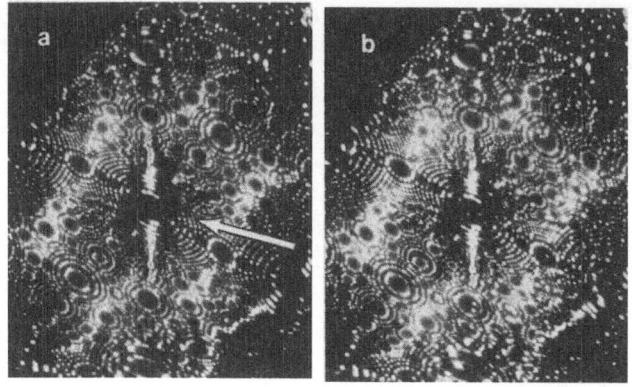

Fig.12.3a,b. Condensation of W atoms on a W substrate at ≈20 K [12.16].
a) Starting surface, with {110} at center. Arrow indicates direction of
vapor stream; b) same surface, after deposition of W from source at ≈3000 K.
The deposit is mostly bounded by the shadow line

complete within less than 100 atomic excursions. The extent to which this
result can be extrapolated to other materials still should be tested, but
this experiment suggests that movement to growth sites is an ordinary thermal process.

The likelihood that atoms will actually reach a lattice step after landing on a smooth terrace will therefore depend primarily upon the particulars
of surface diffusion, and upon the strength of the interactions binding the
atom to the terrace. About the latter, rather little is known. During the
beginnings of field ion microscopy it was hoped that information about binding of adatoms at specific surface sites would become accessible through
studies of field evaporation. If evaporation in a high field occurs over a
Schottky saddle, that is by removal of an ion over a barrier created by the
superposition of the applied field and the image potential between the ion
and the surface, then knowledge of the field at which desorption occurs will
indeed provide information about the strength of binding [12.7]. Such measurements are quite simple and have been made for a number of different adatoms on different planes of tungsten [12.17,18]. However, there is no indication that the Schottky model is applicable [12.19]. Until that can be demonstrated, the values of the binding energy derived in this way must be
suspect.

12.3.2 Atom Diffusion

Much has already been said about the migration of atoms on solids [12.20], as this is the first process about which field ion microscopy revealed any quantitative detail. The field ion microscope provides information about the location of individual adatoms, as in Fig.12.4. By determining the location of an atom before and after an interval t at a high temperature at which diffusion can occur, it is simple to obtain a quantitative measure of the mean-square displacement $<\Delta x^2>$. The Einstein relation

$$<\Delta x^2> = 2Dt \tag{12.1}$$

provides a connection to the diffusivity of the adatoms. If migration occurs by hops of a single atom from one surface site to an adjacent one at a rate Γ, then

$$<\Delta x^2> = \Gamma t \ell^2 \tag{12.2}$$

and

$$D = D_0 \exp(-\Delta E_m/kT), \quad D_0 = \nu \ell^2/2 \quad . \tag{12.3}$$

Here ΔE_m is the barrier opposing motion, ν is a characteristic vibrational frequency, on the order of kT/h, and ℓ is the jump length, which in this model is just the spacing between sites.

That this simple picture describes actual atomic motion on metals reasonably well was already shown by the earliest studies in the field ion microscope [12.22]. We expect jumping of atoms to be activated, and that is in fact what is demonstrated by the diffusion data in Fig.12.5. Also, if migration occurs by the random movement of an adatom, then the mean-square displacement should increase linearly with the length of the diffusion interval t. This dependence was confirmed in early studies of self-diffusion on W{110}. More recently work has focused upon the details of the jump process itself. Do atoms really jump between nearest-neighbor sites, as assumed earlier, or do they make excursions over long regions of the surface? This important question can be settled by examining not just the mean-square displacement, which yields the diffusivity D, but the entire distribution curve describing the probability of finding an adatom at a given distance from its origin after a short time interval. Although only little has been done to measure such distributions, so far all indications are that jumps to nearest-neighbor sites predominate [12.23].

The overall magnitude of the barrier to atomic motion on flats is also noteworthy. For tungsten on W{110} this amounts to only 1/10 the heat of sublimation of the substrate. We might expect the energetics of diffusion to

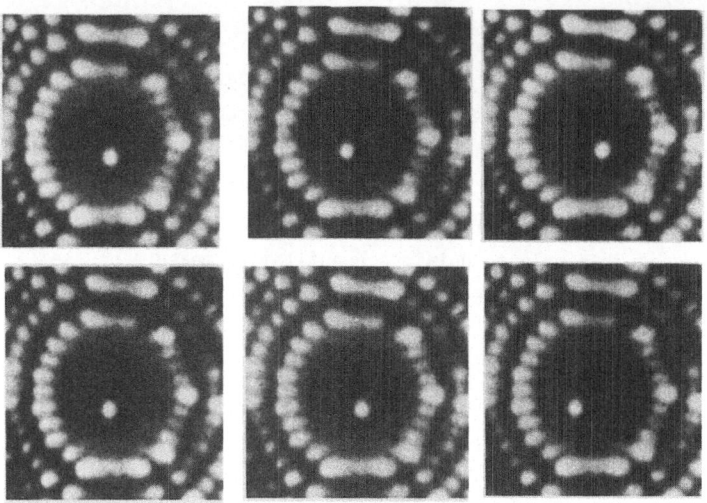

Fig.12.4. Observations of a single Re atom after diffusion on a W{211} plane. Movement occur during 60 s intervals at 327 K, in the absence of applied fields [12.21]

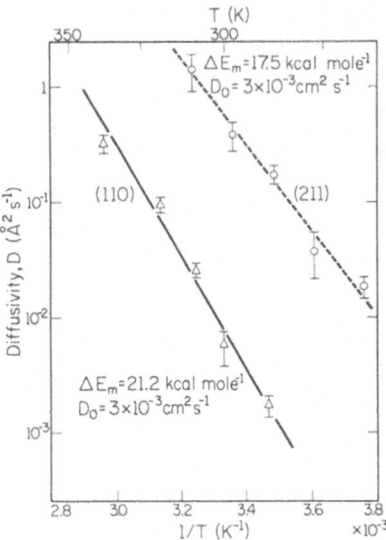

Fig.12.5. Temperature dependence of W adatom self-diffusion on W{110} and W{211} [12.22]

be significantly affected by the atomic arrangement of the substrate, which varies considerably from one plane to the next, as suggested in Fig.12.6. This is not too strikingly shown by the quantitative data on tungsten. Only three planes, the {110}, {211} and {321}, have been examined at all care-fully. The activation energy for atom motion on these amounts to 21, 17, and 20 kcal mole^{-1} respectively [12.20]. However, on rougher planes the diffusion

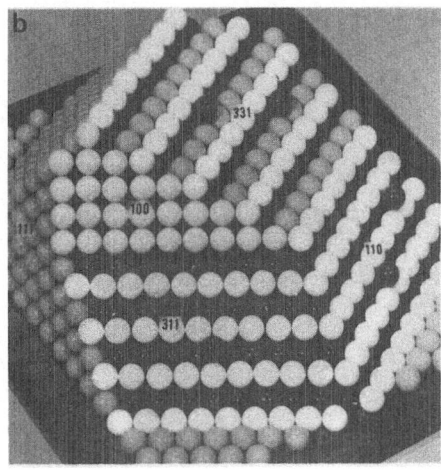

Fig.12.6a,b. Hard-sphere models of bcc (a) and fcc surfaces (b)

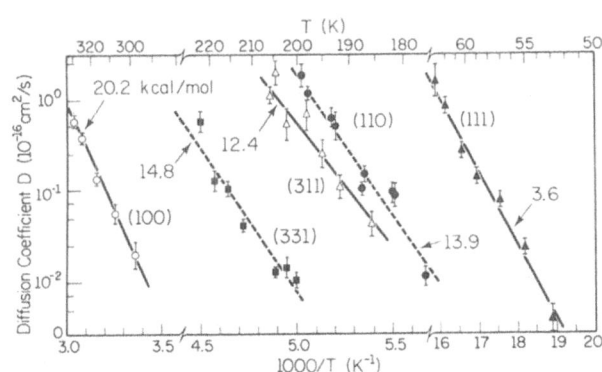

Fig.12.7. Temperature dependence of Rh adatom self-diffusion on different planes of Rh [12.24]

barrier is much higher, so much so that quantitative measurements have not yet been feasible. On fcc metals a larger variety of smooth planes is accessible to study, as is clear from the data in Fig.12.7. For rhodium [12.24], self-diffusion on the close-packed {111} has been found to occur at cryogenic temperatures; on the atomically rougher {100}, atom motion becomes important only around room temperature. Even for the latter plane, however, the barrier for diffusion is only ~1/7 of that for evaporation from the crystal.

A structural effect important in understanding atomic transport toward steps is the strong directionality observed in motion on some surfaces. On a bcc crystal, the {211} plane, for example, consists of channels of close-packed lattice atoms along <111>, apparent in Fig.12.6. Only motion along these channels, not across them, has so far been observed. Unfortunately, the

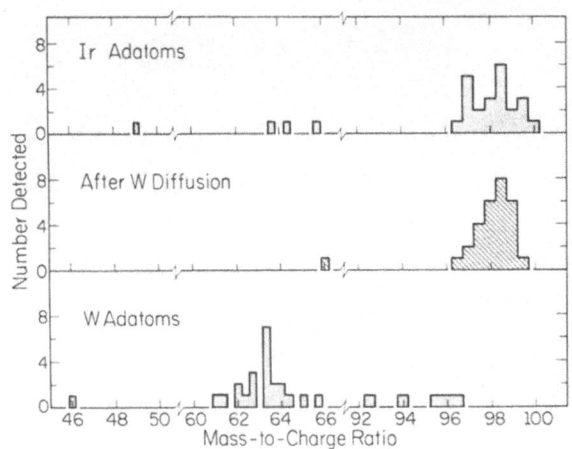

Fig.12.8. Atom probe measurements of cross-channel motion on Ir {110} [12.26]. *Top*: observations of Ir atoms field desorbed from Ir{110} without prior heating. *Bottom*: observations of W atoms under similar conditions. *Center*: after cross-channel motion on Ir{110} with a W adatom on it, atom in adjacent channel is revealed as Ir

geometry of the surface is not always a reliable guide in predicting pre-ferred directions for diffusion, as has become clear from studies on {110} planes of fcc metals. The atomic arrangement of an ideal {110} of an fcc crystal is rather similar to that of {211} in the bcc system, with close-packed rows in the lattice defining the direction along which we might expect preferential diffusion. That has indeed been found in self-diffusion on rho-dium [12.24]; however, on platinum {110} self-diffusion occurs just as often across as along these channels [12.25], and on iridium {110} movement across the channels is strongly preferred [12.26]. Detailed measurements by *Wrigley* for tungsten atoms on iridium {110} using the atom probe [12.7], a combina-tion time-of-flight mass spectrometer and field ion microscope, have revealed that cross-channel motion occurs by a replacement mechanism [12.26]. It fol-lows from the data in Fig.12.8 that a tungsten adatom takes the place of a lattice atom in the channel wall, leaving the lattice atom to carry on the diffusion process. Still unsettled is the question why cross-channel diffu-sion occurs on some fcc metals but not others. There seems to be a correla-tion between cross-channel motion and the reconstruction of the surface, but even that needs clarification. At the moment we can say only that direc-tional effects must be expected in surface transport, but predictions are uncertain.

12.3.3 Adatoms at Surface Steps

The behavior of atoms at steps is of interest as this concerns the critical stage in which the atom is incorporated into the lattice. In the standard view of growth phenomena, incorporation occurs when atoms strike a step regard-

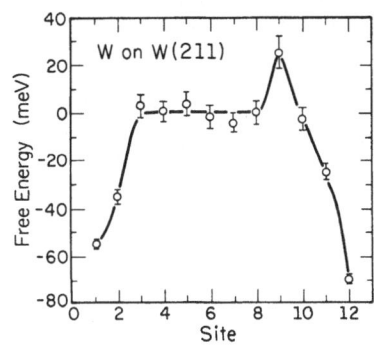

Fig.12.9. Free energy of W adatom bound at different sites in <111> channel on W{211} [12.30]

less of whether it is ascending or descending. Early work with the field ion microscope already showed this picture to be unrealistic. Tungsten atoms, for example, were found to be reflected from descending steps on W{110} and {211} with fair efficiency [12.22]. This is consistent with the view that there is an effective energy barrier at a descending step which most atoms are not able to surmount. Such a barrier is expected from simple pair potential estimates of atom binding [12.27], and attempts have been made to obtain quantitative values of this barrier height by measuring the rate of atom disappearance from W{110} at high temperatures [12.28,29]. So far these experiments have not been entirely satisfactory. Nevertheless they have established the generality of reflection at descending steps.

Rather more is known about atomic behavior at steps on W{211}. The equilibrium distribution of a single tungsten atom over the different sites available on such a plane has recently been measured [12.30]. As shown in Fig.12.9, the free energy varies with position in quite an unexpected way: the two sites next to the step are strongly attractive and there are significant differences in atomic behavior at the two ends of the plane. The latter is expected from symmetry, but the sign and magnitude of the edge effects are not. Interactions are highly specific to the chemical identity of the atom; a rhenium atom shows rather different properties from those of a tungsten atom. These effects must obviously be accounted for in a proper interpretation of the atomic reflection at descending steps — they also suggest that we are dealing with quite a complicated phenomenon.

12.3.4 Clustering of Adatoms

As the concentration of adatoms on a surface increases so does the likelihood of forming a nucleus from which new layers can grow. The field ion microscope has been especially helpful in revealing the properties of these

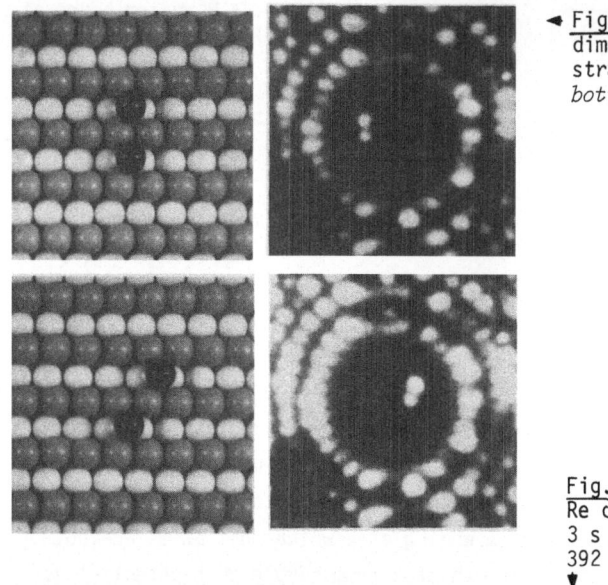

Fig.12.10. Cross-channel dimers on W{211}. *Top*: straight configuration; *bottom*: staggered dimer

Fig.12.11. Observations of Re dimer dissociation [12.31]. 3 s heating intervals at 392 K

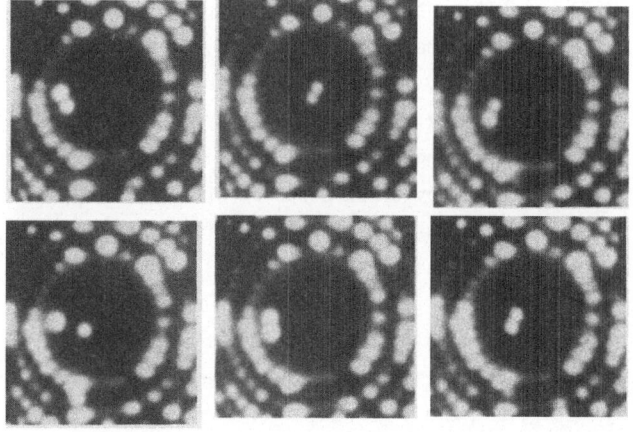

clusters, as well as in providing information about the interactions that are responsible for nucleation. Most detail has been accumulated about rhenium clusters on W{211}. Rhenium dimers with one atom in each of two adjacent rows can exist in two states, straight or 0 and staggered or 1, shown in Fig.12.10. The energetics of these dimers have been probed extensively [12.31]. The staggered state, in which the atoms are further apart than in the straight, is actually more stable. The forces holding the dimer together are rather weak; the energy difference between a dissociated and a staggered dimer amounts only to 3.7 kcal mole^{-1}. Nevertheless, dimers are stable to well over

400 K; that is clear from the observations in Fig.12.11. The fascinating thing about the dimers is that their mobility on the {211} is greater than that of single rhenium atoms [12.21]. This high mobility, however, appears specific to rhenium and has not been found for chemically different clusters [12.32].

All of these properties are also dependent upon the arrangement of the atoms on the substrate. Rhenium dimers formed from two atoms in the same channel are more strongly bound than cross-channel dimers and diffuse more slowly than single atoms [12.33]. When an additional rhenium atom is added to a dimer to form a cross-channel trimer, the properties of this new cluster are similar to those of a cross-channel dimer [12.34]. Addition of a third atom to the channel containing an in-channel dimer eventually leads to a trimer with quite different diffusion properties.

When more and more rhenium atoms are added to the {211} plane of tungsten, a two-dimensional rhenium layer is formed. Atom bonding to such layers has recently been examined by *Fink* [12.35]. From observations of the distribution of a rhenium atom over the different positions in a <111> channel adjacent to such a rhenium layer, the relative strength of binding at these sites has been determined. The fact that binding differs at the two sides of the layer is of course expected from the symmetry of the {211} plane. Quite surprising, however, is the behavior of a cluster with five sites: binding at the central site is weaker than at either of the adjacent sites. This is still another indication that the simple notions of additive interactions which underlie many of the basic ideas in crystal growth are deficient in describing real systems.

Atom clusters have also been studied on other two-dimensional substrates such as W{110}, but in less quantitative detail. Nevertheless, some unexpected behavior has been unearthed here as well. *Bassett* and *Tice* [12.36] have estimated the binding energy of chemically different dimers and found that in the series tantalum, tungsten, rhenium, iridium and platinum, there is a binding energy minimum at rhenium. Rhenium does not form stable dimers. Recent studies in this laboratory have shown, however, that higher clusters of rhenium are stable on this plane [12.37]. Another indication of the complex nature of interactions on W{110} is the behavior of platinum metals. At a low concentration of adatoms these metals form atom chains [12.38], which transform into two-dimensional layers only at higher concentration. These layers are quite loosely packed and their structure has been found to depend sensitively upon the number of adatoms in the layer [12.39]. This number can be adjusted by selective field evaporation; equilibration then

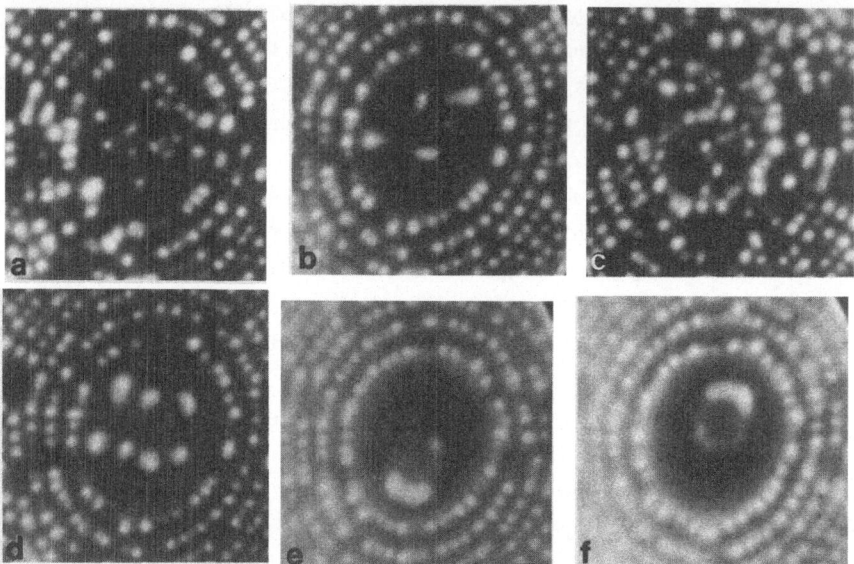

Fig.12.12a-f. Pd clusters on W{110} [12.41]. a) Approximately 20 Pd atoms deposited on W at ≈20 K; b) chains form after heating to 235 K; c) further deposition of Pd; d) two-dimensional clusters formed at 235 K; e) large Pd island, obtained by joining of smaller clusters at 340 K; f) the island sweeps over the surface at 390 K

may produce a different surface mesh and even dissociation into chains. These effects are not limited to the platinum metals—similar loose packing has also been observed with silicon layers on W{110} [12.39,40].

Most interesting for their bearing on layer growth are the dynamic properties of such clusters. The clusters are not stationary entities as originally envisioned. Instead even fairly large aggregates have been found capable of moving over the metal substrate. As shown in Fig.12.12, palladium clusters of more than 50 atoms can sweep over the W{110} surface at temperatures in the vicinity of 400 K [12.41].

12.4 Summary

This has been a brief survey of some insights into layer growth processes obtained by observations on the atomic level using the field ion microscope. It should be evident that such observations have already revealed quite unexpected effects. Most striking is the complexity of the atomic interactions underlying atomic movement and nucleation. Certainly the qualitative notions of atomic bonding which have served as intuitive guides to crystal growth in the past will have to be considerably revised. In addition, these fine-scaled studies have revealed an unexpected richness of phenomena contributing to crystal growth. Despite these complicating factors, it should also be clear that studies of individual atomic behavior are capable of yielding quantitative information not yet attained in any other way, information which is important in building up a sound atomic view of how crystals and overlayers really grow.

Acknowledgments. It is a pleasure to acknowledge the considerable help received from R.S. Chambers, H.-W. Fink, D.A. Reed, K. Stolt, and J.D. Wrigley in the preparation of this review.

References

12.1 For a sketch of the history, see F.C. Frank: Adv. Phys. **1**, 91 (1952)
12.2 Crystal growth theories are reviewed by J.D. Weeks, G.H. Gilmer: Adv. Chem. Phys. **40**, 157 (1979);
H. Müller-Krumbhaar: In *Current Topics in Materials Science,* Vol.1, ed. by E. Kaldis (North Holland, Amsterdam 1978) p.1
12.3 G.H. Gilmer: This volume
12.4 The subject has been examined by R. Kern, G. Lelay, J.J. Metois: In *Current Topics in Materials Science,* Vol.3, ed. by E. Kaldis (North-Holland, Amsterdam 1979) p.131
12.5 J.H. Van der Merwe: This volume
12.6 For further details see G. Ehrlich: In *Proceedings of the 9th International Vacuum Congress and 5th International Conf. on Solid Surfaces,* ed. by J.L. de Segovia (ASEVA, Madrid 1983) p.3
12.7 The technique of field ion microscopy is described by E.W. Müller, T.T. Tsong: *Field Ion Microscopy Principles and Applications* (American Elsevier, New York 1969);
K.M. Bowkett, D.A. Smith: *Field Ion Microscopy* (North-Holland, Amsterdam 1970). For a recent review, see J.A. Panitz: J. Phys. E**15**, 1281 (1982)
12.8 M. Isaacson, D. Kopf, M. Utlaut, N.W. Parker, A.V. Crewe: Proc. Natl. Acad. Sci. USA **74**, 1802 (1977);
A.V. Crewe: Science **221**, 325 (1983)
12.9 K. Takayanagi: Jpn. J. Appl. Phys. **22**, L4 (1983)
12.10 M.S. Isaacson, J. Langmore, N.W. Parker, D. Kopf, M. Utlaut: Ultramicroscopy **1**, 359 (1976);
M. Utlaut: Phys. Rev. B**22**, 4650 (1980)
12.11 K. Takayanagi: Ultramicroscopy **8**, 145 (1982)
12.12 M.J. Yacaman: This volume
12.13 G. Binniq, H. Rohrer, Ch. Gerber, E. Weibel: Appl. Phys. Lett. **40**, 178 (1982); Phys. Rev. Lett. **49**, 57 (1982); Phys. Rev. Lett. **50**, 120 (1983); also
G. Binnig, H. Rohrer: Helv. Phys. Acta **55**, 726 (1982); Phys. Bl. **39**, 16 (1983); Surf. Sci. **126**, 236 (1983)
12.14 R. Young, J. Ward, F. Scire: Rev. Sci. Instrum. **43**, 999 (1972)
12.15 M.W. Roberts, C.S. McKee: *Chemistry of the Metal-Gas Interface* (Clarendon, Oxford 1978) Sect.8.2
12.16 G. Ehrlich: Brit. J. Appl. Phys. **15**, 349 (1964). See also
T. Gurney, Jr., F. Hutchinson, R.D. Young: J. Chem. Phys. **42**, 3939 (1965); and

R.D. Young, D.C. Schubert: J. Chem. Phys. **42**, 3943 (1965)

12.17 G. Ehrlich, C.F. Kirk: J. Chem. Phys. **48**, 1465 (1968)

12.18 E.W. Plummer, T.N. Rhodin: J. Chem. Phys. **49**, 3479 (1968)

12.19 R. Gomer, L.W. Swanson: J. Chem. Phys. **38**, 1613 (1963);
T.T. Tsong, E.W. Müller: Phys. Stat. Solidi A1, 513 (1970);
R.G. Forbes: Surf. Sci. **102**, 255 (1981);
D.R. Kingham: Vacuum **32**, 471 (1982)

12.20 For reviews of the subject, see G. Ehrlich, K. Stolt: Annu. Rev. Phys. Chem. **31**, 603 (1980); also
D.W. Bassett: In *Surface Mobilities on Solid Materials*, ed. by Vu Thien Binh (Plenum, New York 1983) p.63

12.21 K. Stolt, W.R. Graham, G. Ehrlich: J. Chem. Phys. **65**, 3206 (1976)

12.22 G. Ehrlich, F.G. Hudda: J. Chem. Phys. **44**, 1039 (1966); also
W.R. Graham, G. Ehrlich: Thin Solid Films **25**, 85 (1975)

12.23 G. Ehrlich: J. Vac. Sci. Technol. **17**, 9 (1980)

12.24 G. Ayrault, G. Ehrlich: J. Chem. Phys. **60**, 281 (1974)

12.25 D.W. Bassett, P.R. Weber: Surf. Sci. **70**, 520 (1978)

12.26 John D. Wrigley, Jr.: *Surface Diffusion by an Atomic Exchange Mechanism*, Coordinated Science Lab., Univ. of Illinois at Urbana, Report T-115, July 1982;
J.D. Wrigley, G. Ehrlich: Phys. Rev. Lett. **44**, 661 (1980)

12.27 For early work of this type, see M. Drechsler: Z. Elektrochem. **58**, 327 (1954)

12.28 D.W. Bassett: Surf. Sci. **53**, 74 (1975);
D.W. Bassett, C.K. Chung, D. Tice: Vide **176**, 39 (1975)

12.29 S.-C. Wang, T.T. Tsong: Surf. Sci. **121**, 85 (1982)

12.30 H.-W. Fink, G. Ehrlich: 43rd Annual Conference on Physical Electronics, Santa Fe, New Mexico, June 1983

12.31 K. Stolt, J.D. Wrigley, G. Ehrlich: J. Chem. Phys. **69**, 1151 (1978)

12.32 D.A. Reed, G. Ehrlich: Philos. Mag. **32**, 1095 (1975)

12.33 D.A. Reed: *Studies of Surface Diffusion*, Ph.D. Thesis, Univ. of Illinois at Urbana, 1980

12.34 K. Stolt, G. Ehrlich: Abstracts, TMS-AIME Fall Meeting, Milwaukee, WI, September 1979

12.35 H.-W. Fink, G. Ehrlich: 42nd Annual Conference on Physical Electronics, Atlanta, GA, June 1982

12.36 D.W. Bassett, D.R. Tice: In *The Physical Basis of Heterogeneous Catalysis*, ed. by E. Drauglis and R.I. Jaffee (Plenum, New York 1975), p.231. It should be noted that the presence of clusters in FIM images was first noted by D.W. Bassett: Surf. Sci. **23**, 240 (1970)

12.37 H.-W. Fink: *Atomistik der Monolagenbildung*, Ph.D. Thesis, Technical University Munich, 1982

12.38 D.W. Bassett: Thin Solid Films **48**, 237 (1978)

12.39 H.-W. Fink, G. Ehrlich: Surf. Sci. **110**, L611 (1981)

12.40 T.T. Tsong, R. Casanova: Phys. Rev. Lett. **47**, 113 (1981)

12.41 H.-W. Fink: Private communication

13. Ising Model Simulations of Crystal Growth

G. H. Gilmer

With 15 Figures

13.1 Introduction

Kinetic Ising model simulations have elucidated many aspects of crystal growth. For example, studies of the motion of close-packed surfaces of perfect crystals provided evidence for a surface roughening transition, where the two-dimensional nucleation barrier disappears [13.1-3]. Competing mechanisms such as spiral growth and 2d nucleation have been simulated and the relative importance of the processes has been assessed [13.3,4].

In this chapter, I review some of these results and also treat the crystallization of binary systems. In the following section I discuss the model which will be applied to study various crystal growth mechanisms. Section 13.3 treats the equilibrium structures of surfaces. Kinetics of crystal growth are discussed in Sect.13.4. Impurities can have a strong influence on crystal growth kinetics, even when they are present in very small concentrations. Certain impurities segregate at the interface and therefore are concentrated in the region where the atoms crystallize. Binary systems also yield information of the effect of a driving force on the crystal structure and composition. Experimental applications include molecular-beam studies of deposition onto cold substrates, where a very large driving force is present. Also, the results can be applied to the laser annealing of amorphous semiconductors, where dopant concentrations in the crystalline product have been found to exceed the maximum equilibrium values by several orders of magnitude [13.5]. Some conclusions are contained in Sect.13.5.

13.2 Model of Crystal Growth

The spin-1 Ising system provides a relatively simple model that has most of the basic growth mechanisms and properties of monatomic and binary crystals. This model is analogous to a lattice gas, with sites that may be occupied by either of two atomic species, or they may be vacant. It has been applied

to three-component alloys without vacancies; this system exhibits a variety
of phase transitions and critical points [13.6]. The model with vacant sites
has been used previously in studies of surface segregation [13.7].

The Hamiltonian of the binary lattice gas with nearest-neighbor interactions is

$$H = - \sum_{\{i,j\}} \left[c_i^A c_j^A \phi_{AA} + (c_i^A c_j^B + c_i^B c_j^A) \phi_{AB} + c_i^B c_j^B \phi_{BB} \right] - \sum_i (c_i^A \mu_A + c_i^B \mu_B) \quad .$$

$$(13.1)$$

Here the sum over $\{i,j\}$ includes all pairs of nearest neighbor sites. The
$c_{i,j}^{A,B}$ indicate the occupancy of the sites; $c_i^A = 1$ if an A atom is on site i,
otherwise $c_i^A = 0$. The bond energies ϕ represent the energy required to separate pairs of atoms; for example, two neighboring A atoms interact with the
energy $-\phi_{AA}$. Finally, μ_A and μ_B are the chemical potentials of the two species.

The Hamiltonian in (13.1) is easily transformed into that of the magnetic
Ising system. This is instructive, since the effects of a magnetic field are
more transparent than those due to the chemical potentials. We consider first
the case where all sites are occupied. Then the status of site i can be specified by the spin variables $\sigma_i = \pm 1$, (the spin-½ Ising model) with

$$c_i^A = \tfrac{1}{2}(1 + \sigma_i), \quad c_i^B = \tfrac{1}{2}(1 - \sigma_i) \quad .$$

$$(13.2)$$

Replacing the C's by σ's in (13.1), we obtain

$$H = \tfrac{1}{4} \sum_{\{i,j\}} \Omega \sigma_i \sigma_j - \sum_i h_i \sigma_i + H_0 \quad ,$$

$$(13.3)$$

where

$$\Omega = \phi_{AA} + \phi_{BB} - 2\phi_{AB} \quad ,$$

$$(13.4)$$

$$h_i = \tfrac{1}{2}[\Delta\mu_A - \Delta\mu_B] - (Z - Z_i)(\phi_{AA} - \phi_{BB})/4 \quad ,$$

$$(13.5)$$

and H_0 is a constant. In (13.5), $\Delta\mu = \mu - \mu^{(0)}$, where $\mu^{(0)} = -Z\phi/2$ is the chemical potential required for equilibrium with a crystal composed of only one
species [13.8]; Z_i is the coordination of a site, and it may be less than
the bulk coordination number Z if the site is near a free surface. It is
easy to relate the properties of the system to the values for ϕ and μ in this
Hamiltonian. The atoms may cluster with their own kind, or assume an ordered
arrangement of alternating A and B atoms. From (13.3) it is clear that Ω is
the parameter that determines the nature of the ordering at low temperatures:

> 0 clustering alloy

Ω = 0 ideal alloy

< 0 ordering alloy.

The magnetic field h_i determines the type of atom favored at site i; $h_i > 0$ favors A and $h_i < 0$ B. At the fully coordinated sites in the bulk, $h = \frac{1}{2}(\Delta\mu_A - \Delta\mu_B)$. The surface positions have the additional term $-(Z - Z_i)$ $(\phi_{AA} - \phi_{BB})/4$ that favors the more volatile species. If $\phi_{BB} < \phi_{AA}$, this term is negative and assigns a lower energy to the B atoms. This is the familiar "broken bond" driving force for surface segregation. The model does not include a driving force associated with the relative sizes of the atomic species, although this may be significant in some alloys [13.9]. However, the broken-bond effect can produce strong segregation and is therefore adequate to demonstrate growth mechanisms involving surface layers of impurities.

13.3 Equilibrium Surface Structure

The kinetics of crystal growth are largely determined by the equilibrium structure of the close-packed faces. These orientations predominate on the growth morphologies, since they usually have the slowest growth rates. Often is is possible to deposit new layers on a close-packed face only by the nucleation of stable two-dimensional clusters which then expand and cover the entire face. In the following we first discuss the structure of close-packed faces of single-component crystals.

It had been recognized by *Burton* and *Cabrera* [13.10] and by *Jackson* [13.11] that equilibrium roughening of close-packed surfaces would affect both the growth kinetics and the morphology of crystals. At low temperatures the surfaces are essentially flat with only a few adatoms and vacancies. Configurations are improbable if they involve large deviations in the heights of neighboring sections of the surface. At higher temperatures, however, height differences become more likely, and whole regions may be displaced from their "ground-state" position at zero temperature. This is illustrated in Fig.13.1 with some typical surfaces generated by the Monte Carlo simulation procedure [13.12] with the equilibrium chemical potential $\mu^{(0)}$. These results show a definite change in the connectivity of the clusters between $kT/\phi = 0.6$ and 0.667. The properties of the surface were observed to change when the connected clusters appeard. At low temperatures the interface appeared to be pinned at one lattice plane and the average height of the surface never changed by more than a small fraction of the distance between close-packed layers. At the higher temperatures, however, the interface was no longer stationary and the average height was observed to fluctuate through large distances. Clearly the large clusters observed at high temperatures provide stable nuclei when

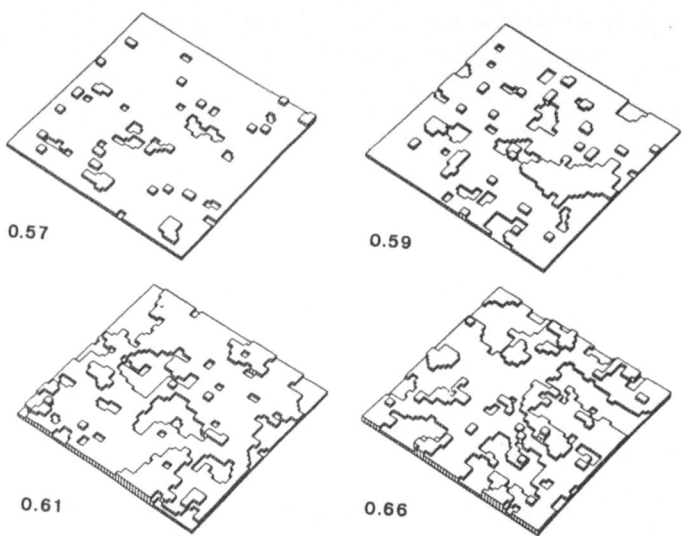

Fig.13.1. Typical configurations of the simple cubic {100} face generated by Monte Carlo simulations for a crystal in equilibrium with its vapor. The numbers adjacent to the figures indicate the ratio kT/φ. Small clusters and atoms with fewer than three bonds are not represented

exposed to a supersaturated vapor, and the face can add new layers continuously without nucleation.

The first theoretical approach in agreement with this picture was given by *Weeks* et al. [13.13]. They pointed out that the interface density profile must vanish when the interface becomes delocalized. A low-temperature series for the profile width was calculated, and series extrapolations indicated that the width diverges at a temperature slightly above $T_c^{(2D)}$, the critical temperature of a single (two-dimensional) layer. This result and the qualitative changes observed in the Monte Carlo studies provided convincing evidence that a phase transition occurred at a defined temperature. This is in contrast to early calculations based on the mean-field theory which predicted only a quantitative increase in surface roughness with increasing temperature [13.14].

The best theoretical evidence is now based on work relating interface models to other systems with confirmed phase transitions. *Chui* and *Weeks* [13.15] obtained a direct relation to the metal-insulator transition in a 2D Coulomb gas, and *van Beijeren* [13.16] related it to the phase transition in the six-vertex model. These results confirm the predicted divergence of the interface profile width and show that thermodynamic qunatities such as

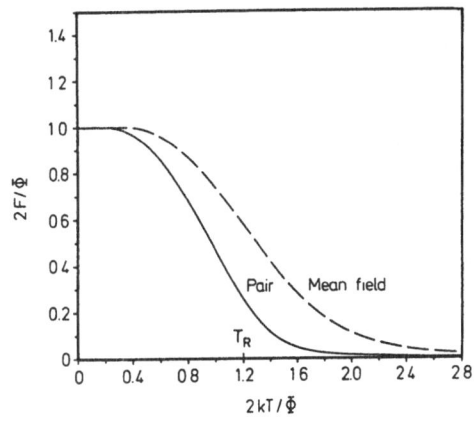

Fig.13.2. Free energy of a step along the <001> direction in a simple cubic {100} face

the specific heat do not exhibit anomalous behavior near the roughening transition.

Another manifestation of the roughening transition, confirmed by the six-vertex model, is the vanishing of the excess free energy associated with an isolated monatomic step on the crystal surface. In Fig.13.2 I show the mean-field and pair approximation results for a <001> step on a simple cubic {100} face [13.17]. Although the Monte Carlo evidence shows that the free energy vanishes at $kT_R/\phi = 0.6$, these approximations predict finite values at all temperatures. Note that the more accurate pair approximation does predict very small values of F at temperatures significantly above T_R. The magnitude of the step free energy is a critical parameter for the crystal growth process, since the free energy of the periphery of a cluster is closely related to that of the monatomic step. A small step free energy implies that clusters can be nucleated readily on the close-packed surface.

The equilibrium surface structure can be altered by the addition of impurities to the system. The surface layer of a clustering alloy may have a composition very different from the interior of the crystal. The clustering alloy has a bulk miscibility gap as shown in Fig.13.3 [13.18]. The dotted line indicates the states traversed by a particular alloy crystal as its temperature is changed. At the temperature T_0 where the dotted line enters the miscibility gap, precipitation of a bulk phase rich in B atoms can begin. At this point the concentration of B is equal to the equilibrium solubility limit of the crystal, and hence $h = 0$. But the sites at the surface are biased toward the volatile B atoms, as discussed above, and the B population of the surface layer must be greater than that in the bulk. Precipitation of the B-rich phase occurs preferentially at the surface.

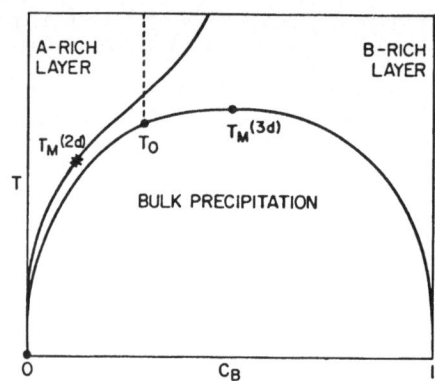

Fig.13.3. Phase diagram of a clustering alloy as calculated by mean-field
theory. The impurity concentration at
the crystal surface is also indicated

(Figure labels: A-RICH LAYER, B-RICH LAYER, $T_M^{(2d)}$, T_0, $T_M^{(3d)}$, BULK PRECIPITATION, T, C_B, 0, 1)

A first-order phase transition may occur in the surface layer [13.19]. At
a temperature above T_0 the population of B atoms in the layer may rise dis-
continuously from a small minority to a majority, while the bulk concentration remains small. A necessary condition is that the broken bond driving
force exceeds that resulting from interactions with A atoms in the layer below. It is also required that T is below the critical mixing temperature
$T_M^{(2D)}$ of A and B atoms in the surface layer [13.20].

The curve on the left in Fig.13.3 indicates schematically the locus of
temperatures (for crystals of various compositions) where the surface B concentration exceeds 50%, and first-order transitions occur for crystals of
concentrations C_B less than that corresponding to the mark at $T_M^{(2D)}$.

The variation of the surface excess of B atoms with temperature is shown
in Fig.13.4; these results were calculated by a mean-field approximation in
which the layer concentrations were chosen to give the minimum total free
energy. A simple cubic (SC) lattice with an exposed {100} face was employed.
In this case a phase transition occurs below a critical temperature
$kT_M^{(2D)}/\Omega \approx 1$, the mean-field value for the critical mixing temperature of an
isolated layer.

Surface segregation can affect crystal growth kinetics in several ways.
First, the free energies of steps may be very different from those on the
pure crystal surface. The energy of a step at low temperatures is given by a
simple argument illustrated in Fig.13.5, where two steps are produced by a
shear displacement of a thin crystal. The interatomic bonds crossing the
dashed line before displacement have an energy (per column) $\Phi = 2\phi_{BB} + N\phi_{AA}$,
where $N + 2$ is the number of layers in the crystal. After displacement the
interactions are $\Phi' = 2\phi_{AB} + (N - 1)\phi_{AA}$, and the edge energy per site of a step
is

$$\varepsilon_s = - (\Phi' - \Phi)/2 = \phi_{BB}/2 + \Omega/2 \quad . \tag{13.6}$$

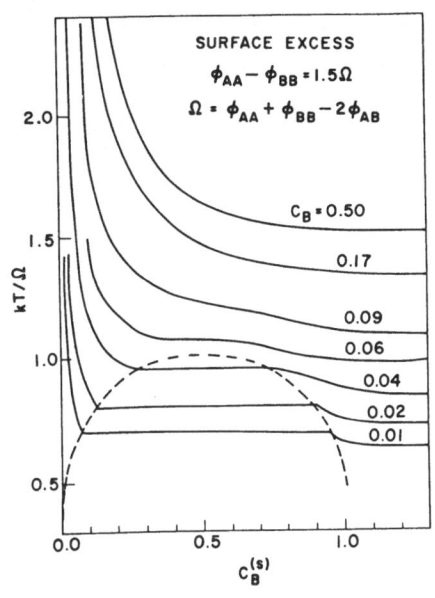

SURFACE EXCESS

$\phi_{AA} - \phi_{BB} = 1.5\Omega$

$\Omega = \phi_{AA} + \phi_{BB} - 2\phi_{AB}$

$C_B = 0.50$

0.17

0.09

0.06

0.04

0.02

0.01

kT/Ω

2.0

1.5

1.0

0.5

0.0   0.5   1.0

$C_B^{(s)}$

Fig. 13.4. Surface excess of impurity $C_B^{(s)}$ at a free surface vs kT for a clustering alloy. The bulk composition is indicated on the curves. The dashed line indicates the region where a discontinuous change in $C_B^{(s)}$ with temperature is predicted by the model

LOW T STEP ENERGY

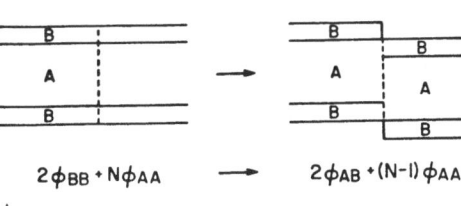

$2\phi_{BB} + N\phi_{AA} \longrightarrow 2\phi_{AB} + (N-1)\phi_{AA}$

Fig. 13.5. Change in energy of a thin crystal with a segregated layer when two steps are introduced

It is apparent that a step on the segregated surface of an ideal alloy ($\Omega = 0$) has a lower energy than does a step on a clean surface, ($\phi_{AA}/2$). The nucleation of 2D clusters on the surface is enhanced in this case, since the boundary of such a cluster is a curved step and also has a lower energy. *Cabrera* was the first to discuss this effect of volatile impurities on the surface energy [13.21].

The ideal alloy impurities are not effective at moderate to high temperatures, however, and the surface roughening point is apparently not much affected. Figure 13.6 is a plot of the step free energies for various crystal compositions of the ideal alloy ($\phi_{BB} = \phi_{AA}/2$, $\phi_{AB} = 3\phi_{AA}/4$). The mean-field approximation used here suppresses the roughening transition as discussed above; however, the relative changes in the step free energies of the different alloys should be correct. These results show that the free energy of the alloy crystal step is reduced by a factor of two at $T = 0$, in agreement with the calculation of the step energy above. But the step energies of the alloy approach those for steps on the pure A crystal at higher temperatures, unless C_B is very large. The small reduction in the system energy achieved by arranging the B atoms at the step edge is offset by the entropy of a random arrangement on bulk and surface lattice sites.

A segregated layer on a clustering alloy is stable at relatively high temperatures. The weak bonds between A and B atoms inhibit the dissolution of the segregated layer in the bulk of the crystal. But the segregated layer

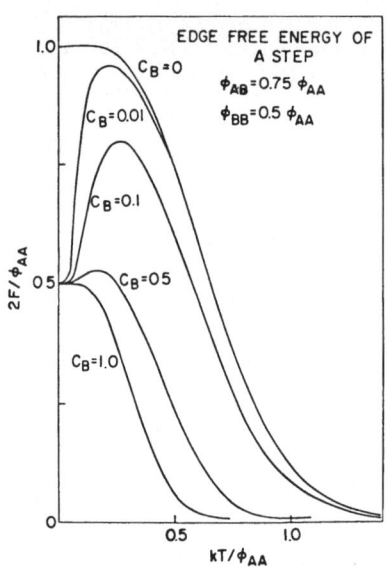

Fig.13.6. Free energy vs T for a step on an ideal alloy crystal surface. Bulk composition of the crystal is indicated on the curves

probably does not enhance the 2D nucleation rate through a significant reduction of the step free energy. At low temperatures the step energy is higher than that of the ideal alloy according to (13.6). Furthermore, few atoms will remain attached very long at the surface because of the weak bonding to the volatile segregated layer. We shall consider some Monte Carlo simulations which indicate that a new mechanism operates under these conditions.

13.4 Crystal Growth Kinetics

Two basic events are used to simulate the crystal growth process in the kinetic Ising model. These correspond to the impingement of an atom from the vapor onto the crystal surface, and the evaporation of an atom from the crystal. An event and the site where it is to take place are chosen using a computer-generated random number, in accordance with standard Monte Carlo procedure [13.2]. The probability of impingement is proportional to the supersaturation in the vapor, and we assume that it is site independent,

$$k^+ = \nu \exp(\mu/kT) \quad , \tag{13.7}$$

where ν has the dimensions of an attempt frequency. The evaporation of an atom occurs at a frequency that is sensitive to the surface site

$$k_A^-(n_A, n_B) = \nu \exp[-(n_A \phi_{AA} + n_B \phi_{BB})/kT] \quad , \tag{13.8}$$

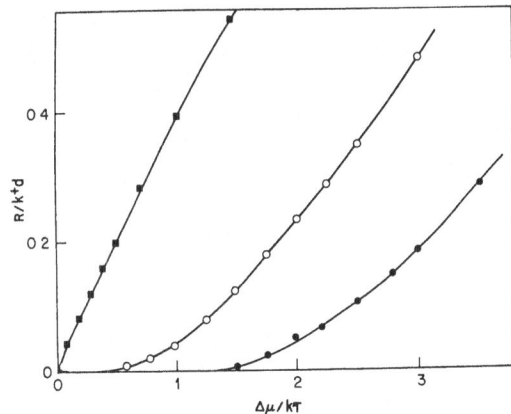

Fig.13.7. Monte Carlo calculations of the normalized growth rates of simple cubic {100} faces on a perfect crystal. Here d is the interlayer spacing. $kT/\phi =$ ■: 0.667; ○: 0.333; ●: 0.250

where n_A and n_B are the numbers of A and B neighbors, respectively. A corresponding equation holds for k_B^-. Crystal growth simulation is described in detail elsewhere [13.3,4].

The steady-state growth rates obtained by the Monte Carlo procedure described above are shown in Fig.13.7. Here I plot the average growth rates versus $\Delta\mu$ for a single-component crystal without lattice defects. Again, the data represent the SC {100} face (Fig.13.1) and temperatures of $kT/\phi = 0.667$, 0.333, and 0.250 correspond to the squares, open circles and solid circles, respectively. Thus, only the data represented by the squares were obtained at a temperature above T_R. Although the roughening transition has little effect on the thermodynamic properties of the surface, the kinetics are changed dramatically. We see that above T_R the growth rate is linear in $\Delta\mu$ at the origin, whereas below T_R there is a region of metastable states at finite $\Delta\mu$ where the rate is essentially zero. These curves are characteristic of rates limited by nucleation, and analysis of the data shows that this is indeed the growth mechanism in the regime where $T < T_R$ [13.3]. Growth above T_R occurs by the continuous addition of atoms to the edges of existing clusters.

Crystals can often grow at low temperatures and small $\Delta\mu$ where the rate of 2D nucleation of clusters should be prohibitively slow. Certain defects and impurities can account for this, and we now discuss some of these growth mechanisms.

Frank pointed out that a screw dislocation intersecting the crystal surface could produce steps that terminate at the intersection. These steps should wind up into a spiral pattern under the influence of a finite driving force, and thus provide an inexhaustible source of edge positions where atoms could attach to the crystal [13.22]. A simulation of this process is shown in Fig.13.8. The equilibrium surface is shown in (a), and various stages of

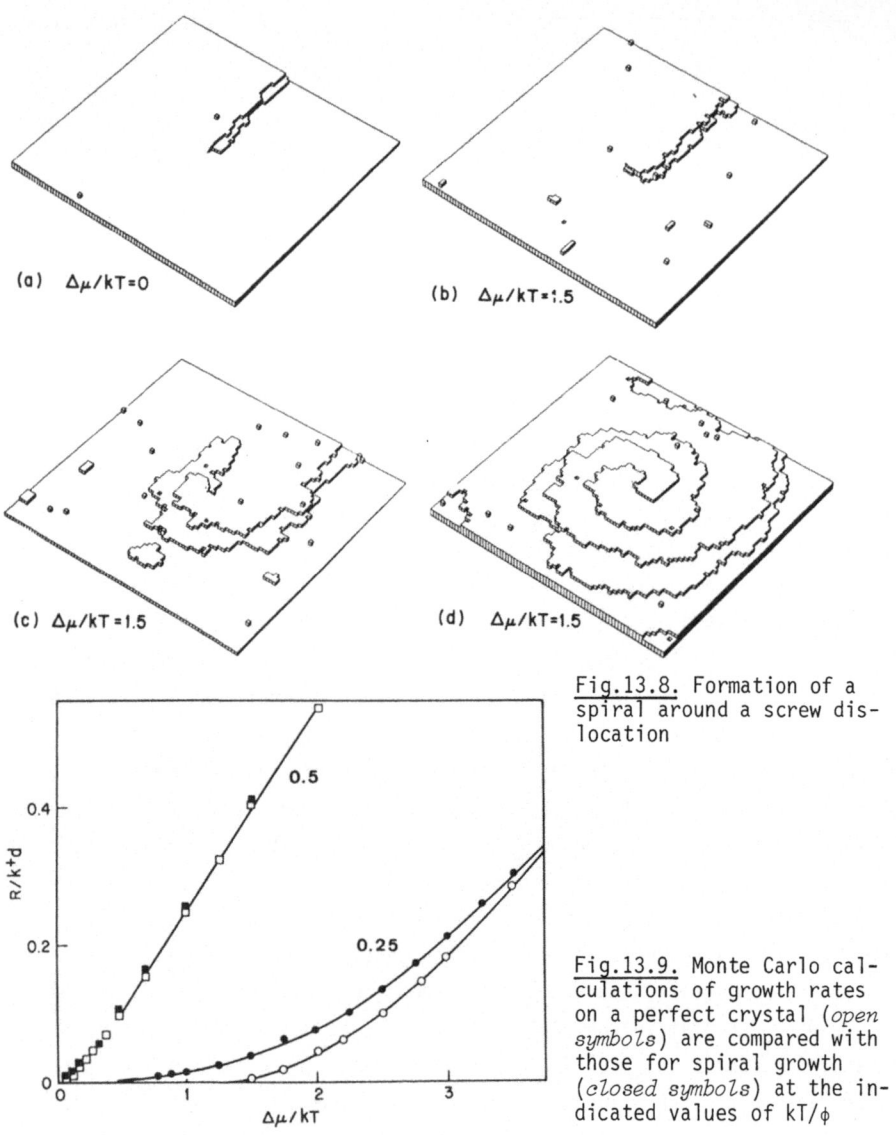

(a) $\Delta\mu/kT=0$

(b) $\Delta\mu/kT=1.5$

(c) $\Delta\mu/kT=1.5$

(d) $\Delta\mu/kT=1.5$

<u>Fig.13.8.</u> Formation of a spiral around a screw dislocation

<u>Fig.13.9.</u> Monte Carlo calculations of growth rates on a perfect crystal (*open symbols*) are compared with those for spiral growth (*closed symbols*) at the indicated values of kT/ϕ

the approach to a steady-state spiral pattern with $\Delta\mu = 1.5$ kT are shown in (b-d). The steady-state rates are shown in Fig.13.9, again for the SC {100} face. At the lower temperature ($kT/\phi = 0.25$) the presence of the screw dislocation does increase the crystal growth rate considerably, especially at small $\Delta\mu$. At the higher temperature, near T_R, the screw dislocation has little effect since the step free energy is quite small and the nucleation of clusters on the perfect surface is fast.

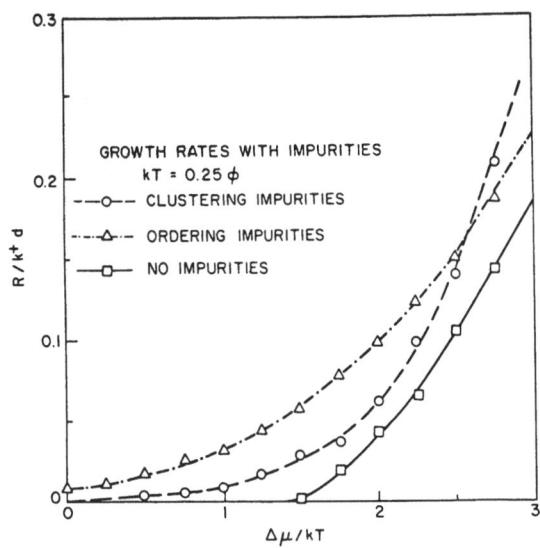

Fig.13.10. Alloy growth rates compared with pure crystal growth rates. The chemical potential of the ordering impurities was maintained at the value $\mu_B = \mu_A - 6$ kT; whereas that of the clustering impurities was fixed at $\mu_B = \mu_B^{(0)} - 0.5$ kT, where $\mu_B^{(0)}$ is the value required for equilibrium with a crystal composed of B atoms. Here $\Delta\mu \equiv \Delta\mu_A = \mu_A - \mu_A^{(0)}$

We now consider the influence of a segregated layer on the kinetics. The presence of a segregated layer can be predicted from the expressions for the rate constants, (13.7,8). For simplicity, we treat a system at a low temperature where the bulk phases are essentially pure. A layer of B atoms can be deposited, in principle, by the addition of atoms at kink sites. If atoms in the layer have m neighbors in the layer of A atoms beneath, then an atom in a kink site has $(Z/2 - m)$ B neighbors within the layer. The evaporation rate from the kink site is $\nu\exp[-m\phi_{AB}/kT - (Z/2 - m)\phi_{BB}/kT]$. For the layer to be stable, the B impingement rate $k_B^+ = \nu\exp(\mu_B/kT)$ must equal this evaporation rate, or μ_B must have the value

$$\mu_B^{(1)} = \mu_B^{(0)} - m(\phi_{AB} - \phi_{BB}) \quad . \tag{13.9}$$

Segregation without bulk precipitation can occur only when $\phi_{AB} > \phi_{BB}$, according to (13.9). A monolayer or more of B atoms will adhere to the surface when $\mu_B > \mu_B^{(1)}$.

The kinetics of the SC{100} face in the presence of such a layer are shown in Fig.13.10. We plot as a function of $\Delta\mu_A$ the steady-state rate of advance R of the crystal surface (or the net rate of deposition of A and B atoms). Here $\phi_{AB} = 0.533\phi_{AA}$, $\phi_{BB} = 0.4\phi_{AA}$, $\phi_{AA} = 4kT$, and $\Delta\mu_B = -0.4kT$. For comparison we have also plotted R for the single-component system (A atoms), in which case a nucleation barrier inhibits measurable growth for $\Delta\mu_A$ below about 1.5 kT. The rate of formation of new layers of the crystal in the pre-

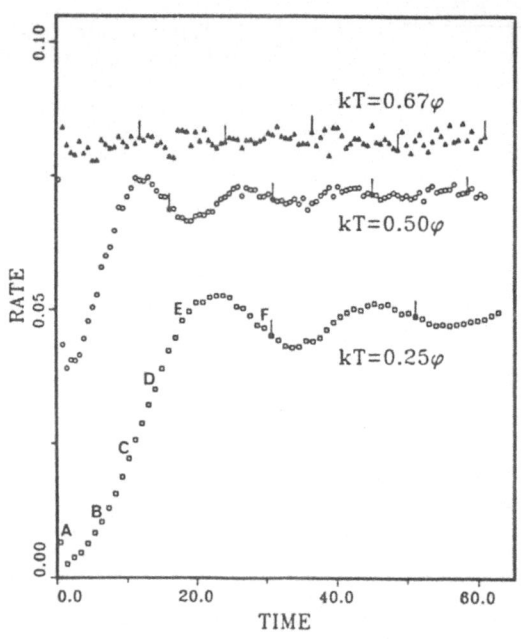

Fig.13.11. Transient growth rates of SC{100} face. The data points plotted are the average deposition rates during the time intervals after the driving force is applied. Each data point represents an average of 50-100 simulations on 60×60 segments. Vertical lines on the data points indicate the completion of new layers. The driving force was chosen in each case to give approximately equal asymptotic growth rates; $\Delta\mu/kT = 2$ when $kT/\phi = 0.25$; $\Delta\mu/kT = 0.6$ when $kT/\phi = 0.5$; and $\Delta\mu/kT = 0.5$ when $kT/\phi = 0.67$

sence of the segregated layer is many orders of magnitude larger than the nucleation rate on the clean surface, at the smaller values of $\Delta\mu_A$.

The segregated layer stabilizes the A atoms at the surface. An atom that impinges directly on a vacant site in the layer of B atoms has lateral interactions that reduce the evaporation probability. Although the nucleation of 2D clusters on top of the segregated layer is very unlikely, the formation of A-rich clusters within the layer may proceed at an appreciable rate. Above $T_M^{(2D)}$ these clusters may form without a nucleation barrier, and when $\Delta\mu_A > 0$ the concentration of A atoms in the layer increases monotonically with time. When the concentration reaches a critical value, a new layer of B atoms can adhere, and the process is repeated.

The unique nature of this mechanism is clearly revealed by growth transients [13.23]. A crystal initially in equilibrium with its vapor is suddenly placed in a supersaturated environment, and the subsequent deposition rate is measured. The average rates of deposition are shown in Fig.13.11 for a crystal composed solely of A atoms [13.24]. Here $t = 0$ is the time the driving force was applied. The curve with the lowest indicated temperature corresponds to the conditions of Fig.13.10. The initial rates are small in this case, since most atoms impinge on sites without lateral neighbors. As the clusters nucleate and spread, the growth rates increase and reach a maximum when about two-thirds of the first layer has been completed. The damped oscillations in-

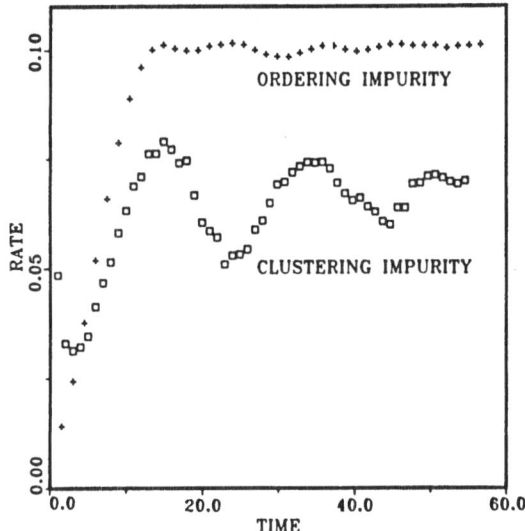

RATE

ORDERING IMPURITY

CLUSTERING IMPURITY

TIME

Fig. 13.12. Transient growth rates of crystals in the presence of ordering and clustering impurities. In both cases $\Delta\mu_A = 2$ kT, but $\mu_B = \mu_A - 6$ kT and $\mu_B = \mu_B^{(0)} - 0.5$ kT for the ordering and clustering impurities, respectively

dicate the deposition of successive layers, and eventually the crystal surface is distributed over several levels and the growth rate approaches the steady-state value. Surface roughening occurs at $kT_R = 0.6\phi_{AA}$, and the absence of oscillations at $kT = 0.667\phi$ confirms that the surface is rough. Above T_R the surface is already distributed over many levels in its initial equilibrium configuration.

Transients measured during the growth of the alloy crystals are shown in Fig.13.12; the conditions are identical to those specified for Fig.13.10. At $t = 0$ the driving force $\Delta\mu_A = 2kT$ was applied to a clustering alloy crystal that was in equilibrium with a vapor containing both species. The presence of oscillations implies that the equilibrium surface was not rough, but was localized to the vicinity of a single level. The amplitude of the oscillations is actually larger than that observed for the pure crystal at any temperature, Fig.13.11. This suggests that the segregated surface remains more localized during growth, as might be expected if the surface layer is gradually transformed into an A-rich layer as described above.

This mechanism of growth is sensitive to the relative magnitudes of the interactions between atoms. The growth rate is limited by the rate at which A atoms impinge on vacant sites, and for this reason a rather volatile impurity is most effective. (In equilibrium, the surface layer modeled for the data of Fig.13.12 consists of 20% A, 75% B, and 4% vacant sites.) It is interesting to speculate on the effect of a less volatile layer. This could reverse the situation and impede growth, provided ϕ_{AB} is small. In that case few vacancies would exist in the segregated layer, and the 2D nucleation rate on the top surface would be slower than that which occurs without the B atoms.

The presence of an ordering alloy species may also have a large effect on the kinetics. The triangular symbols in Fig.13.10 indicate Monte Carlo data [13.25] for a system with strong AB bonds; specifically, $\phi_{AB} = 2\phi_{AA} = 2\phi_{BB}$, and $kT = \phi_{AA}/4$ as before. In this case a relatively small flux $k_B^+ = 2.5 \times 10^{-3}$ k_A^+ caused a large increase in the interface mobility at small $\Delta\mu_A$. The B atoms stabilize small clusters and thus provide many 2D nuclei under conditions where the only stable single-component clusters are very large. At large values of $\Delta\mu$ where the single-component nucleation rate exceeds k_B^+, there is little effect.

Transient growth rates for this system are plotted in Fig.13.12. In this case the vapor did not contain B atoms in the initial state; and at $t = 0$ the vapor pressure of the A component was increased to give $\Delta\mu_A = 2$ kT, and the flux of B atoms was applied. The small initial rates are almost the same as those plotted in Fig.13.11 at $kT/\phi = 0.25$, since the initial states of the crystals are identical. But subsequent oscillations are greatly attenuated with the flux of B atoms present. This indicates that the interface rapidly assumes a multilevel structure. Enhanced nucleation of clusters around B atoms allows portions of the surface to advance quickly to higher levels. B atoms impinge on the tops of small clusters and initiate clusters at the higher levels while the first layer is still only partially completed. Although the steady-state kinetics of the ordering and clustering alloys appear very similar in Fig.13.10, the difference in the mechanisms is readily apparent in the transient curves of Fig.13.12. Transient data are of great value for the identification of crystal growth mechanisms.

The composition of an alloy crystal depends on the conditions during growth, and may differ substantially from the values corresponding to equilibrium phase diagrams. At large values of the driving force, the more volatile component is trapped in excess of the equilibrium concentration [13.26]. This occurs because it has a high impingement rate, (13.7), but the rapid motion of the interface gives insufficient time for equilibration by evaporation of the atoms, and many are trapped in the crystal. This effect is enhanced when a layer of the volatile component segregates at the surface, and rapid growth traps portions of this layer in the crystal.

Simulation data for the quantities of the B species incorporated are shown in Fig.13.13. A steep initial rise in the concentration of B atoms (clustering alloy) with $\Delta\mu$ is apparently the result of the relatively stable monolayer of B atoms at the surface. Also plotted are data for an ideal alloy

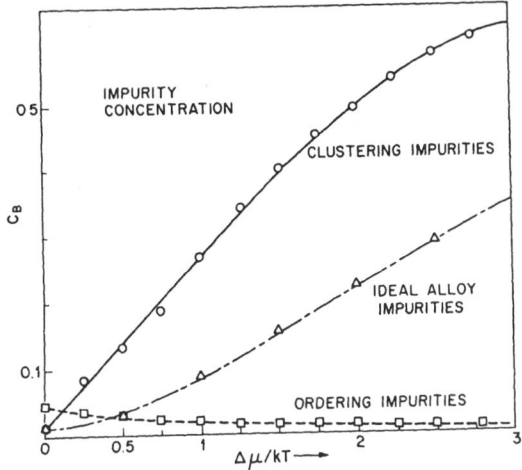

Fig.13.13. The concentration of impurities trapped in the crystal. The square and circular symbols correspond to ordering and clustering impurities, respectively, under the conditions specified for Fig. 13.12

with $\phi_{AB} = \frac{1}{2}(\phi_{AA} + \phi_{BB})$, but with the same value of ϕ_{AB} and approximately the same equilibrium concentration. The initial rise is much more gradual in this case, where segregation does not occur.

The concentration of clustering impurities captured by the crystal can exceed the maximum solubility of the B atoms in the A-rich region. The data of Fig.13.13 indicate that $C_B = 0.13$ at $\Delta\mu_A/kT = 0.5$, whereas the maximum equilibrium concentration at this temperature is $C_B^{(e)} = 0.02$. Thus, the kinetics of the growth process have produced a metastable alloy. Since the B-rich phase can be created only by a nucleation event, the metastable system may remain indefinitely in the supersaturated condition. A small solid-state diffusion coefficient also slows the precipitation process. New techniques of crystal growth that generate very large values of $\Delta\mu$ have produced metastable systems with impurity concentrations that exceed the equilibrium value by several orders of magnitude [13.5]. Techniques of this type can permit the construction of a new class of materials with unusual and perhaps useful properties.

The enhanced rate of impurity capture in the presence of a segregated layer may explain variations in impurity capture on different faces of a single crystal. Certain minerals and artificially grown crystals exhibit sectors of high impurity concentration that correspond to the regions of the crystal deposited on specific low-index faces [13.27]. Since the surface layers along different low-index orientations have different binding energies for impurities, it is apparent that the driving force for segregation may depend on orientation. Thus, some orientations may contain B-rich monolayers, while

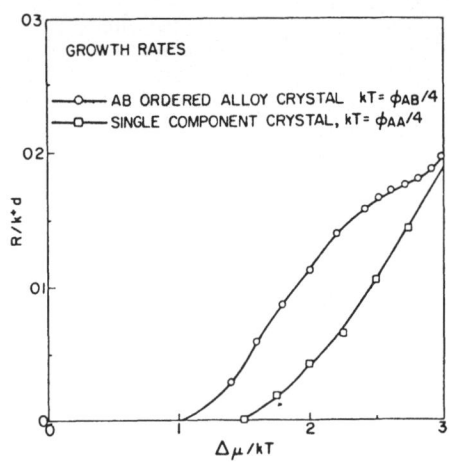

Fig.13.14. The steady-state growth rate of an ordered alloy crystal with $\Delta\mu_A = \Delta\mu_B \equiv \Delta\mu$, circles, compared with the rate of a pure crystal

others do not. These results suggest that the capture of a given species will be significantly greater on surfaces containing the segregated layer.

For the ordering alloy the B atoms are incorporated in the greatest numbers at small driving forces. These atoms are more likely to stick than the A atoms, especially at small values of $\Delta\mu$ where the probability of an A atom sticking is very small. As $\Delta\mu$ is increased, fewer B atoms impinge on a given layer of the crystal and the concentration is reduced. In general, an increase in the driving force causes an increase in the B concentration when $\phi_{AB} < \phi_{AA}$, and a decrease when $\phi_{AB} > \phi_{AA}$.

Our final topic is the growth kinetics and structure of an ordered alloy in which the two species have the same chemical potential, and appear in the crystal in equal concentrations. An ordered pattern of alternating A and B atoms is obtained from low-temperature growth at small $\Delta\mu$. On the simple cubic lattice, the A and B atoms assume a sodium chloride structure.

The growth rate of the ordered alloy is shown in Fig.13.14 for the {100} face and for $kT = \phi_{AB}/4$. (Note that $kT = \phi_{AA}/4$ for the data of Figs.13.10,12.) The perfect ordered alloy contains only AB bonds, and for comparison we have included the kinetics of the single-component crystal with the same ground-state energy ($kT = \phi_{AA}/4$). This alloy has a regime of very slow growth for $\Delta\mu < kT$, suggesting that a 2D nucleation mechanism also applies here. The growth rate is somewhat larger than that of the pure crystal. Although an atom has only a 50% chance of landing on a site that provides the stronger AB bond to the crystal, atoms with the weaker bonding apparently contribute to the growth process in some way.

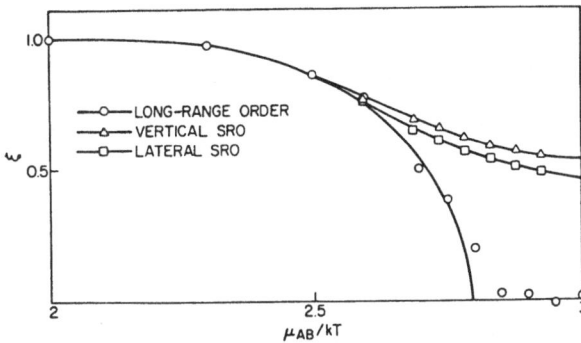

<u>Fig.13.15.</u> The long-range order for the ordered alloy (*circles*) compared with the pair ordering in the vertical and horizontal directions (*triangles* and *squares*, respectively)

The alloy growth rate has an inflection point at $\Delta\mu \approx 2.5$ kT, where many imperfections degrade the ordering. The long-range order parameter of the crystal is plotted versu $\Delta\mu$ in Fig.13.15. Here

$$\xi = \frac{n_A^A - n_A^B}{n_A^A + n_A^B} = \frac{n_B^B - n_B^A}{n_B^B + n_B^A} \quad , \tag{13.10}$$

where n_A^A and n_A^B are the number of A atoms on the A and B sublattices, respectively. The order remains almost perfect for $\Delta\mu < 2$ kT, but drops rapidly to zero above this driving force. This "kinetic phase transition" was discussed by *Chernov* and *Loomis* [13.28] in their simulation study of crystal growth using impingement and evaporation at a kink site. The disorder slows growth, since AB bonds are replaced by the weaker AA and BB bonds, and hence the evaporation rate is increased.

Short-range order is also illustrated in Fig.13.15 for pairs oriented perpendicular (vertical) and parallel (horizontal) to the surface. Appreciable short-range order remains even after the long-range order has vanished. Note that the vertical pairs are more highly ordered than the horizontal pairs. Similar growth-induced anisotropy has been utilized in garnet crystals designed to support bubble domains, where it is essential to have the direction of easy magnetization perpendicular to the plane of the film [13.29].

Growth-induced anisotropy is a direct result of the large lateral growth rate of a cluster. The expansion of a cluster occurs at a much faster speed than R, the normal growth rate. Since the lateral pairs are created during the cluster expansion, more disorder is expected in these directions.

313

13.5 Conclusions

The Ising model exhibits a variety of crystal growth mechanisms. At low temperatures the close-packed planes of a perfect crystal grow by the 2D nucleation mechanism, as expected. Above a well-defined roughening temperature, continuous growth occurs by the addition of atoms at the edges of existing clusters and steps. The spiral growth mechanism has been shown to be most effective at temperatures well below T_R. Near T_R and above, the rate of 2D nucleation is fast and steps and clusters caused by thermal disordering are present, so the additional steps produced by the screw dislocation have little effect. However, the hillock formed by the spiral steps surrounding the screw dislocation continues to exist, even at relatively high temperatures [13.30].

Simulations have also demonstrated the important effects of surface segregation on crystal growth kinetics. Screw dislocations are not necessary for the growth of certain alloy systems at small driving force and low temperatures. Either a segregating volatile species, or small quantities of a species that exerts strong interactions with the host atoms can permit measurable growth under these conditions. In view of the ubiquitous nature of impurities in most types of crystal growth systems, it is likely that these mechanisms are influential under a wide range of conditions. Only at high temperatures during growth on a thermally roughened surface is the effect of the second component of little consequence.

The simulations also show that the concentrations of the two species in the crystal are very sensitive to the conditions that prevailed during its formation. Surface segregation and the relative strengths of the atomic interactions are factors that may have important effects, especially if the crystal is grown at a high driving force. Also, the degree of perfection in ordered alloys is drastically affected by the growth conditions.

The Ising model is well suited to a study of interface kinetics. It is probably the simplest model that can exhibit the basic phenomena considered here. It is equivalent to the Kossel model discussed by *Volmer* and *Stranski* many years ago [13.31]. However, only recently has the computer technology been available that has made it possible to obtain extensive data on its properties. Improvements in the model are needed. Models that allow continuous particle coordinates permit the formation of lattice defects during the growth process. *Broughton* et al. [13.32] used molecular dynamics techniques to study the crystallization of a supercooled Lennard-Jones melt on a fcc {100} orientation, and the occasional formation of twin-planes was observed. Currently crystallization on the {111} plane is being investigated, and frequent twinning is observed, even at relatively small undercooling. Models of this type which impose few restrictions on the crystallization process provide information on defect structures and concentrations, absolute crystal growth rates and preferred crystal lattices.

References

13.1 The surface roughening transition has been reviewed by H.J. Leamy, G.H. Gilmer, K.A. Jackson: In *Surface Physics of Materials I*, ed. by J.B. Blakeley (Academic, New York 1975) p.121; and J.D. Weeks: In *Ordering in Strongly Fluctuating Condensed Matter Systems*, ed. by T. Riste (Plenum, New York 1980) p.293

13.2 G.H. Gilmer, P. Bennema: J. Appl. Phys. **43**, 1347 (1972); S.W.H. de Haan, V.J.A. Meeussen, B.P. Veltman, P. Bennema, C. van Leeuwen, G.H. Gilmer: J. Crystal Growth **24/25**, 491 (1974)

13.3 A review of surface roughening and crystal growth kinetics in general is provided by J.D. Weeks, G.H. Gilmer: Advances in Chem. Phys. **40**, 157 (1979). Also see J.P. van der Eerden, P. Bennema, T.A. Cherepanova: In *Progress in Crystal Growth and Characterization 3*, ed. by B.R. Pamplin (Pergamon, Oxford 1979) p.219

13.4 G.H. Gilmer: J. Crystal Growth **35**, 15 (1976); and R.H. Swendsen, P.J. Kortman, D.P. Landau, H. Müller-Krumbhaar: J. Crystal Growth **35**, 73 (1976)

13.5 J. Narayan, W.L. Brown, R.A. Lemons (eds): *Laser-Solid Interactions and Transient Thermal Processing of Materials* (North-Holland, Amsterdam 1983)

13.6 D. Furman, S. Dattagupta, R.B. Griffith: Phys. Rev. B**15**, 441 (1977)

13.7 F.L. Williams, D. Nason: Surf. Sci. **45**, 377 (1974); V.S. Sundarum, P. Wynblatt: Surf. Sci. **52**, 569 (1975); K. Binder, D. Stauffer, V. Wildpaner: Acta Met. (1978)

13.8 See Ref. [13.1], first citation, for a calculation of $\mu^{(0)}$

13.9 P. Wynblatt, R.C. Ku: Surf. Sci. **65**, 511 (1977); F.F. Abraham, N.-H. Tsai, G.M. Pound: Surf. Sci. **83**, 406 (1979)

13.10 W.K. Burton, N. Cabrera: Disc. Faraday Soc. **5**, 33 (1949)

13.11 K.A. Jackson: In *Liquid Metals and Solidification* (American Society for Metals, Cleveland 1958) p.174

13.12 G.H. Gilmer: Science **208**, 355 (1980)

13.13 J.D. Weeks, G.H. Gilmer, H.J. Leamy: Phys. Rev. Lett. **31**, 549 (1973)

13.14 J.W. Cahn, J.E. Hilliard: J. Chem. Phys. **28**, 258 (1958); J.W. Cahn: J. Chem. Phys. **30**, 1121 (1959); and D.E. Temkin: Sov. Phys. Crystallogr. **14**, 344 (1969)

13.15 S.T. Chui, J.D. Weeks: Phys. Rev. B**14**, 4978 (1976)

13.16 H. van Beijeren: Phys. Rev. Lett. **38**, 993 (1977); and H.J.F. Knops: Phys. Rev. Lett. **39**, 766 (1977)

13.17 G.H. Gilmer, J.D. Weeks: J. Chem. Phys. **68**, 950 (1978)

13.18 Approximation methods for calculating the properties of the Ising model, including the extent of the miscibility gap are discussed by T.L. Hill: In *An Introduction to Statistical Mechanics* (Addison-Wesley, Reading 1960) Chap.14

13.19 A first-order transition was observed first by J.C. Shelton, H.R. Patil, J.M. Blakeley: Surf. Sci. **43**, 493 (1974), for carbon segregation to Ni{111} surface; and by J.C. Hamilton, J.M. Blakeley: J. Vac. Sci. Technol. **15**, 559 (1978), for carbon on Pt{111}. Also

P.C. Bettler, D.H. Bennum, C.M. Case: Surf. Sci. **44**, 360 (1974), noted surface phase changes for carbon and silicon on tungsten

13.20 A simple mean-field theory for segregation that exhibited a first-order transition in the surface layer was presented by
C.R. Helms: Surf. Sci. **69**, 689 (1977); and a diffuse-interface model was described in a paper on critical-point wetting of fluids by
J.W. Cahn: J. Chem. Phys. **66**, 3667 (1977)

13.21 N. Cabrera: Z. Electrochemie **56**, 294 (1952)

13.22 F.C. Frank: Discuss. Faraday Soc. **5**, 48 (1949)

13.23 Reproducible transient growth rates have been measured after the application of a potential during electrodeposition. See
R. Roussinova, E. Budevski: J. Electrochem. Soc. **119**, 1346 (1972)

13.24 G.H. Gilmer: J. Cryst. Growth **49**, 465 (1980)

13.25 G.H. Gilmer: J. Cryst. Growth **42**, 3 (1977)

13.26 A.A. Chernov: Sov. Phys.-Uspekhi **13**, 101 (1970). A model for the trapping of impurities during crystal growth from the melt is given by
K.A. Jackson, G.H. Gilmer, H.J. Leamy: In *Laser and Electron Beam Processing of Materials*, ed. by C.W. White, P.S. Peercy (Academic, New York 1982). Ising model simulations for laser-annealed silicon are described by G.H. Gilmer: [Ref. 13.5], p. 249

13.27 L.S. Hollister: Am. Mineral. **55**, 742 (1970)

13.28 A.A. Chernov, J. Loomis: J. Phys. Chem. Solids **28**, 2185 (1967). Also see A.A. Chernov: [Ref. 13.26]

13.29 E.M. Gyorgy, M.D. Sturge, L.G. van Uitert, E.J. Heilner, W.H. Grodkiewicz: J. Appl. Phys. **44**, 438 (1973);
R. Wolfe, R.C. Le Craw, S.L. Blank, R.D. Pierce: Appl. Phys. Lett. **29**, 815 (1976)

13.30 G.H. Gilmer, K.A. Jackson: In *Crystal Growth and Materials*, ed. by E. Kaldis, H.J. Scheel (North-Holland, Amsterdam 1977) p.80

13.31 M. Volmer: Die Kinetik der Phasenbildung (Steinkopf, Dresden 1939);
N. Stranski: Z. Physik Chem. **136**, 259 (1928)

13.32 J.Q. Broughton, G.H. Gilmer, K.A. Jackson: Phys. Rev. Lett. **49**, 1496 (1982)

14. Phase Transitions on Surfaces

P. Bak

With 20 Figures

14.1 Introduction

14.1.1 2D Physics and Adsorption

Adsorption is one of the most interesting phenomena in surface science. The study of adsorption is of enormous interest in achieving a better understanding of catalysis, semiconductor interfaces, crystal growth, etc.

In addition, adsorbed layers are unique examples of two-dimensional physical systems. Therefore, theories of statistical physics in two dimensions often can be applied directly to and provide insight into the properties of adsorbed systems. Vice versa, the study of surface structures allows for experimental testing of two-dimensional theories of great current interest.

In this chapter, we shall be concerned with phase transitions in monolayers of adsorbed atoms or molecules. Theories for phase transitions and phase diagrams in 2D are quite fascinating and distinct from three-dimensional theories in that they involve topological defects, such as domain walls and dislocations. As we shall see, one can sometimes predict complete global phase diagrams knowing only the symmetry of the various phases involved. The adsorbed monolayer may form an ordered structure which is commensurate with the crystal surface ("substrate"). When the coverage is changed by varying the pressure (or temperature) the monolayer may expand or contract and become incommensurate with the substrate. When the temperature is increased the ordered structure will eventually melt into a two-dimensional liquid with the same symmetry as the substrate.

A possible schmeatic phase diagram with commensurate, incommensurate and fluid phases is shown in Fig.14.1. According to current theories [14.1,2] the transition between the commensurate and the incommensurate phases is brought about by spontaneous formation of domain walls (or rather domain lines) in the C phase. We shall see that due to an amazing transformation of an approp riate statistical model onto the quantum sine-Gordon equation, these walls

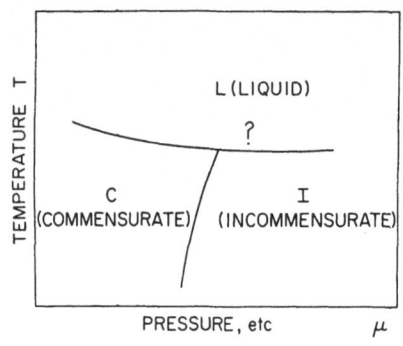

Fig.14.1. Schematic phase diagram with commensurate (C), incommensurate (I), and fluid (L) phases

can be described as "quantum solitons." The melting of the I phase is a Kosterlitz-Thouless [14.3] transition caused by dissociation of dislocation pairs.

Much of the work in the field has been concerned with transitions within rare gas monolayers physisorbed on graphite. In general, the various phases have hexagonal symmetry, and the theories of phase diagrams in such systems are quite complicated [14.3]. Here we shall consider the simpler case with uniaxial, rectangular symmetry, as found for instance in chemisorbed layers of gases or metals on clean, oriented metal surfaces. Although the binding of atoms in chemisorbed monolayers is often much stronger than any thermal energy, corrugation of the surface potential is often in the meV regime and allows for thermally induced transitions between C, I, and liquid (L) phases. The commensurate phase is a $p \times 1$ (or $p \times n$) uniaxial phase, which becomes incommensurate in the "p" direction. We shall call this a commensurate-incommensurate transition of order p. The incommensurate phase is a uniaxial "$(p + \delta) \times n$" phase. Near the C phase the I phase takes the form of a striped soliton or domain wall structure, and with increasing temperature it melts into a 2D "soliton liquid" with 1×1 (or $1 \times n$) symmetry. Surprisingly, it turns out that the phase diagram depends in a crucial way on the order of commensurability, p. Five topologically different phase diagrams may arise for $p = 1,2,3,4$, and 5 [14.4-8] (see Figs.14.15-19 below). We suggest that experimental systems (some of which will be discussed next) be studied to check these diagrams.

14.1.2 Experiments

Several experimental techniques are available for studying surface structures and phase diagrams. The bulk of the work has been performed by means of low energy electron diffraction (LEED) techniques. The interpretation of LEED data is often hampered by multiple scattering associated with strong inter-

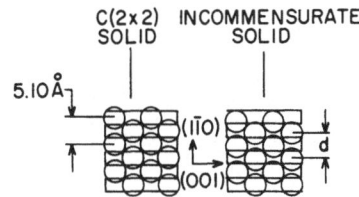

C(2x2) INCOMMENSURATE
SOLID   SOLID

5.10Å

(1̄10)

(001)

d

Fig.14.2. Arrangement of Xe atoms on the Cu{110} surface in the commensurate and incommensurate phases [14.9]. The CI transition is a $p = 2$ transition

action with matter. Most of our understanding of surface structures, nevertheless, stems from LEED measurements. Neutron scattering measurements are useful for light atoms, but a large volume of material is needed so the technique is useful only where huge surface areas can be studied, as for instance in exfoliated graphite. The most promising technique is certainly synchrotron X-ray diffraction. Today it is possible to determine the structure of a single layer, and to analyze the shape of diffraction peaks to determine the nature of correlation functions in the various surface phases. This is important in checking theoretical predictions of phase transitions and phase diagrams. Let us mention a few specific systems.

Jaubert et al. [14.9] studied the phase diagram of a xenon (Xe) monolayer adsorbed on the Cu{110} surface at $T \sim 77$ K. At low pressure the Xe atoms form a commensurate 2×2 phase (Fig.14.2). When the pressure is increased the monolayer contracts along one direction and becomes incommensurate in that direction. The commensurate phase (and the CI transition) is of order $p = 2$. At the transition the diffraction peaks at the commensurate position split, and the splitting is a measure of the incommensurability. It was found that the incommensurability $\delta \sim (P - P_c)^{\frac{1}{2}}$, where P_c is the critical pressure at which the transition takes place.

Diehl and *Fain* [14.10,11] studied a monolayer of N_2 molecules (represented by ellipses in Fig.14.3) adsorbed on graphite, at $T \sim 33$ K. In the commensurate phase the positions of the N_2 molecules are ordered in the commensurate $\sqrt{3} \times \sqrt{3}$ structure. Because of the herringbone orientational order the C phase is a uniaxial phase of order $p = 3$. With increasing pressure the monolayer becomes incommensurate in the b direction (Fig.14.3) such that the average periodicity becomes slightly less than three graphite lattice units. In the LEED pattern the incommensurability shows up as an asymmetric splitting of diffraction peaks.

Kortan et al. [14.12] investigated the ordering of Br_2 molecules in the stage 4 graphite intercalation compound Br_2C_{28} utilizing X-ray scattering techniques. Strictly speaking, the compound is three-dimensional, but it was found that the interlayer couplings are sufficiently small to suppress correlations or ordering between layers so that each layer can be viewed as an in-

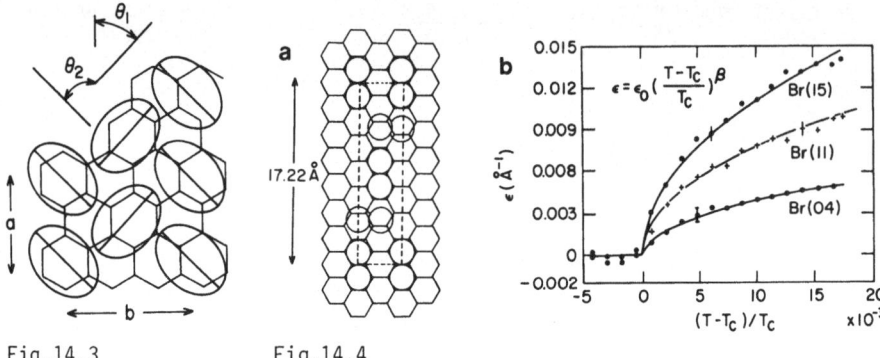

Fig.14.3   Fig.14.4

<u>Fig.14.3.</u> Incommensurate ordering of N_2 molecules on graphite 14.10,11 . In the commensurate phase the periodicity along the b direction is three graphite lattice units, so the C phase is of order p = 3

<u>Fig.14.4a.</u> Ordering of Br_2 molecules in the commensurate phase of the graphite intercalation compound Br_2C_{28} (p = 7). (b) Splitting of Bragg peak vs temperature near the CI transition [14.12]

dependent 2D system. Figure 14.4a shows the ordering of Br_2 molecules in the $7 \times \sqrt{3}$ commensurate phase. At T = 342.2 K there is a CI transition along the 7-fold direction, so p = 7. The CI transition here is driven by temperature alone. Figure 14.4b shows the splitting of selected Br_2 Bragg peaks. The splitting has a square root dependence on the reduced temperature.

The phase diagram of H adsorbed on the Fe{110} surface has been investigated by *Imbihl* et al. [14.13] (Fig.14.5). The phase diagram includes two commensurate structures, of order p = 2 and p = 3, respectively. Inbetween the commensurate phases there seems to be an "antiphase" domain wall structure. We shall see that domain wall structures are characteristic for incommensurate phases near the CI transition.

Tellurium adsorbed on a W{110} surface exhibits several commensurate phases. For instance, *Park* et al. [14.14] discovered a series of p × 2 phases with p = 5, 22, 17, 20 and 2, with incommensurate phases inbetween. Sulfur adsorbed on W{110} has a similarly complicated behavior [14.15]. An X-ray scattering study of Pb adsorbed on Cu{110} [14.16] has revealed a commensurate-incommensurate transition of order p = 5 and a reversible incommensurate melting transition, so all the various phases and transitions to be discussed in the following are represented. *Grunze* et al. [14.17] found a p = 2 CI transition in a monolayer of N_2 adsorbed on Ni{110}. Finally *Lyuksyutov* et al. [14.18] found a 3 × 1 ordered structure of Mg on the Re{1010} surface and a 5 × 1 structure in Ba on the Mo{112} surface. It would be interesting

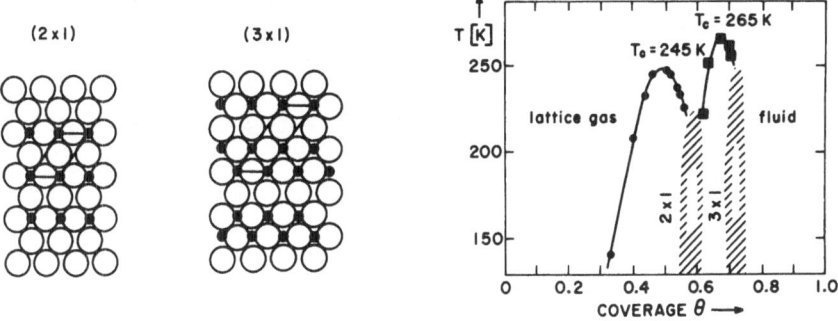

Fig.14.5. *Right*: Phase diagram of the H/Fe{110} system in the temperature-coverage plane obtained by *Imbihl* et al. [14.13]. *Left*: Models of the (2 × 1) and (3 × 1) phases. The large circles represent Fe atoms, the small circles H atoms

to investigate these structures further as a function of coverage to see if I and liquid phases exist in addition to the commensurate ones.

These examples should be sufficient to illustrate that we are dealing with a rather general phenomenon. In Sects.14.2,3, theories of phase diagrams involving uniaxial commensurate, incommensurate and liquid phases will be reviewed. Many authors have contributed to the understanding of various aspects of phase diagrams. A rather general treatment, based on Bethe ansatz solutions of the 1D quantum sine-Gordon equation, and the Kosterlitz-Thouless [14.19] theory of melting, has recently been given by *Haldane* et al. [14.4], and for simplicity we shall generally follow this work. The analysis will take place in two steps. First, in Sect.14.2, the CI transition will be treated using a model in which the melting transition has been impeded by not allowing dislocations to be formed in the C and I phases. The results at T = 0 were first obtained by *Frank* and *van der Merwe* [14.20], and at T > 0 by *Pokrovsky* and *Talapov* [14.21]. Second, in Sect.14.3, the melting transition of the I phase is studied, by including the possibility of having dislocations. The melting is expected to take place through a dislocation-separation mechanism. Phase diagrams for all values of p are derived.

The model which will be studied is a continuum model where the adsorbed monolayer forms a continuously deformable lattice on the periodic surface. Others have applied discrete lattice gas models (chiral Potts models, Ising models, etc.) confining the atoms to specific positions in the surface unit cell to study essentially the same phase diagrams. We believe that the general structure of the phase diagrams does not depend on which type of model is being applied; it is the symmetry of the various phases that matters.

14.2 The Commensurate-Incommensurate Transition and Quantum Solitons

Figure 14.6 illustrates the simple model of an adsorbed monolayer on a uni-axial substrate, on which our discussion on the CI transition will be based. A two-dimensional array of particles, connected by harmonic springs, is situated on a two-dimensional "washboard" potential representing the crystal surface. Since we are interested only in phase transitions involving periodicity in the uniaxial x direction, the potential has been assumed to be smooth in the perpendicular y direction, $V(x,y) = V(x)$. The structure remains commensurate (or incommensurate) in that direction. Any broken symmetry in the y direction can be restored at a separate transition. The particular structure shown is "almost" commensurate with the periodicity of the crystal surface ($p = 1$). For a C phase of general p there would be a particle in every p^{th} potential minimum. The shift in the x direction of the atom (i,j) relative to the commensurate position is denoted $\phi_{i,j}$. The Hamiltonian of the model may be written

$$H = \sum_{<i,j>} \frac{1}{2} (\phi_{i,j} - \phi_{i+1,j} - \mu)^2 + \frac{1}{2} (\phi_{i,j} - \phi_{i,j+1})^2$$

$$+ \lambda \cos\phi p_{i,j} \quad . \tag{14.1}$$

The first two terms represent the elastic interactions in the x and y directions, respectively. The parameter μ is the misfit between the inter-atomic distance favored by the springs, and the distance $2\pi/p$ favored by the periodic potential represented by the last term. We have chosen a cosine potential, but it can be argued [14.2] that the general properties are independent of the choice of potential. In a lattice gas model the cosine term would be replaced by a periodic array of δ functions. The p equivalent commensurate arrangement corresponds to $\phi = 2\pi n/p$, $n = 0,1..., p-1$.

Note that since the particles are connected with harmonic springs the 2d crystal cannot melt. To allow for melting it must be possible to have dislocations which can break up the crystal. If the harmonic interaction in the y direction is replaced by a cosine potential

$$\frac{1}{2} (\phi_{i,j} - \phi_{i,j+1})^2 \to \frac{1}{2} \cos(\phi_{i,j} - \phi_{i,j+1}) \quad , \tag{14.2}$$

the Hamiltonian would be invariant under a shift of one chain in the x direction, $\phi_{i,j} \to \phi_{i+n,j}$ and the model would permit melting. The complete phase diagrams to be discussed here and in Sect.14.3 can in principle be investi-

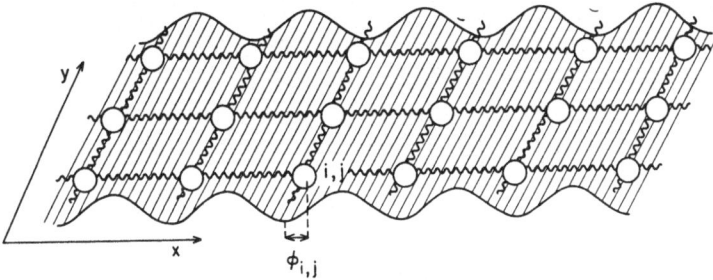

Fig.14.6. Adsorbed monolayer on uniaxial substrate. The monolayer is almost commensurate with p = 1. The "phase" $\phi_{i,j}$ is the displacement of the atom (i,j) in the x direction relative to the pth potential well, so $\phi_{i,j}$ = 0 in the commensurate phase

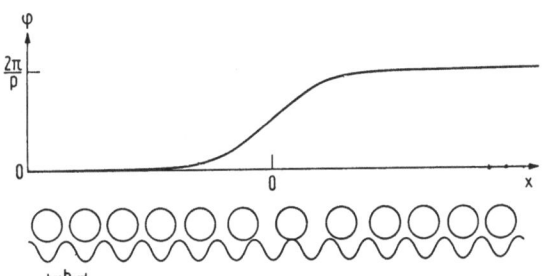

Fig.14.7. Domain wall, or soliton solution to (14.4). The domain wall connects a commensurate regime with $\phi = 2\pi n/p$ and a commensurate regime with $\phi = 2\pi(n+1)/p$. At the domain wall the monolayer is far from registry with the crystal surface

gated by studying the Hamiltonian (14.1) with the substitution (14.2), for instance by molecular dynamics or Monte Carlo techniques.

When distortions of the regular periodic lattice are not too dramatic, the continuum approximation

$$\phi_{i,j+1} - \phi_{i,j} \sim \frac{\partial\phi}{\partial y} \quad , \quad \phi_{i+1,j} - \phi_{i,j} \sim \frac{\partial\phi}{\partial x}$$

can be applied, and the Hamiltonian (14.1) becomes

$$H \sim \int dxdy \left[\frac{1}{2}\left(\frac{\partial\phi}{\partial x} - \mu\right)^2 + \frac{1}{2}\left(\frac{\partial\phi}{\partial y}\right)^2 + \lambda\cos p\phi\right] \quad . \tag{14.3}$$

At T = 0 the equilibrium configuration is given by the function $\phi(x,y)$ which minimizes the integrand in (14.3). The Euler-Lagrange equation is

$$\frac{\partial^2\phi}{\partial x^2} - \lambda p\cos p\phi = 0 \quad , \tag{14.4}$$

which is the pendulum equation. In addition to the simple commensurate solutions $\phi = 2\pi n/p$, the equation has solutions which are regular arrays of phase defects, or walls, or solitons, with distance ℓ. Figure 14.7 shows the posi-

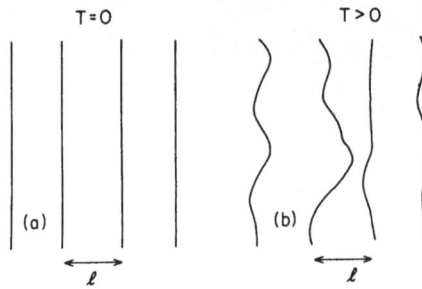

Fig.14.8. Arrays of domain walls. At T = 0 (a) the walls are straight. It costs energy to bend the walls. At T > 0 (b) configurations with meandering walls may be thermally excited and must be included in the partition function

(a)   (b)
ℓ   ℓ

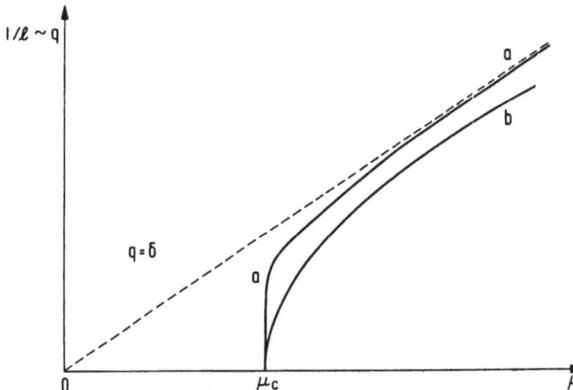

Fig.14.9. Incommensurability vs the misfit μ at (a) T = 0 and (b) T > 0

tions of atoms near a single wall. Between the walls the monolayer is almost commensurate; at the wall the particles are far out of registry with the surface potential. A "light" wall connecting C phases with $\phi = 2\pi n/p$ to the left and $\phi = 2\pi(n+1)/p$ to the right is shown in the figure. A "heavy" wall, or antiwall, connecting C phases characterized by n and n-1 would be favored by a negative μ in (14.3). The width w of the wall depends on the potential, $w \sim 1/\sqrt{\lambda}$. Figure 14.8 shows a regular array of domain "lines" which may be stable at T = 0.

When the misfit is smaller than a critical value $\mu_c = 4/\pi\sqrt{\lambda}$, the C solution $\phi = 0$ minimizes the energy (14.3). Beyond the critical value of μ there are solitons in the ground state. The incommensurability q is proportional to the density of solitons, $q \sim 1/1$. Near the CI transition

$$q = 1/1 \sim - \ln^{-1}(\mu - \mu_c) \quad , \tag{14.5}$$

which is an extremely steep curve (Fig.14.9). This result was first obtained by *Frank* and *van der Merwe* [14.20].

At T > 0 all possible configurations of the Hamiltonian (14.3) should be included with the proper Boltzmann factor in order to calculate the thermodynamic properties of the system. For instance, the walls can wiggle as shown in Fig.14.8b. The free energy F of the system is given by

$$\exp(-\beta F) = \int D\phi(x,y)\, \exp\{-\beta H[\phi(x,y)]\} \quad , \tag{14.6}$$

where $\int D\phi$ indicates a functional integral over all possible configurations (including also, for instance, wall-antiwall loops as shown in Fig.14.3).

Now we shall apply a trick which is often used in the statistical mechanics of 2D systems. The problem of finding the free energy of a 2D system can be transformed into that of finding the lowest eigenvalue of a certain matrix, the *transfer matrix*. The transfer matrix relates configurations on the row $y+1$ with configurations on the row y. In the present case the transfer matrix can be written as $\exp(-\beta H)$ whose H is the *1D quantum sine-Gorden* (SG) Hamiltonian:

$$H = T \int dx \left[\frac{1}{2}\gamma\pi^2 + \frac{1}{2}\gamma^{-1}\left(\frac{d\phi}{dx} - \mu\right)^2 + \frac{\lambda}{T}\cos p\phi \right] \quad , \tag{14.7}$$

with $\gamma = 2\pi T$. Here, π is the conjugate momentum to the phase ϕ, $[\pi(x),\phi(x')] = \delta(x-x')$. The temperature T plays the role of Plancks constant $\hbar$.

Thus, the 2D statistical mechanics problem has been "reduced" to the 1D quantum problem of finding the *ground state* of (14.7) merely by noting that two matrices (the transfer matrix and the 1D Hamiltonian), representing two seemingly very different physical problems, happen to be identical. We can thus immediately take over the field theorists' results on the sine-Gordon Hamiltonian [14.22,23]. The thermal fluctuations in the original model have been replaced by quantum fluctuations in the 1D quantum model. The walls in the classical model are represented by the quantum solitons in the sine-Gordon Hamiltonian, and we shall see that the misfit μ plays the role of a chemical potential for solitons. It is fascinating that a very abstract construction like the quantum sine-Gordon equation (which hardly has any application in particle quantum field theory) can be applied to the down-to-earth problem of calculating surface phase diagrams which can be measured by simple and direct experimental techniques.

The field theorists prefer to introduce a parameter β which in our language is defined by

$$\beta^2 = p^2\gamma = 2\pi p^2 T \quad . \tag{14.8}$$

The quantum SG equation is the Hamiltonian for an array of quantum pendula, connected by the harmonic interaction $(d\phi/dx - \mu)^2$ which favors a chirality, or twist, of the chain. In the C phase the pendula fluctuate around their "down" positions; in the I phase the chain has a net average winding number.

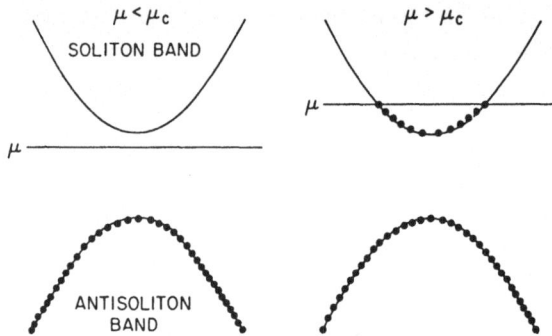

Fig.14.10. "Soliton" fermion bands derived from the quantum sine-Gordon equation. All states up to the chemical potential μ are filled. For $\mu > \mu_c$ there are solitons in the ground state, corresponding to the existence of infinite domain walls in the adsorbed monolayer. The incommensurability is proportional to the number of solitons in the ground state

The quantum solitons can be thought of as fermions. It is possible to transform the Hamiltonian to a fermion Hamiltonian [14.24], the massive Thirring model. At a specific temperature [14.25] ($\beta^2 = 4\pi$, or $T = 2/p^2$) the fermions are noninteracting and can be labeled with the wave vector k. These fermion states are illustrated in Fig.14.10. There is a "valence" band and a conduction band of solitons. The antisolitons (holes in the valence band) correspond to heavy walls, and the solitons correspond to the light walls (Fig.14.7). The misfit μ acts as a chemical potential for the soliton. At $\mu = 0$ the Fermi level is in the middle of the band gap, so the soliton band is empty: the system is commensurate. With increasing μ the Fermi level eventually hits the bottom of the "conduction" band $\mu_c(T)$: this indicates the CI transition. As the chemical potential increases further, the soliton band is filled. Because of the quadratic minimum, the band filling n_s obeys a square root law above the critical value:

$$q \sim \frac{1}{T} \sim n_s \sim (\mu - \mu_c)^{\frac{1}{2}} \tag{14.9}$$

(Fig.14.9b), a result to be compared with the steep logarithmic behavior at $T = 0$ (curve a in Fig.14.9). This result was first obtained by *Pokrovsky* and *Talapov* [14.21]. The physical reason for the square root behavior is an entropy-mediated interaction between meandering walls of the form

$$F_c \sim \frac{T^2}{K l^2} \, ,$$

which arises because of the possibility of the solitons in Fig.14.10b colliding with each other. Configurations involving antiwalls have been effectively included. The fact that there are no antisolitons in the quantum ground state simply means that there are no *infinite* heavy walls in the incommensurate phase for positive μ.

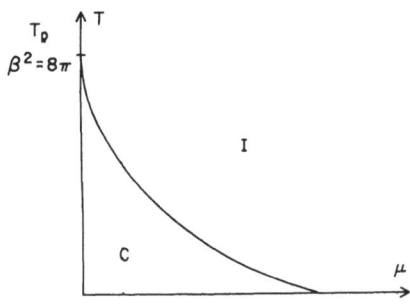

T_ρ
$\beta^2 = 8\pi$

Fig.14.11. Phase diagram of the CI transition. Note that above an upper critical temperature $t_\rho = 4/p^2$ the monolayer does not lock into the C phase. The effective substrate potential is zero above this temperature

Even if the noninteracting fermion picture is valid only at a special temperature, one can argue [14.2] that at other temperatures the qualitative behavior remains the same.

Figure 14.11 shows the resulting phase diagram. Note that above an upper critical temperature T_ρ there is no C phase (except, accidentally, at $\mu = 0$). In the quantum soliton picture the gap (Fig.14.10) reduces to zero at this temperature, i.e., the zero-point energy of the soliton lattice balances the positive soliton formation energy. The effective substrate potential is reduced because the walls fluctuate about their "equilibrium" positions, and above the critical temperature the effective potential renormalizes to zero. The critical temperature is given by

$$\beta^2 = 8\pi \qquad \text{or} \qquad T_\rho = \frac{4}{p^2} \ . \tag{14.10}$$

The square root dependence of the misfit on the pressure found by *Pokrovsky* and *Talapov* has been confirmed experimentally by *Kortan* et al. [14.12] for the $p = 7$ system Br_2C_{28} (Fig.14.4). *Jaubert* et al. [14.9] also observed this behavior but in the following section it will be seen that for $p = 2$ complications are expected to occur because of the possibility of having dislocations and thus a liquid phase interfering with the CI transition.

Before proceeding with the melting transition let us have a closer look at the I phase. In a three-dimensional crystal the positions of the atoms fluctuate around perfectly ordered equilibrium positions, i.e., there is complete long-range order. In a 2D incommensurate system this is not so. Because of the continuous symmetry associated with the position of the incommensurate monolayer on the surface, there can be no real long-range order in this phase [14.26]. The correlations between lattice positions decay algebraically at long distances:

$$<\cos[\phi(0) - \phi(r)]> \sim \cos q \cdot r \times r^{-\eta} \ . \tag{14.11}$$

327

At T = 0, the exponent $\eta = 0$, so there are long-range oscillating correlations given by the wave vector q of the I phase. In the absence of the substrate potential the exponent η characterizing the decay of correlations can be calculated rigorously since the Hamiltonian (14.3) is then harmonic. One finds

$$\eta_{harm} = T \quad . \tag{14.12}$$

A 2D I phase may be called a "floating phase" to illustrate that there is no long-range order, and that it does not generally lock into the surface potential when the wave vector accidentally becomes commensurate. This situation is not to be confused with the fluid phases with exponential decay of position correlations

$$<\cos[\phi(0) - \phi(r)]> \sim \cos \mathbf{q} \cdot \mathbf{r} \times \exp\left(-\frac{|\mathbf{r}|}{\xi}\right) \quad . \tag{14.13}$$

14.3 The Melting Transition and the Phase Diagram-Dislocations and Soliton Liquids

The uniaxial incommensurate phase has the symmetry of the 2D xy models. The phase of the xy spins corresponds to the position ϕ of the monolayer on the surface. The melting transition is thus isomorphic with the Kosterlitz-Thouless transition in the xy model. Note that the symmetry is quite different from that of the 2D melting transition on a smooth substrate discussed by *Nelson* and *Halperin* [14.27], where two coordinates u_x and u_y are needed to describe the position of the system, and there may be two transitions: one involving destruction of positional order and one involving destruction of orientational order. On a periodic substrate, there is generally no need for a transition of the second type. The substrate induces rotational order at all temperatures.

In the 2D xy picture the cos $p\phi$ term corresponds to a p fold spin anisotropy term. The p = 1 term represents a magnetic field. *José* et al. [14.28] have investigated the consequences of such terms for $\mu = 0$ and their phase diagrams must appear as a special limit of our results here. The melting takes place through a dislocation-dislocation mechanism. Let us, therefore, first identify the Kosterlitz-Thouless dislocations in our soliton description of the nearly commensurate incommensurate phase. Figure 14.12 shows 2 walls coming together at p = 2 and 3 walls meeting at p = 3. Since the phase shift imposed by the wall (Fig.14.7) is $2\pi/p$, the total phase shift as the defect is traversed is 2π. Thus, p domain walls coming together form a dislocation. The numbers shown in the figure are n = 1,2...p, representing the p equivalent

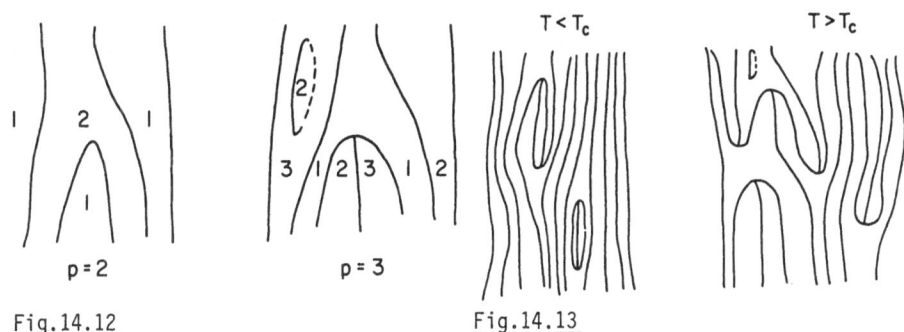

p = 2   p = 3

Fig.14.12   Fig.14.13

Fig.14.12. Dislocations formed by p domain walls coming together for p = 2, and p = 3. The numbers represent the p different equivalent positions of the commensurate monolayer on the surface. The phase ϕ increases 2π as the dislocation is traversed in the positive direction

Fig.14.13a. Bound dislocation pairs below the melting transition. b) Free dislocations above the transition. The resulting structure may be called a "soliton liquid"

commensurate positions. Clearly, it is topologically impossible to have $n < p$ walls meeting, or to have a single wall ending somewhere (unless p = 1).

Below the melting transition the dislocations are bound in pairs (Fig. 14.13a). Above the phase transition the dislocations screen each other and form a plasma of free dislocations, Fig.14.13b. The resulting structure can be characterized as a soliton liquid.

It is difficult to perform explicit analytical calculations on specific models allowing for both CI and melting transitions, such as the one defined by (14.1,2), or the "chiral Potts", or clock models to be defined later. However, in the limit where the energy cost of forming a dislocation E_0 is large (i.e., the low fugacity limit)

$$y = \exp\left(-\frac{E_0}{T}\right) \ll 1 \quad , \tag{14.14}$$

the problem becomes tractable. By studying the properties of the I phase in the *absence* of dislocation, we can find out whether or not dislocations are relevant, if they are permitted to occur. If dislocations are relevant, the I phase is unstable and will melt. If dislocations are irrelevant, the I phase is stable. According to the theory of Kosterlitz and Thouless [14.19] the dislocations are not relevant if the exponent η defined by (14.11) < 1/4; they are relevant when $\eta > 1/4$. The line $\eta = 1/4$ defines the melting line in the limit of vanishing fugacity. The analysis presented below closely follows the work of *Haldane* et al.[14.4].

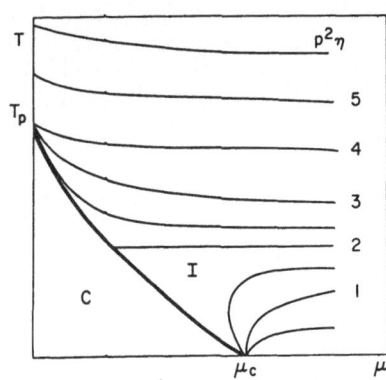 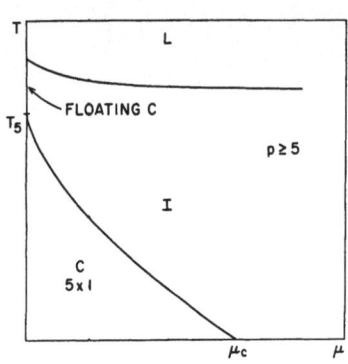

Fig.14.14   Fig.14.15

Fig.14.14. Lines of constant   derived from the quantum sine-Gordon model
(14.7). In the absence of dislocations the lines for different commensurability order p may be scaled onto each other as illustrated

Fig.14.15. Phase diagram for p > 4. There is a floating phase separating the
C and L phases at $\mu = 0$

Haldane [14.23] calculated the constant η, or rather constant $p^2\eta$, trajectories in the I phase by applying the Bethe ansatz to the 1D quantum sine-
Gordon equation in the presence of a finite density of solitons (Fig.14.14).
At large misfits ($\mu \gg \mu_c$) η approaches the harmonic value, $\eta = T$ (or
$\eta = \beta^2/2\pi p^2$). The surface potential plays no role in this case. At the CI
transition, $p^2\eta \to 1/2$: this result was first obtained by *Schulz* [14.29]. Near
the CI transition $p^2\eta$ is given by an equation of the form

$$\left(\frac{1}{2} p^2\eta - 1\right) = 4R(\beta)\sqrt{\mu/\mu_c(\beta) - 1} \quad , \tag{14.15}$$

where $R(\beta)$ is a function of temperature which can be found by solving a complicated integral equation. The lines $p^2\eta < 2$ all end at the point $T = 0$,
$\mu = \mu_c(0)$. The lines $4 > p^2\eta > 2$ end at the point $T = T_p$, $\mu = 0$. Finally, the
lines $p^2\eta > 4$ cross the line $\mu = 0$ above the critical point.

It is now quite straightforward to determine the phase diagrams. For a
given p we find the line $\eta = 1/4$ and identify it with the melting line.

i) p > 4. The phase diagram for p > 4 is shown in Fig.14.15. For all values of
μ there is a floating incommensurate phase separating the C and L phases.
Thus, the liquid phase does not interfere with the CI transition (unless, of
course, there is a first-order transition due to mechanisms not accounted
for here; we shall always implicitly assume that this is not the case). Even
at $\mu = 0$ there is a "massless" accidentally commensurate floating phase separating the KT transition from the critical point of the CI transition. This

result was first obtained by *José* et al. [14.28]. A phase diagram of this type is expected for instance for the $p = 7$ system Br_2C_{28} (Fig.4.12) studied by *Kortan* et al. [14.12]. At a given pressure they found both the CI transition and the melting transition as a function of temperature. It would be interesting to perform experiments with varying coverage to explore the complete diagram. *Kortan* et al. also analyzed the line shape of the diffraction peaks to determine η. Their findings are consistent with having η increasing from $2/p^2 = 0.04$ at the CI transition to $1/4$ at the melting transition as predicted by theory [14.19,29]. In passing, we note again that the dislocations and walls introduced here cannot destroy the ordering in the perpendicular direction. If the transitions in Br_2C_{28} take place through the mechanism discussed here, the resulting "liquid" would still have $p = 2$ fold commensurability in the perpendicular direction. It would be a *1 × 2* phase (not a "smetic" phase as suggested by *Kortan* et al.). The fact that the monolayer is rather dilute in the "1" direction does not change its symmetry classification. An equivalent situation in freon adsorbed on graphite has been discussed by *Bak* and *Bohr* [14.30]. The 1×2 phase could melt through an Ising transition at a higher temperature.

ii) $p = 4$ (Fig.14.16). For $p = 4$ the CI line and the melting line meet at the singular point $\mu = 0$, $T = T_p$. The transition at this particular point has the symmetry of the 2D xy model in a cubic anisotropy field [14.31], or the Ashkin-Teller model. The critical behavior is expected to be nonuniversal, i.e., there are no specific predictions for exponents. For $\mu \neq 0$ there is an I phase between the C and L phases.

iii) $p = 3$. The phase diagram for $p = 3$ is shown in Fig.14.17. Again, the transition lines meet at a singular point p on the $\mu = 0$ line. The transition when passing through this point is expected to belong to the universality class of the three-state Potts model (or clock model).

The phase diagram for $p = 3$ is quite controversial. Several authors suggest the existence of a Lifshitz point LP where the transition from the C phase changes from commensurate-incommensurate to commensurate fluid [14.32-36]. The corresponding phase diagram is shown by a broken line in the figure. The transition along the CL line could then either be of the three-state Potts type [14.32-33], or it could be of a different "chiral" type, the properties of which are not known [14.36].

When $\mu \neq 0$ the system has "chirality": the spins (in the xy analogy) prefer one way of rotating with respect to the other. *Huse* and *Fisher* [14.36] apply scaling laws near the Potts point and estimate that the chirality μ is a re-

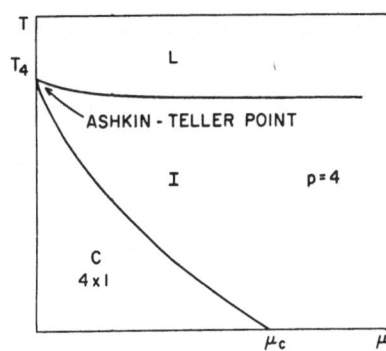
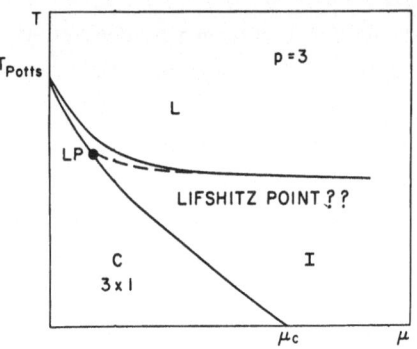

Fig.14.16.   Fig.14.17

Fig.14.16. Phase diagram for p = 4. The melting line and the CI line meet at a singular point at $\mu = 0$. The phase transition at this point, when approached along the $\mu = 0$ line, is expected to be of the "2d xy with cubic anisotropy" universality class

Fig.14.17. Phase diagram for p = 3. The melting line and the CI line meet at the Potts point at the $\mu = 0$ line. The alternative phase diagram, including a Lifshitz point (LP) separating fluid, commensurate and incommensurate phases, which has been suggested by other authors, is indicated by the broken melting line

levant perturbation, i.e., the nature of the phase transition for nonzero μ is expected to be different from that at $\mu = 0$. This is perfectly consistent with a phase diagram having no Lifshitz point. However, *Huse* and *Fisher* take this result to indicate a line of critical points between the Potts point and a Lifshitz point with a different critical behavior which they call "chiral". This interpretation seems less natural since such "chiral" transitions do not exist in other situations where chirality is relevant: within mean field theory one can always eliminate relevant chiral terms (which exist whenever the Lifshitz condition [14.26] is not fulfilled) by expanding the free energy around a wave vector which is displaced from the commensurate one [14.37]. The chiral term leads to a transition into a chiral (incommensurate) phase, not to a transition of "novel" first-order type as originally suggested by *Landau* and *Lifshitz* [14.26]. Renormalization group arguments near D = 4 dimensions lead to the same conclusions [14.38]. Moreover, the calculations presented here leading to the phase diagram without a Lifshitz point are rigorous, so the Huse-Fisher interpretation is incorrect in the limit of small fugacity of dislocations.

Schulz [14.5], who independently obtained many of the results in this section, has applied renormalization group arguments to a fermionic model with a finite fugacity of dislocations, and also found a phase diagram with no Lifshitz point.

The discrete "chiral clock" or "asymmetric Potts" model has the same symmetry as the model discussed here. It is defined by the Hamiltonian [14.34]

$$H = - J \sum_{ij} \cos[2\pi(n_i - n_j - \mathbf{\Delta} \cdot \mathbf{R}_{ij})/p] \qquad (14.16)$$

where $\mathbf{\Delta} = (\Delta, 0)$, and $\mathbf{R}_{ij}$ is the unit vector connecting nearest-neighbor sites. The states $n = 1, \ldots p$ represent the p degenerate commensurate phases. *Howes* et al. [14.32] studied a "self-dual" version of this Hamiltonian, which is probably in a different universality class (it has no I phase), and found almost rigorously a Lifshitz point. *Howes* [14.33] has studied the 1D transfer matrix (or quantum Hamiltonian) derived from (14.16) by a series expansion technique and also found a Lifshitz point. *Selke* and *Yeomans* [14.35] studied the Hamiltonian using a standard Monte Carlo method. The phase transition lines were identified as maxima in the specific heat, and again a Lifshitz point was found. As they correctly pointed out, this method of locating transition lines could be incorrect for models with xy symmetry. *Gehlen* and *Rittenberg* [14.39] argue that the Lifshitz point is a finite-size effect and would not exist for an ideal infinite system. *Houlrik* et al. [14.40] studied the Hamiltonian by Monte Carlo renormalization group methods. The existence of the floating phase was confirmed, but the transition does not take place at the maximum of the specific heat. The transition temperature was found to be a factor 6 higher than that found by Howes. No conclusions concerning the existence of the Lifshitz point were drawn.

Hydrogen adsorbed on the Fe{110} surface has a 3×1 uniaxial phase [14.13], Fig.14.5. It would be interesting to study this system further, for instance by synchrotron X-ray methods, to obtain more information on the phase diagram. *Selke* et al. [14.41] constructed a lattice gas model which produces both the 2×1 and the 3×1 phase observed in this system. *Kinzel* [14.42] studied this model by transfer matrix scaling, finding an anisotropic transition into an incommensurate phase with continuously varying wave vector, in agreement with [14.4,5]. However, the claim was made that the I phase has true long-range order, in disagreement with the arguments presented in Sect.14.2 which predict an I phase with algebraic decay of correlations.

iv) $p = 2$. The phase diagram for $p = 2$ is shown in Fig.14.18. The IL line penetrates all the way down to $T = 0$, so there is always a liquid phase between the C and I phases. This result was first derived by *Villain* and *Bak* [14.6] from a study of the 2D axial next-nearest-neighbor Ising (ANNNI) model, which is in fact a lattice gas model with the proper symmetry.

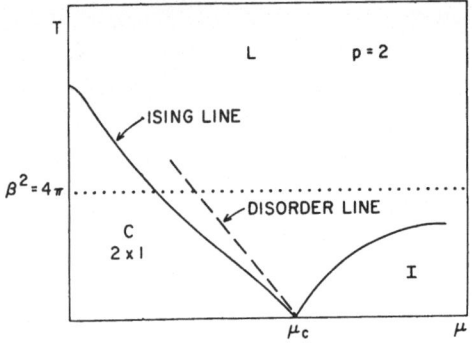

Fig.14.18. Phase diagram for p = 2. There is a liquid phase between the I and C phases even at the lowest temperatures. At the disorder line the wave vector q in the liquid phase defined by (14.13) goes continuously to zero. The dotted line indicates the line $\beta^2 = 4\pi$ where the sine-Gordon Hamiltonian including a dislocation term can be solved rigorously [14.8]

Coppersmith et al. [14.43] noted that this result can be derived directly by combining Schulz's result that $\eta = 2/p^2$ at the CI transition with the Kosterlitz-Thouless result that $\eta < 1/4$ in the I phase. Since the C phase has a simple up-down symmetry the CL transition becomes an Ising transition.

It is possible to include the dislocations explicitly in the Hamiltonian for p = 2. *Bohr* [14.7] analyzed a discrete model including both walls and dislocations and indeed found an Ising transition with the proper critical indices. *Bohr* et al. [14.8] investigated the quantum sine-Gordon model with an extra dislocation operator. They were able to solve the problem analytically along with line $\beta^2 = 4\pi$ (the dotted line in Fig.14.18). They found an Ising transition, but no I phase at this temperature. As can be seen from Fig.14.18, this is because the I phase exists only at lower temperatures.

It would be interesting to repeat the experiment on Xe absorbed on Cu{110} [14.9], and also the experiment on H on Fe{110} near the 2 × 1 phase, to look for the fluid phase between the C and I phases. Recently, *Grunze* et al. [14.17] reported the discovery of such a liquid phase in N_2 adsorbed on Ni{110}. The transition in this case is complicated by a shear distortion which seems to take place simultaneously with the CI transition.

It has been shown by *Peschel* and *Emery* [14.44], and by *den Nijs* [14.45], that a disorder line is expected in the liquid phase. A disorder line separates a regime with oscillating, but exponentially decaying correlations [$q \neq 0$ in (14.13)] from a regime with monotonically decaying correlations ($q = 0$). *Den Nijs* suggested that in the experiment on Xe on Cu{110} the disorder line could have been mistaken for a CI transition since at both transitions the wave vector goes continuously to zero.

v) p = 1. Finally, for p = 1 the melting transition of the I phase also preempts the CI transition (Fig.14.19). At T > 0 the melting transition is a KT transition; as T → 0 it becomes a CI transition with a vanishing soliton

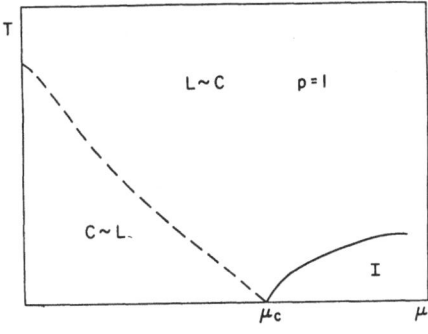

Fig.14.19. Phase diagram for p = 1. The liquid phase penetrates down to T = 0 as for p = 2. The CL transition is washed out by dislocations. Again, a disorder line which could be mistaken for a CI transition line in an experiment is expected to exist in the liquid phase

density. Similar conclusions have been reached by *Schaub* and *Mukamel* [14.46]. However, dislocations have another interesting effect in this case: the CL transition is completely wiped out. This agrees with the xy analogy because the $\cos\phi$ term acts as a magnetic field on the CI transition. The p = 1 "CL" transition can be described by a 1D quantum spin 1/2 Hamiltonian:

$$H = \sum_n S_n^x S_{n+1}^x + S_n^y S_{n+1}^y + \mu S_n^z + y S_n^x \quad , \tag{14.17}$$

where the y term gives the effects of dislocations. For y = 0 there is a transition at $\mu = \mu_c = 1$ where the ordering in the xy plane vanishes. For finite y there is no transition since there will always be an ordered component of the spin in the xy plane. The p = 1 C phase and the L phase have the same 1 × 1 symmetry. One can go continuously from an "almost commensurate" to an "almost L" phase by forming more and more vacancies, without ever crossing a phase transition line. A p = 1 CI transition in Pd adsorbed on Nb{110} has been observed by *Sagurton* et al. [14.47]. The transition is strongly first order and cannot be described by the theory presented here.

As the coverage is varied over a wide range in a real physical system, it could well be that the wave vector locks in at several p × n structures. Figure 14.20 shows a possible phase diagram for such a system. At T = 0 the discrete model (14.1) has an infinity of C phases which may or may not have I phases between them [14.1]. Most of these phases exist only up to a temperature T_p which is less than the melting temperature. The phases with p = 3 and p = 4 are expected to be stable all the way up to the melting temperature. The I phases will join together to form a floating phase which may be accidentally commensurate. If an experiment is performed, say at roughly constant temperature with varying concentration as indicated by the broken line in Fig.14.20, one would expect to see only a finite number of C phases with incommensurate and liquid phases between. *Park* et al. [14.14]

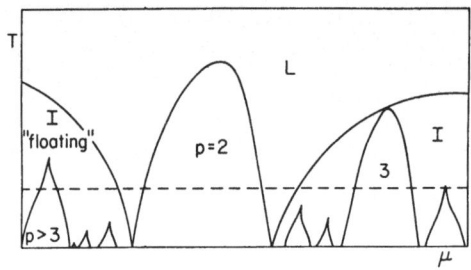

Fig.14.20. Possible general phase diagram with several C phases. An experiment performed along the broken line would show a finite number of C phases separated by regimes of I and L phases

studied tellurium (Te) adsorbed on the W{110} surface. They found a sequence of commensurate phases $p \times 2$ with $p = 4$, 20, 17, 22 and 5, corresponding to coverages of 1/4, 3/10, 6/17, 4/11 and 2/5, respectively. These 5 C phases are separated by I phases. In principle, there could be many more C phases in the coverage range in which the 11×2 structures occur; for instance, the structures 10×2 and 11×2, corresponding to concentrations 3/10 and 4/11 could exist but were not observed. Their findings are in agreement with a phase diagram like Fig.14.20.

Acknowledgements. First of all, I should like to thank T. Bohr and F.D.M. Haldane for pleasant collaboration in carrying out the work on which a significant part of this review is based. I am grateful to H.J. Schulz, M.P.M. den Nijs, S. Howes, W. Selke, A. Erbil, P. Kleban, and E. Bauer for illuminating discussions on theoretical and experimental aspects of adsorbed monolayers.

This work was supported by the Division of Materials Sciences U.S. Department of Energy under contract DE-AC02-76CH00016.

References

14.1 For a review see P. Bak: Rep. Prog. Phys. **45**, 587 (1982)
14.2 V.L. Pokrovsky, A.L. Talapov, P. Bak: In *Solitons*, ed. by V.E. Zakharov, S. Trullinger (North-Holland, Amsterdam, to be published)
14.3 For a review see J. Villain, M.B. Gordon: Surf. Sci. **125**, 1 (1983), and Ref.14.1
14.4 F.D.M. Haldane, P. Bak, T. Bohr: Phys. Rev. B**28**, 2743 (1983)
14.5 H.J. Schulz: Phys. Rev. B28, 2746 (1983)
14.6 J. Villain, P. Bak: J. Physique (Paris) **42**, 657 (1981)
14.7 T. Bohr: Phys. Rev. B**25**, 6981 (1982)
14.8 T. Bohr, V.L. Pokrovsky, A.L. Talapov: Pis'ma Zh. Eksp. Teor. Fiz. **35**, 165 (1982)
14.9 M. Jaubert, A. Glachant, M. Bienfait, G. Boato: Phys. Rev. Lett. **46**, 1679 (1981)
14.10 R.D. Diehl, S.C. Fain, Jr.: Phys. Rev. B**26**, 4785 (1982)
14.11 R.D. Diehl, S.C. Fain, Jr.: Surf. Sci. **125**, 116 (1983)

14.12 A.R. Kortan, A. Erbil, R.J. Birgeneau, M.S. Dresselhaus: Phys. Rev. Lett. **49**, 1427 (1982); Phys. Rev. B**28**, 6329 (1983)
14.13 R. Imbihl, R.J. Behm, K. Christmann, G. Ertl, T. Matsushima: Surf. Sci. **117**, 257 (1982);
F. Boszo, G. Ertl, M. Grunze, M. Weiss: Appl. Surf. Sci. **1**, 103 (1979)
14.14 Ch. Park, E. Bauer, H.M. Kramer: Surf. Sci. **119**, 251 (1982)
14.15 G. Popov, E. Bauer: Surf. Sci. **122**, 433 (1982)
14.16 W.C. Marra, P.H. Fuoss, P.E. Eisenberger: Phys. Rev. Lett. **49**, 1169 (1982)
14.17 M. Grunze, P.H. Kleban, W.N. Unertl, F.S. Rys: Phys. Rev. Lett. **51**, 582 (1983)
14.18 I.F. Lyuksyutov, V.K. Medvedev, I.N. Yakovkin: Zh. Eksp. Teor. Fiz. **80**, 2452 (1981)
14.19 J.M. Kosterlitz, D.J. Thouless: J. Phys. C**6**, 1181 (1973);
J.M. Kosterlitz: J. Phys. C**7**, 1046 (1974)
14.20 F.C. Frank, J.H. van der Merwe: Proc. R. Soc. (London) A**198**, 205 (1949)
14.21 V.L. Pokrovsky, A.L. Talapov: Phys. Rev. Lett. **49**, 1169 (1982); Zh. Eksp. Teor. Fiz. **75**, 1151 (1978) Engl. transl.:[Sov. Phys.-JETP **51**, 134 (1980)]
14.22 R. Dashen, B. Hasslacher, A. Neveu: Phys. Rev. D**10**, 4114 (1974)
14.23 F.D.M. Haldane: J. Phys. A**15**, 507 (1982)
14.24 S. Mandelstam: Phys. Rev. D**11**, 3026 (1975)
14.25 A. Luther, V.J. Emery: Phys. Rev. Lett. **33**, 589 (1974)
14.26 L.D. Landau, E.M. Lifshitz: *Statistical Physics* (Pergamon, London 1959)
14.27 D.R. Nelson, B.I. Halperin: Phys. Rev. B**19**, 2457 (1979)
14.28 J.V. José, L.P. Kadanoff, S. Kirkpatrick, D.R. Nelson: Phys. Rev. B**16**, 1217 (1977)
14.29 H.J. Schulz: Phys. Rev. B**22**, 5274 (1980)
14.30 P. Bak, T. Bohr: Phys. Rev. B**27**, 591 (1983)
14.31 E. Domany, M. Schick, J.S. Walker, R.B. Griffiths: Phys. Rev. B**18**, 2209 (1978)
14.32 S. Howes, L.P. Kadanoff, M.P.M. den Nijs: Nucl. Phys. B**215**, 169 (1983)
14.33 S. Howes: Phys. Rev. B**27**, 1762 (1983)
14.34 S. Ostlund: Phys. Rev. B**24**, 398 (1981)
14.35 W. Selke, J.M. Yeomans: Z. Physik B**46**, 311 (1982)
14.36 D.A. Huse, M.E. Fisher: Phys. Rev. Lett. **49**, 793 (1982)
14.37 D. Mukamel, S. Krinsky: Phys. Rev. B**13**, 5065 (1976)
14.38 A. Aharony, P. Bak: Phys. Rev. B**23**, 4770 (1980)
14.39 G.V. Gehlen, V. Rittenberg: Preprint
14.40 J.M. Houlrik, S.J. Knak Jensen, P. Bak: Phys. Rev. B**28**, 2883 (1983)
14.41 W. Selke, K. Binder, W. Kinzel: Surf. Sci. **125**, 51 (1983)
14.42 W. Kinzel: Phys. Rev. Lett. **51**, 996 (1983)
14.43 S.N. Coppersmith, D.S. Fisher, B.I. Halperin, P.A. Lee, W.F. Brinkman: Phys. Rev. Lett. **46**, 549 (1981)
14.44 I. Peschel, V.J. Emery: Z. Physik B**43**, 241 (1981)
14.45 M.P.M. den Nijs: Private communication
14.46 B. Schaub, D. Mukamel: J. Phys. C**8**, L225 (1983)
14.47 M. Sagurton, M. Strongin, F. Jona, J. Colbert: BNL 33388 Preprint (1983)

15. Finite Size Effects, Surface Steps, and Phase Transitions

P. Kleban

With 7 Figures

15.1 Introduction

15.1.1 Goals

This chapter is not a review in the usual sense. It rather comprises an outline of several aspects of the interplay between finite-size effects on surfaces —especially as induced by ordered arrays of steps —and phase transitions or related phenomena.

Some of the effects we consider have already been observed experimentally,
others remain to be seen in the laboratory. All of them concern ordering or
(equilibrium thermodynamic) phase transition behavior on stepped surfaces or
related systems. In discussing them we make use of the language of phase
transitions in adsorbed layers, however, most of the results described also
apply to (reversible) reconstruction. Our purpose is as much to show what
has been learned in this area as to motivate further experimental and theoretical studies.

There has been considerable work on phase transitions in physisorption
systems, especially on grafoil (or other basal graphite) substrates [15.1].
Research on chemisorption systems, with several notable exceptions [15.2],
has tended to concentrate on electronic or other surface properties. Thus,
for these systems, the experimental study of phase transitions, and especially the modification of phase transitions or ordering by introducing surface
steps is much more of a frontier area, with many rewards awaiting the successful investigator. This is especially true since much of the appropriate
theory, i.e., the effects of finite size on ordered phases and phase transitions, is under very active study [15.3-11] and has matured to the point
where considerable significant contact with experiment can be made. We hope
to illustrate this in what follows, by outlining the general theory and applying it to explain some experimental results and predict others. We concentrate on what we believe are the main ideas and consider only a few simple
examples that we have found of interest. Thus, we do not pretend that this

treatment is either comprehensive or exhaustive, but rather hope it will serve as an introduction, guide and inspiration to other researchers interested in this area.

15.1.2 Phase Transitions on Surfaces

Most of the examples we are concerned with here are adsorption systems that can be modeled as a "lattice gas." This means that the adsorbed molecules sit (most of the time) in a lattice of adsorption sites that are fixed with respect to the substrate, and the system is "closed," i.e., the adsorbate does not diffuse irreversibly into the bulk or go into a new bonding state, such as occurs in oxide formation. We also consider only systems that are in thermodynamic equilibrium, which means that an adsorbed molecule can hop (diffuse) from one site to another at a rate sufficiently rapid for equilibrium to be established in the time available for an experiment. Note that if the hopping is activated with energy E, the rate is proportional to $\exp(-E/T)$, so that equilibrium will not be established when the temperature T is too low. The assumption of fixed adsorption sites also implies that the lateral interactions between adatoms (adatom-adatom, or AA, interactions) are small compared to the adatom substrate interaction and the energy barrier between adsorption sites. This situation is realized in many (but not all!) physisorption [15.1] and chemisorption [15.12] systems for coverages in the submonolayer regions.

Another reason why we consider lattice gas models is that they are comparatively easy to handle theoretically and that many results are at hand. The presence or absence of an adatom at adsorption site i is simply specified by an occupation number $n_i = 1$ or $n_i = 0$, respectively. Further, the discrete lattice of sites simplifies many techniques, e.g., Monte Carlo calculations. The adsorption system thermodynamics is, in this case, specified by a lattice of sites, a set of AA interactions, the temperature T and coverage θ (or, equivalently, using thermodynamics, by T and the chemical potential μ). Many of the features of these models are discussed elsewhere [15.2,13]. Most of the results we mention below apply to more general systems, but it is helpful to restrict oneself to lattice gas models for pedagogical reasons.

The AA interactions give rise to the possibility of various ordered configurations of the overlayer. A rich variety of such orderings has been observed [15.1,14] via LEED (Low-Energy Electron Diffraction) and other scattering and thermodynamic techniques. The adlayer can form ordered solid-like structures that give rise to extra Bragg spots (adlayer beams) or disordered fluid-like structures that contribute to the diffuse scattering and may be observable by thermodynamic and other measurements.

It is worth mentioning that the fluid-like structures in a lattice gas model are *not* fluids in the usual sense. The adatoms are still restricted to the lattice of adsorption sites. What distinguishes this phase from a solid-like structure is the lack of long-range order. Because of thermal effects, the adatoms do not form a periodic lattice with unit mesh larger than the adsorption site lattice, hence no extra Bragg peaks are observed. This kind of disordered state can exist in a given system at higher coverage (lattice liquid) or lower coverage (lattice gas), with the possibility of the two phases becoming identical at a critical point, just as for ordinary three-dimensional substances. Note the unfortunate confusion of language — the term "lattice gas" is used to refer to the discrete adsorption site model in general *and* to a low density fluid phase.

The AA interactions determine the phase diagram, i.e., the types of phases present and their boundaries as a function of T and θ or T and μ. The latter set of thermodynamic variables is convenient for physisorbed or weakly chemisorbed systems in equilibrium with a dilute gas above the surface so that, aside from an additive function of T alone and a constant due to the adsorption energy, $\mu \propto k_B T \ln P_{gas}$. For strong chemisorption, P_{gas} is too small to measure, so θ is the only experimentally accessible variable.

The phase boundaries are lines or regions along which one phase melts, freezes or transforms (say by increasing θ) into another. The phase transitions that occur are classified as either first-order to second-order (strictly speaking, the latter should be called higher order or continuous). The distinction refers to whether or not the order parameter, the thermodynamic variable that specifies the ordering in the phase in question, vanishes abruptly (first-order phase transition) or goes continuously to zero (second-order) as one moves from the more ordered to less ordered phase. A first-order transition generally has a coexistence region — a range of θ values over which two adlayer phases are in thermodynamic equilibrium. Second-order transitions are of particular interest. The continuous vanishing of the order parameter (e.g., the adlayer Bragg intensity in a solid phase) means that the two thermodynamic phases become identical at the transition point. Hence, near this point, fluctuations in either phase become very pronounced, and their range (the correlation length ξ, see below) diverges at the transition. This latter effect is responsible, for example, for critical opalescence in three-dimensional systems. These long-range effects have a universal nature. The divergence of ξ (and the behavior of other thermodynamic singularities) as T approaches the transition temperature T_c is characterized by critical exponents: e.g. $\xi \sim |T - T_c|^{-\nu}$ with $\nu > 0$. A second-order transition is characterized by a set of critical exponents (Sect.15.3.3) which are in general independent of the details of the AA interactions. They are determined in most cases by the symmetry of the adsorption site lattice and the ordered phase. Thus, all possible second-order transitions fall into only a few universality classes, i. e., sets of values of the critical exponents. Knowing the universality class, on the other hand, determines much of the thermodynamics over a range which may extend quite far from the phase transition itself, including (as we shall see), the finite-size behavior of the system.

Since the concepts of order parameter and correlation length are important here, let us give a working definition of them. Consider an overlayer ordering characterized by a Bragg spot at a reciprocal lattice vector $\mathbf{k}_0$. If there are N adsorption sites in the system, we can define an order parameter p via

$$p \propto (1/N)|<\rho_{\mathbf{k}_0}>| \tag{15.1}$$

where

$$\rho_{\mathbf{k}_0} = \sum_{i=1}^{N} \exp(i\mathbf{k}_0\mathbf{R}_i)n_i \quad . \tag{15.2}$$

In (15.1) $\rho_{\mathbf{k}_0}$ is the $\mathbf{k}_0$ Fourier component of the overlayer density, defined in (15.2), and the thermal average is taken over all properly weighted configurations (of occupation numbers) of the system. The AA interactions enter through the average, according to the laws of statistical mechanics. In (15.2), $\mathbf{R}_i$ locates adsorption site i, the occupation number $n_i = 1(0)$ for an occupied (empty) site, and p is generally normalized so $p = \pm1$ for perfect order. Hence in the ordered phase, $p = O(1)$ but generally $|p| < 1$ due to thermal effects (e.g., random extra sites in the perfectly ordered lattice occupied or random filled sites unoccupied). In the disordered phase, $p = O(1/N)$, hence $p = 0$ for an ideal, infinite system (i.e., in the thermodynamic or bulk limit). In this case, p jumps discontinuously to zero at a phase transition if it is first order, but vanishes continuously for a second-order transition. An explicit example of this type of order parameter is given in Sect.15.3.1.

The scattering intensity is closely related to the order parameter p for overlayer ordering characterized by wave vector $\mathbf{k}_0$. In single-scattering (kinematic) approximation, the *total* scattering intensity at scattering (momentum transfer) vector $\mathbf{k}_0$ (within a multiplicative constant) is

$$I(\mathbf{k}_0) = <\rho_{\mathbf{k}_0}\rho_{-\mathbf{k}_0}> \quad . \tag{15.3}$$

When one is well into the ordered phase, in many situations (but not always, Sect.15.3), fluctuations in ρ_{k_0} are small and, therefore,

$$I(k_0) \propto <\rho_{k_0}\rho_{-k_0}> = |<\rho_{k_0}>|^2 \propto N^2 p^2 \quad . \tag{15.4}$$

This N dependence in (15.4) is the familiar result for the size dependence of the Bragg peak intensity. It is of interest to examine the extent to which (15.4) does *not* hold. This is measured by the difference between the left- and right-hand sides

$$\chi_{k_0} = 1/N(<\rho_{k_0}\rho_{-k_0}> - |<\rho_{k_0}>|^2) \quad . \tag{15.5}$$

By Fourier transformation, it is easy to write (15.5) as the integral over all space of a function involving the correlation between occupied adsorption sites separated by **R**. If this correlation has some finite range ξ, called the correlation length, $\chi_{k_0} = O(1)$ in the bulk limit and the correction to (15.4) is small, $O(N)$. This is the case which occurs generally when the over-layer is well ordered or disordered, say, by a few random defects. The corre-lation length ξ can also be understood as the distance over which a small fluctuation is felt. As one approaches a second-order transition, however, ξ approaches infinity. The fluctuations responsible for this also drive the order parameter p to zero, cause χ_{k_0} to diverge, and change the scattering intensity $I(k_0)$ from $O(N^2)$ to $O(N)$ (i.e., from a Bragg peak to the level of diffuse scattering). The divergence or vanishing of these quantities are all related in the modern theory of critical phenomena. The divergence of ξ at the critical point is the central physical concept, since it implies the lack of a length scale.

It is important to note that in the disordered phase, ξ is the (short) dis-tance over which an ordered region extends, since the range it measures re-fers to fluctuations about configurations with no long-range ordering (no overlayer Bragg peaks). In the ordered phase, by contrast, the correlation length refers to the range of fluctuations about an already ordered state, e.g., if we disturb a $c(2 \times 2)$ ordering at point i, it is the distance from i over which that disturbance will be felt. This is not the same as the size of the ordered region itself (island size), which is a different quantity sometimes called the "coherence length."

There has been a revolution in the understanding of second-order transi-tions in the last decade. Scaling and renormalization group theory have pro-vided a theoretical basis for understanding fluctuations and calculating their properties. A variety of review articles [15.15-21] are available. Two-dimensional systems are of particular interest [15.2,13,22,23], since second-order phase transitions are more generally encountered there than in higher dimensions, apparently because for many systems two dimensions is a border-line case. In one dimension, long wavelength fluctuations preclude the possibility of any phase transition, except in special cases, e.g., at $T = 0$. In two dimensions fluctuations tend to be large, favoring second-order transitions.

The original aim of scaling and renormalization group theory was to describe fluctuations near second-order phase transitions. This would seem a limited, albeit extremely interesting goal, since critical points are often isolated and appear to be somehow "special", e.g., in three-dimensional liquid-gas systems. However, as hinted at above, the theory has turned out to be useful for describing thermodynamics over a wide range of conditions. This is especially true of certain of its manifestations, such as the position space renormalization group technique [15.24], which is capable of calculating entire phase diagrams. Further, and this is rather surprising, it appears that first-order phase transitions (transitions with a discontinuous jump of the order parameter) can be understood as a limiting case of second-order transitions [15.6-10,25]. This was unexpected since the theory of second-order transitions [15.15-21] is centered around the divergence of the correlation length ξ at the transition point (lack of a length scale), a type of divergence that apparently does not generally occur for first-order transitions. This suggests that the theory is not yet in its final form, which should encompass both types of transitions in a single unified picture.

What does this have to do with stepped surfaces? One of the important effects of steps is to limit the size of the region in which overlayer ordering can occur. This finite-size effect "washes out" a phase transition by broadening and shifting the singularity. For second-order transitions, when the correlation length ξ equals the step size, the system "feels" its finiteness and these effects set in. Similar changes occur for first-order transitions. However, according to scaling theory, ξ is the fundamental quantity and making it finite because the system is spatially limited has a similar effect to making it finite by moving away from a transition in an infinite system. Thus the theory of the approach to the critical point in an ideal, infinite system implies a theory of size effects in real, finite systems. For experimentalists, this means that the possibility of creating an ordered array of steps of well-defined, controllable widths by cutting a crystal at a small angle to a low index plane allows one to study these effects in a systematic way. For other types of substrates grain sizes have also been determined [15.26], but in general a continuous range of sizes is not available.

The effects we consider here apply to two broad types of variables, thermodynamic quantities and correlation functions. Both are averages taken in thermodynamic equilibrium. The former refers to "classical" thermodynamic quantities such as specific heat, pressure, order parameter, etc. The latter refers to static spatial correlations, as measured by scattering experiments —electrons, X-rays, or neutrons in the cases of interest to surface science. The correlation functions are in fact also thermodynamic quantities, since, in general, they are averaged. As will become apparent below, there has been considerably more work on the theory of finite-size effects for thermodynamic quantities than for correlation functions.

In describing scattering we limit ourselves to kinematic (single-scattering) effects. Multiple scattering is also affected by statistical disorder, but there appears to have been little work on this point.

15.2 Overlayer Ordering and Adatom Binding Energy at Step Edges

In this section we consider certain effects of stepped surfaces on ordered overlayer phases [15.27-31]. Thus, we shall apparently not be concerned with phase transitions per se. However, we shall see in Sect.15.3 that, in fact, our main result is arguably an effect of finite size on a first-order phase transition.

The change in overlayer scattering from the flat to stepped surface is the effect of interest. We consider an ordered array of steps with a well-defined average terrace width D (Fig.15.1). This can be prepared in a controlled way by cutting a single crystal at an angle to a low index plane. Here (and in Sect.15.3) we assume that interterrace effects are negligible, i.e., the ordering is determined by forces inside a single terrace. Hence we omit the case of long-range AA interactions, such as those due to dipole forces in the adsorption of alkali metals, where one might expect the adatoms on one terrace to strongly influence those on the next. We also assume the system can be modeled as a lattice gas, although some of our results should be more general (for an incommensurate overlayer [15.33], for instance, there should be overlayer lattice constant shifts due to binding energy changes at the terrace edges).

Probably the most interesting result of this study is that the *sign* of the change in adsorption energy δE_B at the terrace edge sites can be determined by a simple comparison of experimental results for overlayer scattering versus coverage on stepped and flat surfaces. How this can be done is described be-

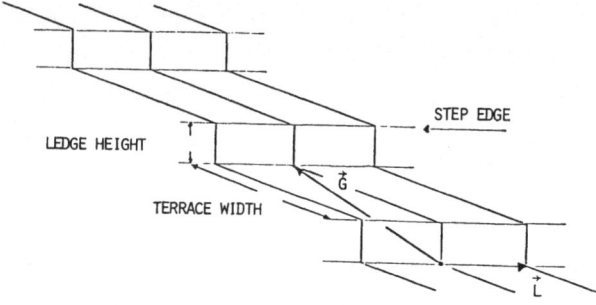

Fig.15.1. Regularly stepped surface. The vector **G** connects closest equivalent sites on adjacent terraces. L denotes the terrace length, and Da the
terrace width, where a is either a substrate or adsorption lattice constant.
Surfaces of this sort can be prepared and characterized [15.32]. Possible
kinks occurring along the edge are not shown

low. The *result* is that one can tell whether the adatoms bind more or less
strongly at terrace edge sites than in the terrace interior (or on the corresponding flat surface).

Now in physisorbed systems where the binding is mainly determined by van
der Walls forces [15.34], one expects the binding energy E_B to depend mainly
on the number of nearby substrate atoms. Hence it should be greater at edge
sites in the step "elbow" and (somewhat) less on the open edge. (This has
been demonstrated for Xe physisorbed on a stepped-kinked Ru{001} surface
[15.35,36].) In chemisorption systems, by contrast, it is not at all obvious
how E_B will change at terrace edge sites. In general, one expects for a given
adsorbate and given substrate a maximal E_B for some particular local arrangement of substrate atoms (e.g., a fourfold hollow). If the terrace interior
allows this arrangement, E_B at the edge sites will be less. If not, it can
be less or greater (δE_B positive or negative), depending on the circumstances.
So in general, for a given substrate and adsorbate, δE_B depends on the terrace
orientation, the terrace edge direction, and whether one considers the "elbow"
or open side of the terrace.

The total scattering intensity (for LEED, X-rays or neutrons) is, in kinematic (single-scattering) approximation, given by (15.2,3). Note that the
average in (15.3) is taken over all overlayer configurations at the coverage
in question, weighted by the appropriate Boltzmann factor. If the configurations on different terraces are independent, (15.3) may be written as a sum

$$I(\mathbf{k}) = N_s(<|\rho_k|^2> - |<\rho_k>|^2) + h(N_s, \mathbf{k} \cdot \mathbf{g})|<\rho_k>|^2 \quad . \tag{15.6}$$

Here the average refers to configurations on a single terrace only, N_s is the
number of terraces, **g** a vector between equivalent points on adjacent terraces

(Fig.15.1), and

$$h(N_s,x) = \sin^2(N_s x/2)/\sin^2(x/2) \quad . \tag{15.7}$$

The first term in (15.6) is due to the effects of statistical disorder on a single terrace. If only one adlayer configuration is important (e.g., for no overlayer, $\theta = 0$), this term vanishes, but in general it is finite. The second term is due to interterrace interference, i.e., the fact that overlayers on different terraces can have fixed phase relations. If only one configuration is important, it is nonzero. This term gives rise to the splitting of (flat surface) Bragg spots due to the regular array of steps, such as is observed for a clean stepped surface. In cases when several configurations count, it can be suppressed or even vanish, e.g., by antiphase effects.

Using (15.6) the intensity $I(k)$ may be calculated for a given adsorbate system, and various choices of the δE_B, if one has a reasonable model for the AA interactions. We assume that δE_B is much larger than any AA interaction, as is reasonable for strong chemisorption [15.12]. This implies that the edge sites will remain either empty ($\delta E_B > 0$) or filled ($\delta E_B < 0$) while the interior terrace sites are filled with adatoms. Explicit calculations have been performed [15.28,29] for a model appropriate for oxygen on W{110}, and a variety of one- [15.30] and two-dimensional [15.31] models at $T = 0$. The details of the results vary from one case to another but one feature is common to them all. If both δE_B are positive (less attractive adsorption sites), there is less room available for ordering, and, therefore, the maximum overlayer scattering intensity I_{max} occurs at a coverage θ_{max} that is *smaller* than on the flat surface. If both δE_B are negative, θ_{max} is increased, and if there is one of each sign, θ_{max} is close to the flat surface value. Consider a (2×1) overlayer with the dense direction parallel to the step edge (dark circles in Fig.15.2) and a terrace width of D adsorption sites. We approximate θ_{max} as follows. For both edges attractive (filled) and D even, the maximum intensity occurs when exactly half the interior sites are filled, i.e., $\theta_{max,even} = (2 + (D - 2)/2)/D = 1/2 + 1/D$. For D_{odd} it occurs when one fills half the interior sites less one, so $\theta_{max,odd} = (2 + (D - 3)/2)/D = 1/2 + 1/2D$. The other possibilities for the δE_B can be calculated similarly. The value of I_{max} varies somewhat according to whether D is even or odd, but it is reasonable to take θ_{max} to be the average of the two cases. The result is

$$\theta_{max}^{++} \cong \frac{1}{2} - \frac{3}{4D} \tag{15.8a}$$

$$\theta_{max}^{+-} \cong \frac{1}{2} - \frac{1}{4D} \tag{15.8b}$$

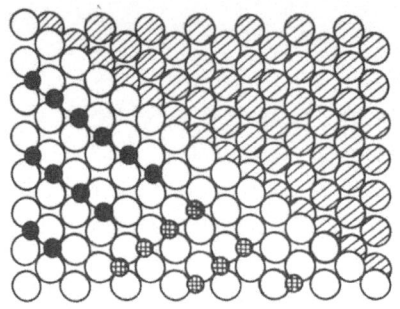

Fig.15.2. Stepped W{110} surface studied in [15.28,37]. Large circles represent W atoms, small circles adsorbed O. The step edge runs from upper left to lower right. There are two possible (2 × 1) overlayer domains, indicated by the dark and cross-hatched circles on the upper terrace. Scattering from the former is strongly enhanced. The evidence indicates that O atoms are less strongly bound at either terrace edge in this system

$$\theta_{max}^{--} = \frac{1}{2} + \frac{3}{4D} \ , \tag{15.8c}$$

where the + and - signs refer to the signs of δE_B and we have assumed equal numbers of odd and even width terraces.

Thermal effects in the case tested (oxygen/stepped W{110}) do not appear to affect (15.8) significantly. This is reasonable since one is looking at scattering where I is a maximum and the overlayer is most ordered. Here one expects temperature to play only a minor role.

Note that the $\frac{1}{2}$ in (15.8) is just θ_{max} for a flat surface (D → ∞). Hence the *shift* in θ_{max} with respect to the flat surface value can be used to determine the change in the terrace edge site binding energies E_B. This requires only a comparison of experimental results, and can be done without detailed modeling.

One caveat must be added here — if one is near a second-order phase transition (in the corresponding flat system) fluctuations could modify (15.8) — this point has yet to be investigated. Otherwise, results similar to (15.8) should apply to other types of overlayer ordering and step edge directions.

There are experimental results for oxygen adsorbed on flat [15.38,39] *and* stepped [15.37] W{110}, Fig.15.2. Figure 15.3 illustrates the AA interaction model [15.40]. A detailed comparison leads to the conclusion that for the step direction considered, both δE_B are positive (15.8a), i.e., both edge sites are less favorable. This conclusion is buttressed by the trend observed in oxygen poisoning of N_2 dissociation on stepped W{110} [15.28,29,44]. However, perhaps the most important point is that such conclusions can be drawn in such a straightforward manner.

Another point worth mentioning is the overlayer Bragg spot splitting found on stepped surfaces. The amount of this depends on the relative size of the two terms on the rhs of (15.6). This, in turn, is *very* sensitive [15.27,28] to the coverage and also the sign of δE_B. For $\theta \leq \theta_{max}$, and one $\delta E_B < 0$, for

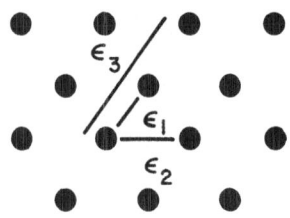

Fig.15.3. Adatom-Adatom (AA) interaction energies for oxygen/W{110} used [15.40] in the stepped surface study [15.28] illustrated in Fig.15.2. Values used were $\varepsilon_1 = -0.072$ eV, $\varepsilon_2 = +0.08$ eV, $\varepsilon_3 = -0.049$ eV, where $\varepsilon > 0$ indicates two adatoms at the specified displacement repel, $\varepsilon < 0$ attract. The interactions respect the substrate symmetry. Note that these values imply a (2×1) ordering is the most stable state at low T

instance, the ordered region will anchor and there will be little if any statistical disorder. Hence the first term will be small and splitting large. For smaller values of θ and both $\delta E_B > 0$, one generally has less splitting since the ordered islands are free to move perpendicular to the step edges, destroying the phase relation between different terraces. The behavior of the splitting as a function of θ thus checks the results of an analysis using (15.8).

15.3 Phase Transitions and Finite-Size Effects

15.3.1 First-Order Transitions: Ordered Overlayers

In this section we consider some of the effects that regions of finite size, especially those due to surface steps, have on phase transitions. To begin, we consider first-order phase transitions. As an example, consider a (2×2) ordered state on a square lattice with lattice constant a as shown in Fig. 15.4. The ordering here may be specified by the order parameter $p = (1/2N)$ $\langle \rho_{k_0} \rangle$, (15.1), where $k_0 = (2\pi/\sqrt{2}a)(\hat{i} + \hat{j})$. Now the system will order in state A or B at low enough T due to the AA interactions. A nearest-neighbor repulsion or second-neighbor attraction will result in this ordered structure. Such an ordered state is in fact a coexistence region. The coexisting phases A and B are differentiated by the sign of $\langle \rho_{k_0} \rangle$. For such an ideal system one generally expects that most of the important configurations are, aside from some short-range thermal fluctuations (remember $\xi < \infty$), either entirely phase A or entirely phase B. In other words the probability of an antiphase boundary (domain wall) is small. Hence, the scattering intensity $I \propto p^2 N^2$ as anticipated in (15.4), i.e., there is a Bragg peak at scattering vector k_0. This size dependence of I is just the familiar result for the size dependence of the overlayer beam. It can, in fact, be shown to be rigorously true for $T < T_c$ in the two-dimensional Ising model and various arguments for it are made in other cases [15.6].

Note that this argument implies a block geometry, i.e., we have implicitly assumed the system to be a square or other compact shape. For a long strip,

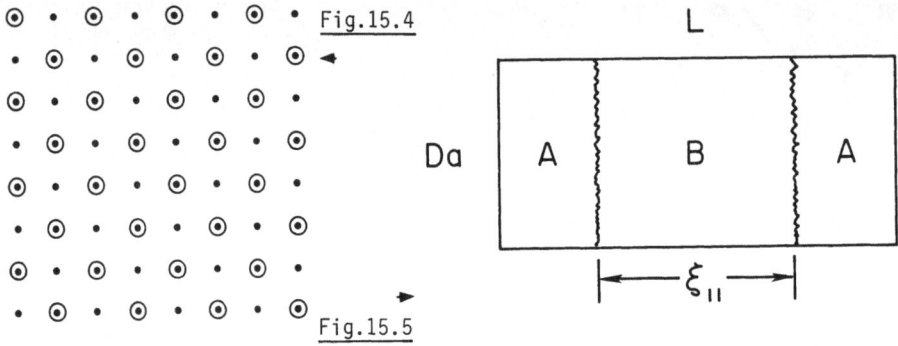

Fig.15.4

L

Da

A   B   A

$\xi_{\parallel}$

Fig.15.5

Fig.15.4. Perfect $c(2 \times 2)$ ordering of adatoms (large circles) in a square lattice of adsorption sites (small circles). The adatoms occupy the A sublattice, remaining sites constitute the B sublattice. The order parameter may be defined as $p = (<N_A> - <N_B>)/2N$, where $<N_A>$ = average number of adatoms in the A sublattice and N = total number of sites; here p = +1

Fig.15.5. Strip of dimension $L \times Da$, e.g., a single terrace from Fig.15.1. The A and B domains refer to an ordered overlayer, e.g., the $c(2 \times 2)$ structure in Fig.15.4. Domain walls with energy per unit length Σ are shown as wiggly lines. The domain size $\xi_{\parallel}$ is the average distance between walls

on the other hand, one expects that antiphase boundaries (also called domain walls, solitons or interfaces) will be present, as shown in Fig.15.5. The average distance between these walls depends on the domain wall energy (interface free energy or surface tension) per unit length, Σ. If the strip width is Da, the average wall separation is [15.45,46]

$$l_0 = a \, \exp(Da\Sigma/k_0 T) \quad . \tag{15.9}$$

Here the overlayer beam-scattering intensity from *one* strip of length L with N_1 adsorption sites will be approximately the number of domains multiplied by the scattering from a single domain, i.e.,

$$I_1 \propto (L/l_0)[N_1/(L/l_0)]^2 p^2 \tag{15.10}$$

$$= \frac{N_1^2 p^2}{(L/l_0)} \quad .$$

Now suppose we have two samples of a given adsorption system of size $L \times L$ (or a larger system, where L is the coherence length of the incoming radiation), one sample being flat and the other stepped with constant terrace width Da and length (assumed to be) L. Each has the same number N adsorption sites. For the stepped surface there are L/Da steps, each with $N_1 = N/(L/Da)$ adsorption sites. Assuming the scattering from different steps to be incoherent, we find by (15.9,10) and the above

$$I_{flat} \propto N^2 p^2$$

$$I_{stepped} \propto (L/Da)I_1$$

$$= N^2 p^2 / [(L/Da) \cdot (L/L_0)]$$

$$I_{stepped} / I_{flat} = (Da^2/L^2) \cdot \exp(Da\Sigma/k_B T) \quad . \tag{15.11}$$

The exponential terrace width dependence in (15.11) has, to our knowledge, never been observed. Note that one must have $Da < L$ and $l_0 < L$ for (15.11) to hold. Hence the lhs of (15.11) is less than unity. If $l_0 > L$ one must replace l_0 by L in (15.11), resulting in a ratio linear in D. Hence the exponential behavior requires

$$D \frac{a\Sigma}{kT} < \ln(L/a) \quad . \tag{15.12}$$

Now if the coexistence region in question terminates with a second-order transition at temperature T_c, $\Sigma(T_c) = 0$ since the interface energy vanishes (the A and B phases become identical). Thus, there is always a range of temperatures near T_c where (15.12) is satisfied. For the Ising model $a\Sigma/k_B T_c \cong 4J/k_B T_c(T_c - T)/T_c$ for T near T_c, where 2J is the AA interaction energy. Substituting this in (15.12) and using the exact value for T_c gives

$$\frac{T_c - T}{T_c} < (0.57)(\ln L/a)/D \quad . \tag{15.13}$$

The value for L for surfaces in current use is the surface grain size, i.e., the length of a well-ordered surface region. This depends on the material and how it is prepared. Reported values are approximately in the range 50-1000 Å. Taking a = 3 Å as a typical value, and D between 5 and 20, gives

$$0.08 < \frac{T_c - T}{T_c} < 0.66 \quad . \tag{15.14}$$

In most cases it is no problem to attain and control temperatures within 10% of T_c. Thus this exponential shape effect should be easily accessible by experiment.

Note also that the *shape* of the Bragg peaks should be quite different in the flat and stepped surface cases, with an anisotropic, exponentially varying linewidth in the latter.

A general theory for the finite-size behavior at first-order phase transitions as the *shape* of the system is varied [15.8] is reviewed in Sect. 13.3.4.

Explicit formulas for the crossover from block to strip geometries are given. Up to the present, however, this has been done only for thermodynamic quantities and (with one exception [15.7]) for systems with ideal (e.g., periodic) boundary conditions. General results for correlation functions, needed to describe scattering experiments, and the effects of various boundary conditions have yet to be determined.

15.3.2 First-Order Phase Transitions: Liquid-Gas Coexistence

A second effect of interest is the finite-size behavior of the compressibility in a (surface) liquid-gas transition. This is related to the slope of the isotherm in the coexistence region, as shown in Fig.15.6. Here the order parameter is proportional to the coverage difference, $\Delta\theta$, between the two phases, and the isothermal compressibility K_T per site in fact acts like the scattering intensity I per site in an ordered system. For block geometry, using the analog of (15.4) and thermodynamics

$$K_T = \left(\frac{\partial\theta}{\partial\mu}\right)_T$$

$$K_T = [(\Delta\theta)^2/4k_BT]N \quad , \tag{15.15}$$

where $\Delta\theta$ is to be evaluated in the thermodynamic limit and N is the number of adsorption sites. Now K_T is just the slope of an isotherm in the θ,μ plane. Hence this slope is a measure of the number of adsorption sites in a surface grain [15.41]. Now, if the isotherm has a straight region, as is observed [15.42], with chemical potential difference $\Delta\mu$ corresponding to the coverage difference $\Delta\theta$, making the approximation $(\Delta\theta/\Delta\mu) \cong \Delta\theta/\Delta\mu$ and using the ideal

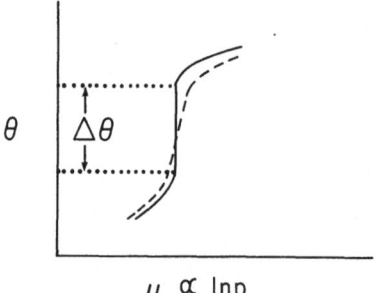

$$\mu \propto \ln p$$

Fig.15.6. Isotherm ("riser") in a liquid-gas coexistence region. Solid line: ideal system with two coverages coexisting at the same chemical potential; dashed line: finite grain or crystallite size N induces a finite isotherm slope $\propto$ N, (15.15)

gas form for μ, (15.15) is then reduced to the appealingly simple expression

$$N = 1/\Delta\theta \cdot \ln(p_2/p_1) \quad , \tag{15.15a}$$

where p_1 and p_2 are the (gas) pressures corresponding to the end points of the straight part of the isotherm. This result has yet to be tested extensively. However, for CH_4 on graphite [15.42,43] at 84.69 K one finds $\Delta\theta = 0.026$, $\ln p_2/p_1 \cong 0.05$ which gives $N = 770$, in agreement with what is expected for the grain size on grafoil [15.26].

For a strip geometry of width Da by length L, domains as shown in Fig. 15.5 are again important. Here the explicit approximate form for the compressibility [15.8] may be used

$$K_T = [2(\Delta\theta)^2/k_B T] \cdot D \cdot \xi_\| \quad , \tag{15.16}$$

where $\xi_\|$ measures the domain size (coherence length) along the strip, and can be evaluated using results for the two-dimensional Ising model as above for $T < T_c$, but with T large enough so that exponential effects are important:

$$K_T \simeq (1.89/k_B T_c) \, (\Delta\theta)^2 (T_c/T_c - T)^{1/2} D^{3/2} \exp[1.76 \, D \, (T_c - T)/T_c] \quad . \tag{15.17}$$

The conditions for (15.17) to be valid are [15.8]

$$\ln D \gtrsim 2J/kT \quad ,$$
$$D > 2.4 \quad \text{at} \quad T_c \tag{15.18}$$

and

$$D \gtrsim 1.76 \, (T_c - T)/T_c \quad . \tag{15.19}$$

The first of these conditions is required for the general form used for $\xi_\|$ to be valid. It also ensures that $l_0 < L$ at low T so that one has the situation shown in Fig.15.5. The second ensures that critical fluctuations are not important, i.e., that one is not too near T_c.

We do not believe that (15.17) has been tested experimentally.

15.3.3 Finite-Size Scaling Theory: Second-Order Transitions

Let us now briefly review the main points of finite-size scaling theory for systems that may be represented by discrete spin models (we do not consider commensurate-incommensurate transitions, superfluid He layers, etc.) in two dimensions. The theory treats the effects of finite sizes on second [15.3-5] and first-order [15.6-11] phase transitions. For the former, a well-developed body of results exists, that has recently been summarized [15.5]. We consider it first.

A second-order transition in an infinite system is characterized by singularities in thermodynamic quantities and correlation functions at the critical point T_c. For instance, the correlation length diverges according to

$$\xi = \xi_0 t^{-\nu} ,$$ (15.20)

where the reduced temperature

$$t = \left| \frac{T - T_c}{T_c} \right|$$ (15.21)

measures the deviation from the critical point, ξ_0 is a constant, and ν is a critical exponent. The free energy per site also has a singular term

$$f_s = A\, t^{2-\alpha} .$$ (15.22)

Here A is a constant and α another critical exponent. Differentiating (15.22) twice with respect to the temperature shows that the specific heat per site is also singular.

$$C \propto t^{-\alpha}$$ (15.23)

Generally $\alpha > 0$ so C diverges at T_c. There are other various critical exponents associated with other thermodynamic quantities and also whether the critical point is approached by varying the temperature or the ordering field (see below). They are related to each other (for an ordinary critical point, only two of them may be chosen independently), and their values are determined by the universality class of the phase transition.

A finite system is characterized by a length L. In two dimensions one generally considers either a system with a block geometry, e.g., an L × L square, or a system with strip geometry similar to Fig.15.5. To determine the problem fully one must also specify the boundary conditions, e.g., in an adsorption problem, whether the adsorbed molecules are attracted to or repelled from the edge sites. It should be noted that most theoretical work has been done with periodic boundary conditions which are of course not realized in nature.

Now consider a thermodynamic quantity P that is singular at the critical point in the infinite system. For finite L, $P_L(t)$ is the quantity of interest. Finite-size scaling asserts that near T_c (for small t), the deviation of P from its infinite system value is measured by the ratio of L to the (infinite system) correlation length

$$P_L(t)/P_\infty(t) = f[L/\xi(t)]$$ (15.24)

as long as L is large compared to a lattice spacing. The function f is universal, i.e., the same for any system in a given universality class, but it does depend on geometry, boundary conditions, and of course the quantity P. There is an adjustment that must be made to (15.24) to complete the theory, but before considering it, let us analyze the significance of (15.24).

First, it is clear that the scaling function $f(x)$ must satisfy

$$f(y) \to 1$$
$$y \to \infty \qquad\qquad (15.25)$$

since $P_L \to P_\infty$ as $L \to \infty$ at fixed t. Secondly, if $P_\infty(t) \cong A\, t^{-\rho}$ near T_c (for small t)

$$f(y) \to \left(\frac{C}{A}\right) x^{\rho/\nu}$$
$$y \to 0 \quad, \qquad\qquad (15.26)$$

where C is a constant, since P_L is not singular for small t at fixed L (in general a finite-size system in two dimensions cannot have a phase transition).

Note that the scaling variable $y = \xi_0 L t^\nu$ (sometimes called the scaled temperature) controls the crossover from regular or noncritical behavior (15.25) to singular behavior in the critical region (15.26). Thus, it sets the scale of the finite-size induced *rounding* of the critical region. For $y \gg 1$ we are out of the critical region, for $y \ll 1$ inside it, and the boundary is given by $y \approx 1$. Since $\nu > 0$, $y \approx 1$ defines a temperature region that narrows about T_c as the size L of the finite system grows.

A second important consequence of (15.24) is the behavior of P_L right at the bulk system transition temperature $T = T_c$ (t = 0). Combining (15.24,26) gives

$$P_L(0) = \lim_{t \to 0} P_\infty(t) f(y)$$
$$P_L(0) = CL^{\rho/\nu} \quad . \qquad\qquad (15.27)$$

Equation (15.27) shows that (for $\rho > 0$) the divergence in the *finite* system quantity is governed by the *infinite* system singularity, i.e., the bulk system critical exponent. We have in fact already seen an example of this: (15.4 or 15) may be shown [15.6] to reflect the corresponding effect for first-order phase transitions.

The third important effect of finite size on second-order phase transitions is a *shift* of the position of the maximum of the quantity in question. This is somewhat less well understood than the rounding or finite-size divergence described above. If one considers the position $T_m(L)$ of the maximum of a singular quantity, e.g., the specific heat, this will approach T_c according to

355

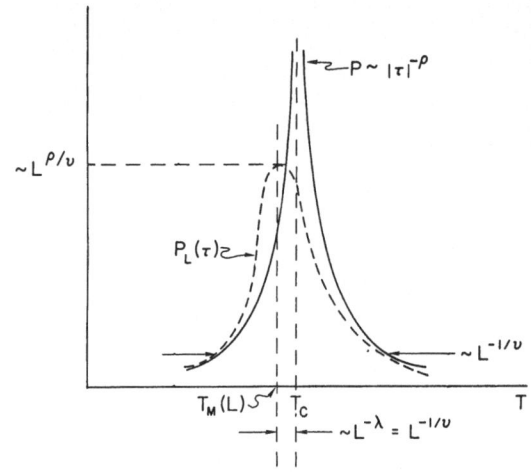

Fig.15.7. The three main features of finite-size scaling theory for second-order transitions: rounding, shift and divergence. Solid curve: bulk thermodynamic quantity diverging as $P \sim |t|^{-\rho}$, $\rho > 0$ at the critical temperature T_c. Dashed line: the same quantity P_L for a finite-size system characterized by length L

$$\overset{\bullet}{t} \equiv [T_m(L) - T_c]/T_c = bL^{-\lambda} \quad , \tag{15.28}$$

where λ is the shift exponent. If one assumes that the only criterion important for finite-size effects on the critical region is $y \approx 1$, i.e., the correlation length equaling the system size, one can argue [15.5] that $\lambda = 1/\nu$.

One finally may incorporate (15.24,28) by writing

$$P_L(t) \sim L^{\rho/\nu} Q(L^{1/\nu} \overset{\bullet}{t}) \quad . \tag{15.29}$$

Equation (15.29) is, aside from the shift, equivalent to (15.24) if $Q(x) \to C$ for $x \to 0$, $Q(x) \to Ax^{-\rho}$ for $x \to \infty$.

The finite-size induced rounding, shift and divergence at a second-order phase transition are all illustrated in Fig.15.7.

The influence of geometry on finite-size effects has not been extensively investigated. However, it has recently been shown [15.47,48] that as one passes from block to strip geometry, domains similar to those found in a coexistence region (Fig.15.5) become important, and govern the shape dependence of the specific heat peak shift. The difference here from the first-order case for $T < T_c$, Sects.13.3.1,2, is that at fixed y, since T approaches T_c as $L \to \infty$, the total domain wall energy remains constant (the domain wall energy per unit length vanishes as 1/L). As a result, the distance between domain walls goes up only as L, instead of exponentially with L as occurs in a coexistence region, (15.9). This result has been shown for the two-dimensional Ising model with periodic boundary conditions, but general arguments [15.47] indicate that it remains valid for a wide class of critical points and boundary conditions. It clearly has consequences for the behavior of the scattering intensity I($\mathbf{k}$) similar to those due to the influence of domain walls for $T < T_c$ (15.11). However, the effects are much more subtle in this case and remain to be calculated.

The fact that similar domain wall configurations occur in these two cases is more evidence for the connection between first- and second-order phase transitions. Its full significance in this regard is not yet clear, however.

15.3.4 Finite-Size Scaling Theory: First-Order Transitions

A partial theory also exists for the effects of finite size in first-order transitions [15.6-11]. We have already seen some of its consequences in (15.10,11) that describe the divergence of I(k) in a coexistence region for different finite geometries. The main result of Sect.15.2, (15.8), may also be understood as a finite-size shift effect, as we shall argue below.

To outline the theory, we must first introduce the idea of an ordering field h. This is the quantity thermodynamically conjugate to the order parameter p. For a magnetic system, p may be taken as the magnetization per site, and h the external magnetic field. For a surface liquid-gas transition, where p is proportional to $\Delta\theta$ (the difference in coverage between the two states), h is the chemical potential μ. For the case of overlayer ordering, p is proportional to $<\rho_{\mathbf{k}_0}>/N$, as discussed in Sect.13.1.2. Here, h is understandable as a potential that modulates the adsorbate binding energy with wave vector $\mathbf{k}_0$. Any finite amount of h will, therefore, cause the system to choose one of the possible coexisting ordered states, e.g., A or B in Fig.15.4, depending on the phase of h. Such a field is generally not physically accessible, but it is necessary for an understanding of the physics of the situation.

One important theoretical result [15.8] describes the behavior of the free energy as a function of ordering field h and geometry at given temperature $T < T_c$. If f is the free energy per site one has

$$f_s(h,T) \equiv f(h,T) - f_\infty(0,T)$$
$$= - (k_B T/N) \ln 2 \cosh\{(phN/k_B T)^2 + (L/2\xi_\parallel)^2\}^{\frac{1}{2}} \ . \qquad (15.30)$$

In (15.30) $f_\infty(0,T)$ is the bulk zero-field free energy per site, which is not described by the theory. The geometrical factors Da and L define the system width and length, as in Fig.15.5. For a strip $\xi_\parallel$ is the domain size in the long direction, as in (15.16). Since a is the lattice spacing (which we assume for simplicity to be square), D enters (15.30) via $N = Da L/a^2$.

The full consequences of (15.30) are spelled out elsewhere [15.8]. We confine ourselves to recapitulating a few of them. Note that for large fields it reduces to

$$f_s = -p|h| \tag{15.31}$$

so that $\partial f_s/\partial h = \mp p$ in the bulk limit, giving the correct relation between the free energy and magnetization. For a block geometry, $L \ll \xi_{||}$, so that

$$f_s = - (k_B T/N) \ln 2 \cosh(phN/k_B T) \tag{15.32}$$

which is just the free energy of a *single* domain of net magnetization $\pm pN$, i.e., domain walls do not contribute. Here the generalized susceptibility at $h = 0, \chi = \partial p/\partial h = -\partial^2 f/\partial h^2$ is easily seen to diverge as N in the bulk limit, in accord with the divergence of I(k) seen above and in (15.15). For $L \gtrsim \xi_{||}$, on the other hand, the second term in the argument on the rhs of (15.30) becomes important, and one finds an exponential divergence for χ as in (15.10). Note that these results are the analog of (15.27) for the second-order case. Here, however, we see explicitly the effects of geometry on the divergence.

It is worth noting that (15.30) may be rewritten in terms of the two natural variables

$$y_V = phN/k_B T \tag{15.33a}$$

$$y_A = phD(\xi_{11}/a)/k_B T \quad , \tag{15.33b}$$

which are just the bulk ordering energy divided by the thermal energy and the ordering energy of a single domain divided by the thermal energy respectively. Here we see a difference from the formulation of the theory for the second-order case, where the natural variable is L/ξ, the system size divided by the bulk correlation length. The variables y_V and y_A also determine the rounding of the transition as a function of ordering field h and geometry. Further details on this point are given in [15.8].

There are some conditions imposed on the validity of (15.30) that sould be mentioned. First, it applies to first-order transitions, so that one must stay away from the critical point. The condition for this is that $Da \gg \xi$, i.e., the bulk correlation length must not be able to "cross" the smaller dimension. A second, more technical and weak condition, is $L \gg \xi \ln D$, which ensures that the system does not break up into domains in the direction along which D is measured.

Equation (15.30) does not consider the effects of boundary conditions that might break the symmetry about the usual physical ordering field value $h = 0$, i.e., prejudice the system into one of the coexistence phases *at* $h = 0$, and therebey induce a finite-size shift in the ordering field. Now consider the example discussed in Sect.15.2, especially (15.8), which predicts a shift in *coverage* location of the maximum of the scattering $\Delta\theta_{max} \propto 1/D$, where D is the strip width. This occurs for boundary conditions of the symmetry-breaking type due to adsorption energy changes at the terrace edges. If the shift is governed by a variable like y_V or y_A, one must consider the energy change induced by the altered boundary conditions. This is $\Delta E \sim L$, assuming block geometry or strip geometry with $\xi_{11} >> L$, as will be the case for T sufficiently less than T_c. The corresponding change in ordering energy is [15.7]

$$\Delta hN \sim L \tag{15.34}$$

so the shift induced in h becomes

$$\Delta h \sim L/N \sim 1/D \quad . \tag{15.35}$$

Now the coverage θ_{max} at which the overlayer scattering $I(\mathbf{k}_0)$ reaches a maximum value I_{max} corresponds to an ordering field $h = h_{max}$ and a chemical potential $\mu = \mu_{max}$. The change of boundary conditions may generally be expected to induce a shift $\Delta\mu \sim 1/D$ in the chemical potential as well as in h.

Now assume, consistent with finite-size scaling in second-order transitions, that the intensity I' with new boundary conditions is the same function of the shifted variables as with the original boundary conditions (both $\delta E_B = 0$)

$$I'(h,\mu) \cong I(\dot{h},\dot{\mu}) \tag{15.36}$$

where

$$\dot{h} \cong h - 1/D$$
$$\dot{\mu} \cong \mu - 1/D \quad . \tag{15.37}$$

Hence by (15.36) I_{max} will occur at

$$\dot{h}_{max} \cong h_{max}, \quad \dot{\mu}_{max} \cong \mu_{max} \tag{15.38}$$

or

$$h \sim h_{max} + 1/D \tag{15.39a}$$

$$\mu \sim \mu_{max} + 1/D \quad . \tag{15.39b}$$

In general, the ordering field h is not physically accessible, hence (15.39a) will not be satisfied. But in most cases, maximizing I will still induce a shift in μ of the type given in (15.39b). Since

$$\theta \sim \partial f/\partial\mu \quad , \tag{15.40}$$

the coverage at this shifted chemical potential value becomes

$$\dot{\theta}_{max} = \theta(\mu_{max} + 1/D) \sim \partial f(\mu_{max} + 1/D)/\partial\mu$$

$$= \partial f/\partial\mu + (1/D)\partial^2 f/\partial\mu^2$$

$$\sim \theta_{max} + 1/D \quad , \tag{15.41}$$

recovering the D dependence of (15.8). Finally, we note that if μ itself is the ordering field, as for liquid-gas coexistence, this line of argument indicates the possibility of large coverage shifts, since $\partial^2 f/\partial\mu^2$ diverges, invalidating the expansion in (15.41).

15.4 Summary, Conclusions and Future Directions

15.4.1 Summary

We have briefly reviewed the main points of finite-size scaling theory for first- and second-order phase transitions in discrete spin (e.g., Ising) systems in two dimensions. In addition we have pointed out several experimental results for physisorbed and chemisorbed systems that can be understood with the help of the theory. It should be clear to the reader by now that as promised in Sect.15.1, if this work is a review of anything it is the theoretical and experimental research that is just over the horizon. Therefore, we conclude by pointing out the areas where progress can be made, first for the systems considered here, and then for some related problems of interest.

15.4.2 Future Directions

15.4.2.1 Theory

We believe the most fundamental problem in this area, as emphasized recently elsewhere [15.5,49] is an understanding of the connection between finite-size behavior for first- and second-order phase transitions. A comprehensive theory of both would also no doubt go a long way toward improving the *general* theory of first-order transitions, which are clearly related [15.25] to the second-order case.

On a more prosaic level, the existing scaling theory [15.6-11] for first-order transitions needs to be more fully developed before it reaches the level of the theory for the second-order case [15.3-5]. For *either* type of transition there are several rather glaring omissions, however. First and foremost, for those interested in surface phenomena, is the fact that almost all the results apply to thermodynamic quantities. A detailed description of scattering intensities (i.e., correlation functions), including line-shape analysis, is lacking. Clearly, the main lines of the theory will resemble those for thermodynamic quantities, but the results would be of great value [15.50] in understanding experimental results and predicting interesting measurements.

In addition, there are various "loose ends" in the theory that need to be cleaned up. In particular it is important to clarify the influence of nonperiodic boundary conditions. Also the question of finite-size induced shifts in second- and especially first-order transitions is not yet fully answered.

In summary, great progress has been made in the theory, and there is significant contact with experiment, but there is still much research to be done and even a world or two to conquer.

15.4.2.2 Experiment

Finite-size scaling theory for second-order transitions is apparently generally well established [15.5]. However, there are few if any definitive tests in the interesting case of submonolayer adsorption systems. The careful specific heat results for adsorption in grafoil [15.51,52] clearly exhibit finite-size induced rounding; however, this is not a detailed test of the theory. One reason for this is that the size of the adsorbing region is fixed by the substrate, and cannot be varied at will. It follows that stepped surfaces, in which the terrace width D can be controlled (and characterized), provide a way to get around this problem and test some of the predictions of finite-size scaling directly. In doing this one will necessarily be restricted to adsorption on a single surface, in contrast to a layered system such as grafoil, which will make accurate thermodynamic measurements difficult. Hence, the technique of choice is scattering. The results of such measurements may be very interesting as a test of and inspiration for the theory, since as mentioned in Sect.15.4.2.1 the necessary fully detailed results (line-shape analysis) for scattering intensities are not yet at hand.

There are several adsorption systems that one can suggests as candidates for this kind of study. For strong chemisorption, perhaps the most suitable is oxygen/Ni{111}, since the flat surface phase transition behavior has been extensively studied [15.53-55]. Other systems include those reviewed in [15.2].

15.4.2.3 Related Areas of Interest

There are some other very interesting areas involving finite-size effects and two-dimensional systems that we have not dealt with here, in part for pedagogical reasons. Recent theoretical results [15.56-59] for models of the commensurate-incommensurate transition in two dimensions [15.33] show what are apparently [15.60] rather strong finite-size effects. These effects may well be observable in laboratory systems.

Finally, there are some technologically very important and interesting effects of finite particle size on reaction rates in heterogeneous catalysis (e.g., steam reduction). In some cases these may be due to thermodynamic effects as modified by finite size. For instance, there is a very interesting general prediction [15.61] for the slowing down of reactions near critical points in infinite systems that may be relevant. This work would complement the many interesting studies using stepped single-crystal surfaces as model systems for surface defect effects on catalytic reactions and adsorption [15.32].

Acknowledgements. We should like to thank M. Grunze, M.E. Fisher, V. Privman, and M.N. Barber for helpful comments and discussions. This work was supported in part by the Office of Naval Research.

References

15.1 J.G. Dash: *Films on Solid Surfaces* (Academic, New York 1978); O.E. Vilches: Ann. Rev. Phys. Chem. **31**, 463 (1980)
15.2 See the excellent recent review by L.D. Roelofs: Appl. Surf. Sci. **11/12**, 425 (1982)
15.3 M.E. Fisher: In *Critical Phenomena*, Proc. Enrico Fermi Int. School of Physics, Vol.51, ed. by M.S. Green (Academic, New York 1971)
15.4 M.E. Fisher, M.N. Barber: Phys. Rev. Lett. **28**, 1516 (1972)
15.5 M.N. Barber: In *Phase Transitions and Critical Phenomena*, Vol.8, ed. by C. Domb, J.L. Lebowitz (Academic, New York, in press)
15.6 P. Kleban, C.-K. Hu: Submitted to Phys. Rev.
15.7 M.E. Fisher, A.N. Berker: Phys. Rev. B**26**, 2507 (1982)
15.8 V. Privman, M.E. Fisher: J. Stat. Phys. **33**, 385 (1983)
15.9 J.L. Cardy, P. Nightingale: Phys. Rev. B**27**, 4256 (1983)
15.10 H.W.J. Blöte, M.P. Nightingale: Physica **112A**, 405 (1982)
15.11 Y. Imry: Phys. Rev. B**21**, 2042 (1980)
15.12 T.L. Einstein: In *Chemistry and Physics of Solid Surfaces II*, ed. by R. Vanselow (CRC, Boca Raton 1979)
15.13 T.L. Einstein: In *Chemistry and Physics of Solid Surfaces IV*, ed. by R. Vanselow, R. Howe (Springer, Berlin, Heidelberg, New York 1982)
15.14 G. Somorjai: Surf. Sci. **34**, 156 (1973)
15.15 K.G. Wilson, J. Kogut: Phys. Reports **12C**, 77 (1974)
15.16 S.K. Ma: *Modern Theory of Critical Phenomena* (Benjamin, Boston 1976)
15.17 M.E. Fisher: Rev. Mod. Phys. **46**, 597 (1974)
15.18 P. Pfeuty, G. Toulouse: *Introduction to the Renormalization Group and to Critical Phenomena* (Wiley, New York 1977)
15.19 C. Domb, M.S. Green (eds.): *Phase Transitions and Critical Phenomena*, Vol.6 (Academic, New York 1976)
15.20 M.N. Barber: Phys. Reports **29C**, 1 (1977)
15.21 D.J. Wallace, R.K.P. Zia: Rep. Prog. Phys. **41**, 1 (1978)
15.22 See the review by M.N. Barber: Phys. Repts. **59**, 375 (1980)
15.23 For recent work, see the articles in Surf. Sci. **125**, 1-50 (1983); M.W. Cole, F. Toigo, E. Tosatti (eds.): *Statistical Mechanics of Adsorption*
15.24 A.N. Berker: In *Ordering in Two Dimensions*, ed. by S.K. Sinha (North-Holland, NY 1980), and references therein
15.25 B. Nienhuis, M. Nauenberg: Phys. Rev. Lett. **35**, 477 (1975)
15.26 See, for example R.J. Birgeneau, P.A. Heiney, J.P. Pelz: Physica (Utrecht) **109,110**, 1785 (1982)
15.27 P. Kleban: Surf. Sci. **103**, 542 (1981)

15.28 P. Kleben, R. Flagg: Surf. Sci. **103**, 552 (1981)
15.29 P. Kleban: In *Ordering in Two Dimensions*, ed. by S.K. Sinha (North-Holland, New York 1980)
15.30 R. Roy, P. Kleban: Bull. Am. Phys. Soc. **26**, 59 (1980)
15.31 Y. Shen, P. Kleban: Unpublished
15.32 H. Wagner: In *Solid Surface Physics*, Springer Tracts in Modern Physics, Vol.85, ed. by G. Höhler (Springer, Berlin, Heidelberg, New York 1979)
15.33 See the article by P. Bak, this volume
15.34 L.W. Bruch: Surf. Sci. **125**, 194 (1983)
15.35 J. Küppers, K. Wandelt, G. Ertl: Phys. Rev. Lett. **43**, 928 (1979)
15.36 K. Wandelt, J. Hulse, J. Küppers: Surf. Sci. **104**, 212 (1981)
15.37 T. Engel, T. von dem Hagen, E. Bauer: Surf. Sci. **62**, 361 (1977)
15.38 T. Engel, H. Niehus, F. Bauer: Surf. Sci. **52**, 237 (1975)
15.39 G.-C. Wang, T.-L. Lu, M.G. Lagally: Phys. Rev. Lett. **39**, 411 (1977); J. Chem. Phys. **69**, 479 (1978)
15.40 W.Y. Ching, D.L. Huber, M.G. Lagally, G.-C. Wang: Surf. Sci. **77**, 550 (1978); see also
E. Williams, S. Cunningham, W.H. Weinberg: J. Chem. Phys. **68**, 4688 (1978)
15.41 F.A. Putnam: Private communication
15.42 Y. Larher: J. Chem. Phys. **68**, 2257 (1978)
15.43 A. Thomy, X. Duval: J. Chem. Phys. **67**, 1101 (1970)
15.44 K. Besocke, H. Wagner: Surf. Sci. **87**, 457 (1979)
15.45 L. Onsager: Phys. Rev. **65**, 117 (1944)
15.46 M.E. Fisher: J. Phys. Soc. Japan Suppl. **26**, 87 (1969)
15.47 P. Kleban, G. Akinci: Phys. Rev. Letters **51**, 1058 (1983)
15.48 P. Kleban, G. Akinci: Phys. Rev. B**28**, 1466 (1983)
15.49 M.E. Fisher, D.A. Huse: In Proc. of Ninth Midwest Solid State Theory Symposium on "Melting, Localization and Chaos", to be published
15.50 R.J. Birgeneau: *Exhortations to Theorists*, unpublished
15.51 M. Bretz: Phys. Rev. Lett. **38**, 501 (1977)
15.52 M.J. Tejwani, O. Ferreira, O.E. Vilches: Phys. Rev. Lett. **44**, 152 (1980)
15.53 L.D. Roelofs, A.R. Kortan, T.L. Einstein, Robert L. Park: Phys. Rev. Lett. **46**, 1465 (1981)
15.54 L.D. Roelofs, A.R. Kortan, T.L. Einstein, Robert L. Park: J. Vac. Sci. Technol. **18**, 492 (1981)
15.55 L.D. Roelofs, A.R. Kortan, T.L. Einstein, Robert L. Park: In *Ordering in Two Dimensions*, ed. by S.K. Sinha (North Holland, New York 1980)
15.56 W. Selke, M.E. Fisher: Z. Phys. B**40**, 71 (1980)
15.57 M.N. Barber, P.M. Duxbury: J. Phys. A**14**, L251 (1981)
15.58 W. Selke: Z. Phys. B**43**, 335 (1981)
15.59 S. Redner: Unpublished
15.60 M.N. Barber, P.M. Duxburg: J. Phys. A**15**, 3219 (1982)
15.61 H. Procaccia, M. Gitterman: Phys. Rev. Lett. **46**, 1163 (1981)

16. Recent Developments in the Theory of Epitaxy

J. H. van der Merwe

With 7 Figures

16.1 Introduction

16.1.1 Nomenclature and Techniques

Epitaxy has become a very topical field of research, not only because of its fundamental interest [16.1-3] but also because of its vital importance in solid-state device fabrication. The phenomenon owes its technological importance to its unique role in the fabrication of single crystals with certain desired properties [16.4-6].

A single crystal which grows in a unique orientation on a single crystalline substrate is said to grow epitaxially. When the substances are the same the phenomena is called homoepitaxy, and otherwise heteroepitaxy. Unless specified otherwise, "epitaxy" will refer to the case in which the growing crystal and original substrate are different. Precipitates in the bulk are often also oriented with respect to the matrix and then referred to as endotaxial. The oriented growth, stimulated by an artificial microrelief pattern on an amorphous substrate, is named artificial epitaxy or more specifically, topotaxy or diataxy [16.7,8].

The first systematic investigation of epitaxy in the laboratory was carried out by *Royer* [16.9]. Progress in fundamental understanding was rather slow until the development of sophisticated ultrahigh vacuum (UHV) techniques for growth and characterization [16.10]. In the absence of ultrahigh vacuum, impurities are ever present and almost invariably have drastic detrimental influences [16.11,12] on the growth and properties of thin epitaxial overlayers.

Of the different techniques of crystal growth, the results obtained by evaporation in UHV lend themselves more uniquely than others to theoretical interpretation. Since the *present considerations are directed mainly towards fundamental understanding,* we shall limit ourselves to such results. The deliberations have, nevertheless, usually a wider range of validity.

Because of its importance a considerable amount of work [16.10] has been and is being done on epitaxy. However, limited space and time prohibit a detailed discussion of even a fraction of the results of interest, although an attempt will be made in this section to present a rudimentary picture of the field. A more detailed description of some important recent developments on growth modes and epitaxy at {111} fcc and {110} bcc interfaces will be given respectively in Sects.16.2,3. No attempt will, however, be made to give a complete bibliography here; the references given are meant to represent typical examples only.

16.1.2 Deposition

Evaporation [16.13] usually consists of heating the relevant substance to sufficiently high temperatures to yield a vapor which is commonly modeled with an ideal gas at temperature T and (vapor) pressure P(T). Extreme care [16.12] is taken to minimize the presence of impurities in the vapor.

Contamination, emanating from the substrate [16.11] held at temperature T_s, may also be very detrimental and must be averted. Apart from its ability to adsorb [16.14] (desorption energy E_a) vapor molecules by dissipating at impact a sufficient fraction of their kinetic energies, the most important property of the substrate in the present context is its crystallinity, its periodicity, which the adatom sees as a periodic potential [16.15] when translating on the substrate surface. The favorable adsorption sites (potential wells) accordingly form a surface lattice.

An adatom migrates (diffuses) [16.14,16] along the surface by jumping over the activation energy barriers E_D at the saddle points between neighboring adsorption sites. It may reevaporate after a mean stay time τ_s, depending on T_s and E_a, or migrate a distance λ_s. If the latter is large enough (depending on T_s and E_D) then during its stay it can meet another adatom or cluster of adatoms. If the vapor is supersaturated, there will be a net flux towards the surface and critical nuclei will form and grow. In order for the nuclei to grow crystalline, the arrival rate of adatoms at a cluster must be small enough (low supersaturation) to allow them sufficient time (high T_s) to take up appropriate positions in the overlayer lattice. Clearly, T_s controls a delicate balance: it must be low enough for realistic growth rates (supersaturation) and to prohibit undesirable interdiffusion and compound formation or even melting, while, on the other hand, it must be large enough to allow adatoms to reach appropriate positions in the existing regular cluster rather than joining up with other single adatoms.

The behavior of the adsorbate during the premonolayer stage [16.3,17] (the stage at which gas, fluid and solid phases may exist — the latter with several superstructures), the nature of the transitions between phases, and the dependence on T_s, coverage, and substrate lattice, has recently been the subject of considerable interest. It forms the theme of another review. For the present, the phenomenon of orientational epitaxy [16.18] will also be regarded as falling within this category.

16.1.3 Governing Principles and Models

From the foregoing considerations it becomes abundantly clear that the phenomena under consideration range from "practically" equilibrium phenomena on the one hand to "practically" nonequilibrium ones on the other [16.14]. The processes involved are activated, and the equilibration time may accordingly be anywhere between real short and real long, depending strongly on the substrate temperate T_s.

Mathematically it is vastly more difficult to describe nonequilibrium than equilibrium phenomena. Equilibrium approaches use the formalisms of equilibrium thermodynamics and statistical mechanics, specifically, the minimum free energy criteria, which often are approximated by minimum energy criteria [16.19,20]. In the approximation, the temperature is only an implicit parameter, in so far as it is assumed that the atomic processes are fast enough to ensure quasiequilibrium. Nonequilibrium cases are usually treated equilibriumwise in terms of the equilibrium concentrations of critical nuclei [16.21], or using a master equation approach [16.14]. To perform the mimimization, models need be constructed of the overlayer-substrate system.

In all these approaches, at least up to monolayer coverage, the substrate is modeled as a *rigid* periodic potential energy surface with the symmetry and periodicity of the substrate surface lattice [16.15,22] (lattice parameter a_s) and represented by a truncated Fourier series

$$V(\mathbf{r}) = \sum_{\mathbf{k}} V_{\mathbf{k}} \exp(i\mathbf{k} \cdot \mathbf{r}) \quad , \tag{16.1}$$

where the $\mathbf{k}$'s are the reciprocal lattice vectors of the substrate surface lattice. The term "lattice parameter" is used at this stage in a collective sense as comprising both "dimensions" and "symmetry" [16.23] of the unit cell. For coverages exceeding one monolayer the "rigidity" approximation is poor and alternate approaches are needed, as indicated below. The effects of surface reconstruction on the form of $V(\mathbf{r})$ are usually neglected in applications.

In the submonolayer regime the adsorbate-substrate system is usually modeled by a lattice gas [16.24] and at higher coverages by a two-dimensional harmonic (elastic) solid [16.19,25] with natural lattice parameter a_0, which usually is approximated by the corresponding bulk value [16.26].

For coverages exceeding one monolayer, both overlayer and substrate are treated as solids in the harmonic approximation and the interfacial interaction in (16.1) is either generalized to Peierls-Nabarro type [16.27] stresses or replaced by a Volterra dislocation model [16.20].

It may be anticipated that the degree to which the two relevant surface lattices match or do not match plays a significant role in the realization of epitaxy; mismatch refers to differences in both dimensions and to lattice symmetry. The importance of such a role [16.11,19] has been a matter of controversy [16.28,29] ever since it was suggested by *Royer* [16.9].

16.1.4 Growth Modes

From the foregoing one might conclude that monolayer coverage is equivalent to the existence of a two-dimensional solid covering the entire substrate surface. This is simply not so. The initial growth mode may be either two-dimensional, or three-dimensional, or a mixture of the two. *Bauer* [16.30] has suggested useful criteria for the different growth modes, based on equilibrium principles, and they may be regarded as zero-order approximations. Some refinements of the criteria, taking into account supersaturation, substrate electronic influence, misfit accommodation and equilibration rate, are discussed more fully in Sect.16.2. The relevance of "weak" and "strong" bonding [16.23,31] is also briefly dealt with. The growth modes are of great importance, both fundamentally and technologically; the latter because 3D growth is normally accompanied by a higher density of detrimental defects.

16.1.5 Orientation

Complete specification of the orientation of an epitaxial overlayer not only comprises the azimuthal orientation but also the overlayer crystal plane in contact with the substrate. If the adatom-substrate (a-s) bonding is strong, one may speculate that the contact plane will depend strongly on the existence of a plane (in the normal overlayer lattice) whose atomic configuration resembles that of the substrate; the overgrowth is then a quasicontinuation of the substrate [16.19]. If, on the other hand, the a-s bonding is relatively weak, a macroscopic approach suggests that the contact plane will be largely determined by minimum surface free energy considerations. In this case, the island is three-dimensional, and in order to make a prediction one needs to know the relations between the various specific surface and interfacial free energies.

Nucleation theory suggests that the choice of contact plane is a matter of size and form of the critical nucleus [16.32]. For example, the {001} fcc metal plane grows parallel to a {001} NaCl plane if the critical nucleus consists of four atoms, while the {111} fcc plane grows in this fashion if the critical nucleus consists of three atoms. The question of contact plane clearly needs to be considered more fully.

It has been almost unanimously accepted ever since it was suggested by *Volmer* and *Weber* [16.33] that the azimuthal orientation is determined primarily by minimum interfacial free energy, specifically the (azimuthal) orientational dependence of the well (minima) depths. Clearly, the parameters which determine the depths (epitaxial tendencies) are adatom-substrate and adatom-adatom bond strengths, mismatch and size. Nucleation and growth are seen as secondary factors.

In their concept of an "accommodation center", introduced to explain the epitaxial growth of metals (contact plane {001}) on {001} NaCl surfaces exposed to water vapor, *Henning* and *Vermaak* [16.34] have in principle included all the above-mentioned primary parameters. Their criterion for epitaxy was that a (rigid) intact square array of (hard sphere), adatoms matches the corresponding square array of (hard sphere) substrate cations, and, furthermore, that the radius of the (hard sphere) anion, or its substitute, is small enough to allow close approach between the two square arrays; the configuration thus obtained yields strongest bonding, which is possible only when it is allowed by the geometry.

The statements given above implicitly assume that equilibrium is realizable within the time scale of the observations, and that the overgrowth has macroscopic dimensions, which is not necessarily the case [16.3,35]. The fact that a misorientation exists implies that the relevant configuration is metastable. The system can break away from it only by a fluctuation, the probability $p(T_s, \Delta E)$ being in the form of (16.2), where ΔE now is the energy barrier in the metastable configuration. The equilibration time τ could be very long and clearly can be speeded up by annealing at higher substrate temperature T_s.

For islands of microscopic dimensions, there may be finite fluctuations, and the probability p that the island be misfitted [16.30] (translationally and/or azimuthally) satisfies a relation of the form

$$p \propto \exp(-\Delta E/kT) \quad , \tag{16.2}$$

where ΔE is the interfacial free energy change of the island in the fluctuation [16.36]. A model for determining ideal minimum free energy configurations will be dealt with more fully in Sect.16.5. The magnitude of ΔE depends in a complicated manner on bond strengths, misfit (including misorientation), and island size. This dependence will also be discussed more fully in Sect. 16.3. For the present it suffices to note that there will be a distribution [16.37,38] of orientations of small islands in which the spread may be quite prominent in the case of weak a-s bonding and that usually ΔE increases in proportion to the island size. The latter has the implication that the probability of a fluctuation of a given misorientation decreases drastically with size.

This is also the appropriate point to mention that the curve of ΔE versus mismatch (misfit and misorientation) develops a cusped minimum with vertical tangent at zero mismatch as the island height (overlayer thickness) becomes large [16.39]. This probably contributes significantly to the fact that homoepitaxy (the growth of a pure perfect crystal) can be perfect under ideal conditions. Calculations have shown that an entire hierarchy of cusps exists associated with the order of commensurability of the interfacial lattices. The character of the hierarchy may be illustrated in the case of one-dimensional mismatch [16.40]; the zero-order cusp is at zero mismatch, the first-order ones when

$$na_0 = (n \pm 1)a_s \quad ; \quad n = 2,3..., \tag{16.3}$$

and higher-order hierarchies satisfy more complicated relations. The depths of minima decrease drastically with the order of the minima. The existence of cusps clearly is important in large misfit [small n in (16.3)] systems, where the concept of a misfit vernier, supplemented by (cusped) coincidences [16.16, 20], rather than misfit accommodation by misfit dislocations (Sect.16.1.6) becomes appropriate.

It has already been mentioned in connection with the choice of the contact plane that growth may emanate from a (nonequilibrium) nucleation mechanism [16.32]. Macroscopic (capillarity) and microscopic (atomic) nucleation theories have been put forward with some success to show that the orientation may originate in a nucleation phenomenon in which the interfacial energy, or its equivalent, enters into the predictions, as ΔE in (16.2), apart from other terms in the exponent. Clearly, the smaller ΔE the more probable the relevant critical nucleus. An implicit assumption of the theory is that the island continues to grow in the nucleated orientation. The azimuthal orientations of the {001} and {111} fcc contact planes on {001} NaCl will be such as to minimize the corresponding ΔE's.

370

While a change of contact plane requires a rather large fluctuation involving recrystallization, a change in azimuthal orientation may be highly
probable in the early history of the island [16.26,37]. Thus, one would expect
to see a distribution of orientations. The spread of orientations may be enhanced in view of the fact that at a given supersaturation islands with different unrelated orientations, e.g., the {001} and {111} fcc orientations on
NaCL, may nucleate simultaneously, usually in different proportions determined by their critical free energies of formation. Whichever dominates subsequently depends on their growth modes and growth rates [16.41].

Another important property of islands is their mobility [16.26,37,42,43],
which, according to (16.2) depends critically on ΔE and T_s. If they were approximately free to translate (small ΔE and large T_s), adatoms which would be
even more free would behave almost like a 2D gas. The islands accordingly behave in a Brownian-like way in two dimensions. The equipartition theorem requires that

$$\frac{1}{2} M\langle v^2\rangle = kT_s \quad , \tag{16.4}$$

where M is the mass and v the speed of the particle. It follows that at T_s
= 300 K the mean $U \equiv \langle v^2\rangle^{\frac{1}{2}}$ is about 16 m/s for an island containing 100 atoms
of mass number 200 each, and that it takes roughly 10^{-10} s to move a distance
equal to its width.

If viewed somewhat differently, u represents the component of velocity
(parallel to the substrate surface) of a molecule that collides (directly
from the vapor) with an island, the number thus colliding per second with velocity component exceeding u is given by

$$n_u = \frac{\beta SP(T)}{(2\pi mkT)^{\frac{1}{2}}} \exp\left\{-\frac{mu^2}{2kT}\right\} \quad , \tag{16.5}$$

where P(T) is the vapor pressure, S the island-substrate contact area and
$\beta(>1)$ a geometrical factor near unity. The prefactor of the exponential is the
total number of hits per second. A fraction e^{-1} of hits are such that the
horizontal component of velocity transmitted to the island in a "head-on" collision exceeds the mean $V = \frac{1}{n} (2kT/m)^{\frac{1}{2}}$, obtained by equipartition and momentum
conservation estimates; the island is assumed to consist of n atoms and to
move freely over the surface. When n = 100 and m is equal to 200 mass units,
$V \simeq 2$ m/s. For monolayer coverage of an island comprising 100 atoms, the number
of these hits experienced during the build up is of the order of 5. The exercise shows that apart from Brownian-like motion, also direct hits from molecules in the vapor may contribute to erratic motion. Note also that the island

may experience an impulsive torque with rotational effects if the impact is not truly "head on". The evidence that islands are mobile is overwhelming. The considerations above demonstrate the theoretical credibility of the mobility phenomenon.

An important consequence of such motion is that the islands not only coalesce as a result of growth, but also in mutual collisions. In the latter case the violence of the impact may stimulate favorable reorientation changes, recrystallization, and even melting [16.44]. An important feature of coalescence of islands with relative azimuthal misorientations is that they are separated by low-angle grain boundaries [16.41,45]. Free energy minimization constitutes a driving force eliminating such boundaries. If the azimuthal misorientations are of opposite sign, the boundary induces a torque which aids a favorable reorientation. Once the neck between two islands is eliminated the boundary may sweep rather rapidly through the smaller island. If the misorientations are of the same sign they contribute to stabilizing the relevant misorientation. When more than two touch, the situation becomes more complicated. Since originally the misorientations are completely uncorrelated, one would expect that coalescence, at appropriate substrate temperatures, will induce an improvement of the overall perfection of azimuthal orientation.

It is also of interest that islands exert forces on each other even when separate [16.46]. Although knowledge is still incomplete, there is strong theoretical and experimental evidence that islands interact elastically [16.47,48] and electronically via the substrate, and that for insulating and semiconducting substrates the islands may become charged with consequential Coulombic interaction [16.49]. The elastic interaction normally has a repulsive nature [16.48]. By opposing or favoring close approach and coalescence, these forces may influence the rate of reorientation.

16.1.6 Misfit Accommodation

We adopt the convention that "misfit" f specifically refers to quantification of dimensional differences, including differences in thermal expansion, and that "mismatch" be used in a collective sense, i.e., to include also misorientation and differences in symmetry [16.23,26]. Also, we shall speak of one-dimensional (1D) misfit or matching when we compare the spacings of corresponding parallel atomic rows in the two interfacial lattices [16.26]. Some of the contents of this section will also be dealt with elsewhere. For the sake of clarity and continuity we summarize the relevant essentials here.

Consider an overlayer consisting of a finite but large monolayer on a much larger substrate surface. We adopt the model — a harmonic 2D crystal on rigid crystalline substrate (equivalent to a periodic 2D potential) governed by equilibrium principles — defined above [16.19]. The considerations show that there are three basic modes of misfit accommodation resulting from the competition between the harmonic adatom-adatom forces and the adatom-substrate potential [16.19,20]. When the former is weak compared to the latter, the monolayer is misfit-strained to match the substrate completely. As the former increases and/or the latter decreases a critical misfit f_c is attained, whereafter the so-called pseudomorphic (coherent) monolayer becomes unstable. It is still metastable though, and only when f exceeds a value $f_1 > f_c$ will it break away spontaneously. For $f > f_c$, equilibrium requires the misfit to be accommodated by misfit strain (MS) e and misfit dislocations (MD's) jointly, such that

$$f = e + \bar{f} \quad , \tag{16.6}$$

where the part $\bar{f}$ of f is accommodated by MD's. The partition of f between e and $\bar{f}$ under equilibrium requirements has been expressed in terms of elliptic integrals [16.19]. The so-called commensurate-incommensurate (CI) phase transition at $f = f_c$ is continuous with logarithmic singularity [16.17].

Two-dimensional coherency also occurs with different symmetries. Thus some ultrathin {111} fcc metal layers grow coherently on {110} bcc substrates [16.3], requiring the deformation of both angles and distances. Some of these revert at higher thicknesses to 1D matching and appear to lead to the well-known Nishiyama-Wassermann (NW) and Kurdjumov-Sachs (KS) orientations [16.50]. The critical conditions for coherence in such systems are considered in some detail in Sect.16.3.

When the adatom-substrate interaction becomes weak compared to the adatom-adatom forces, the influence of the former diminishes and vanishes in the limit. The relative bond strength where the influence becomes negligible (the crystals being regarded rigid) also depends on the mismatch and the thickness of the crystals [16.26]. Such a bicrystal system is said to be simulated by a "rigid model", and the mismatch between them, to be accommodated by a misfit vernier (MV), the third mode of misfit accommodation.

Epitaxial layers, which started off as subcritical monolayers, clearly must become unstable at some critical thickness h_c [16.20,51]; the MS energy continues to increase in proportion to the number of monolayers, while the substrate interaction energy of a coherent layer remains practically constant. Various models based on equilibrium principles have been used to estimate h_c, as well as the distribution of misfit between MD's and MS. Often the calculated contribution of MS is greatly underestimated, as compared to the observed contribution. This discrepancy can almost certainly be ascribed to the equilibrium basis of the theory, an ideal which is often not accomplished in practice, because the acquisition of the necessary MD's may constitute a major barrier [16.20].

The generation, motion, and nature of MD's is a subject in itself [16.20]. It suffices to mention here that MD's may be injected along the interface from the island perimeter, or climb from the overlayer free surface towards the interface, or originate from glide dislocations in the overlayer, and that it is a discrete process with discrete effects on MS and location of islands of finite lateral extent. Apart from the generation barrier, there are the barriers to motion, e.g., temperature-activated mass transport in the case of climb, and the Peierls stresses in the case of glide [16.20].

The Burgers vectors of the MD's acquired by glide along the interface, and some obtained by climb, lie in the interface; these dislocations are named perfect MD's. The Burger's vectors and glide planes of other glide dislocations are almost always inclined to the interface and less efficient (imperfect MD's) in accommodating misfit [16.20]; only their projections on the interface accommodate misfit. A special case is the one in which a threading dislocation — a dislocation which is a continuation of a substrate dislocation into the overlayer — is swept by the MS-induced stresses into the interface, to constitute an imperfect MD [16.20].

The equilibrium free energy criteria for intermixing, and the (diffusion) kinetic conditions for its realization, are dealt with more fully in Sect. 16.2. The basic effect of intermixing on misfit accommodation is the smearing out [16.3,20,52] of the discontinuity in physical properties, such as bonding, lattice parameters, and thermal expansion coefficients. The dependence of lattice parameter on concentration is usually approximated by Vegard's law [16.5,20,53], and one may anticipate that this law will also represent a first approximation for the dependence of the other physical properties.

A well-established consequence of lattice parameter smearing is the smearing of misfit, the redistribution of the MD's, and their rearrangement into walls normal to their glide planes [16.20,52]. The number of MD's is conserved in this redistribution process until the diffusion zone reaches the crystal surface where MD's can escape by climb and glide.

An important application of the dependence of physical properties on intermixing is the fabrication of dislocation free layers [16.5]. Both MD's and threading dislocations are usually detrimental to the operation and life times of solid state devices [16.4-6]. While threading dislocations may be eliminated by misfit strains [16.20], the need for MD's can be avoided by

growing [16.5] either graded layers, in which the gradient of lattice parameter is too low to attain the critical thicknesses at which MD's are needed,
or lattice-matched systems (using ternary and quaternary layers of III-V
compounds) in which both lattice parameters and thermal expansion coefficients
are matched.

Another form of misfit accommodation is the one in which the atoms at the
interface rearrange so that the new positions of given atoms are not necessarily a quasicontinuation of those in the parent crystal. This concept has
been dealt with in detail by *Aleksandrov* [16.53] in his considerations of
transition layers in semiconductors.

It is known that many crystal surfaces, particularly in semiconductors,
reconstruct in vacuum. The effects of such reconstruction on epitaxial growth
are not completely understood as yet. There are several phenomena, specifically defects, which are believed [16.54] to have their origin in the mismatch due to reconstruction of the substrate surface. This subject clearly
needs to be explored in more detail.

Polimorphism [16.41,55] and tetragonal distortion [16.56] may either be
regarded as MS type modes of misfit accommodation or as direct consequences
of MS. Iron is known to grow on Cu and Au with fcc structure (γ phase) in
the initial stages, which seems to be coherent, and revert to the stable
bcc structure (α phase) at a thickness of about 15 Å. Similarly epitaxial
Ge layers on GaAs are tetragonally distorted, the lateral dimensions being
homogeneously deformed to match the atomic arrangements at the interface.

Up to now it has been tacitly assumed that the contact plane is a simple
crystallographic plane. Small deviations from such orientations have been
observed and interpreted as a means of accommodating misfit at the interface
[16.57-59].

16.2 Growth Modes

16.2.1 Adsorption-Desorption

It should be remembered that we limited ourselves to simple growth from vapor.
Also, for our purpose, the following elementary considerations are adequate.
We approximate the vapor by an ideal gas satisfying the Maxwell distribution
of velocities. Accordingly, the flux of vapor molecules, i.e., the number
striking unit substrate area per unit time, is given by [16.16]

$$\phi_1(T) = \frac{P(T)}{(2\pi mkT)^{\frac{1}{2}}} \quad .$$
(16.7)

Only a fraction α_v depending mainly on the nature of the surface is adsorbed, and thus defines the deposition rate

$$R(T) = \frac{\alpha_v P(T)}{(2\pi mkT)^{\frac{1}{2}}} \quad . \tag{16.8}$$

Kinetic equilibrium between vapor and substrate of temperature T_s exists when the evaporation rate equals the deposition rate

$$R_e(T_s) = \frac{\alpha_v P_e(T_s)}{(2\pi mkT_s)^{\frac{1}{2}}} \tag{16.9}$$

There will be a net flux

$$\Phi = R(T) - R_e(T_s) \tag{16.10}$$

towards the surface, and consequently growth, if the vapor is supersaturated, i.e., $P(T) > P_e(T_s)$. Defining the saturation ratio ζ as P/P_e one obtains

$$\frac{N_1}{N_{1e}} \simeq \frac{R(T)}{R_e(T_s)} = \frac{P(T)}{P_e(T_s)} \left\{ \frac{T_s}{T} \right\}^{\frac{1}{2}} \simeq \frac{P(T)}{P_e(T_s)} = \zeta \quad , \tag{16.11}$$

where the first approximation is based on relations of the form in (16.13) below, N_1 and N_{1e} being the existing and equilibrium densities of adatoms, and the second approximation is based on the fact that P is very much more sensitive to temperature changes than $T^{\frac{1}{2}}$; see, for example, (16.9,13).

Atoms adsorbed on the surface have a mean stay time τ_s given by

$$\tau_s = \nu^{-1} \exp(E_a/kT) \quad , \tag{16.12}$$

where $\nu(\sim 10^{-13}\ \text{s}^{-1})$ is the vibration frequency of an atom on an adsorption site. For complete condensation one accordingly obtains [16.60]

$$R_e(T_s) \simeq \frac{N_0}{\tau_s} = N_0 \nu \exp(-E_a/kT_s) \quad , \tag{16.13}$$

where N_0 is the number density of adsorption sites on the substrate surface and E_a the desorption energy.

For incomplete condensation, adatoms will be scattered on the surface, and will migrate freely by consecutive jumps between adsorption sites, with a jump frequency

$$\omega = \nu' \exp(-E_D/kT_s), \quad \nu' \sim \nu \quad , \tag{16.14}$$

where E_D is the activation energy for surface migration. It follows from random walk theory that the mean square of its surface displacement during its stay time τ_s is given by

$$<r^2> = <x^2> + <y^2> = 2\omega\tau_s a^2 \quad , \tag{16.15}$$

where a is the elementary jump distance. It follows from (16.12,14,15) that its range (diffusion length) in time τ_s is given by

$$\lambda_s = (<r^2>)^{\frac{1}{2}} = \sqrt{2}a \, \exp[(E_a - E_D)/2kT_s] \quad . \tag{16.16}$$

Since $E_a > E_D$ this implies that the range increases sharply with decreasing temperature, because of the exponential increase in τ_s, while on the other hand the rate of equilibration drops sharply, because of a decrease in the jump frequency ω.

16.2.2 Bauer's Criteria

A diffusing adatom may impinge on and join a single adatom or a cluster of adatoms and thus bring about growth. The growth mode is usually classified [16.30] as Volmer-Weber (VW = 3D) growth when initiated from the three-dimensional islands, as Frank-van der Merwe (FM = 2D) growth when it grows by the successive addition of extensive (two-dimensional) monolayers, and as Stranski-Krastanov (SK) when initial FM growth is followed by VW type growth. The growth mode is also technologically important because of its influence on perfection.

Since *Bauer* [16.30] first suggested criteria for the three growth modes, several improvements have been suggested. Bauer's criteria are based on equilibrium considerations and follow most easily from minimum free energy principles. In Fig.16.1 each step is supposed to be of monatomic height and σ_0, σ_s and σ_i are the specific surface free energies of the overlayer, the substrate and the overlayer-substrate interface, respectively. If the free energy of the overlayer-substrate system is less when an additional adatom is added at A rather than at B, two-dimensional growth will be preferred. The growth criteria thus obtained, hereafter referred to as *Bauer's criteria*, may be summarized as follows:

$$\Delta\sigma = \sigma_0 + \sigma_i - \sigma_s \begin{array}{ll} \leq 0 & ; \quad 2D \equiv FM \quad , \\ > 0 & ; \quad 1D \equiv VW \quad , \end{array} \tag{16.17}$$

where the equality representing, for example, the growth of a substance on itself, is classified as FM.

Clearly, Bauer's criteria represent a zero-order approximation [16.3]. They contain, nevertheless, the dominating variables, but neglect the influences of intermixing, mismatch, anisotropy and supersaturation. They ignore the range of substrate electronic influences in the overlayer and attribute macroscopic properties to microscopic systems, e.g., monolayers.

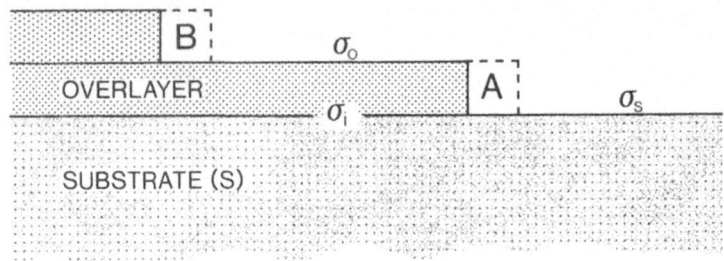

Fig.16.1. Substrate(s) and two growing monolayers terminating at A and B.
Symbols σ_0, σ_s and σ_i are specific surface free energies as shown

These influences are particularly important at the 2D-3D transition ($\Delta\sigma = 0$),
where a small change in any of the σ's and/or the involvement of the neglected
quantities may introduce drastic deviations from the zero-order prediction.

The criteria in (16.17) can be written usefully in terms of (macroscopic)
adhesion energies. In view of the fact that two surfaces are created when a
crystal is separated into two halves, it follows that [16.23]

$$\sigma_0 = \frac{1}{2}\,\varepsilon_{00} \quad , \quad \sigma_s = \frac{1}{2}\,\varepsilon_{ss} \quad , \qquad\qquad (16.18)$$

where, for example, ε_{00} is the work needed per unit area (adhesion energy)
in performing the separation reversibly at constant temperature and pressure.
Likewise, the interface energy is defined in terms of the work needed to create
pairs of 0-0 and s-s surfaces and to rejoin them into pairs of 0-s surfaces;

$$\sigma_i = \frac{1}{2}(\varepsilon_{00} + \varepsilon_{ss} - 2\varepsilon_{0s}) \quad . \qquad\qquad (16.19)$$

In view of (16.18,19) Bauer's criteria may be expressed as

$$\Delta\sigma = \varepsilon_{00} - \varepsilon_{0s} \begin{array}{l} \le 0 \quad ; \quad \text{FM} \quad , \\ > 0 \quad ; \quad \text{VW} \quad , \end{array} \qquad (16.20)$$

which explicitly states that FM growth is obtained when the (adatom) overlayer-
substrate bonding is relatively strong. This situation is often referred to
as "complete whetting".

The criteria in (16.17,20) apply to the interface between the first layer
and substrate. To the same approximation, these relations reduce to ($\varepsilon_{0s} \rightarrow \varepsilon_{00}$)

$$\Delta\sigma = 0 \qquad\qquad (16.21)$$

for the interface between the first and second monolayers. One would accordingly expect the corrections to the Bauer criteria to be particularly significant for continued growth on top of a first FM layer. It is certainly not
surprising that a change of growth mode is usually observed in this case.

On the basis of the schema adopted in Fig.16.1, it may be anticipated that an overlayer that has commenced in the 3D mode at the interface will be bounded by rather steep sides because of the constraints at the edges on the substrate, unless influenced otherwise by significant surface free energy anisotropies.

16.2.3 Young's Equation

It is appropriate at this point to consider briefly the application of Young's equation [16.16]

$$\sigma_s = \sigma_i + \sigma_0 \cos\theta \tag{16.22}$$

in deducing Bauer's criteria. This relation is the condition that a macroscopic cap-shaped overgrowth, radius of curvature r and contact angle θ, has a minimum surface free energy

$$G = 2\pi r^2 (1 - \cos\theta)\sigma_0 + \pi r^2 \sin^2\theta(\sigma_i - \sigma_s) \tag{16.23}$$

at constant volume

$$V = \frac{1}{3}\pi r^3 (2 - 3\cos\theta + \cos^3\theta) \ . \tag{16.24}$$

The requirement that physically meaningful solutions for θ from (16.22) be real yields the criteria in (16.17). In addition, it yields the criterion

$$\sigma_s + \sigma_0 - \sigma_i \leq 0 \tag{16.25}$$

for homogeneous, completely nonwhetting nuclei. Otherwise, for $\sigma_i \leq \sigma_s \leq \sigma_0 + \sigma_i$, i,e., $\pi/2 \geq \theta \geq 0$, it may be taken as a measure of the deviation from two-dimensionality.

In the previous derivation of Bauer's criteria, using a simple cubic type model, the VW islands will have vertical sides. In real crystals the side inclinations are determined by surface free energy an isotropies.

Evidently, the derivation of Bauer's criteria from Young's equation suffers from much the same shortcomings as the previous one. Furthermore, with these criteria, there is the additional constraint that the island be cap-shaped, an assumption that is probably good enough for a liquid drop with isotropic specific surface free energy. It also neglects the line tension at the perimeter island base which is important when the base radius becomes small and probably is also inducive to VW growth. The importance of line tension effects in nucleation phenomena has been considered by *Navascués* and *Tarazona* [16.21].

16.2.4 Supersaturation

Markov and *Kaischew* [16.61] analyzed the modification of the criteria in (16.17), as a nucleation phenomenon, under conditions of supersaturation. Their calculations were addressed towards nuclei of fcc crystals with {100} and {111} faces parallel to the substrate. For the present purpose it is sufficient to demonstrate the essentials of their arguments using an island with a simple square base of $n \times n$ atoms and vertical sides [16.60] (Fig.16.2) ν atoms high. Anisotropy is taken into account by letting the specific surface free energies σ_1 of the sides be different from that at the top and the specific area s_a per atom in the horizontal plane be different from that, namely s_1, on the vertical plane.

The first fundamental assumption of *Markov* and *Kaischew* was that the shape of the island, the n/ν ratio in this case, will be thermodynamically stable, determined by minimizing the free energy

$$G = \Delta\sigma s_a(n^2 + 4\alpha n\nu) \quad ; \quad \alpha = s_1\sigma_1/\Delta\sigma s_a \tag{16.26}$$

at constant volume (constant number N of atoms)

$$V = n^2\nu v = N v \quad , \quad N = n^2\nu \quad , \tag{16.27}$$

yielding the equilibrium values

$$n_0 = (2\alpha N)^{1/3} \quad , \quad \nu_0 = n_0/2\alpha \quad , \tag{16.28}$$

$$G_0 = 3\Delta\sigma s_a(2\alpha N)^{2/3} \quad . \tag{16.29}$$

Let the particle be in the presence of a supersaturated ($\zeta > 1$) vapor, the saturation ratio ζ defined by (16.11). When a molecule is adsorbed from the vapor its chemical potential decreases by

$$\Delta\mu = \mu - \mu_e = kT \ln\zeta \quad . \tag{16.30}$$

Accordingly, the Gibbs free energy of formation of a nucleus containing N atoms (at constant temperature and pressure) is given by

$$\Delta G = 3\Delta\sigma s_a(2\alpha N)^{2/3} - N\Delta\mu \quad . \tag{16.31}$$

When $\Delta\sigma < 0$ and the first monolayer is in FM mode, then ΔG will continue to decrease with increasing N. This implies continued FM growth in the first monolayer until the substrate is completely covered.

Of particular interest is the case in which the bonding to the substrate is relatively weak, (16.20), and $\Delta\sigma$ is positive. As the cluster size N increases ΔG at first increases up to a maximum to define a critical nucleus of height given by the Gibbs-Thomson equation [16.60]

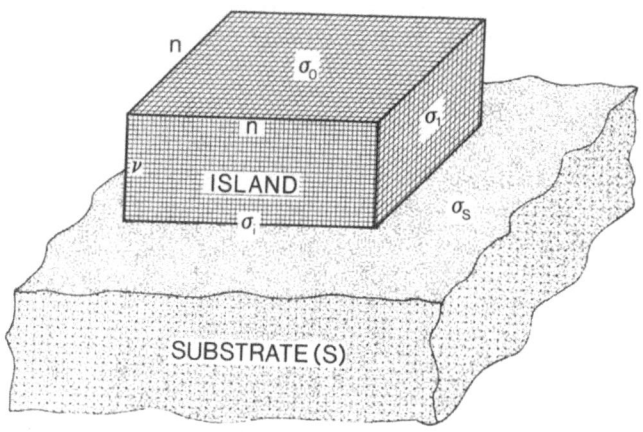

Fig.16.2. Rectangular island ($n \times n \times \nu$ atoms) on a substrate. The specific surface free energies σ_0, σ_s, σ_1 and σ_i are as shown

$$\nu_0^* = \frac{2s_a \Delta\sigma}{\Delta\mu} \tag{16.32}$$

and free energy of formation

$$\Delta G^* = 16 \Delta\sigma s_a \left(\frac{s_1 \sigma_1}{\Delta\mu}\right)^2 . \tag{16.33}$$

Markov and *Kaischew* [16.61] proposed that the initial growth mode is determined by the shape of the critical nuclei at the prevailing supersaturation. As the supersaturation and hence $\Delta\mu$ increases, the nucleus height ν_0^* decreases from a multiple value typical of VW growth, to unity which defines FM growth [16.60]. The transition occurs at $\Delta\mu = \Delta\mu_t$, where $\Delta\mu_t$ is given by

$$\Delta\sigma = \Delta\mu_t/2s_a . \tag{16.34}$$

Above $\Delta\mu_t$ the nuclei will be 2D, yielding a FM growth mode. This implies a modification of the Bauer's criteria in (16.17) to

$$\Delta\sigma - \Delta\mu_t/2s_a \begin{array}{l} \leq 0 \quad ; \quad \text{FM} , \\ > 0 \quad ; \quad \text{VW} . \end{array} \tag{16.35}$$

It follows from the analysis that the critical nucleus side length is given by

$$n_0^* = 2 \frac{s_1 \sigma_1}{s_a \Delta\sigma} \nu_0^* , \tag{16.36}$$

where $\nu_0^* = 1$ for $\Delta\mu \geq \Delta\mu_t$. Since the analysis is specifically important close to the limit $\Delta\sigma = 0$, the lateral extent of the critical 2D nucleus may be anticipated to be large compared to its height, at least when $\Delta\sigma \sim 0$.

It follows from (16.11,30,34) that the transition will occur at a deposition rate

$$R(T) = R_e(T) \exp(2s_a \Delta\sigma/kT) \quad . \tag{16.37}$$

If we assume that in "equilibrium" the substrate will be completely covered when the vapor is supersaturated, the equilibrium deposition rate $R_e(T)$ is given by [16.60] [cf. (16.13)]

$$R_e(T) = \frac{1}{s_a \tau_s} = s_a^{-1} \nu \exp(-E_a/kT) \quad . \tag{16.38}$$

This may be substituted into (16.37) to define R(T) in terms of more fundamental quantities.

Two interesting observations may be made at this point. Firstly, with reference to (16.35) one might speculate that continued FM growth, after the first FM monolayer, may be stimulated in subsequent layers provided the supersaturation is properly controlled, and the conditions are otherwise not very opposing. Secondly, if one switches from supersaturation to undersaturation ($\Delta\mu < 0$), it may be speculated that complete evaporation will occur if $\Delta\sigma > 0$ and incomplete evaporation otherwise; the nucleus size will stabilize where ΔG is minimum and the nucleus height defined by (16.32) (both $\Delta\sigma$ and $\Delta\mu$ negative). The height will decrease with increasing undersaturation, becoming a monolayer when $\Delta\mu = \Delta\mu_t$, as defined in (16.34).

16.2.5 Electronic Influence of the Substrate

To this category, of which we know least, belongs the influence of the substrate on the electronic configurations in the overlayer [16.3,23]. One may speculate that strong overlayer-substrate bonding at the interface induces drastic changes in the overlayer electronic structure, that the distance in which the change decays depends on the bonding type, and that a reduction of bond strength is the main characteristic of the change. This would imply that the relation $E_{00} - E_{0s} \leq 0$ at the overlayer-substrate interface be followed by

$$E_{0,n+1} - E_{0n} > 0 \tag{16.39}$$

for the interface between the nth FM monolayer and the next, n being small. This is exactly the condition for SK growth. Unless other opposing influences prevail, this kind of argument suggests that strong overlayer-substrate bonding and a few FM atomic layers are characteristic of SK growth. For metals n appears to be of the order of one to three [16.3]. The reverse, i.e., the possibility that VW growth reverts to FM growth, is inhibited by the constraint at the substrate. The theme of overlayer islands merging into continuous films, and the 2D growth appearance [16.35] of high island densities, will be dealt with elsewhere.

An indirect influence of the substrate on bonding may emanate from substrate-induced misfit strains [16.16,61]. For example, if one makes the zero-order approximation that interface bonding is proportional to the number density of interfacial atoms, then

$$\varepsilon_{00}(e) = \varepsilon_{00}(0)(1 + e)^{-2} \simeq \varepsilon_{00}(0)(1 - 2e) \quad , \tag{16.40}$$

where e, $\varepsilon_{00}(e)$, and $\varepsilon_{00}(0)$ are the 2D strain, the adhesion in a strained overlayer and the adhesion in an unstrained overlayer, respectively. In view of the form of Bauer's criteria in (16.20), misfit strains may have significant consequences on growth modes. If the supposition in (16.40) is valid, one would expect opposite effects for misfit strains of tension and compression, and one may further speculate that the other geometrical influences of misfit strain will enhance or diminish such an asymmetry. The direct influence of misfit strain energy and misfit dislocation energy will be dealt with below.

16.2.6 Intermixing

Under intermixing are subsumed all material combinations which are miscible and form alloys, being either simple solid solutions or compounds. The free energy criterion for the miscibility of two metallic components A and B is

$$\Delta G = G_{AB} - (G_A + G_B) \leq 0 \quad . \tag{16.41}$$

The Hume-Rothery rules of thumb that a given combination possibly complies with (16.41) are that (i) the atomic radii must be within 15% of each other; (ii) the crystal structures must have the same symmetry; (iii) the chemical valencies may not differ by more than one; and (iv) the electronegativities should be nearly equal if compound formation is to be prohibited too.

The role of miscibility in the occurrence of different growth modes has been considered in some detail by *Bauer* [16.3]. He stressed the fact that the realization of solid solutions for metal pairs, which comply with the miscibility criterion, is still a temperature-dependent rate process, involving volume interdiffusion governed by relations analogous to (16,14,16) valid for surfaces. Little work has been done with the special aim of studying the effects of intermixing at epitaxial crystal interfaces. There appear to be two main consequences [16.3] of intermixing; firstly, it seems to smooth out electronic effects of strong interface-localized substrate-overlayer bonding which favors SK growth, and secondly, it similarly smooths out the abrupt change of lattice parameters at the interface, which is responsible for the presence of misfit dislocations or/and a discontinuity in (misfit) strain. The effects of the latter are briefly dealt with in the next section. The former may possibly be interpreted in terms of either a smoothing of the discontinuity in surface free energy values, or a relaxation of the "electronic" hold on the overlayer by strong bonding to the substrate.

Bauer's considerations clearly show that for complete understanding of the role of miscibility and strength of overlayer-substrate bonding a considerable amount of research is still required.

16.2.7 Misfit Accommodation

It has been argued previously [16.3,20,23] that the accommodation of misfit by misfit strain (MS) and misfit dislocations (MD's) may make significant contributions to the energy balance which determines the growth mode. If, for example the first layer is coherent in FM mode and the second one completely MS relaxed, Bauer's criteria in (16.17) are modified approximately to

$$\Delta\sigma + E^e - E_D \leq 0 \quad ; \quad \text{FM} \quad ,$$
$$> 0 \quad ; \quad \text{VW} \quad , \tag{16.42}$$

where E^e is the energy gain due to the strain relaxation in the second layer and E_D the energy associated with the introduction of misfit dislocations between the first and second layer. If it is true that strong overlayer-substrate bonding induces weak bonding between, say, the first and second monolayers, then E_D will be small. It implies a positive contribution in (16.42) because E^e will be relatively large. This is in addition to the effect on $\Delta\sigma$ as explained with reference to (16.20). The combined effect of bonding and misfit accommodation is clearly a subject which justifies further investigations.

16.2.8 Kinetic Constraints

Until now we have implicitly assumed that the relevant atomic processes are fast enough to ensure quasiequilibrium at every growth stage. For example, in deriving the relations in (16.17,28) it has been tacitly assumed that the atoms needed for shape equilibration would be available on the spot. Most of them would, however, arrive by surface migration from elsewhere. Accordingly, there would be an equilibration time lag at sufficiently high supersaturations (low surface temperatures T_s). This problem has been dealt with by *Venables* and co-workers [16.35] in their considerations of SK growth of Ag on {111} Si.

Like *Markov* and *Kaischew* [16.61], *Venables* and co-workers also proposed that the growth mode is a nucleation phenomenon. They further assumed that the critical 2D nuclei are rather large at high temperatures. This assumption is supported by their UHV-SEM observations and is also in agreement with the speculations based on (16.32). In this case existing nucleation theories predict that the number density of critical nuclei would be approximately given by [16.35]

$$N = C_{\infty}(R/\nu) \exp[(E - E_a + E_D)/kT] \quad , \tag{16.43}$$

where E is the bulk Ag sublimation energy per atom and E_D the diffusion energy per atom; the other symbols have been defined previously. Using estimated values of the E's, the authors concluded that the relation in (16.43) gives a reasonable description of the trend of their observations, thus justifying the supposition that the growth mode is a nucleation phenomenon.

They then dealt with the kinetic barrier by estimating the maximum sizes of islands that could maintain equilibrium shapes at the given deposition conditions. The equilibration time τ by surface diffusion, assuming no difficulties with incorporation into the islands, has been approximated by [16.62]

$$\tau \simeq Cr^4/B \quad , \tag{16.44}$$

where the mass transfer coefficient B is proportional to the surface diffusion coefficient D, r is the radius of the particle, and C a constant. The time t it takes to deposit the material needed for the growth of N pill-box islands is given by

$$N\pi\omega^2 h/4\Omega = Rt \quad , \tag{16.45}$$

where Ω is the atomic volume, h the island height and ω, which may be approximated by r, their width. *Venables* and co-workers [16.35] adopted the condition

$$\tau < t \tag{16.46}$$

as a suitable criterion for equilibration. In view of (16.44,45), criterion (16.46) may be written in the form

$$h/\omega > C'(R\omega/ND) \quad , \tag{16.47}$$

where N and D have opposing temperature dependences, and C' is another constant.

The energies concerned were estimated at $E_D \simeq 1$ eV and $E - (E_a - E_D) \approx 1.11$ eV. The equilibrium was accordingly subtly balanced. Equilibrium was estimated to break down for r [cf. (16.45)] between 0.3 and 1 μm at $T_s = 400^\circ$C and between 0.1 and 0.3 μm at $T_s = 300^\circ$C. This is in qualitative agreement with their observation that the ratio h/ω decreases with increasing temperature. This, together with the evidence that the continuous film grown at temperatures below 200°C is unstable, supports their view that apparent 2D growth at low temperatures is a kinetic phenomenon, which influences the growth mode to such an extent that basic SK growth has an FM appearance.

16.3 Epitaxy of {111} fcc Metals on {110} bcc Metal Substrates

16.3.1 Model

The purpose of this section is to apply the model formulated in Sect.16.1 to the growth of monolayers of fcc metals on {110} bcc metal substrates [16.3, 50,63]. We shall assume [16.26,64,65] that the adatoms have a stable {111} fcc structure (approximated by the bulk one) on a substrate with modulation free interaction potential and specifically address the problems (i) of predicting the ideal epitaxial configurations — the relationship between the corresponding dimensions and orientations of the interfacial lattices for which the equilibrium minimum (free) energy criteria are ideally satisfied — when the modulations of the adatom-substrate interaction are introduced; and (ii) of the equilibrium distribution of the misfit between the different accommodation modes when the configuration is not ideal. The first part of the problem is dealt with by using the rigid model (rigid lattices) [16.26] and the second part by considering an elastic monolayer on a rigid substrate [16.26, 64-66].

It is convenient to analyze the matching at the interface in terms of the corresponding rhombic unit cells in Fig.16.3, which also illustrates the meaning of many of the relevant geometrical quantities. The shorthand notations

$$s = \sin\alpha, \quad c = \cos\alpha, \quad s_\beta = \sin\beta, \quad c_\beta = \cos\beta, \quad \bar{s} = s/s_\beta, \quad \bar{c} = c/c_\beta,$$
$$a_x = 2as, \quad a_y = 2ac, \quad b_x = 2bs_\beta, \quad b_y = 2bc_\beta,$$
$$d_a = 2asc, \quad d_b = 2bs_\beta c_\beta, \quad r = b/a \quad, \tag{16.48}$$

in which a and b are nearest-neighbor distances in the substrate and monolayer, respectively, the a_i and b_i the rhombic unit cell diagonals, and the d_i their widths, help to shorten the analysis. (Note that for simplicity we deviate from the convention of designating substrate and overlayer by means of subscripts s and 0, respectively). The misfits [16.26]

$$f = \frac{b - a}{a} = r - 1 \quad, \quad f_{an} = 2(\beta - \alpha) \quad,$$

$$f_x = \frac{b_x - a_x}{a_x} = \frac{r}{\bar{s}} - 1 \quad; \quad f_y = \frac{b_y - a_y}{a_y} = \frac{r}{\bar{c}} - 1 \quad, \tag{16.49}$$

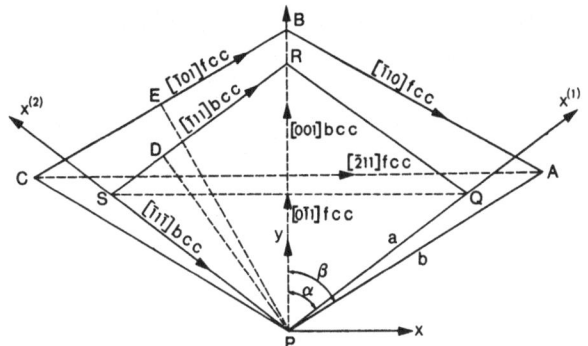

Fig.16.3. Rhombic surface unit cell structures, related crystal directions of an fcc {111} (β = 60°) PABC overlayer on a bcc {110} (α≈54.74°) PQRS substrate and reference systems (x,y) and $(x^{(1)}, x^{(2)})$. The unit cell side lengths are PQ = a and PA = b; the long diagonals SQ = a_x, CA = b_x; the short diagonals PR = a_y, PB = b_y and the widths PD = d_a and PE = d_b. Two important observed epitaxial configurations [16.50] are (i) Nishiyama-Wassermann (NW): fcc [0Ī1]//[001] bcc, ideal configuration; either $f_x = 0$ or $f_y = 0$, and (ii) Kurdjumov-Sachs (KS): fcc [Ī10]//[Ī1Ī] bcc or fcc [Ī01]//[Ī11] bcc, ideal configuration $f_d = 0$. (Van der Merwe [16.26], by permission of Taylor and Francis, Ltd.)

$$f_d = \frac{d_b - d_a}{d_a} = \frac{r}{sc} - 1$$

in which f_x, f_y and f_d refer to matching of corresponding atom rows, are useful in interpreting the results. It should also be noted that the definitions in (16.49) are convenient in the present analysis and differ from those employed by *Matthews* in analyzing misfit accommodation in thickening overlayers [16.20]. In the latter, b rather than a is taken as the reference lattice parameter.

The Fourier representation of the adatom-substrate interaction in (16.1), appropriated to the {110} bcc substrate symmetry, is [16.67]

$$V(x,y) = \sum_{\substack{h,k=0 \\ (h+k)\ even}} A_{hk}\left\{\cos\left[2\pi\left(\frac{hx}{a_x} + \frac{ky}{a_y}\right)\right] + \cos\left[2\pi\left(\frac{hx}{a_x} - \frac{ky}{a_y}\right)\right]\right\} \quad . \quad (16.50)$$

This potential, truncated at second-order terms, may be written in the form [16.26]

$$V(x,y) = A_0\left\{1 + A_1 \cos\left[2\pi\left(\frac{x}{a_x} + \frac{y}{a_y}\right)\right] + A_2 \cos\left[2\pi\left(-\frac{x}{a_x} + \frac{y}{a_y}\right)\right]\right.$$

$$\left. + A_3 \cos\left(4\pi\frac{x}{a_x}\right) + A_4 \cos\left(4\pi\frac{y}{a_y}\right)\right\} \quad , \quad A_1 = A_2 \quad . \quad (16.51)$$

This has previously been written in terms of skew coordinates

$$\frac{x^{(1)}}{a} = \frac{x}{a_x} + \frac{y}{a_y} \quad , \quad \frac{x^{(2)}}{a} = -\frac{x}{a_x} + \frac{y}{a_y} \quad . \quad (16.52)$$

With the origin of coordinates located at a potential minimum, which itself is chosen zero,

$$A_0(1 + A_1 + A_2 + A_3 + A_4) = 0 \quad \text{with} \quad A_1 = A_2 \quad . \tag{16.53}$$

One obtains an additional relation between the coefficients by equating the height of the saddle point at $(\frac{1}{4} a_x, \frac{1}{4} a_y)$ with the activation energy Q of surface migration

$$-A_0(A_1 + A_2 + 2A_3 + 2A_4) = Q \quad . \tag{16.54}$$

Further relations between the coefficients can be obtained only by the appropriate atomic calculations. Such estimates, using simple potentials, are at present in progress [16.68]. Previously, additional relations had been assigned using hand-waving arguments and intuition. These rather crude assignments are [16.26]

$$A_0 = W, \; A_1 = A_2 = -0.4, \; A_3 = -0.12, \; A_4 = -0.08,$$
$$Q = 1.2 \, W \quad . \tag{16.55}$$

A significant feature of these assignments is that the magnitudes of A_1 and A_2 are large compared to those of A_3 and A_4.

The quantity Q is usually assumed to be related to the strength of adatom-substrate interaction (desorption energy E_a) by [16.26]

$$Q = \kappa E_a \quad , \tag{16.56}$$

where κ depends on the type of adatom-substrate bonding and is estimated to lie in the range 0.1 to 0.3.

In the harmonic approximation the strain energy per atom, ε, is written in terms of macroscopic quantities [16.69]

$$E = W l^2 r^2 (e_x^2 + e_y^2 + 2P e_x e_y + R e_{xy}^2)$$

$$P = \frac{C_{12} C_{33} - C_{13}^2}{C_{11} C_{33} - C_{13}^2} \quad , \quad R = \frac{C_{66} C_{33}}{C_{11} C_{33} - C_{12}^2} \quad , \quad l^2 = \frac{\Omega(C_{11} C_{33} - C_{13}^2)}{2 W r^2 C_{33}}$$

$$2C_{11} = c_{11} + c_{12} + 2c_{44} \; , \quad 6C_{12} = c_{11} + 5c_{12} - 2c_{44} \; ,$$

$$3C_{13} = c_{11} + 2c_{12} - 2c_{44}$$

$$3C_{33} = c_{11} + 2c_{12} + 4c_{44} \; , \quad 6C_{66} = c_{11} - c_{12} - 4c_{44} \; , \tag{16.57}$$

where the e's are the strain components, the C's the {111} fcc transformed values of the elastic constants (c_{11}, c_{12}, c_{44}) for cubic crystals, and Ω the atomic volume in the overlayer.

The total interaction energy needs to be calculated for various translational (x_0, y_0) and rotational θ configurations of the overlayer. These configurations are obtained from the coordinate transformations [16.26]

$$x = x_0 + x'\cos\theta - y'\sin\theta \quad ,$$

$$y = y_0 + x'\sin\theta + y'\cos\theta \quad , \tag{16.58}$$

where the Cartesian system (x', y') is fixed in the overlayer lattice with its origin on an atom (lattice point).

16. 3. 2 Ideal Epitaxial Configurations

The supposition [16.26,65,66] is made that the ideal epitaxial configurations, the configurations for which the interfacial energy has minima, may be obtained by minimization with respect to r and θ, and that the ideal configurations may be predicted on the basis of the rigid (rigid overlayer and substrate) model.

The total adatom-substrate interaction energy V_G of a monolayer with rectangular boundaries [16.26] containing $2M + 1$ adatoms on a side, and hence

$$G = (2M + 1)^2 + (2M)^2 \tag{16.59}$$

adatoms all together is obtained by summing over all G adatoms

$$V_G = W\left\{ G + \sum_{i=1}^{4} A_i C_i(x_0, y_0) K(M; p_i q_i) \right\} \quad . \tag{16.60}$$

The epifunction K is defined by [16.26]

$$K(M; p, q) = \frac{1}{\sin p \, \sin q} \{\sin[(2M + 1)p]\sin[(2M + 1)q] + \sin 2Mp \, \sin 2Mq\} \quad ,$$

and

$$p_{1,2} = [\pi r s_\beta \cos(\alpha \mp \theta)]/sc, \qquad q_{1,2} = [\pi r c_\beta \sin(\alpha \mp \theta)]/sc,$$

$$p_3 = 2\pi r s_\beta s^{-1} \cos\theta, \qquad q_3 = 2\pi r c_\beta s^{-1} \sin\theta,$$

$$p_4 = 2\pi r s_\beta c^{-1} \sin\theta, \qquad q_4 = 2\pi r c_\beta c^{-1} \cos\theta \quad , \tag{16.61}$$

where the $C_i(x_0, y_0)$ in (16.60) are the cosine factors of (16.51), but with (x, y) replaced by (x_0, y_0).

Relations of the form in (16.61) are well known in diffraction theory. The most significant consequences of the relation in (16.60), some of which are vividly illustrated in the perspective representation of the mean interacting energy per atom

$$v_G = V_G/G \tag{16.62}$$

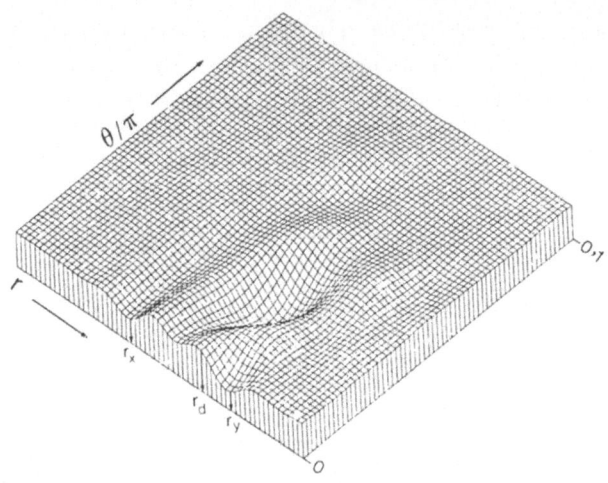

Perspective representation of the mean energy per atom V_G/GWA_0 = $v_G(r,\theta)/WA_0$ for a rectangular island, containing (with $M = 5$) $G = 221$ atoms. The representation is in a space spanned by the ratio $r = b/a$ of the rhombic unit mesh side lengths and the misorientation θ (in radians). The sharp minima at (r,θ) equal to $(r_x,0)$, $(r_d,\beta - \alpha)$ and $(r_y,0)$ define ideal configurations, where respectively f_x, f_d and f_y are zero. Not shown in the figure are other ideal configurations and an extensive plateau (with small undulations), which is removed from the near neighborhood of the above-mentioned ideal points. (Van der Merwe [16.26], by permission of Taylor and Francis, Ltd.)

in Fig.16.4, may be summarized thus [16.26]:

(i) Ideal epitaxial configurations are defined by the criteria

$(x_0,y_0) = (0,0)$, uneven atom number $2M + 1$ on sides,

$$p_i = \pi m_i \text{ and } q_i = \pi n_i; \; m_i + n_i = \text{even}, \; m_i,n_i = \text{integers} \tag{16.63}$$

for $i = 1,2,3$ and 4 independently. These criteria are equivalent to 1D (row) matching.

(ii) The depths of the minima, which may be taken as some measure of the epitaxial "strengths" of the ideal configurations, are given by

$$v_i = -WA_0A_i \; . \tag{16.64}$$

This result lends great importance to reliable estimates of the Fourier coefficients in the representations in (16.50,51).

(iii) For $\beta = \alpha$, all four $(i = 1,...4)$ criteria are satisfied simultaneously for $r = 1$ and $\theta = 0$ (coherency) with a (largest) epitaxial strength of

$$v_G = -WA_0 \sum_{i=1}^{4} A_i \; . \tag{16.65}$$

This result suggests that the tendency for complete coherency is relatively strong.

(iv) For values of (r,θ) significantly different from the ideal values (r_i,θ_i) the undulations in the v_G surface become mere ripples as compared to A_0; so much so, that now even the total energy V_G is of the order A_0 only. The implication [16.26,36] of this is that in rigid translation, and/or rotation an overlayer will experience barriers of order $A_0 G$ when $r \simeq r_i$ and only of order A_0 when r is significantly different from all r_i. The anticipation that this is true also for elastic monolayers has been confirmed recently [16.43].

(v) A growing (increasing M) monolayer with fixed $\theta = \theta_i$ and small $|r - r_i|$ will experience an oscillating torque with increasing M, or if free to rotate will experience an oscillating orientation about $\theta = \theta_i$. (This is only partly due to the island shape.) The implication of the oscillation is that the equilibrium orientation of a growing island will oscillate, and thus contribute to the spread of island orientation.

(vi) The considerations predict [16.26] two well-known orientations: the Nishiyama-Wasserman (NW) orientation [16.50], for which either

$$m_3 = 2, n_3 = 0; \quad r_3 = \sin\alpha/\sin\beta \simeq 0.943; \quad f_x = 0,$$
or
$$m_4 = 0, n_4 = 2; \quad r_4 = \cos\alpha/\cos\beta \simeq 1.155; \quad f_y = 0 \quad, \tag{16.66}$$

and the Kurdjumov-Sachs (KS) orientation [16.50], for which either

$$m_1 = n_1 = 1, \quad r_1 = \sin\alpha\cos\alpha/\sin\beta\cos\beta \simeq 1.089; \quad f_d = 0,$$
or
$$m_2 = n_2 = 1, \quad r_2 = r_1, \quad f_d = 0 \quad. \tag{16.67}$$

The 1D matching for these two cases is illustrated in Fig.16.5a,b. One-dimensional matching is also an important concept in the resonance approximation [16.24].

16.3.3 Accommodation of Misfit

16.3.3.1 Background

Having established the ideal conditions for epitaxy using the rigid model, it remains to find the conditions to realize the different modes of misfit accommodation of an elastic monolayer. The important variables in such an exercise are the misfit and the relative bond strengths [16.26]. The analysis shows that the effect of the latter is carried by the configurational para-

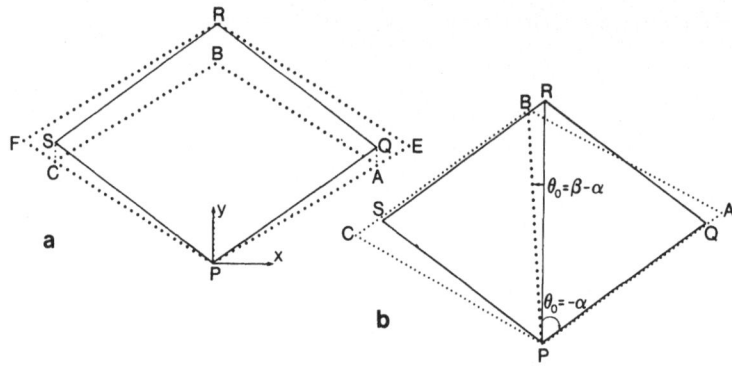

Fig.16.5a,b. Matching of atomic rows for some elementary ideal epitaxial configurations of dissimilar rhombic meshes ($\beta \neq \alpha$). The substrate unit cell PQRS
is drawn in full lines and overlayer unit cell in dashed lines. (a) Overlayer
cell PABC matches rows parallel to PR and PERF matches rows parallel to SQ;
N-W orientations in Fig.16.3. (b) Overlayer cell PABC matches rows parallel
to PQ and PS; K-S orientations in Fig.16.3. (Van der Merwe [16.26], by permission of Taylor and Francis, Ltd.)

meter 1, defined in (16.57). Estimates [16.26] of 1 for metals on metals,
using data on diffusion and elastic constants, yield values for 1 of about
4. This is indicative of relatively strong adatom-substrate bonding. The quantity W in these equations may now simply be regarded as a scale factor for
the energies concerned.

In the rigid model the misfit is completely accommodated by 2D misfit ver-
niers (MV's) or a 1D vernier if there is perfect matching in one direction
[16.70-72]. Such configurations are approached by an elastic monolayer when
1 becomes large [16.26], i.e., the ratio of adatom-adatom to adatom-substrate
bond strength becomes large. The limiting ratio at which this is an appropriate approximation decreases with increasing misfit. As the rigidity is relaxed (1 decreases) and the effect of the substrate increases, the adatoms
take up equilibrium positions closer to their respective nearest-neighbor
substrate potential minima, while still conserving their average spacings.
The total energy of the system decreases in this process from say, E_1 to E_2.
However, the overlayer normally does not acquire a (stable) minimum energy
(E_3) configuration without the monolayer being misfit strained to accomplish
better overall matching. In the extreme case ($1 = 1_c$) the coherent configuration in which the misfit is accommodated by MS only is the stable one [16.26].
Let E_4 be the energy for which the coherent configuration is just stable.
Clearly these energies depend on 1 and the mismatch.

If l increases beyond l_c the configuration of complete coherence becomes unstable. If there were perfect matching in one direction [16.26,62,73], say $f_x = 0$, Fig.16.5a, then presumably the matching in this direction (1D matching) will be conserved in the process of attaining the stable noncoherent configuration [16.26]. If the misfit f is changed, then evidently at some critical value $f_x^c(l)$ this 1D coherency will be lost too with the introduction of MD's parallel to the originally matched atomic rows. The corresponding 1D misfit will now be accommodated by MD's and MS jointly. As the misfit increases, the MD structure gradually assumes a vernierlike appearance [16.70-72]. Similar remarks are applicable [16.26] to misfits f_y and f_d, Fig.16.6a,b. Clearly, as the ratio $r = b/a$ increases from below 0.943 $(f_x = 0)$ through 1.089 $(f_d = 0)$ to above 1.155 $(f_y = 0)$, all possible combinations of misfit accommodation modes within the categories MS, MD's and MV's come and go. These modes may also be regarded as different 2D solid phases with boundaries dependent on f_i and l [16.26].

The strains appearing in the strain energy per atom in (16.57) are most conveniently expressed [16.64] in terms of the linear and angular deformations of the diagonals PB and CA of the monolayer unit cell in Fig.16.3, thus

$$e_x = \frac{u(A) - u(C)}{b_x} \quad , \qquad e_y = \frac{v(B) - v(P)}{b_y} \quad ,$$

$$e_{xy} = \frac{1}{2}\left[\frac{u(B) - u(P)}{b_y} + \frac{v(A) - v(C)}{b_x}\right] \quad ,$$

(16.68)

where u and v are respectively the x and y components of the elastic displacements of the lattice points P, A, B and C.

16.3.3.2 Complete Coherency

An exact mathematical analysis of even the rather simplified model of complete coherency is still very complicated. Accordingly, its treatment usually involves additional approximations, two of which are considered briefly. In the first [16.26], the forms of the phase boundaries have been estimated by assuming (i) that the energy ΔE that can be gained per atom from the substrate in the process of attaining coherency is proportional to the epitaxial strengths defined in (16.64,65)

$$\Delta E = \eta V \quad ; \quad \eta = (1 + P)^{-1} \quad ,$$

(16.69)

where the approximate value $(1 + P)^{-1}$ of the proportionality (scale) factor η has been inferred from semiexact calculations [16.25] on systems with simple rectangular symmetry; and (ii) that it may otherwise be identified with the minimum elastic energy ε_m needed for coherency, thus

$$\Delta E = \varepsilon_m \quad . \tag{16.70}$$

While for complete coherency the final configuration is unique, for 1D coherently this is not so. For example, for a given unit cell diagonal length a_x (spacing of corresponding substrate atomic rows) and any arbitrary ratio $\bar{r} = \bar{b}/a$ ($\bar{b}$ is the MS value of b) an angle $\bar{\beta}$ can always be selected such that $2\bar{b}\sin\bar{\beta} = a_x$. The appropriate values of $\bar{r}$ and $\bar{\beta}$ are assumed to be those which minimize the strain energy [16.26]. A simple mathematical description of these considerations is appropriate.

Consider the NW orientation with f_x near zero and $\bar{f}_x = 0$. Assume that

$$\delta = \frac{\bar{r} - r}{r} , \qquad \delta\beta = \bar{\beta} - \beta \tag{16.71}$$

are small quantities. Substituting these in $2\bar{b}\sin\bar{\beta} = 2a\sin\alpha$, one obtains (to first order)

$$\delta\beta = (\gamma_x - \delta)\tan\beta , \qquad \gamma_x = f_x\bar{s}/r \quad . \tag{16.72}$$

In terms of these, the MS strain components in (16.68) become

$$\bar{e}_x = \gamma_x, \quad \bar{e}_y = (\delta - \gamma_x s_\beta^2)c_\beta^{-2}, \quad \bar{e}_{xy} = 0 \quad . \tag{16.73}$$

If these are introduced into the relation in (16.57), and minimization with respect to δ and $\delta\beta$ is performed, the result

$$\varepsilon_m = Wl^2\gamma_x^2 r^2(1 - p^2) \tag{16.74}$$

is obtained, which, together with (16.70), yield the phase boundaries

$$r = \bar{s} \pm \left[\frac{|A_3|}{(1 - P)(1 + P)^2}\right]^{\frac{1}{2}} l^{-1} \quad . \tag{16.75}$$

The phase boundaries of the configurations with $\bar{f}_y = 0$, $\bar{f}_d = 0$ and $\bar{f} = 0$ may be obtained analogously [16.26]. The results are illustrated in Fig.16.6 for values of the A's given in (16.55). The results are qualitatively in agreement with expectation; as l decreases (adatom-substrate bond strength increases) misfit accommodation by complete coherency begins to dominate, whereas the boundaries narrow down towards the respective limits defined by $f_i = 0$, as l increases.

In this section only misfit accommodation by coherence has been considered. Furthermore, the estimate of the energy gained from the substrate in attaining coherency is rather crude. The main merits of the analysis are its simplicity, and the useful guidelines it provides [16.26].

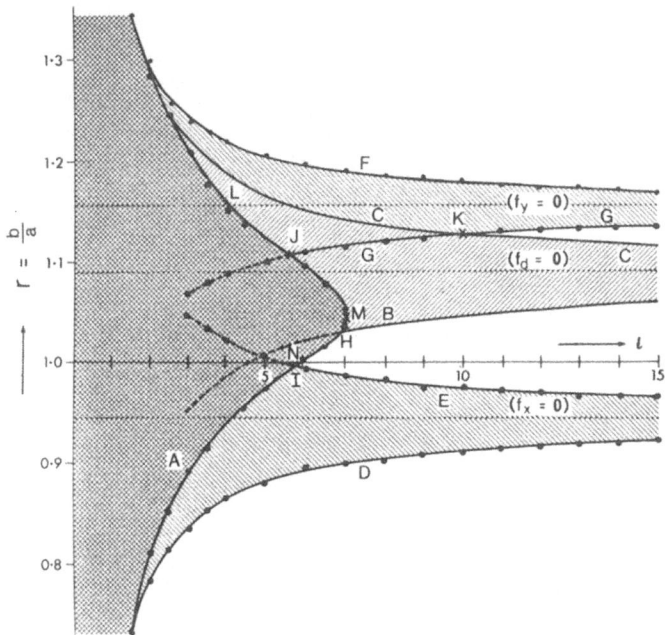

Fig.16.6. "Phase" boundary lines in space (l,r) between regions where transitions to ideal epitaxial configurations are energetically favorable and regions where they are not (taking R = 1/2 and P = 1/3). The region in this figure and the ideal matching are respectively: (i) region (XX) between r axis and Curve A, $\bar{f}$ = 0 (complete coherence); (ii) region (//) between Curves B, C and A, $\bar{f}_d$ = 0; (iii) region (\\) between Curves D, E and H, $\bar{f}_x$ = 0 and (iv) region (\\) between Curves G, C and F, $\bar{f}_y$ = 0. (Van der Merwe [16.26], by permission of Taylor and Francis, Ltd.)

16.3.3.3 Misfit Dislocations (MD) and Misfit Verniers (MV)

In this section the analysis is refined, but also at a price. To make the governing equations tractable, the two-dimensional discrete problem is reduced to an approximate one-dimensional continuous one [16.64-66]. The equations for the equilibrium positions of the adatoms are 2G second-order nonlinear difference equations in two dimensions [16.64,66]. They reduce to two second-order nonlinear differential equations in a 2D continuum by approximating the differences by differentials on the supposition that the atomic displacements vary slowly from atom to atom. In the second approximation, the 2D problem is reduced to a 1D problem by assuming that in a perpendicular direction, the x direction say, the misfit is accommodated by a vernier. In view of the small values of l both simplifying approximations are rather crude.

The governing equation, thus simplified [16.66], is a sine-Gordon type equation

$$\frac{d^2\bar{v}}{d\bar{k}^2} = \frac{\pi}{2\bar{L}_y^2} \sin(2\pi\bar{v}) \quad , \tag{16.76}$$

$$\bar{L}_y^2 = \bar{L}^2\bar{c}^2 = -\frac{31\bar{c}^2\bar{c}^2}{2A_4} \quad ; \quad \bar{k} = \frac{y}{ac} \quad , \quad \bar{v} = \frac{v}{ac} = \frac{2v}{a_y} \quad .$$

The solution of this equation has been expressed in terms of elliptic integrals (K complete and F incomplete of first kind)

$$\frac{\pi\bar{k}}{\kappa\bar{L}y} = \pm F\left[\kappa, \pi\left(\bar{v} \mp \frac{1}{2}\right)\right] \quad ,$$

$$\bar{P}(\kappa) = \frac{2\kappa\bar{L}_y K(\kappa)}{\pi} \quad , \tag{16.77}$$

for MD's of displacement vectors $\Delta v = \frac{1}{2} a_y$ and spaced at "distances" $\Delta\bar{k} = \bar{P}(\kappa)$ apart. The modulus κ of the integrals accordingly determines the MD spacing. The upper and lower sings apply to positive ($\bar{b}_y > a_y$) and negative ($\bar{b}_y < a_y$) MD's, respectively. In such a configuration the MD's accommodate misfit $\bar{f}_y = \pm\bar{P}^{-1}$ in the presence of misfit strain $\bar{e}_y = (\bar{f}_y - f_y)\bar{c}r^{-1}$. The partition of f_y between $\bar{f}_y$ and $\bar{e}_y$ is determined by

$$\frac{2E(\kappa)}{\pi\kappa\bar{L}_y} \mp P^2\bar{f}_y\bar{c} \mp (r - \bar{c})(1 - P^2) = 0 \quad , \tag{16.78}$$

which is the condition for minimum energy, $E(\kappa)$ being a complete elliptic integral of second kind. The critical condition for stability of the coherent ($\kappa = 1$) configuration, i.e., the phase boundary, is given by

$$r^c_{(y)} = \bar{c} \pm \frac{2}{\pi\bar{L}_y(1 - P^2)} \quad . \tag{16.79}$$

The phase boundaries [16.66] associated with the ideal configurations characterized by $\bar{f}_x = 0$, $\bar{f}_d = 0$ and $\bar{f} = 0$ have been similarly obtained; for $\bar{f}_x = 0$ simply replace $\bar{c}$ by $\bar{s}$ in (16.79) and A_4 by A_3 in (16.75). The near identity of the result with (16.75) is clear. The results of these calculations are demonstrated in Fig.16.7, in which

$$\lambda = 1^2(rs_\beta c_\beta)^{-1} \quad . \tag{16.80}$$

The calculations apply to separate semi-ideal cases and do not make provision for the continuous transition from one to another which would have rounded off the corners at intersections of the estimated phase boundaries for complete 2D coherence. The correspondence between the diagrams in Figs. 16.6 and 7 is evident. Quantitatively they differ, though. For example, the

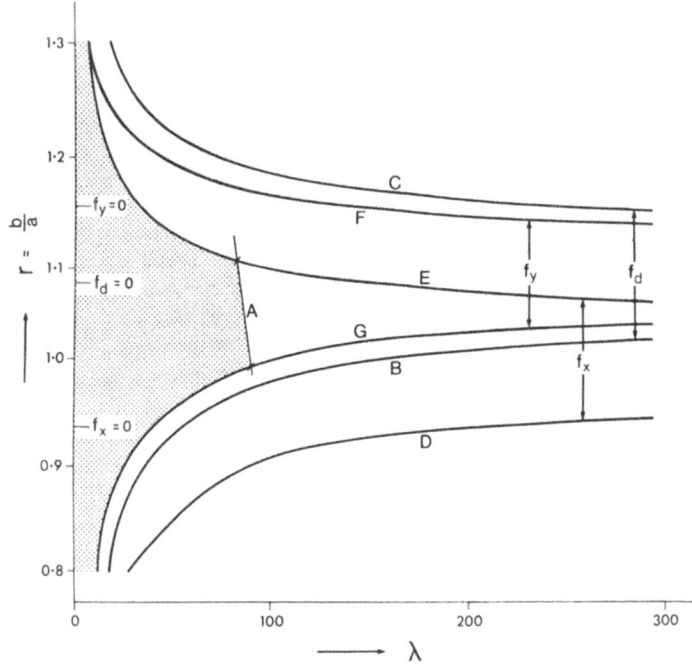

Fig.16.7. "Phase" boundaries in space $r = b/a$, $\lambda = 1^{c}/rs_{\beta}c_{\beta}$ for regions of different coherencies. (i) Complete 2D coherence to left of GAE; (ii) 1D coherence in NW orientation, (a) Curves D and E for $\bar{f}_x = 0$ and (b) Curves G and F for $\bar{f}_y = 0$; (iii) 1D coherence in KS orientation, Curves B and C for $\bar{f}_d = 0$

boundaries for 1D coherence in the KS model enclose those for 1D coherence in the NW mode; at least on the part of the diagram shown in the figure. These discrepancies must be due to the different and drastic approximations made in the two cases.

16.4 Concluding Remarks

The primary goals of studies on epitaxy are fundamental understanding and technological innovation, for the former meaning primarily the prediction of epitaxial orientations, for the latter, crystal perfection, and for both, suitable material combinations. That the orientation under equilibrium conditions is governed mainly by minimum free energy principles, and that the relevant criteria depend strongly on the strength of adatom-substrate and adatom-adatom bonding on matching interfacial symmetries and dimensions, and on growth modes, is undeniable. Although it is possible that one or more of these factors may play on overriding role in a given case, none of them can be singled out as being of sole importance. While a substantial amount of work in predicting azimuthal orientations has been done, the choice of contact plane requires elucidation. Nonequilibrium considerations are much more difficult, but of vital practical importance.

The first step in a theory is to construct a reliable mathematical model. The harmonic approximation for adatom-adatom interaction in a monolayer, in which the stable configuration is identified with the corresponding one in the bulk, is an extreme model, the other extreme being the lattice gas model. Clearly, the reality lies somewhere inbetween, including perhaps indirect electronic influences via the substrate. Electronic influences are possibly also of paramount importance for the interaction between consecutive adatom monolayers, particularly when the adatom-substrate bonding is strong. It has been indicated that this concept may play a significant role in the growth modes, particularly in Stranski-Krastanov growth, i.e., the transition from FM to VW growth. This propagation of the substrate influence should, and probably will, constitute a field of great research interest, both experimentally and theoretically.

The mechanism of intermixing as a means of propagating influences both ways and of smearing out the abrupt transitions between crystals is regarded as a matter of great fundamental and technological importance. Its effects on both growth mode and adhesive strength are desirable features which should be exploited.

The other vital component of the model is the substrate, which is modeled in terms of adatom-substrate interaction competing with the adatom-adatom forces for the configurational order of the adatoms. The most important feature of this interaction is the periodic potential, having the periodicity and symmetry of the substrate. Many important predictions are simply a consequence of this periodicity and symmetry. It has been shown, however, that epitaxial tendencies and modes of misfit accommodation depend vitally on the numerical values of the relevant Fourier coefficients. Very little is known about such coefficients and calculations are at present in progress with the view of obtaining estimates.

Recent work on predicting epitaxial configurations has been reviewed: it is shown that ideal azimuthal orientations are analytically predictable on the basis of the rigid model, in which both overlayer and substrate are assumed rigid for given interfacial lattice symmetries, provided the Fourier coefficients in the adatom-substrate interaction potential are known.

These concepts have been applied with success in the case of epitaxy at {111} fcc - {110} bcc interfaces, and for the first time occurrence of the

Nishiyama-Wassermann and the Kurdjumov-Sachs epitaxial orientations is understood on the basis of sound physical principles. As a profitable by-product, interesting predictions have been obtained regarding the mobility of epitaxial islands. When the geometrical parameters and orientation are close to ideal for epitaxy, motion is prohibited by rather high potential barriers. The barrier heights increase approximately in proportion to adatom-substrate bond strength and contact area. When the parameters and orientations differ significantly from the ideal, the barrier heights are but of the order of the surface migration energy of a single adatom, independent of contact area. Also the dependence of the modes of misfit accommodation on bond strength and mismatch has been estimated for deformable monolayers and the (phase) boundaries between the various modes displayed in two-dimensional interaction-misfit space. Though the estimates are crude they are seen as valuable guidelines for topical theoretical and experimental research.

Acknowledgement. Financial support for research from the South African Council for Scientific and Industrial Research is gratefully acknowledged.

References

16.1 J.L. Robins: Surf. Sci. **86**, 1 (1979)
16.2 J. Villain: Surf. Sci. **97**, 219 (1980)
16.3 E. Bauer: Appl. Surf. Sci. **11/12**, 479 (1982); in *Interfacial Aspects of Phase Transformations*, ed. by B. Mutaftschiev (Reidel, Boston 1982) pp.411-431
16.4 G.H. Olsen, C.J. Nuese, R.T. Smith: J. Appl. Phys. **49**, 5523 (1978)
16.5 C.A.B. Ball, J.H. van der Merwe: In *Dislocations in Solids*, Vol.6, ed. by F.R.N. Nabarro (North Holland, Amsterdam 1983) Chap.27, pp.121-141
16.6 L.L. Chang: In *Handbook of Semiconductors*, Vol.3, ed. by T.S. Moss, S.P. Keller (North Holland, Amsterdam 1980) Chap.4, pp.563-597
16.7 C. Weissmäntl: Thin Solid Films **92**, 55 (1982)
16.8 H.G. Schneider, V. Ruth, Karmany (eds.): *Advances in Epitaxy and Endotaxy* (Elsevier, Amsterdam 1976)
16.9 L. Royer: Bull. Soc. Fr. Mineral. Crystallogr. **51**, 7 (1928)
16.10 J.W. Matthews (ed.): *Epitaxy* (Academic, New York 1975)
16.11 M.I. Bîrjega, N. Popescu-Pogrion, V. Topa: Thin Solid Films **94**, 67 (1982); Rev. Roum. Physique **25**, 697 (1980)
16.12 J.M. Gibson, J.C. Bean, J.M. Poate, R.T. Tung: Thin Solid Films **93**, 99 (1982)
16.13 S. Mader: In *Epitaxial Growth*, ed. by J.W. Matthews (Academic, New York 1975) Part A, Chap.2, pp.29-36
16.14 J.A. Venables, G.L. Price: In *Epitaxial Growth*, ed. by J.W. Matthews (Academic, New York 1975) Part B, Chap.4, pp.382-432
16.15 J. Frenkel, T. Kontorowa: Phys. Z. Sowjetunion **13**, 1 (1938)
16.16 R.W. Vook: Internat. Met. Rev. **27**, 209 (1982)
16.17 P. Bak: Rep. Prog. Phys. **45**, 587 (1982)
16.18 A.D. Novaco, J.P. McTague: Phys. Rev. Lett. **38**, 1286 (1977)

16.19 F.C. Frank, J.H. van der Merwe: Proc. R. Soc. **198**, 205, 206 (1949)
16.20 J.W. Matthews: In *Epitaxial Growth*, ed. by J.W. Matthews (Academic, New York 1975) Part B, Chap.8, pp.560-607
16.21 G. Navascués, P. Tarazona: J. Chem. Phys. **75**, 2441 (1981)
16.22 V.L. Pokrovskii, A.L. Talapov: Zh. Eksp. Teor. Fiz. **78**, 134 (1980)
16.23 J.H. van der Merwe: In *Chemistry and Physics of Solid Surfaces*, Vol. II, ed. by Ralf Vanselow (CRC, Boca Raton 1979) pp.129-153
16.24 M. Schick: Prog. Surf. Sci. **11**, 245 (1981)
16.25 J.H. van der Merwe: J. Appl. Phys. **41**, 4725 (1970)
16.26 J.H. van der Merwe: Philos. Mag. A**45**, 127, 145, 159 (1982)
16.27 J.H. van der Merwe: Proc. Phys. Soc. (London) A**63**, 616 (1950)
16.28 M. Smolett, M. Blackmann: Proc. Phys. Soc. (London) A**64**, 683 (1951)
16.29 P.W. Pashley: In *Epitaxial Growth*, ed. by J.W. Matthews (Academic, New York 1975) Part B, pp.2-24
16.30 E. Bauer: Z. Kristallogr. **110**, 423 (1958)
16.31 E. Bauer, H. Poppa: Thin Solid Films **12**, 167 (1972)
16.32 D. Walton: Philos. Mag. **7**, 1671 (1962)
16.33 M. Volmer, A. Weber: Z. Phys. Chem. **119**, 277 (1926)
16.34 C.A.O. Henning, J.S. Vermaak: Philos. Mag. **22**, 281 (1970)
16.35 J.A. Venables, J. Derrien, A.P. Janssen: Surf. Sci. **95**, 411 (1980)
16.36 W.A. Jesser, D. Kuhlmann-Wilsdorf: Phys. Stat. Sol. **21**, 538 (1967)
16.37 H. Reiss: J. Appl. Phys. **39**, 5045 (1968)
16.38 J.J. Metois: Appl. Phy.. Lett. **29**, 134 (1976)
16.39 N.H. Fletcher, K.W. Lodge: In *Epitaxial Growth*, ed. by J.W. Matthews (Academic, New York 1975) Chap.7, pp.530-557
16.40 J.C. du Plessis, J.H. van der Merwe: Philos. Mag. **11**, 43 (1965)
16.41 G. Honjo, K. Tagayanagi, K. Kobayashi, K. Yagi: Phys. Status Solidi **55**, 353 (1979)
16.42 A. Masson, J.J. Metois, R. Kern: Surf. Sci. **27**, 463 (1971)
16.43 I. Markov, V.D. Karaivanov: Thin Solid Films **61**, 115 (1979); **65**, 361 (1980)
16.44 J.J. Metois, G.D.T. Spiller, J.A. Venables: Philos. Mag. A**46**, 1015 (1982)
16.45 C. Ghezzi, C. Paorici, C. Pelosi, M. Servidori: J. Cryst. Growth **41**, 181 (1977)
16.46 M. Harsdorf, G. Reiners: Thin Solid Films **85**, 267 (1981)
16.47 J.J. Metois, R. Kern: Thin Solid Films **57**, 231 (1979)
16.48 L.C.A. Stoop, J.H. van der Merwe: J. Crystl. Growth **24/25**, 289 (1974)
16.49 R.B. Marcus, W.B. Joyce: Thin Solid Films **10**, 1 (1972)
16.50 L.A. Bruce, H. Jaeger: Philos. Mag. A**38**, 223 (1978)
16.51 J.H. van der Merwe: In *Treatise on Materials Science and Technology*, Vol.2, ed. by H. Herman (Academic, New York 1973) Chap.1, pp.1-103
16.52 J.S. Vermaak, J.H. van der Merwe: Philos. Mag. **10**, 785 (1964); **12**, 463 (1965)
16.53 L.N. Aleksandrov: Surf. Sci. **86**, 144 (1978)
16.54 J.A. van Vechten: J. Vac. Sci. Technol. **14**, 992 (1977)
16.55 W.A. Jesser, J.W. Matthews: Philos. Mag. **15**, 1097 (1967)
16.56 W. Hagen: J. Cryst. Growth **43**, 739 (1978)
16.57 M. Harsdorf: Fortschr. Mineral. **42**, 250 (1966)
16.58 Y. Kawamura, H. Okamoto: J. Appl. Phys. **50**, 4457 (1979)
16.59 G.H. Olsen: J. Cryst. Growth **31**, 223 (1975)
16.60 R. Kaischew, S. Stoyanov, D. Kashchiev: J. Cryst. Growth **52**, 3 (1981)
16.61 I. Markov, R. Kaischew: Krist. Tech. **11**, 685 (1976); Thin Solid Films **32**, 163 (1976)
16.62 H. Mykura: In *Molecular Processes on Solid Surfaces*, ed. by E. Drauglis, R.D. Gretz, R.I. Jaffee (McGraw-Hill, New York 1969) p.135
16.63 K. Takayanagi, K. Yagi, G. Honjo: Thin Solid Films **48**, 137 (1978)
16.64 J.H. van der Merwe: Thin Solid Films **74**, 129 (1980)

16.65 L.C.A. Stoop: Thin Solid Films **94**, 353 (1982)
16.66 L.C.A. Stoop, J.H. van der Merwe: Thin Solid Films **91**, 257 (1982);
   94, 341 (1982); **98**, 65 (1982)
16.67 M. Braun: Private communication
16.68 P.M. Stoop: Private communication
16.69 J.H. van der Merwe: In preparation
16.70 K. Takayanagi, K. Yagi, K. Kobayashi, G. Honjo: J. Cryst. Growth **28**,
   343 (1975)
16.71 K. Yagi, K. Takayanagi, G. Honjo: Thin Solid Films **44**, 121 (1977)
16.72 F.A. Ponce: Appl. Phys. Lett. **41**, 371 (1982)
16.73 L. Bicelli, B. Rivolta: Surf. Technol. **12**, 361 (1981)

17. Angle-Resolved Secondary Ion Mass Spectrometry

N. Winograd

With 13 Figures

17.1 Introduction

The response of a solid to energetic ion bombardment has been of interest to
scientists since the initial observation by *Grove* [17.1] that cathodes in gas
discharge tubes were subject to erosion. The reason for this erosion is that
when a keV ion strikes a solid, there is sufficient momentum transferred to
the target atoms to initiate considerable nuclear motion, Fig.17.1. A fraction of the moving atoms may obtain a component of momentum oriented into the
vacuum such that they are able to overcome the surface binding force and
eject from the solid. Ion/solid interactions of this type are studied today
from a wide variety of perspectives. Some of the particles emerge directly
as positive or negative ions, while others are ejected as either ground or
excited state neutral atoms. The emission process and the associated radiation damage in the solid are referred to as sputtering — an often critized
term [17.2,3] which, however, appears destined to remain with us for some
time.

  In this chapter I shall focus on the measurement of the angular distributions of sputtered ions as an approach to characterize surface structure.
The discussion is divided into three parts. General background information
about the field is given in Sect.17.2. The theoretical developments which
make interpretation of the experimental results possible are given in Sect.
17.3, and finally a number of specific applications are reviewed in Sects.
17.4,5.

17.2 Background Information

Measurement of the intensity of the secondary ions forms the basis of secondary ion mass spectrometry (SIMS). The secondary positive (negative) ion
yield $Y^{+(-)}$ in this experiment is defined as

Fig.17.1a,b. Atoms positions. a) Before the primary ion (the lone sphere above the solid) has struck. b) Consequences of a single ion impact. The two atoms ejecting to the left form a dimer. For graphical clarity, only a selected group of atoms is shown, with arbitrary size

$$Y^+ = R^+ Y \quad , \tag{17.1}$$

where R^+ is the probability that the ejecting neutral species will leave the solid in an ionized state. A similar expression could be written for negative ions. It is possible that R^+ is a function of energy so that

$$Y^+ = \int_0^{E_m} R^+(E) Y(E) dE \quad , \tag{17.2}$$

where $Y(E)dE$ is the differential secondary neutral particle yield, and E_m is the maximum energy that any of the ejected particles have attained. For many systems, the value of Y^+ may be large enough to permit detection of very low concentrations of surface species. Much of the early interest in secondary ion emission in fact stemmed from this potential analytical application. Evaluation of the secondary ion intensity I^+ (counts/s) recorded during bombardment of the sample is more complex than for the neutral species as determined by weight loss, but it is easy to see the potential sensitivity of the approach. For this case, ignoring angular effects [17.4]

$$I^+ = i_0 \gamma_M \beta(M^+) T(E_0,R) \int_{E_1}^{E_2} Y^+(E') dE' \quad , \tag{17.3}$$

404

where i_0 is the flux of primary ions, γ_M is the isotope abundance of element M, $\beta(M^+)$ is the detector efficiency which depends on the mass and velocity of the ion [17.5] and $T(E_0,R)$ is the transmission of the ion optical system which depends upon energy position $E_0 + \Delta E/2$, width $\Delta E = E_2 - E_1$ of the energy window and the mass resolution of the mass spectrometer. For typical values of $i_0 = 10^{-9}$ amps of Ar^+ ions spread over 10^{-3} cm^2 (6×10^{13} ions s^{-1} cm^{-2}), $\gamma_M \beta T = 10^{-3}$, $E_0 = 10$ eV, $\Delta E = 5$ eV and the integral in (17.3) equal to approximately 10^{-2}, $I^+ = 6 \times 10^8$ counts s^{-1}. With a detection limit of ~1 cps, it is clear that measurements in the parts per million (ppm) atomic percent range can be completed. In addition, the sensitivity suffices to detect ions ejected from a very small area.

There are two major difficulties associated with the measurement of secondary ions. First, it turns out to be nearly impossible to obtain quantitative ion yield measurements. This is due to the difficulty in evaluating the product of β and T and to the large variations found in $y^+(E)$ in (17.3) [or R^+ in (17.1)] with the sample matrix. For example, the yield of Ni^+ from Ni is 10^3 times smaller than the yield of Ni^+ from NiO [17.6] under otherwise identical conditions. To understand these results it is necessary to have a theoretical understanding of the mechanism of secondary ion formation. As shown in Sect.17.3, there are a number of possible approaches to explaining the ionization process, although none have yet proven to be very satisfactory.

A second major problem with SIMS as presented above is that although the primary ion current is small, the primary ion flux is extremely large, typically 10 to 100 μamp cm^{-2}, or 10^{14} to 10^{16} incident ion s^{-1}. If we assume there are roughly 10^{15} atoms cm^{-2} on a typical metal surface and the sputtering yield Y is ~1, then the sample is eroded away at the rate of 0.1 to 10 monolayers s^{-1}. The magnitude of the sample damage is clearly quite large and the prospects of performing a chemical analysis on a molecular system appear dim.

During the late 1960s, this problem was ingeniously overcome by *Benninghoven* and co-workers [17.7,8] who simply proposed to expand the size of the primary-ion beam and reduce the total dose of ions to less than ~10^{13} atoms cm^{-2}. By using incident currents of ~10^{-9} to 10^{-10} amps cm^{-2}, they could study the chemical nature of a surface for minutes without altering the surface composition. The erosion rate was effectively reduced to $10^{-5} - 10^{-6}$ monolayer s^{-1}. In addition, since a wide area was being imaged into the mass spectrometer, they were able to enhance sensitivity considerably by utilizing a quadrupole mass filter instead of a magnetic sector [17.9]. This implementation was also important since the energy distribution of the ejected species extends over a range of several eV and quadrupoles are generally insensitive to ion energy. The equipment can easily be made compatible with ultrahigh vacuum (UHV, 10^{-10} Torr) conditions for surface studies.

Molecular cluster ions are commonly observed to be present in static SIMS spectra. It is of interest to see if these species can be used to learn something about the chemistry of a surface. *Benninghoven* [17.10] first proposed that for a number of different metals exposed to oxygen, observed clusters of the type $M_m O_n^{\pm}$ were formed from contiguous atoms in the solid and that the solid structure could therefore be reconstructed by an appropriate analysis of the cluster ion intensities. This idea was further reinforced by the work of *Buhl* and *Preisinger* [17.11] who studied ZnS crystals and found mainly polyatomic ions such as ZnS^+, ZnS^-, Zn_2S^+ and ZnS_2^-. Other polyatomic ions such as Zn_2^+ or S_2^- which were not originally present in the crystal were observed only in very low intensities. Meanwhile, it became clear that static SIMS could become a useful tool for the detection of many nonvolatile organic and inorganic molecules. *Benninghoven* et al. first showed that when a thin film of an amino acid on a silver foil was bombarded with Ar^+ ions, a cluster ion equal to the molecular weight of the organic compound with a proton attached was ejected from the solid intact [17.12]. Since that discovery many groups have exploited this observation and now such delicate molecules as proteins with molecular weights above 2000 amu can be observed.

The generality of this assessment has been questioned both theoretically and experimentally. For example, *Rabalais* and co-workers found that CsCl diluted into KCl still produced significant numbers of Cs_2^+ cluster ions and that recombination of ejecting monomers could occur over several hundred angstroms [17.13]. Thus, it appears that under certain circumstances rearrangement is observed, but under others the clusters are ejected from the sample intact. We shall see below how these observations can be collated to develop a more unified theory of the cluster emission process.

Of further interest was the proposal that static SIMS could be utilized to characterize monolayers formed during chemisorption of small molecules onto solid surfaces. For example, *Barber* and co-workers studied chemisorption of CO on clean polycrystalline Ni surfaces [17.14]. They found cluster ions of the type Ni_2^+, $NiCO^+$ but none of the type NiC^+, NiO^+ or Ni_2O^+, indicating that CO adsorbs in a molecular state on Ni. Further, by comparison to vibrational spectra for this system they were able to make a correlation between the Ni_2CO^+ ion and bridge bonded CO, and the $NiCO^+$ ion and linear bonded CO. Their rationale for this observation was that the molecular cluster was ejected as a fragment of the surface without rearrangement. Although the interpretation of many of the above results has been questioned in recent years, there is no doubt that the static SIMS approach coupled with the study of the cluster ions offers a unique method of surface analysis.

With the possible advantages of SIMS clearly in mind, there should still remain a nagging doubt about its ability to reveal something new about surfaces. Sputtering phenomena are very complex and theoretical models are hard to come by. It has been the objective of our group over the last few years to concentrate on an analysis of the angular distributions of secondary ions. There have been two primary reasons for this objective. First, it was hoped that by controlling as many variables as possible, i.e., angle, that more detailed knowledge could be obtained about the sputtering process itself. Secondly, early work by *Wehner* [17.15] clearly showed that the angular distributions of neutral atoms reflected the symmetry of the crystal surface. In his experiment, he placed a flat plate collector an arbitrary distance above a Cu single crystal which produced the pattern shown in Fig.17.2. Could this approach, then, coupled with static SIMS, provide structural information about single-crystal surfaces? In the remaining sections, I shall outline the theoretical advances that encouraged us to proceed to answer this question, with a number of representative examples of how angle-resolved SIMS can provide surface structural information.

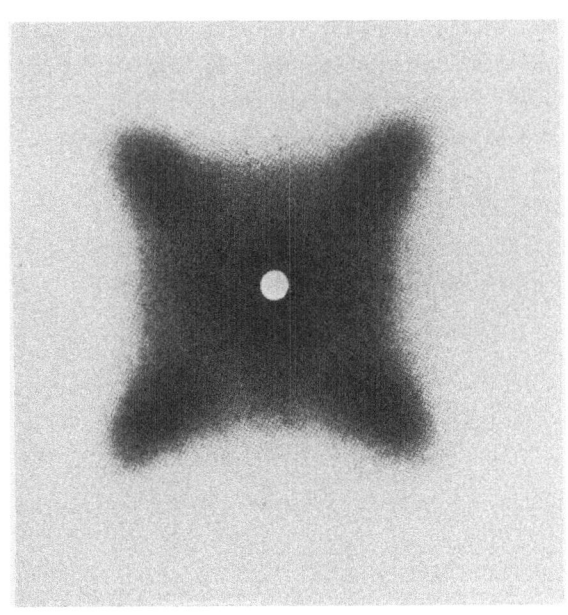

Fig.17.2. Angular distributions of atoms ejected from Cu{001} [17.2]

17.3 Theory of Ion/Solid Interactions

There are essentially two aspects to the successful interpretation of SIMS spectra, and more specifically to angle-resolved SIMS spectra. First, we must be able to determine how the nuclear positions of the atoms or molecules in the sample change with time in response to the impact of the primary ion. Secondly, we must have a quantitative description of the ionization processes which cause certain species to leave the sample surface as ions. In this section, a possible approach to better understanding the SIMS phenomena is presented, along with the basic concepts needed to develop angle-resolved SIMS as a surface characterization tool.

17.3.1 Momentum Dissipation and Classical Dynamics Calculations

The impact of the primary ion induces considerable motion of the atoms in target crystals. Although there have been a number of attempts to develop equations which provide the number of particles that eject from the surface [17.16], none has been generally applicable. It has been our opinion that a molecular dynamics procedure applied to a large ensemble of atoms to compute actual nuclear positions as a function of time can provide a valuable theoretical basis for many experiments. Classical dynamics calculations have, of course, been very successful in explaining trajectories in atom-diatom scattering [17.17], properties of liquids [17.18] and even the solvation of large molecules like dipeptides [17.19]. For describing the sputtering process this approach has the distinct advantages of utilizing many fewer approximations than required for the analytical theories. On the other hand, no simple equation falls out of the calculations, although important concepts may emerge from the resulting numbers. In general, these calculations require considerable computer time, but with the recent surge in computer efficiency, it would appear this is becoming less of a difficulty.

The computation of classical trajectories using the molecular dynamics procedure rests on Newton's equation of motion. For a particle i of mass m_i, the equation of motion is

$$m_i \ddot{r}_i = F_i \quad , \tag{17.4}$$

where r_i is the position vector, $\ddot{r}_i$ is the acceleration and F_i the force. For N particles there will be 3N coupled second-order differential equations that need to be solved. Computationally, it is more convenient to utilize Hamilton's form of the equations of motion as

$$m_i \dot{v}_i = F_i \tag{17.5}$$

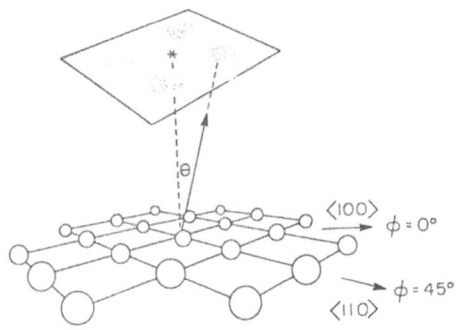

Fig.17.3. Coordinate system used
in determining angular distributions

$$r_i = v_i \quad , \tag{17.6}$$

where v_i is the particle's velocity, since there are now 6N coupled first-order differential equations. These equations provide the time dependence of the position and the velocity and are easier to solve than the second-order equation. For an arbitrary value of F_i the atomic motion can be determined using a finite difference algorithm [17.20].

In all calculations to date, the initial coordinates of the atoms are set at their equilibrium position with zero initial velocity. As shown in Fig. 17.3, the primary ion is given a specific kinetic energy E, with polar angle θ and azimuthal angle ϕ such that

$$v = (2E/m)^{\frac{1}{2}} = (v_x^2 + v_y^2 + v_z^2)^{\frac{1}{2}} \tag{17.7}$$

$$v_z = -v \cos\theta \tag{17.8}$$

$$v_x = v \sin\theta \cos\phi \tag{17.9}$$

$$v_y = v \sin\theta \sin\phi \quad . \tag{17.10}$$

The force is calculated from the gradient of the potential energy V as

$$F_i = - \nabla_i V(r_1, r_2 \ldots r_n) \quad . \tag{17.11}$$

For a many-body solid, we would need a multidimensional potential surface that accounted for all possible positions of the particles. With our present knowledge of the solid state, this potential surface is unknown but is usually approximated using a sum of pairwise additive potentials as

$$V = \sum_{i=1}^{N-1} \sum_{j>i}^{N} V_{ij}(r_{ij}) \quad , \tag{17.12}$$

where r_{ij} is the distance between particle i and j.

The subsequent atomic motion due to the primary ion impact is computed for enough time so that no particles have sufficient kinetic energy to escape the solid. For a metal such as copper, we have found that after the most energetic particle has less than 2 eV of kinetic energy, integration for longer times does not produce any more ejected particles. Approximately 200 integration steps requiring 200 fs are typically required to achieve this condition.

The major difficulty with this model, or any model, of the ion/solid interaction is that we have only limited knowledge of the forces acting between the atoms. For a single-component metal surface such as Cu, the Cu-Cu interaction is often represented using the interaction potential suggested by *Harrison* [17.21]. This potential utilizes a Born-Mayer exponential function to approximate the high-energy collisions ($\overset{\sim}{>}$ 20 eV) and a Morse potential with parameters determined by *Anderman* [17.22] to approximate the low-energy ($\overset{\sim}{<}$ 5 eV) collisions. A cubic spline is used to connect the two parts.

There are several approaches to describe the interaction between the primary ion and the solid. An accurate description of this interaction is important only at fairly high energies since most of the momentum is imparted to the solid after only a few collisions. Probably the most popular form of this interaction is the Moliere potential [17.23]

$$V_{ij} = Z_1 Z_2 e^2 / r_{ij} \sum_{n=1}^{3} \alpha_n \exp(-B_n r_{ij}/a) \quad , \tag{17.13}$$

where $Z_1 e$, $Z_2 e$ are the nuclear charges of the projectile and target atoms, a is a screening length, $\alpha = (0.35, 0.55, 0.10)$ and $B = (0.3, 1.2, 6.0)$. This sum of three exponentials is a fit to the Thomas-Fermi potential. Values of the screening length are calculated from the Firsov formula

$$a = (9\pi^2/128)^{1/3} a_0 (Z_1^{1/2} + Z_2^{1/2})^{-2/3} \quad , \tag{17.14}$$

where a_0 is the Bohr radius. It has been proposed that a value of a which is 20% of the value calculated in (17.14), gives slightly better agreement with experiment.

Due to the uncertainty concerning the accuracy of these potentials, it is important to restrict their application to situations where the results are not terribly sensitive to the potential parameters. The angular distribution of particles ejected from single-crystal surfaces represents such a case [17.24]. The calculation of yields, on the other hand, is very sensitive to the chosen depth of the Morse potential well [17.25].

410

17.3.2 Secondary Ion Formation

To correlate the measurement of secondary ion intensity with the calculated observables of the neutrals using the classical dynamics scheme, it is important to understand as much as possible about the ionization mechanism and to be aware of differences in the trajectories of charged and uncharged species. The theory of ionization in SIMS has been the subject of intense debate over the past decade. The difficulty began when it was observed that there was an empirical correlation between log R^+ and the difference between the ionization potential (or electron affinity for negative ions) and the surface work function (IP - ϕ) for a number of elements [17.26]. This observation was confirmed quantitatively by *Yu* [17.27], who measured log R^+ and ϕ simultaneously during deposition of submonolayers of Cs on Mo. The result was first interpreted by *Andersen* and *Hinthorne* [17.26] in terms of a local thermal equilibrium (LTE) model, which treated the region about the impact point as a high-temperature plasma of ions, electrons, and neutral species. Although the resulting equation has the correct form, it has not been possible to see the physical significance of such a model, especially since dynamical calculations clearly show that only a few collisions lead to particle ejection and that there is nothing approaching a thermal environment in this region.

A more satisfying quantum-mechanical approach is under development. In this model the electronic levels of the ejecting atom are allowed to mix with the manifold of states within the solid. The value of R^+ is then determined by solving the time-dependent Schrödinger equation, using an appropriate coupling interaction between the species leaving and the solid. *Blandin* et al. [17.28] and *Norskov* and *Lundqvist* [17.29] have generalized this model by assuming a continuum of levels in the solid and allowing for the tunneling of electrons from occupied states below the Fermi level. The expression for R^+ is

$$R^+ = 2/\pi\{\exp[- C_1\pi(IP - \phi)/\hbar\lambda v]\} \quad , \tag{17.15}$$

where $\hbar$ is Planck's constant divided by 2π, λ is a coupling parameter, v is the particle velocity, and C_1 is a constant and is of the same form as suggested by the LTE model.

In a major advance, *Sroubek* et al. [17.30,31] developed a similar model that allows for nuclear motion within the solid. Although their model is similar to that described above, they find a much weaker dependence of R^+ on velocity. This prediction has been tested in two recent experiments. *Yu* [17.32] found that for TiO_2 substrates the velocity dependence of R^+ followed (17.1) at oxygen ion energies greater than ~15 eV, but that R^+ was nearly independent of velocity at very low kinetic energies. This observation has been attributed

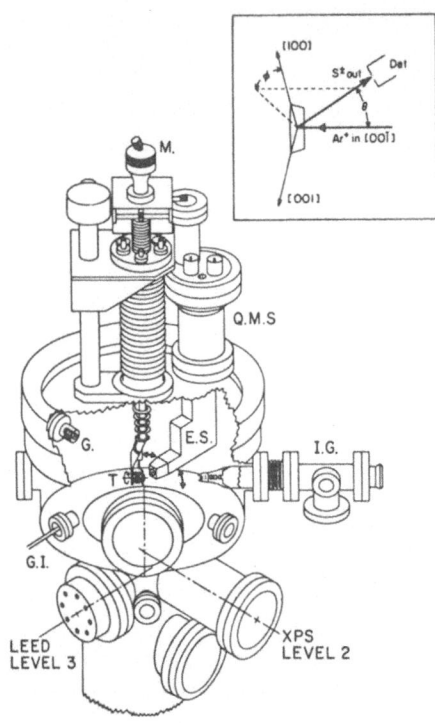

Fig.17.4. Schematic view of spectro-meter. The components illustrated include M, crystal manipulator; Q.M.S., quadrupole mass spectrometer; I.G., primary ion source; E.S., energy spec-trometer; G., Bayard-Alpert gauge; T, crystal target; and G.I., gas in-let. Auxiliary components are omitted for clarity. The SIMS experimental geometry and coordinate system are defined in the inset [17.37]

to surface binding energy effects [17.33,34]. This model also predicts that there is only a very weak dependence of ionization probability on azimuthal takeoff angle [17.33].

Gibbs et al. [17.35,36] measured the polar angle distributions of Ni^+ ejected from a CO-covered Ni{001} surface. The measurements were performed using a new SIMS apparatus aimed at determining ion yields at all θ and ϕ ejection angles for a normally incident beam [17.37]. As shown in Fig.17.4 this measurement was carried out by utilizing a quadrupole mass spectrometer that could be rotated under UHV conditions with respect to the incident ion beam. In addition, a medium resolution energy selector consisting of a 90° spherical sector enabled examining how the angular distributions changed with secondary ion energy. The idea behind designing this apparatus was that the ion trajectories could be directly compared to those calculated for the neu-trals using the molecular dynamics treatment. Good agreement with calculated values could be obtained for the Ni^+ ion yield if a simple image force correc-tion was included. The force arises since as the particle becomes an ion, it will experience a potential

$$E_{image} = e^2/4z = 3.6 \text{ eV}/a_0 \quad , \tag{17.16}$$

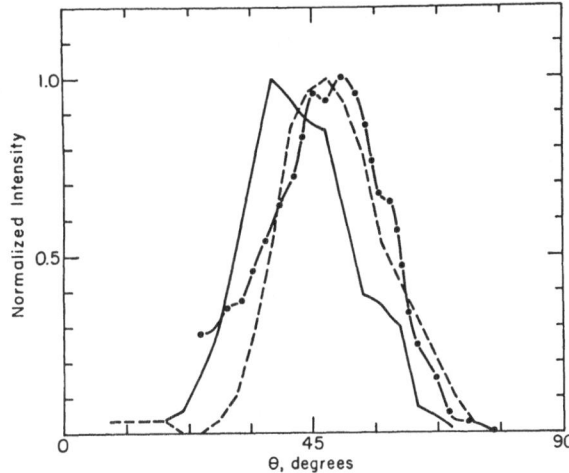

Fig.17.5. Polar angle distributions for Ni ejected from Ni{001}c(2 × 2) CO. The upper set of curves is recorded for a secondary ion energy of 7 ± 2 eV while the lower set of curves is taken at 22 ± 2 eV [17.36]

where a_0 is the height in Å of the particle above the jellium step-edge at the instant of ionization. The force is expected to influence ions near metal surfaces due to the polarization of the metallic electrons and will act to bend the ions closer to the surface plane. The consequences of this correction for the CO/Ni{001} system are shown in Fig.7.5. The observation implies that R^+ is only weakly dependent on kinetic energy. Thus it appears that neglect of ionization processes in the classical dynamical calculations should not be a serious problem when attempting to describe the basic ejection mechanisms of the ions observed in SIMS.

17.3.3 Cluster Formation

It is intriguing to speculate about the origin of the molecular cluster ions observed in SIMS. If they arise from contiguous surface atoms, then their presence could provide key information regarding the local atomic structure of complex surfaces such as alloys and supported metal catalysts. They may also yield information regarding surface bonding geometries. The Ni_2CO^+ cluster, for example, has been proposed to originate from a bridge-bonded CO complex on Ni, while the $NiCO^+$ cluster has been associated with a singly or linearly bonded Ni-CO complex [17.20].

The classical dynamics model is quite helpful in this regard. After termination of a calculation, the particles that have ejected may be tested for possible multimer formation [17.38,39]. To check for the formation of multimers, one merely computes the relative kinetic energy E_r, plus the potential energy V_{ij}, for all pairs of ejected atoms i and j. If the total energy of the dimer

$$E_{tot}^{dimer} = E_r^{dimer} + V_{ij} \qquad\qquad (17.17)$$

is negative, then the tested dimer is considered to be bound. Further, it is often found that several bound dimers are formed above the surface. If some of these have common atoms, i.e., if the dimers are linked or overlapping, then higher multimers may possibly exist. If this condition is found, $E_{tot}^{cluster}$ where

$$E_{tot}^{cluster} = E_r^{cluster} + \sum_{i=1}^{n-1} \sum_{j<i}^{n} V_{ij} \qquad\qquad (17.18)$$

with n being the number of atoms in the cluster, is recalculated for all the atoms in the linkage. As in the dimer analysis, if $E_{tot}^{cluster}$ for the atoms in the linkages is less than zero, then the atoms are considered to be a cluster.

There are other possible definitions of what constitutes a cluster. The requirement that each atom in the multimer be bound to another atom may be an overly stringent requirement for cluster stability. Rigorously, any collection of n atoms with $E_{tot}^{cluster} < 0$ is considered temporarily bound. Since, as we shall see, there are uncertainties in V_{ij}, it is prudent to use the conservative definition of a cluster. Further, for n > 2 it is possible that the cluster will decompose before reaching the detector. For the higher clusters, then, the model really only tests if there are significant numbers of ejecting atoms that are in a favorable spatial position with low enough relative kinetic energy to experience binding interactions.

In general, these calculations tell us that there are three basic mechanisms of cluster formation. First, for clean metals or metals covered with atomic adsorbates, the ejected atoms can interact with each other in the near-surface region above the crystal to form a cluster via a recombination type of process [17.39-41]. This description would apply to clusters of metal atoms and of metal-oxygen clusters of the type M_nO_m observed in many types of SIMS experiments. For this case the atoms in the cluster do not need to arise from contiguous sites on the surface, although we do find that in the absence of long-range ionic forces most of them originate from a circular region of radius ~ 5 Å. This rearrangement, however, complicates any straightforward deduction of the surface structure from the composition of the observed clusters. A second type of cluster emission involves molecular adsorbates like CO adsorbed onto Ni. Here, the CO bond strength is ~ 11 eV, but the interaction with the surface is only about 1.3 eV. Our calculations [17.42] tell us that this energy difference is sufficient to allow CO to eject molecularly, although we do find that $\sim 15\%$ of these molecules can be dissociated by the ion beam or by energetic metal atoms. Clearly, for these molecular systems, it is easy to infer the original atomic configurations of the molecule and to determine the surface chemical state. If CO were dissociated into oxygen and carbon atoms, for example, our calculations suggest that the amount of CO observed should drop dramatically. This type of process undoubtedly applies to the adsorption of organic molecules on surfaces, since the strong carbon framework can soak up excess energy from violent collisions [17.43]. The final mechanism for cluster ejection is essentially a hybrid mechanism between the first two. For CO on Ni, again we find that the observed NiCO and Ni_2CO clusters form by recombining ejecting Ni atoms with ejecting CO molecules. There is apparently no direct relationship between these moieties and linear and bridge-bond surface states. A similar mechanism ought to apply to the formation of cationized organic species. The organic molecule ejects intact, but interacts with an ejecting metal ion to form a new cluster species [17.43].

17.4 Angular Distributions from Clean Single-Crystal Surfaces

Using the general experimental and theoretical approach described in the previous sections, it is now our goal to see what type of structure-sensitive information exists in the angular distributions of the secondary ions. As Wehner showed many years ago, the distributions of the neutrals are highly anisotropic and very clearly reflect the surface symmetry. There have been many attempts to explain these distributions. One such explanation is that the ejection occurs along close-packed lattice directions which extend deep within the crystal [17.44]. This idea conveniently explained the peaks in the angular distributions but required that there be quite a bit of long-range order in the solid even during the impact event. As shown in Fig.17.1, that requirement seems a bit hard to swallow. Although controversy existed concerning these "focusons" for many years, the molecular dynamics calculations of Harrison clearly showed that the ejection was dominated by near surface collisions rather than those from beneath the surface [17.21].

17.4.1 Ni^+ Ion Angular Distributions from $Ni\{001\}$

The results of calculations are displayed schematically in Fig.17.3, where a {001} crystal face is given as an example. Here, each atom's ultimate fate is plotted as a point on a plate high over the solid. Atoms that are ejected perpendicular to the surface ($\theta = 0°$) are plotted in the center of the plate.

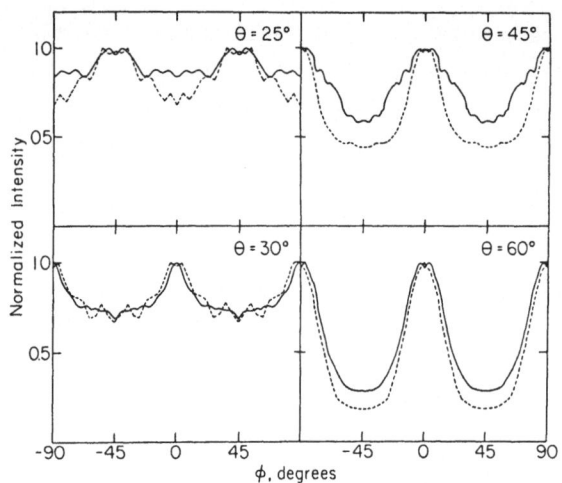

The molecular dynamics calculations yield a clear picture of the scattering mechanisms that give rise to these angular anisotropies, particularly for the higher kinetic energy atoms. Most of the ejected particles arise from within two or three lattice spacings from the impact point and suffer only a few scattering events. The spacings between the surface atoms exert a strong directional effect during ejection. Note that most particles are ejected along $\phi = 0°$, since there are no atoms in the surface to block their path. The nearest-neighbor atom along $\phi = 45°$ inhibits ejection in this direction. It is possible, using the apparatus shown in Fig.17.4, to compare the measured angular distributions of secondary ions to the calculated distributions for a clean Ni{001} single-crystal surface. The results of this comparison are shown in Fig.17.6 [17.36]. Each panel represents an azimuthal angle scan at a particular polar angle. The calculated curves have been corrected for the presence of an image force which, as discussed earlier, tends to bend the secondary ions toward the surface plane. The agreement between the two curves is excellent under all conditions. Note that in accord with the schematic presentation in Fig.17.3, the secondary ion intensity maximizes at $\phi = 0°$ and minimizes at $\phi = 45°$ for $\theta \geq 45°$. Thus it appears that in this simple situation, ion angular distributions behave similarly to the neutrals and are well predicted by theory.

17.4.2 Ni$_2^+$ Ion Angular Distributions from Ni{001}

A real advantage of the static SIMS method over other surface analysis techniques is that molecular cluster ions may be produced which are characteristic of the surface chemistry. It would be most interesting, then, to be able to

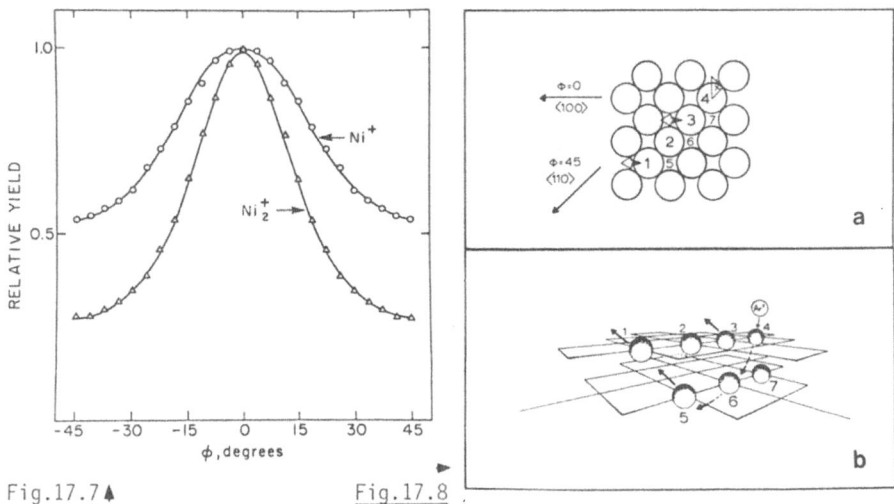

Fig.17.7▲   Fig.17.8

Fig.17.7. Experimental angular distributions of Ni^+ and Ni_2^+ ejected from Ni {001} at a polar angle of $45° \pm 5°$. The center-of-mass kinetic energy of the particles is between 10 and 50 eV. Both curves are fourfold averages of the original data. The incident Ar^+ ion has 2 keV of kinetic energy and is at normal incidence. The solid is at room temperature. The peak counts are ~900 and ~500 counts/s for the Ni^+ and Ni^+_2 distributions, respectively. The [100] azimuthal directions correspond to $\phi = 0$ [17.45]

Fig.17.8a,b. Mechanism of formation of the Ni_2 dimer which preferentially ejects in the [100] directions, contributing the majority of intensity to the peak in the angular distribution. a) Ni 001 showing the surface arrangements of atoms. The numbers are labels while the X denotes the Ar^+ ion impact point for the mechanism shown in Fig.17.8b. Atoms 1 and 3 eject as indicated by the arrows forming a dimer, which is preferentially moving in a [100] direction. b) Three-dimensional representation of a Ni_2 dimer formation process. The thin grid lines are drawn between the nearest-neighbor Ni atoms in a given layer. For graphical clarity, only the atoms directly involved in the mechanism are shown [17.45]

obtain the angular distribution of these species to see if they too contain information about surface structure. As shown in Fig.17.7, the experimental angular distributions for Ni_2^+ ions actually exhibit a sharper azimuthal anisotropy than the Ni^+ ions [17.45]. This result is also observed in the classical dynamics calculations [17.45].

Of particular interest in this case is that fact that the mechanisms giving rise to this increased anisotropy can be ascertained from the theory. As it turns out, most of the dimers formed at $\phi = 0°$ and $\theta = 45°$ originate from a similar set of collision sequences as illustrated in Fig.17.8. The Ar^+ ion strikes target surface atom 4, initiating motion in the solid that eventually ejects atoms 1 and 3 into the vacuum. Both of these atoms are channeled through the fourfold holes in the $\phi = 0°$ direction, are moving parallel to each other,

and are fairly close together. Note that the two atoms that form the dimer
do not originate from nearest-neighbor sites on the surface.

There is an important ramification of the concept that the dimers that
give rise to the maxima in the angular distribution are formed primarily
from constituent atoms whose original relative location on the surface is
known. If this result were extrapolated to alloy surfaces such as $Ni_3Fe\{111\}$,
the relative placement of the alloy components on the surface would be determined. For example, for the $Ni_3Fe\{111\}$ spectra there should be no nearest-
neighbor Fe atoms on a perfect $\{111\}$ alloy surface, yet an Fe_2^+ peak is observed [17.46].

17.5 Angular Distributions from Adsorbate–Covered Surfaces

The channeling phenomena observed from clean surfaces should also be found
in more complex systems such as metals covered with a chemisorbed layer. For
these cases, there are various ways in which one might envisage the angle to
be important. Examples of azimuthal anisotropies have already been seen for
clean metals where surface channeling and blocking give rise to the observed
effect. This situation should also apply to adsorbate-covered surfaces. Other
possibilities include the study of anisotropies in the polar angle distributions as well as in the yield of particles due to changes in the angle of
incidence of the primary ion.

Considerable progress in quantitatively describing the ejection of chemisorbed atoms and molecules from metals has been made using molecular dynamics
calculations. The main difficulty in describing any situation like this is to
develop appropriate interaction potentials which describe the scattering
events. Since little is known about these potentials, early calculations have
utilized pairwise additive potentials for adsorbates which have the same form
as for the substrate, but with different mass. The exact form of the potential
is not as critical as the atomic placement of the adsorbate atom. Thus, in
the calculation, the geometry and coverage of the adsorbate may be varied
over a wide range to test how these quantities influence ejection mechanisms
and ultimately the angular distributions. In this section, examples of how
several different experimental configurations can be utilized will be reviewed.

17.5.1 Atomic Adsorbates

The first application of angle-resolved SIMS to determine the surface structure of chemisorbed layers was for oxygen adsorbed on the $\{001\}$ face of Cu
[17.47]. In this situation, the oxygen overlayer forms a $c(2 \times 2)$ structure

as determined by LEED. Classical dynamics calculations indicate that the oxygen should be ejected in the $\phi = 0°$ direction if it is originally bonded above the copper atom, because it is directly in the path of the ejected substrate species. However, if the oxygen is in a hole site, bonded to four substrate atoms, its predicted angle of ejection is $\phi = 45°$. Experimental studies have confirmed that the oxygen resides in a fourfold bridge site because it is ejected in the $\phi = 45°$ direction [17.47].

There are a number of complications associated with this simple interpretation. First, the magnitude of the azimuthal anisotropies are dependent upon the kinetic energy of the desorbing ion. For the very low energy particles, there has been sufficient damage to the crystal structure near the impact point of the primary ion so that the channeling mechanisms are no longer operative. On the other hand, at higher kinetic energies, say greater than 10 eV, the desorbing ion leaves the surface early in the collision cascade while there is still considerable order in the crystal. The channeling mechanisms are much stronger and the angular anisotropies are larger.

A second complication involves the determination of the height of the adsorbate atom above the surface plane. Calculations have been performed where this bond distance has been varied over several angstroms to find the best fit with experiment [17.47,48]. These studies have also shown that the polar angle distribution is sensitive to the effective size of the adsorbed atom. Thus, it is important to know more about the scattering potential parameters if this distance is to be determined accurately. It appears, however, that the type of adsorption site may be determined in a reasonably straightforward manner.

17.5.2 Adsorption of CO on Ni{001}

The response of a surface to ion bombardment covered with a molecularly adsorbed species is mechanistically distinct from the atomic adsorbate case. For CO on Ni{001}, for example, the strong C-O bond of 11.1 eV and the weak Ni-CO bond of 1.3 eV allow the CO molecule to leave the surface without fragmentation. In the experimental studies, the main peaks in the SIMS spectra for a Ni{001} surface exposed to a saturation coverage of CO are Ni^+, Ni_2^+, Ni_3^+, $NiCO^+$, Ni_2CO^+, and Ni_3CO^+. All ions show a smooth increase in intensity with CO adsorption and reach saturation after 2-L CO exposure (0.5 monolayer coverage). The yields of C^+, O^+, NiC^+ and NiO^+ are all less than 0.01 of the Ni^+ intensity.

The classical dynamics treatment of CO on Ni{001} yields results in qualitative agreement with these findings. Approximately 80% of the CO molecules that eject are found to eject intact, without rearrangement. The formation of NiCO and Ni$_2$CO clusters has been observed over the surface via reactions of Ni atoms and CO molecules. No evidence has been found for NiC or NiO clusters in the calculations. The ion-bombardment approach, then, is a very sensitive probe for distinguishing between molecular and dissociative adsorption processes.

A number of workers have attempted to identify structural relationships found using other techniques such as LEED and vibrational spectroscopy to cluster yields in SIMS, e.g., the correlation of Ni$_2$CO$^+$ to bridge-bonded CO and NiCO$^+$ to linear-bonded CO. As it happens, the calculations clearly show that the mechanism of cluster formation is not consistent with this picture since the clusters form over the surface via atomic collisions. Furthermore, recent combined LEED/SIMS results indicate that the cluster ion yields are not directly related to the adsorbate/substrate geometry [17.49]. The c(2 × 2) structure of CO on Ni{001} with all the molecules in the atop site gave the same Ni$_2$CO$^+$/NiCO$^+$ ratio as the compressed hexagonal LEED structure which must have both atop and bridge-bonded CO molecules.

On the other hand, it is clear that angular distributions for atomic adsorbates are very sensitive to the surface structure so it is not unreasonable to anticipate similar effects for the Ni/CO system. Extensive calculations using the molecular dynamics procedure [17.36] have been completed for the atop and twofold bridge-bonding configurations but statistical considerations have restricted the analysis to only the Ni atoms. As shown in Fig. 17.9, when the CO is in the atop geometry, the calculated Ni distributions

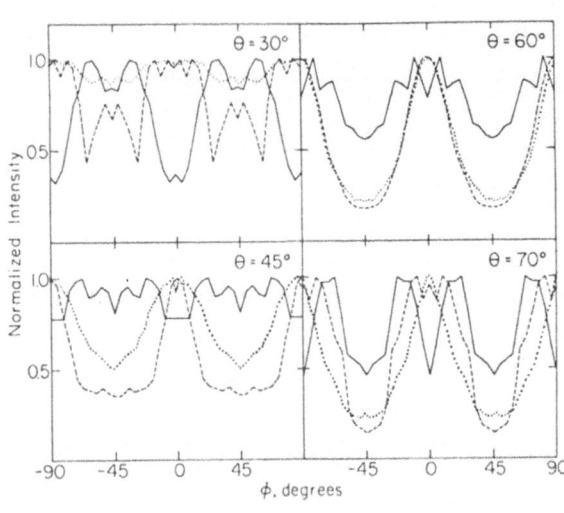

<u>Fig.17.9.</u> Predicted azimuthal dependence of the Ni$^+$ ion yield for Ni{001}c(2 × 2)-CO for CO adsorbed in atop (----) and twofold bridge sites (——). Only those particles with kinetic energies of 3 ± 3 eV were counted. Experimental points (····)

peak along azimuthal directions which are similar to the clean surface. For twofold bridges, however, the CO overlayer tends to scatter the ejecting Ni atoms randomly producing a very different pattern. The predictions for the atop bonding geometry, when corrected for the presence of the image force, are in quite good agreement with experiment, and are consistent with the wide range of other experimental data available for the system.

17.5.3 Adsorption of CO on Ni{7911}

Since the azimuthal angle distributions are sensitive to subtle differences between surface structures, it is of interest to examine the role of larger surface irregularities such as surface steps on the measured quantities. For example, suppose the orientation of the primary ion beam in the SIMS experiment is fixed at different azimuthal angles with respect to the step edge. If the ejection process is structure sensitive, then changes in yield and cluster formation probabilities should be observed as the ion bombards "up" or "down" the steps. In addition, the desorption of chemisorbed molecules should be influenced by their proximity to the step edge.

Carbon monoxide chemisorption on Ni{7911} represents an interesting case with which to check these concepts since comparable studies have been performed on Ni{001} and Ni{111} and since a number of other experimental methods have been applied to this system. Electron energy loss studies performed at 150 K suggest that the initial adsorption occurs at threefold and twofold bridge sites along the step edge. Beyond this point, the CO molecules begin to occupy terrace sites [17.50]. Thus, the low-temperature adsorption of CO on Ni{7911} presents a realm of interesting structural phases which should be sensitive to the azimuthal angle of incidence of the primary-ion beam.

The experimental results for the NiCO$^+$ ion yield as a function of angle is illustrated for this system in Fig.17.10 and the angles are defined in Fig.17.11. Note that the cluster ion yields are higher at $\phi = 180^\circ$ than at $\phi = 0^\circ$, with the most significant variations occurring at intermediate angles. At 0.2 L exposure, the NiCO$^+$ ion signal shows a broad peak which appears at $\phi = 115^\circ$. This peak shifts slightly to 105° and sharpens somewhat at an exposure of 0.4 L. By 0.6 L exposure the peak has become very intense and is only 10° wide, centered at $\phi = 110^\circ$. As the CO coverage increases this peak becomes very broad. At the saturation exposure of 2.8 L the NiCO$^+$ ion intensity displays a broad maximum between $\phi = 40^\circ$ and $\phi = 160^\circ$. An exposure of 0.6 L corresponds almost exactly to the exposure at which the EELS results indicate that all the CO molecules were bound to adsorption sites near the step edge

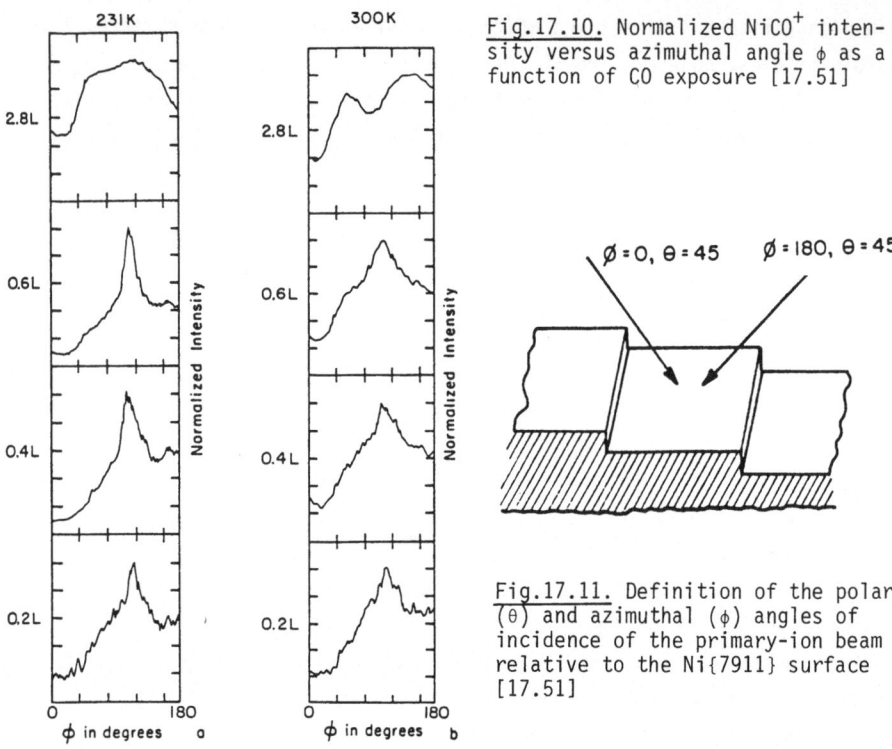

Fig.17.10. Normalized NiCO+ intensity versus azimuthal angle ϕ as a function of CO exposure [17.51]

Fig.17.11. Definition of the polar (θ) and azimuthal (ϕ) angles of incidence of the primary-ion beam relative to the Ni{7911} surface [17.51]

and that all the edge sites were occupied [17.50]. Apparently, the specific bonding site of the CO next to the step edge is responsible for the sharp peak in the NiCO+ ion signal at $\phi = 110°$. At saturation, the peak loses this definition completely presumably since the CO molecules occupy several sites. At CO exposures performed at room temperature the azimuthal plots do not exhibit such sharp features, as illustrated in Fig.17.10. Calculations performed for the twofold bridge step-edge adsorption geometry corresponding to the 0.6 L exposure point successfully reproduce the sharp feature at $\phi = 120°$, although it has not yet been possible to identify the specific collision mechanisms that cause it to occur [17.51].

17.5.4 Angle-Resolved SIMS Studies of Organic Monolayers

We have seen how the angular distributions reflect the bonding geometry of adsorbates through analysis of the azimuthal anisotropies and by varying the angle of incidence of the primary ion. The next possibility is to see if there are channeling mechanisms which act perpendicularly to the surface and which manifest themselves in the polar angle distributions. The model systems which illustrate this effect are benzene and pyridine adsorbed on Ag{111} at 153 K.

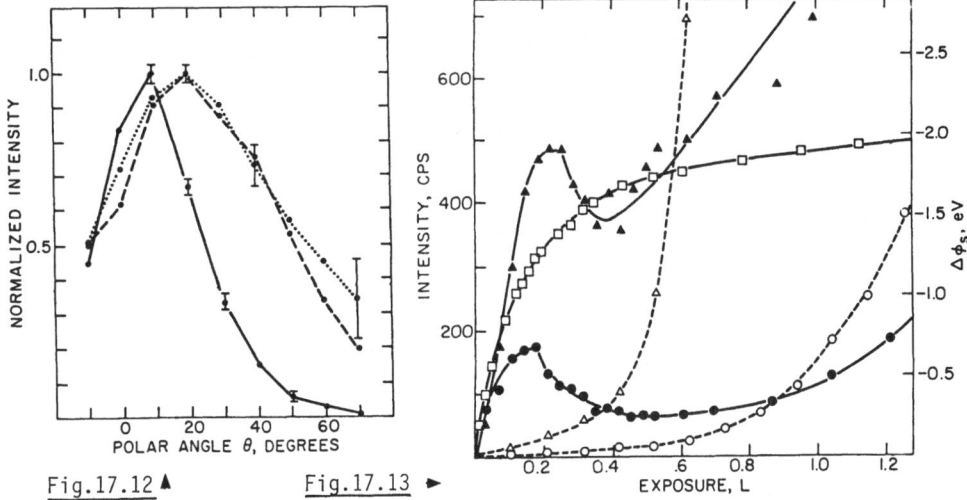

Fig.17.12. Normalized polar angle distribution of molecular ion yields for 4.5 L pyridine [——, $(M+H)^+$], 0.15 L pyridine [——, $(M+H)^+$] and 2.5 L benzene [···, $(M-H)^+$] on Ag{111} at 153 K [17.55]

Fig.17.13. Ion yields of $C_6H_5^+$ (o) and $AgC_6H_6^+$ (△) from benzene on Ag{111} and ion yields of $C_5H_5NH^+$ (●) and $AgC_5H_5N^+$ (▲) from pyridine on Ag{111} as a function of exposure at 153 K. The Ar^+ primary ion was incident perpendicular to the surface, and the polar collection angle θ was 45° relative to the surface normal; work function change (□) for pyridine on Ag{111} as a function of exposure at 153 K. Bombardment conditions: 1 keV, 2 nA, Ar^+ [17.55]

These model systems are of interest for a number of reasons. (i) The molecules are similar in size and shape and should behave in a closely related fashion under the influence of ion bombardment. (ii) Classical dynamics calculations have been performed on these molecules adsorbed on Ni{001} where dramatic differences in the molecule yield are predicted to occur with molecular orientation [17.52]. (iii) Electron energy loss spectroscopy indicates that pyridine on Ag{111} initially adsorbs in a π-bonded configuration but undergoes a compressional phase transition to a σ-bonded configuration as the coverage is increased [17.53]. Benzene, on the other hand, is believed to remain in the π-bonded configuration at all coverages [17.54].

The polar angle distributions for a number of different monolayer structures are shown in Fig.17.12 [17.55]. Note that for monolayer benzene and for π-bonded pyridine where the molecules are believed to lie flat on Ag{111}, the polar angle distributions of $(M-H)^+$ (benzene) and $(M+H)^+$ (pyridine) are broad with a peak at θ = 20°. At the onset of the compressional phase transition, however, the polar angle distribution of the $C_5H_5NH^+$ ion sharpens dramatically, and the peak moves to θ = 10°.

A further point is that the molecular dynamics calculations for benzene and pyridine ejection from Ni{001} predict that there should be a large decrease in yield for molecules bonded perpendicularly to the surface rather than parallel to the surface. This observation is also seen experimentally for pyridine at the 0.2 L exposure point as shown in Fig.17.13, but not for benzene. The 0.2 L point is the same exposure point where the compressional phase transition is observed by EELS [17.53]. These results strikingly confirm how the classical dynamics model eludicates organic SIMS spectra. This confirmation is particularly satisfying in view of the approximations inherent in the classical calculations and the overall complexity of the momentum dissipation process.

17.6 Perspectives

Ion-beam methods are becoming increasingly important in solving the surface characterization problem. The angle-resolved SIMS approach described here is of interest because of its very high sensitivity, its ability to detect molecular cluster ions from more complex surface structures and its sensitivity to adsorption bonding geometries. The method should be complementary to ion-scattering spectrometry where atomic locations can be determined with much higher precision [17.56].

We believe there will be many more applications of these techniques discovered in the near future. Of special note is that it now appears that angle-resolved measurements of the neutral particles should be feasible using multiphoton resonance ionization techniques [17.57]. These types of measurements also ought to provide detailed information about ion fractions — numbers urgently needed for further theoretical developments.

Acknowledgement. I should like to thank a number of my colleagues for helping me to gain the necessary perspective in this field. These include Barbara Garrison who has developed the thrust of our theoretical approach and has suggested many key experiments, and Don Harrison who got us originally involved in this field and gave generously of his knowledge about the ion-bombardment process. Many graduate and post-doctoral students have also helped to develop the approach presented here, including Rick Gibbs, Karin Foley, Dae Won Moon and Roger Bleiler.

I am also most appreciative of the help and cooperation provided by a variety of funding agencies — The National Science Foundation, The Air Force Office of Scientific Research, The Office of Naval Research, The Petroleum Research Foundation administered by the American Chemical Society, as well as The Pennsylvania State University.

References

17.1 W.R. Grove: Trans. R. Soc. (London) **142**, 87 (1852)
17.2 N. Winograd: Prog. Solid State Chem. **13**, 285 (1982)
17.3 W.J. Moore: Am. Sci. **42**, 109 (1960)
17.4 K. Wittmaack: In *Inelastic Ion-Surface Collisions*, ed. by N.H. Tolk, J.C. Tully, W. Heiland, C.W. White (Academic, New York 1977) p.153
17.5 G. Staudenmaier, W.O. Hofer, H. Liebl: Int. J. Mass Spectrom. Ion Phys. **21**, 103 (1976)
17.6 T. Fleish, G.L. Ott, N. Winograd, W.N. Delgass: Surf. Sci. **78**, 141 (1978)
17.7 A. Benninghoven: Phys. Status Solid. **34**, K169 (1969)
17.8 A. Benninghoven: Surf. Sci. **28**, 541 (1971)
17.9 A. Benninghoven, E. Loebach: Rev. Sci. Instrum. **42**, 49 (1971)
17.10 A. Benninghoven: Surf. Sci. **53**, 596 (1975)
17.11 R. Buhl, A. Preisinger: Surf. Sci. **47**, 344 (1975)
17.12 A. Benninghoven, D. Jaspers, W. Sichtermann: Appl. Phys. **11**, 35 (1976)
17.13 F. Honda, Y. Fukuda, J.W. Rabalais: J. Chem. Phys. **70**, 4834 (1979)
17.14 M. Barber, J.C. Vickerman, J. Wolstenholme: Faraday Trans. I, **72**, 40 (1976)
17.15 G.K. Wehner: Adv. Electron. Electron Phys. **7**, 239 (1955)
17.16 P. Sigmund: In *Inelastic Ion-Surface Collisions*, ed. by N.H. Tolk, J.C. Tully, W. Heiland, C.W. White (Academic, New York 1977) p.121
17.17 See for example, D.G. Truhlar, J.T. Muckerman: In *Atom Molecule Collision Theory*, ed. by R.B. Berstein (Plenum, New York 1979)
17.18 See for example, P. Lykos (ed.): *ACS Symp. Ser.*, No. 86 (1978)
17.19 D.J. Rossky, M. Karplus: J. Am. Chem. Soc. **101**, 1913 (1979)
17.20 D.E. Harrison, Jr., W.L. Gay, H.M. Effron: J. Math. Phys. **10**, 1179 (1969)
17.21 D.E. Harrison, Jr., W.L. Moore, H.T. Holcombe: Radia. Eff. **17**, 167 (1973)
17.22 A. Anderman: AFCRL-66-88 Atomics International, Canoga Park, CA, unpublished. See also Ref. [17.20]
17.23 M. Hou, M.T. Robinson: Appl. Phys. **18**, 381 (1979)
17.24 N. Winograd, B.J. Garrison, D.E. Harrison, Jr.: Phys. Rev. Lett. **41**, 1120 (1978)
17.25 B.J. Garrison: In *Potential Energy Surfaces and Dynamics Calculations for Chemical Reactions and Molecular Energy Transfer*, ed. by D.G. Truhlar (Plenum, New York 1981) pp.843-856
17.26 C.A. Andersen, J.R. Hinthorne: Science **175**, 853 (1972)
17.27 M.L. Yu: Phys. Rev. Lett. **40**, 574 (1978)
17.28 A. Blandin, A. Nourtier, D. Hone: J. Physique (Paris) **37**, 369 (1976)
17.29 J.K. Norskov, B.I. Lundqvist: Phys. Rev. B**19**, 5661 (1979)
17.30 Z. Stroubek, K. Zdansky, J. Zavadil: Phys. Rev. Lett. **45**, 580 (1980)
17.31 B.J. Garrison, A.C. Diebold, J.-H. Lin, Z. Sroubek: Surf. Sci. **124**, 461 (1983)
17.32 M.L. Yu: Phys. Rev. Lett. **47**, 1325 (1981)
17.33 J.-H. Lin, B.J. Garrison: J. Vac. Sci. Tech. A**1**, 1205 (1983)
17.34 N.D. Lang: Phys. Rev. B, in press
17.35 R.A. Gibbs, S.P. Holland, K.E. Foley, B.J. Garrison, N. Winograd: Phys. Rev. B**24**, 6178 (1981)
17.36 R.A. Gibbs, S.P. Holland, K.E. Foley, B.J. Garrison, N. Winograd: J. Chem. Phys. **76**, 684 (1982)
19.37 R.A. Gibbs, N. Winograd: Rev. Sci. Instrum. **52**, 1148 (1981)
19.38 D.E. Harrison, Jr., C.B. Delaplain: J. Appl. Phys. **47**, 2252 (1976)
19.39 B.J. Garrison, N. Winograd, D.E. Harrison, Jr.: J. Chem. Phys. **69**, 1440 (1978)
19.40 B.J. Garrison, N. Winograd, D.E. Harrison, Jr.: Phys. Rev. B**18**, 6000 (1978)

17.41 N. Winograd, D.E. Harrison, Jr., B.J. Garrison: Surf. Sci. **78**, 467 (1978)

17.42 N. Winograd, B.J. Garrison, D.E. Harrison, Jr.: J. Chem. Phys. **73**, 3473 (1980)

17.43 B.J. Garrison: J. Am. Chem. Soc. **102**, 6553 (1980)

17.44 R.H. Silsbee: J. Appl. Phys. **28**, 1246 (1957)

17.45 S.P. Holland, B.J. Garrison, N. Winograd: Phys. Rev. Lett. **44**, 756 (1980)

17.46 R.J. Bleiler, A.C. Diebold, N. Winograd: J. Vac. Sci. Tech. A1, 1230 (1983)

17.47 S.P. Holland, B.J. Garrison, N. Winograd: Phys. Rev. Lett. **43**, 220 (1979)

17.48 S. Kapur, B.J. Garrison: J. Chem. Phys. **75**, 445 (1981)

17.49 H. Hopster, C.R. Brundle: J. Vac. Sci. Technol. **16**, 548 (1979)

17.50 W. Erley, H. Ibach, S. Lehwald, H. Wagner: Surf. Sci. **83**, 585 (1979)

17.51 K.E. Foley, N. Winograd, B.J. Garrison, D.E. Harrison, Jr.: J. Chem. Phys., submitted

17.52 B.J. Garrison: J. Am. Chem. Soc. **104**, 6211 (1982)

17.53 J.E. Demuth, K. Christmann, P.N. Sando: Chem. Phys. Lett. **76**, 201 (1980)

17.54 C. Friend, E.L. Muetterties: J. Am. Chem. Soc. **103**, 773 (1981)

17.55 D.W. Moon, R.J. Bleiler, E.J. Karwacki, N. Winograd: J. Am. Chem. Soc. **105**, 2916 (1983)

17.56 P. Eisenberger, L.C. Feldman: Science **214**, 300 (1981)

17.57 F.M. Kimock, J.P. Baxter, N. Winograd: Surf. Sci. **124**, 41L (1983)

18. Determination by Ion Scattering of Atomic Positions at Surfaces and Interfaces

W. M. Gibson

With 20 Figures

18.1 Introduction

Knowledge of the atomic configuration is important to understanding chemical, electronic and mechanical processes at crystal surfaces and interfaces. During the past ten years Ion-Scattering Spectroscopy (ISS) has emerged as a powerful and direct tool to get such information. Application and analysis of ion scattering divides more or less naturally into three energy regimes, Low-Energy Ion Scattering (LEIS) (1-20 kev); Medium-Energy Ion Scattering (MEIS) (20-200 kev); and High-Energy Ion scattering (HEIS) (200 kev-2 Mev). This review will concentrate on the physics, techniques and applications of HEIS with particular attention to studies of atomic structure at *interfaces* — an application for which it is virtually unique. Although for completeness there will necessarily be some overlap of this review with previous reviews [18.1-3] including one in the present series [18.4], the extensive discussion of the principles and details of the technique in those reviews will permit greater concentration here on recent applications and results.

18.1.1 Shadow Cone

One of the most powerful features of the ion-scattering technique is the simplicity of its interpretation. For many surface and interface studies the application is based on the formation of a 'shadow cone' behind an atom because of Coulomb repulsion of positive ions in a parallel beam by the positive nuclear charge of the atom, as shown schematically in Fig.18.1. Atoms lying in the shadow cone cannot be struck by ions in the beam and therefore do not contribute to the yield of any process that requires a close collision between an atom of the solid and ions in the incident beam such as any nuclear reaction, scattering of beam particles through large angles, excitation of inner shell electrons, etc. Therefore, if the ion beam is aligned with low index rows of atoms in the crystal, the beam will scatter from the first atom in each row and not from other atoms in the row which fall inside

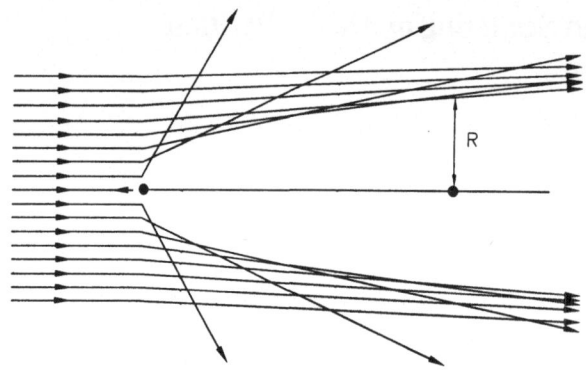

<u>Fig.18.1.</u> Formation of a shadow cone behind an atom exposed to a parallel ion beam. The quantity R is the sha- dow cone radius at the position of the second atom for a pair of atoms aligned with the incident beam

the shadow cone behind the first atom. Even those ions that scatter from the first atom are nearly all scattered through very small angles so that they penetrate far into the crystal before they move laterally to the adjacent row where they may have a chance to interact closely with a lattice atom and scatter again (or make a nuclear reaction, etc.). Even so, most of the ions will be gently deflected and kept away from close interactions. This is called ion channeling, discussed briefly below.

Because of these effects, ions that do not scatter from atoms at the sur- face will, in general, penetrate tens or hundreds of angstroms into the crys- tal before they scatter. These 'bulk scattered' ions can be differentiated from those scattered at the surface because they lose energy penetrating in- to and then out of the crystal. The energy spectrum of scattered particles under the conditions that the incident ion beam is aligned with a low index atomic row in the crystal typically shows a 'surface peak' as in Fig.18.2. After a small-background correction, the area under the surface peak (SP) for a given incident ion fluence (ions/cm^2 incident on the crystal) can be used directly to determine the number of atoms per atomic row (atoms/row) contributing to the scattering. For an ideal crystal one might expect to ob- tain a value of 1.0 atoms/row from such a measurement. In practice the SP intensity frequently corresponds to values greater than 1.0. This can arise from thermal vibration of the atoms in the atomic rows relative to each other which causes deeper lying atoms to spend part of their time at displacements (relative to the surface atom) larger than the shadow cone radius and there- fore exposed to the incident beam ions. For a simple two-atom model in which Coulomb scattering is assumed and the thermal motion is represented as a Gaussian distribution with width ρ, the surface peak intensity is [18.5]

$$I_c = 1 + (1 + R_c^2/2\rho^2)\exp{-(R_c^2/2\rho^2)} \qquad (18.1)$$

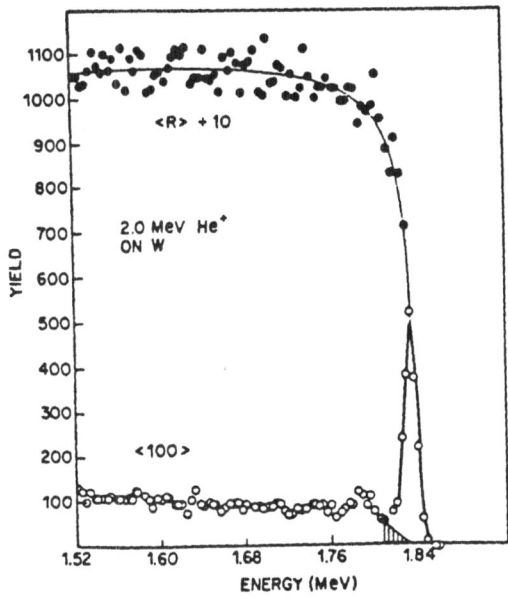

Fig.18.2. Typical scattering spectrum for 2.0 MeV He+ ions incident on a W{001} single crystal in a random (nonchanneling) direction ([R]) and channeling ([100]) direction. Note that the random spectrum is divided by 10 so that the reduction in yield due to channeling is about a factor of 100

where R_c is the radius of the shadow cone at the position of the second atom which for this simple 2-atom case is

$$R_c = 2\left(\frac{Z_1 Z_2 e^2 d}{E}\right)^{\frac{1}{2}} , \qquad (18.2)$$

where Z_1 and Z_2 are the ion and atom charges respectively, d is the atomic spacing and E is the ion energy. The first term in (18.1) represents the unit contribution from the first atom and the second the variable contribution from the second atom. This two-atom Coulomb estimate is inadequate for most cases but it suggests that the surface peak intensity should depend only on one parameter, ρ/R.

For most cases of practical interest, the surface peak yield corresponds to scattering from several atoms per row. Also, a screened Coulomb potential should be used rather than the Coulomb potential. The number of atoms involved is large enough to make analytical calculations formidable, but small enough so that computer simulation using Monte Carlo techniques is easily tractable. In this approach [18.6], the trajectories of a large number of incident ions are followed as they penetrate the crystal with scattering of the individual ions with individual atoms of the solid treated as sequential independent binary collisions. An appropriate screened Coulomb potential such as the Moliere approximation to the Thomas-Fermi potential is normally used. For most initial comparisons or estimates the atoms of the crystal are as-

429

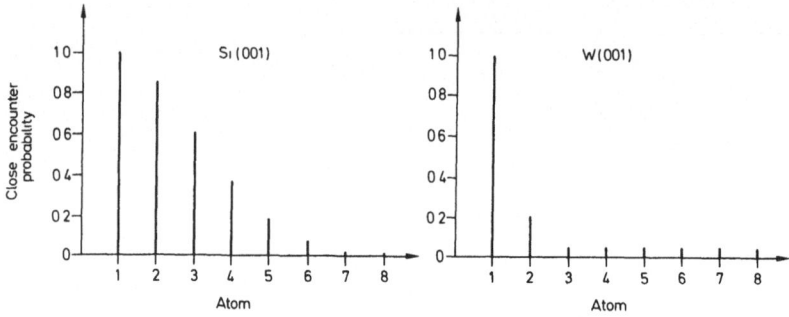

<u>Fig.18.3.</u> Calculated normalized backscattering probability from successsive layers of Si{001} crystal (left) and W{001} crystal (right) for 1.0 MeV He+ incident normal to the crystal surface [18.10]

sumed to vibrate independently with a vibration amplitude determined from a measured Debye temperature. The effects of atomic displacements, enhanced surface vibratio s and correlations in atomic vibration are expected to change the surface peak intensity from the idealized value. Systematic investigation of these effects [18.6-9] has been reported. These will not be reviewed here in detail but will be considered in connection with applications to specific systems or problems. Figure 18.3 shows the relative contribution to the surface peak from successive atoms along an atomic row for 1.0 MeV He+ scattered from a Si{001} and W{001} crystal normal to the surface. It can be seen that for Si as many as eight atoms in the row contribute to the surface peak, whereas for W only two atoms in the row contribute appreciably. This difference arises from both the higher Z of W, which according to (18.2) gives a larger shadow cone radius and from the smaller vibration amplitude for W. The parameter ρ/R_m for Si is about three times larger than that for W in this case.

The number of atoms/row contributing to the surface peak for a variety of different crystals, crystallographic directions, incident ions and ion energies is plotted against ρ/R_m for each case in Fig.18.4. The results shown on this figure were obtained from computer simulations with ideal lattice structure assumed. It can be seen that when plotted this way all results fall on a universal curve. This result forms the basis for quantitative application of High-Energy Ion Scattering (HEIS) since deviations from the universal curve can arise from atomic displacements from the ideal lattice or by changes in the vibrational amplitude, correlation or isotropy assumed for the ideal lattice case. Since either displacement or vibration effects could produce a given deviation, an inherent amibguity arises in such studies. In principle this ambiguity can be resolved by studies at different temperature,

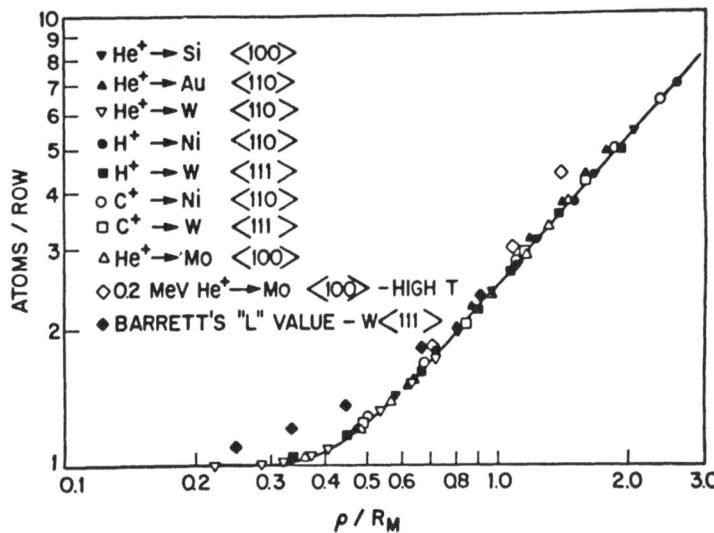

Calculated normalized backscattering probability expressed as the effective number of atoms/row contributing to the scattering as a function of ρ/R_m, where ρ is the thermal vibrational amplitude and R_m is the radius of the shadow cone at the location of the second atom in the row [18.10]

measuring the surface peak intensity for different low index incicent directions and by measurements for ion incidence over a small range of incidence angles relative to the atomic row or plane direction in the bulk lattice. Relatively few measurements have been made over a large range of sample temperature, but those that have been done show that important details of surface vibrational properties can be obtained [18.11]. The use of ion direction and angular scanning studies will be illustrated in the examples. In practice it is most common to calculate the surface peak including the energy, angle or temperature dependence expected for proposed models of surface structure or dynamics and compare the result with measurements of the surface peak intensity. In this way the ion-scattering technique is frequently used to test the validity of models derived from Low-Energy Electron Diffraction (LEED) studies, photo emission studies or theoretical calculations. In some cases the ion-scattering results have been the basis for evolution of models, especially for interface studies where other types of information are not available. The interplay between different structure-sensitive techniques is particularly important since each looks at the surface in a different way and so can contribute to understanding the atomic structure.

Another important aspect of the HEIS technique is the ability to discriminate between different species at the surface or interface. This can be done by measuring the energy of scattered ions or the yield of a specific nuclear reaction between ions in the incident beam and the atoms of interest. For ions scattered 90° or 180° from the incident direction, the relationship between the scattered ion energy and the incident energy E_0 takes a particularly simple form

$$E(90^{\circ}) = E_0\left(\frac{M_2 - M_1}{M_2 + M_1}\right)$$

$$E(180^{\circ}) = E_0\frac{(M_2 - M_1)^2}{(M_2 + M_1)^2} \quad , \tag{18.3}$$

where M_1 and M_2 are the incident ion and the struck atom mass respectively. Therefore, scattering from heavy atoms gives high-energy scattered ions and scattering from light atoms give low-energy ions. If the species of interest is concentrated at the surface or interface, a peak in the energy spectrum is observed, the intensity of which can give directly the absolute number of atoms per unit area at the surface or interface. Likewise, the yield of a nuclear reaction gives a direct and absolute measurement. This ability to determine surface concentrations directly and absolutely (with an accuracy of 3%) is important and differentiates ion scattering from most other surface characterization techniques.

Application of the surface peak to measurement of atomic displacements in the plane of the surface (reconstruction) or normal to the plane of the surface (relaxation) is straightforward. The general principles are indicated on Fig.18.5. If the surface atoms are displaced by an amount corresponding to the shadow cone radius they will no longer effectively shadow the underlying atoms and the surface peak intensity will increase. For 1 MeV He$^+$ ions incident along low index directions in medium mass crystals the shadow cone radius is about 0.1 Å so displacements as small as 0.03 Å can be measured in a good case. When the displacement becomes appeciably larger than the shadow cone radius, the surface peak increase becomes insensitive to the detailed position. By changing the ion energy, the shadow cone radius can be controlled as shown in (18.2). Therefore, the displacement range accessible to quantitative determinations is from a few hundredths to a few tenths of an Angstrom. Displacements of atoms normal to the crystal surface are determined by choosing a low index incident direction inclined to the surface normal. In this case the amount and direction of the relaxation is usually investigated by measurement of the SP as a function of small angular variations about the off-normal direction. Asymmetry in this angular scan can be

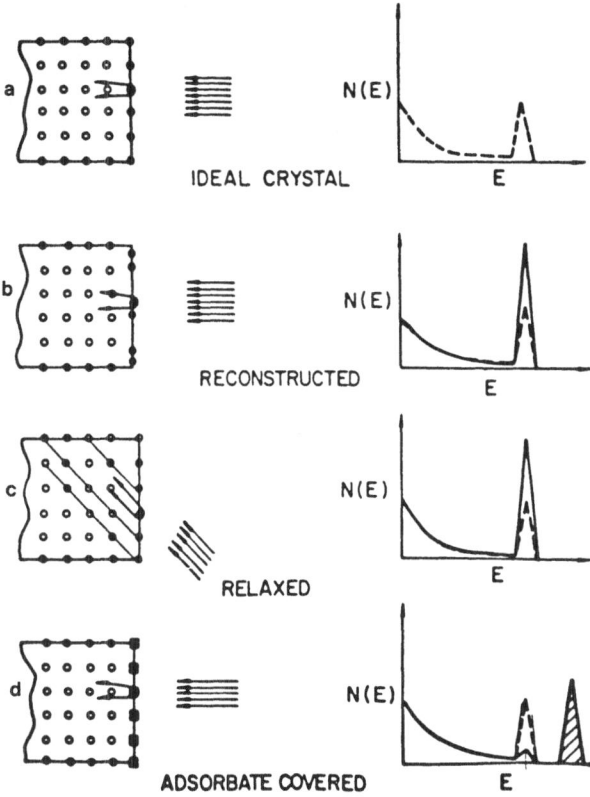

Fig.18.5a-d. Dependence of SP intensity on different crystal surface structures: a) the ideal crystal SP from "bulk-like" surface; b) enhanced SP observed in normal incidence for a reconstructed surface; c) enhanced SP observed for incidence in a nonnormal channeling direction for a relaxed surface; d) reduced SP observed in normal incidence for a registered adlayer.
For (d) an enhanced SP can result from interactions with the adlayer that
moves substrate atoms out of alignment (not pictured)

used to determine the sign of the relaxation (inward or outward) and the
magnitude.

An adsorbate on the surface can affect the substrate surface peak in at
least two ways. If the adsorbate is in good registry with the substrate lattice it can produce a shadow cone that reduces scattering from even the
outermost layers of the substrate, causing the SP to decrease. This is particularly useful for studying epitaxial crystal growth. On the other hand,
the adsorbate can sometimes interact with atoms of the crystal causing them
to be displaced from bulk lattice positions which causes the SP to increase.
It is possible that both effects are present at the same time, leading to
ambiguity. An approach to resolving this type of ambiguity will be discussed

below. As noted previously, the amount of the adsorbate present can be determined directly from the intensity of the adsorbate peak in the scattered
particle energy spectrum. Variations in this intensity for aligned versus
random incidence can also give information on self-shadowing of adsorbate
atoms. This is again particularly useful in detailed studies of epitaxial
crystal formation.

18.1.2 Channeling

The energetic ion beams penetrate several microns into the crystal substrate.
When the ion beam is parallel to a low index atomic row or plane most of the
incident ions are captured in channeling trajectories in which they scatter
gently between rows (strings) or planes of atoms and are kept from close
interaction with lattice atoms. Therefore, the yield of all processes, such
as large-angle scattering, that require close collisions, is very much reduced
This is the reason for the large reduction in scattering yield for aligned
incidence relative to nonaligned (random) incidence shown in Fig.18.2. This
reduction makes observation of the surface peak possible. Ion channeling has
been used extensively for studies of crystals and for other purposes [18.12].
As the ion beam penetrates the crystal the channeled ions oscillate between
rows and planes of atoms with a 'wavelength' hundreds of atoms long. Eventually the phases of the different trajectories are mixed and the flux distribution can be described by statistical equilibrium techniques [18.13]. This
makes analytical solutions tractable. For precise alignment with atomic rows
the flux distribution is

$$f(r) = \ln\left[\frac{Nd}{Nd - \pi r}\right] \quad , \tag{18.4}$$

where r is the perpendicular distance from the atomic row, d is the atomic
spacing along the row and N is the volume density of atoms inside the crystal, so that Nd is the area of a unit cell. This distribution is sharply
peaked at the center of the channel (far from the atomic rows) and decreases
as r^2 to zero at the atomic sites. This flux peaking effect can be used for
surface or interface studies by transmitting the beam through a thin (~1
micron) crystal in order to establish the flux distribution and then observing the interaction of the transmitted particle beam with substrate atoms or
adsorbate atoms displaced from bulk lattice positions at the exit surface or
at an interface near the exit surface of the crystal as shown in Fig.18.6.
This type of measurement is analogous to channeling measurements of impurity
ion location in bulk crystal samples [18.12]. It differs from the determi-

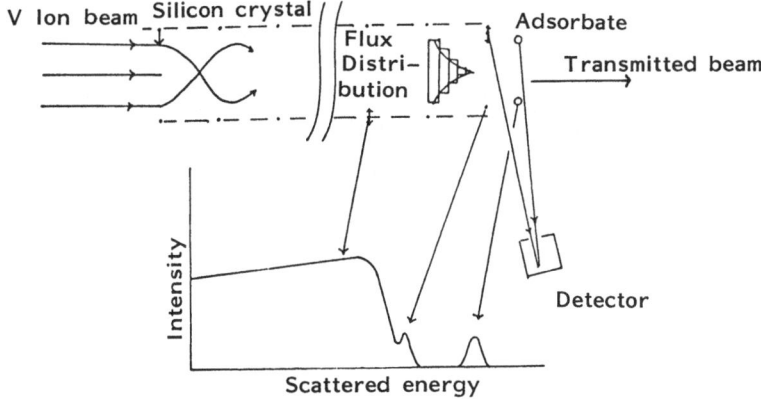

Fig.18.6. Basic principles of the use of ion channeling through thin crystals
for studying substrate disorder and adsorbate atom position at crystal surfaces and interfaces. The initial impact parameter R_{in} describes the limits
of the transverse region within the channel through which channeled ions can
wander. After traversing 1 μ, the flux distribution attains statistical equilibrium which allows a simple characterization of the probability of an ion
and a displaced lattice or adsorbate atom on the exit side of the crystal.
Flux peaking may be visualized as a stacking of allowed transverse areas for
ions which enter the crystal with an initially uniform distribution of impact parameters. The pulse height spectrum from the semiconductor detector
is shown with the various contributions from the crystal and adsorbate
pointed out. The substrate surface peak (SP) is due to scattering from displaced substrate atoms at the exit surface. This appears at higher energy
because of the lower stopping power of channeled relative to nonchanneled
ions. The detector is placed at a grazing exit angle to increase surface
sensitivity

nations described in Sect.18.1.1 in a number of ways. The position of the
adsorbate atoms and displaced substrate atoms are determined independently
and are referenced to the bulk lattice and not to nearby atoms. Therefore,
the ambiguity noted previously that arises from registry of the adsorbate
accompanied by displacement of substrate atoms does not exist. By using
channeling directions inclined to the surface normal of the crystal it is
possible to determine the atomic position relative to the surface normal as
well as the lateral position. In this measurement there is no background
from scattering or reaction with surface atoms, so for perfect alignment the
substrate surface or impurity peak can become very small (equal to the relative yield of scattering from the bulk atoms which is ~1% of the yield for
random incidence for a good case). The range of displacement sensitivity for
this measurement is also different. It is relatively insensitive for displacements up to ~0.3 Å and is particularly sensitive to displacements near
the center of the channel which may be several Angstroms from the atomic
rows. Although this is a useful technique and is virtually the only one pos-

435

sible for measuring the position of important low mass adsorbates (by use of specific nuclear reactions), there have been few reported studies of this kind [18.14-16] because of the complication of preparing [18.17] and cleaning highly perfect, uniform thickness thin crystals.

18.2 Clean Crystal Structure Studies

18.2.1 Semiconductors

a) Silicon

The atomic arrangement of clean silicon crystal surfaces has been a source of much work and speculation over many years. Both Si{100} and Si{111} surfaces have been studied by HEIS, e.g., HEIS measurements of the reconstructed 2×1 surface of Si{100} [18.18] confirmed the presence of several layers of strain proposed from LEED measurements and are generally in agreement with conclusions of those and other studies. However, HEIS measurements of the Si{111} 7×7 surface [18.19] remain to be reconciled with surface models proposed from other studies. The ion channeling measurements of the SP for normal incidence yield a result expected for nearly bulk-like surface while for nonnormal incidence the SP is larger than the expected bulk-like result by nearly two full monolayers. This could be interpreted as two full monolayers that are displaced at least 0.4 Å perpendicular to the surface but are displaced less than 0.15 Å parallel to the surface. Such a large change in bond length is chemically unreasonable and is inconsistent with numerous LEED and photoelectron spectroscopy results which indicate much less vertical displacement. This apparent contradiction has led to a proposal that the surface layers of Si{111} 7×7 have a major rearrangement with a stacking fault in the plane of the surface [18.20]. Figure 18.7 shows the atomic arrangement for the bulk lattice and for three different stacking fault possibilities. It can be seen that these arrangements expose from 1 to 3 additional atomic layers to the ion beam in nonnormal [111], or [001] channeling directions without increasing the number of atoms exposed along the [111] normal direction and without changing the silicon-silicon bond length. Quantitative comparisons favor the structures shown in Figs.18.7b or c. Final resolution of the complex and controversial Si{111} 7×7 surface structure has not yet been achieved but whatever the final answer it will need to accommodate the ion-scattering observations and may well involve atomic arrangements of the type shown in Fig.18.7. In any case, this example illustrates the importance of combining and correlating different kinds of measurements.

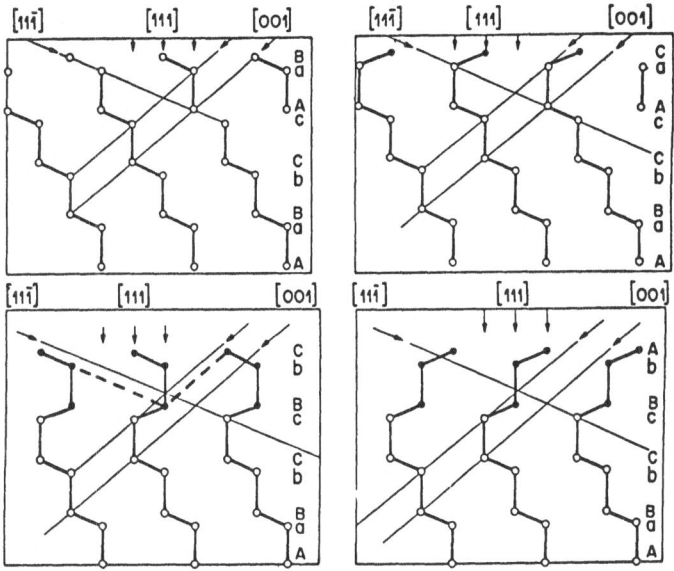

Fig.18.7. Stacking sequence of atom layers for the bulk structure (upper left) and different faulted configurations in a side view of the (110) plane for a Si{111} surface. Also shown are the channeling directions for the incident beam [18.20]

b) GaAs

Nearly all of the semiconductor surface structure studies by ion scattering have concentrated on silicon. The first systematic study of a compound semiconductor surface, GaAs{110}, has recently been reported [18.21,22]. Figure 18.8a shows the measured angular dependence of the SP for normal incidence. The observed SP is incompatible with suggested models of the surface where the GaAs in the surface layer is tilted about $27°$ from the bulk lattice bonding direction parallel to the surface [18.24,25], but is consistent with a $7°$ reconstruction recently proposed by *Duke* et al. [18.23]. To date, as experience and confidence has been developing with the ion-scattering technique, the most usual mode of operation has been to use it to test and possibly choose between or eliminate structure models suggested from other measurements. The work on GaAs also points out the necessity to carry out studies carefully and systematically of radiation domage effects, surface-cleaning procedures and any other things that can possibly influence the SP determination. A preliminary report on GaAs showed apparent agreement with two of the surface models (Meyer and Tong) which are shown in Fig.18.8 to give a too high SP expectation. This was a result of using too low a temperature to anneal surface damage after sputter ion bombardment cleaning of

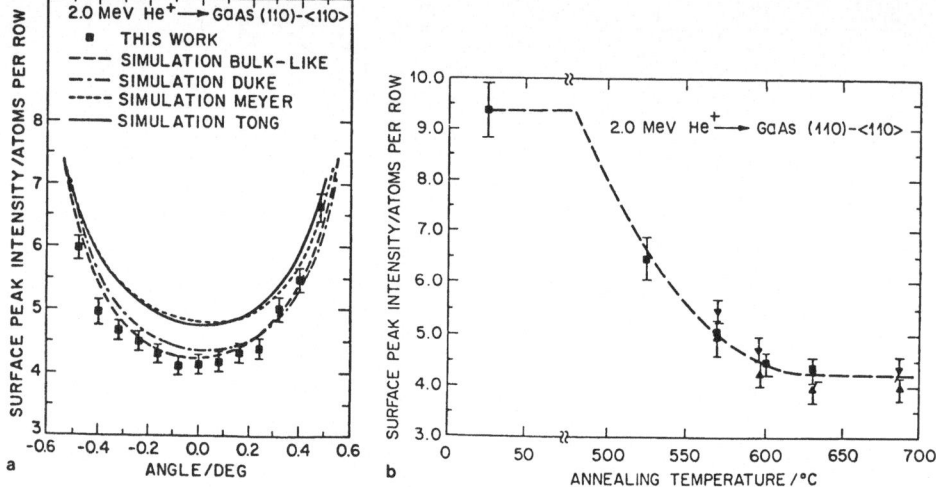

<u>Fig.18.8a.</u> Experimental (squares) and computer-simulated angular scans around the [110] direction for a clean GaAs{110} surface. Simulation results are shown for a bulk-like terminated surface [18.22], the 7° reconstruction proposed by Duke et al. [18.23] and the 27° reconstructions of Meyer et al. [18.24] and Tong et al. [18.25]. b) Dependence of the surface peak intensity on the annealing temperature after sputtering with 1 keV Ar+ ions at a dose of $\approx 10^{20}$ m^{-2}. Different symbols denote different runs. Data points were taken for beam incidence along [110]

the GaAs surface. A temperature of 550°C was used as suggested in the literature. As shown in Fig.18.8b, subsequent measurements showed that this is lower than necessary for full recovery of the surface structure.

18.2.2 Metal Surfaces

In several respects metal surfaces are more congenial to HEIS studies than semiconductors. They are in general less subject to radiation damage perturbations of the SP intensity and generally exhibit less subsurface strain that otherwise complicates the interpretation of this and other surface-structure measurement techniques. A considerable number of HEIS measurements of surface-structure effects have been reported. These include detailed studies of clean and hydrogen-saturated W{001}[18.26,27], surface relaxation and surface vibration studies of Pd{111} [18.28,29], Pt{001} [18.30] and Pt{110} [18.31,32], studies of vibrational properties of Au{110} [18.33], a study of Ag{111} [18.35] and the reconstruction of the Au{100} surface from a (5 × 20) to the (1 × 1) pattern [18.34]. For illustration we shall use a HEIS study of Ni{111} which illustrates some of the approaches to analysis of ion-scattering results and forms a basis for discussing in the next section ad-

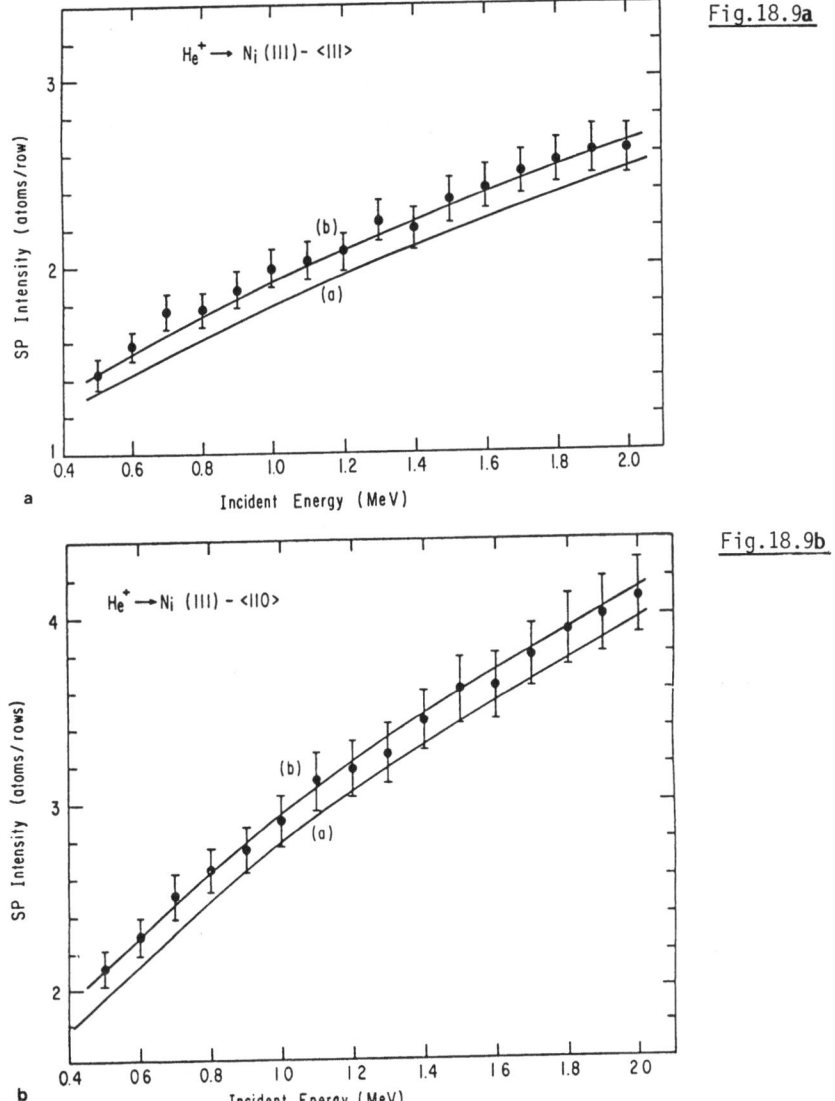

Fig.18.9**a**

Fig.18.9**b**

Fig.18.9a,b. Energy dependence of the SP intensity for a clean Ni{111} 1×1 surface: a) [111] normal incidence; b) [110] incidence. The solid curves in both figures are the results of computer simulations. Simulation (a) assumes a bulk-like structure and isotropic bulk-like thermal vibration (one-dimensional rms amplitude of 0.068 Å); b) assumes a bulk-like structure and an enhanced surface thermal vibration (one-dimensional rms amplitude of 0.084 Å for the first layer, 0.077 Å for the second, 0.072 Å for the third and 0.068 Å for the fourth and deeper layers) [18.36]

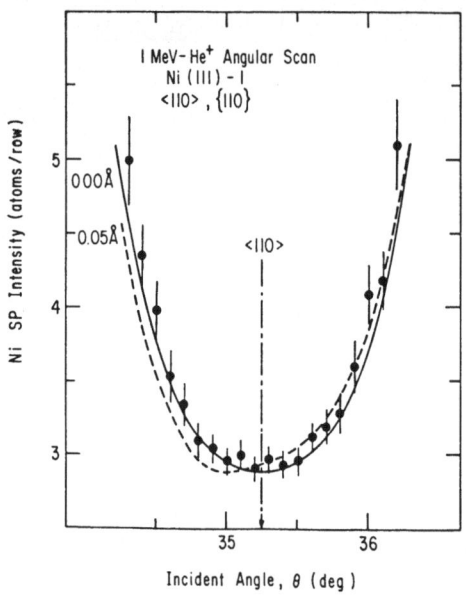

I MeV - He+ Angular Scan
Ni (III) - I
<110>, {110}

Ni SP Intensity (atoms / row)

5

000Å

0.05Å

<110>

4

3

35   36

Incident Angle, θ (deg)

Fig.18.10. Angular dependence of the [110] SP intensity for 1 MeV He+ ions incident on a clean Ni{111} 1 × 1 surface. Solid and dashed curves show the results of computer simulations in which a bulk-like structure with enhanced surface thermal vibration (Fig. 18.9) and a uniform expansion of the first layer by 0.05 Å are assumed, respectively [18.36]

sorbate-induced surface-structure changes. Figure 18.9 shows the energy dependence of the SP intensity as a function of He+ ion energy for Ni{111} in (a) the normal [111] direction and (b) a [110] direction at $35.5°$ from the normal [18.36]. The sputter-cleaned and annealed surface showed a sharp (1 × 1) electron diffraction pattern. LEED measurements on this system indicate a bulk-like structure with no appreciable reconstruction or relaxation [18.37,38]. Curve (a) in both parts of Fig.18.9 shows the expected SP for a bulk lattice and is seen to fall below the observed SP intensity. This difference can arise from either atomic displacement or from enhanced thermal vibration surface atoms, an effect indicated from the LEED studies [18.37, 38].

If the difference between the observed and bulk lattice SP for the nonormal, [110], channeling measurements is due to relaxation, there should be an asymmetry in the angular distribution measured along the {110} plane normal to the surface. For contraction (relaxation inward) the asymmetry should be toward larger angles and for expansion (relaxation outward) toward smaller angles measured relative to the surface normal. The angular dependence of the SP for this case is shown in Fig.18.10. The dotted line shown in this figure is the expected angular distribution for relaxation (expansion) large enough to give the observed SP change. It is clear from the measurement that the observed angular distribution is symmetric about the bulk lattice [110] direction. From this it is concluded that the clean Ni{111} surface has a

bulk-like atomic structure with reconstruction or relaxation less than
0.02 Å and with ~20% nearly isotropic enhancement of thermal vibrational
amplitude. Curve (b) in both parts of Fig.18.9 shows the expected SP intensity versus ion energy for this case. This result confirms the LEED studies
in a straightforward and somewhat more quantitative manner.

18.3 Adlayer-Induced Reconstruction and Relaxation

There have been a number of HEIS studies reported in which adsorbed layers
induce changes in the substrate surface atomic structure. For the strained
silicon surfaces (Si{001}, Si{110}), adsorption of hydrogen or almost anything reduces and even removes the reconstruction and relaxation [18.39,40].
Similar effects are found for the W{001} surface [18.26,27] when saturated
with hydrogen and the Pd{111} surface [18.41] when covered with hydrogen or
carbon monoxide. It is important to repeat that in each case it has been
possible to determine the absolute coverage of the adsorbate by Rutherford
backscattering analysis as noted previously or by Nuclear Microanalysis (NMA)
techniques through measurements of the yield of specific nuclear reactions
such as $^{12}C(d,p)^{13}C$, $^{16}O(d,p)^{17}O$ or $D(^3He,p)^4He$ [18.42].

In most of the cases noted above, strain in the surface was relieved by
the adsorbate. It is interesting to consider oxygen adsorption on Ni{111}
which has been much studied by LEED and exhibits changes from the (1×1)
pattern of the clean surface to a $p(2 \times 2)$ pattern then to a $(\sqrt{3} \times \sqrt{3})R30°$
pattern as the oxygen adsorption increases, with final disappearance of the
LEED pattern at about 10 Langmuirs ($1L = 10^{-5}$ Torr sec of oxygen exposure indicating formation of an amorphous film. Measurement of the Ni SP intensity
as a function of oxygen exposure for ion channeling along the [111] direction
normal to the crystal surface shown by the solid points in Fig.18.11 shows
no change until at 10 Langmuirs exposure, the SP rapidly increases and saturates at 1 atom/row higher value. This coverage corresponds to 3 monolayers
of nickel in the amorphous oxide surface film. For ion incidence in a [110]
channeling direction at 35.5° to the surface normal, the result is dramatically different as shown by the crosses of Fig.18.11. The SP begins to increase upon oxygen exposure, saturates in the region of oxygen coverage corresponding to formation of the $p(2 \times 2)$ pattern, increases again with a plateau at 5L oxygen exposure corresponding to the $(\sqrt{3} \times \sqrt{3})R30°$ pattern, then
undergoes a sudden increase at 10 L exposure as the amorphous surface film
is formed. The SP saturates at 3 atoms/row which in this direction corres-

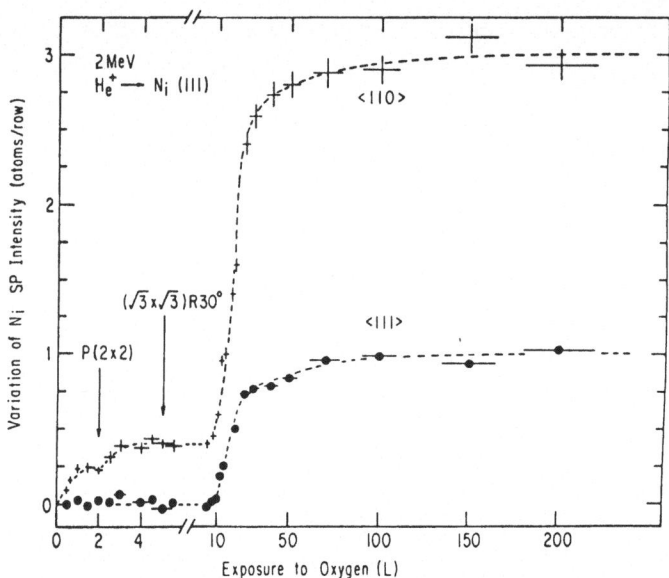

<u>Fig.18.11.</u> Variation of [111] and [110] surface peak intensities for 2 MeV
He$^+$ ions incident on Ni{111} surface as a function of exposure to oxygen at
25°C. Exposures which gave the most intense electron diffraction superstructure patterns are indicated by arrows [18.36]

ponds to 3 monolayers of nickel in agreement with the measurement for normal
incidence.

Measurement of the amount of oxygen in the saturated surface oxide by
integration of the oxygen peak in the channeled particle energy spectrum
gave a value of 2.8 ± 0.4 monolayers, which taken together with the nickel
SP increase indicates that the amorphous oxide film has an NiO stoichiometry. The oxygen coverage measurement by ion scattering at the monolayer
level was used to calibrate the Auger electron spectrometer which was then
used to determine the amount of oxygen present at lower coverages where the
ion scattering was less sensitive. These measurements showed that at 5 L
oxygen exposure (corresponding to the $(\sqrt{3} \times \sqrt{3})R30°$ structure) the coverage
is 0.31 ± 0.05 monolayers and at 2 L exposure (corresponding to the $p(2 \times 2)$
structure), the coverage is 0.23 ± 0.04 monolayers.

The curves of Fig.18.11 show immediately that (a) adsorption of oxygen
does not change appreciably the enhanced thermal vibration of the surface
nickel atoms and, (b) that adsorption of oxygen induces relaxation of the
substrate nickel atoms in stages corresponding to the well-known diffraction
pattern changes. Detailed investigation of the angular distribution of the
SP intensity shown in Fig.18.12 for the two plateaux in the SP change at 2

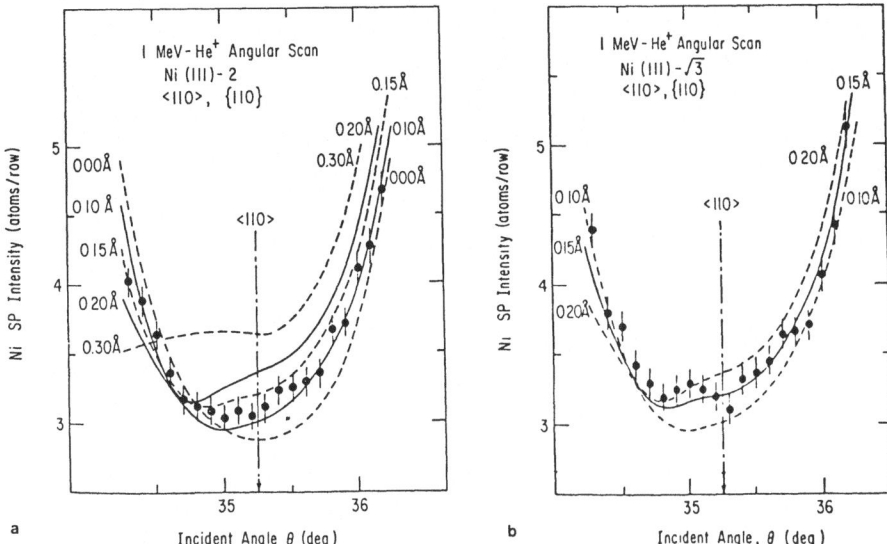

Fig.18.12a,b. Angular dependence of [110] SP intensity for 1 MeV He$^+$ ions
incident on a) Ni{111} p(2 × 2)-O and b) Ni{111} ($\sqrt{3} × \sqrt{3}$)R30°-O structure.
Simulation curves assume enhanced surface thermal vibration and uniform expansion of the first monolayer perpendicular to the surface by various
amounts indicated on the figure. The exact [110] direction was determined
from simultaneous measurements of the bulk yield [18.36]

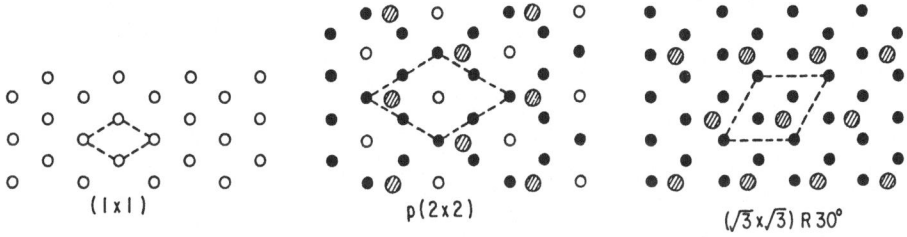

Fig.18.13. Atomic structure models for Ni{111} - (1 × 1), -p(2 × 2)-O, and
-($\sqrt{3} × \sqrt{3}$)R30°-O structures. Large shadowed circles represent oxygen atoms
adsorbed on 3-fold symmetry sites. Small closed and open circles represent
the expanded and nonexpanded Ni atoms in the first monolayer respectively
[18.36]

and 5 L. Analysis of the results shown in Fig.18.12b, which shows the
measured angular distribution for a surface with 1/3 monolayer of oxygen
coverage and corresponding to the ($\sqrt{3} × \sqrt{3}$)R30° pattern, indicates that the
first monolayer of nickel atoms is relaxed outwards by 0.15 Å. The best fit
of the results of Fig.18.12a, which is for an oxygen coverage of 1/4 monolayer and corresponds to the p(2 × 2) pattern, is for 3/4 of the nickel atoms
in the first monolayer relaxed outward by 0.15 Å and the remaining 1/4 at

bulk lattice sites, although relaxation of the entire layer by 0.12 Å could
not be eliminated.

All of these results taken together were used to develop a detailed surface structure which is shown in Fig.18.13. This example not only demonstrates
the sensitivity and usefulness of ion scattering for measuring adsorbate-
induced structural changes but shows the constructive interplay of a number
of techniques (HEIS, LEED, AES) to arrive at a clear result.

18.4 Solid-Solid Interface Structure Studies

As useful as it has proven to be for clean surface and layer-induced structure studies, HEIS is perhaps even more uniquely useful for studying the
structure of buried interfaces since most of the other techniques used to
study surface structure such as LEED, angle-resolved photoemission, atomic
diffraction, or LEIS cannot be used. Three types of interfaces will be used
to illustrate the applicability of HEIS for such studies; amorphous overlayer/
crystalline interfaces such as SiO_2/Si{111}, metal semiconductor interfaces
such as Au/Si{100} or Pd/Si{111} and epitaxial interfaces such as Au/Ag{111}
or Ge/Si{111}.

18.4.1 SiO_2/Si {111}

The silicon oxide/silicon interface is perhaps the most important of a class
of amorphous overlayer/semiconductor interfaces that plays such an important
role in microcircuit manufacture and operation. Understanding the structure
of this interface may be important to understanding and controlling its electrical and mechanical properties.

Using techniques developed by *Feldman* and co-workers [18.43-45], *Haight*
and *Feldman* studied the SiO/Si{111} interface [18.46]. Figure 18.14 shows
the areal density of silicon in the oxide film and at the interface obtained
from the silicon SP versus the oxygen areal density from the oxygen peak intensity in the channeling spectrum as an initially thick (500 Å), thermally
grown SiO_2 film was made thinner by successive chemical etching in dilute
HF solution. The slope of the line obtained gives directly the stoichiometry
of the silicon oxide film which was accurately found to be SiO_2 down to the
interface (an upper limit of one monolayer of SiO could be placed on the film
composition change). Extrapolation of this experimental line to zero oxide
yields an intercept along the silicon axis which gives information on the
amount of disturbed Si at the interface. The measurements shown in Fig.18.14
are for incidence along a [100] direction 54.7° to the surface normal. The

444

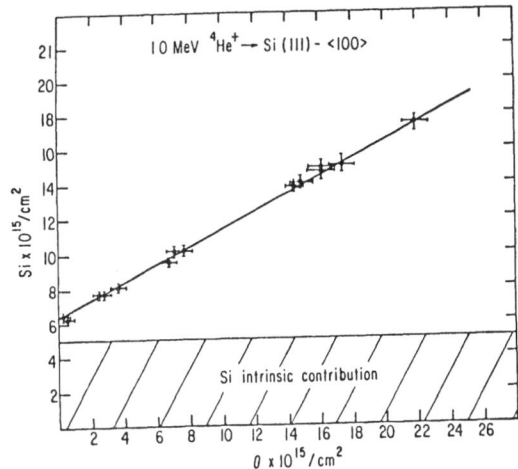

Fig.18.14. Silicon areal density as a function of oxygen areal density for a 1.0 MeV He+ ion beam incident in a [100] direction on an oxide-covered Si{111} surface. The solid line has a slope corresponding to that for SiO$_2$. The extrapolated intercept value on the Si axis is $6.4 \pm 0.5 \times 10^{15}$ Si/cm^2. The hatched region is the intrinsic contribution for a bulk-silicon lattice and corresponds to 4.84×10^{15} Si/cm^2 [18.46]

intercept corresponds to $(6.4 \pm 0.5) \times 10^{15}$ Si/cm^2. From this must be removed the intensity of the SP that would be expected from an unstrained bulk lattice at room temperature (4.84 ± 10^{15})Si/cm^2 leaving an excess of $(1.56 \pm 0.5) \times 10^{15}$ Si/cm^2. This corresponds to 2 monolayers of displaced silicon. Measurements were also made for normal incidence which gave a silicon excess of $(1.0 \pm 0.5) \times 10^{15}$/cm^2 (~1.3 monolayer). A summary of the various silicon SP measurements from extrapolation to zero oxide thickness is shown in Fig. 18.15.

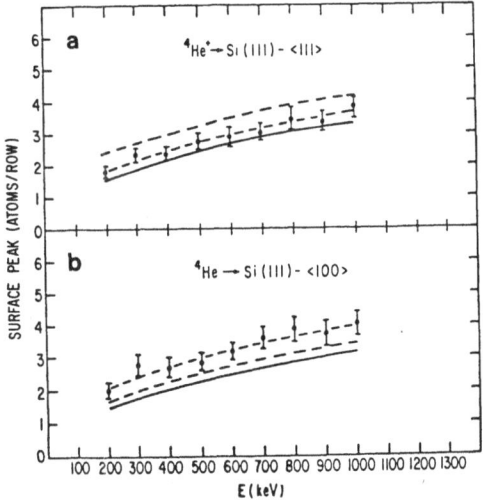

Fig.18.15a,b. Silicon SP intensity for He+ ions incident on a Si{111} oxide covered surface after extrapolation to zero oxide thickness for incidence: a) in a surface normal [111] direction; and b) in a [100] direction 57.2° from surface normal. In (a) the dashed curve, agreeing well with experimental data, is from a computer simulation which assumes two layers of silicon in the substrate crystal are displaced 0.1 Å in the plane of the surface. The uppermost dashed curve corresponds to the 0.3 Å lateral displacement in two layers. The solid curve is for bulk-like lattice termination. For (b) the dashed curve that fits the data corresponds to 0.22 Å vertical displacement inward and outward with respect to the bulk lattice position for the first two layers of the substrate respectively. The 0.1 Å lateral displacements from (a) have been included in the simulation. The dashed curve in the middle corresponds to 0.1 vertical displacement only. The solid curve corresponds to a bulk-like lattice termination

445

The conclusion from this study was that the interface is well described by a stoichiometric amorphous SiO_2 film up to the interface with a region of distorted silicon consisting of two monolayers. The major displacement of the silicon atoms in the two monolayers is in the vertical direction, neglecting any contribution from enhanced vibration of silicon atoms at the interface. Lateral displacements of the interface layers indicated by the simulations is 0.1 Å and vertical displacements are 0.22 Å with one layer displaced inward, the other outward. An alternative explanation which would fit the observations would be one monolayer of SiO at the interface with no laterally displaced silicon but with small (0.1 Å) residual vertical displacements of two layers. In any case, it is clear from these studies that the interface is extremely abrupt and remarkably free of strain. Indeed, it contains much less strain than the atomically clean Si{111} surface.

18.4.2 Au/Si{100} Interface

Metal-semiconductor interfaces are of great importance since they are used extensively for electrical contacts and Schottky barrier formation. In a study of the interface during deposition of gold on clean silicon surfaces it was found that dramatic structural changes occur during the initial deposition stages [18.47]. Figure 18.16 shows the silicon SP intensity as a function of gold coverage at room temperature. At first the SP decreases followed by an increase at ~1 monolayer coverage and a second increase at ~4 monolayers average coverage. These changes are accompanied by changes in the electron diffraction pattern from the clean surface (2×1) first to (1×1) and then to loss of the pattern indicating an amorphous surface layer. The initial decrease could be due to either of two effects; a) removal or reduction of reconstruction in the initial surface, or b) registration of the Au atoms with the substrate lattice with attendent shadowing of Si atomic rows. Measurements to be described below indicate that the former explanation is the correct one. This is supported by the observation that the magnitude of the decrease is the same as that obtained by H saturation of a Si{001} surface and by the change of the diffraction pattern to 1×1. It is noted that only a fraction of a monolayer of Au appears to be sufficient to remove the reconstruction. Therefore, the atomic bonding model for this effect that applies to adsorbed hydrogen removal of the reconstruction would not seem to apply in this case.

The increase at ~1 monolayer and especially at ~4 monolayer coverage can perhaps be seen more clearly for Au deposited on a Si{111} surface since the latter has no large reordering effect, Fig.18.17. It is necessary in

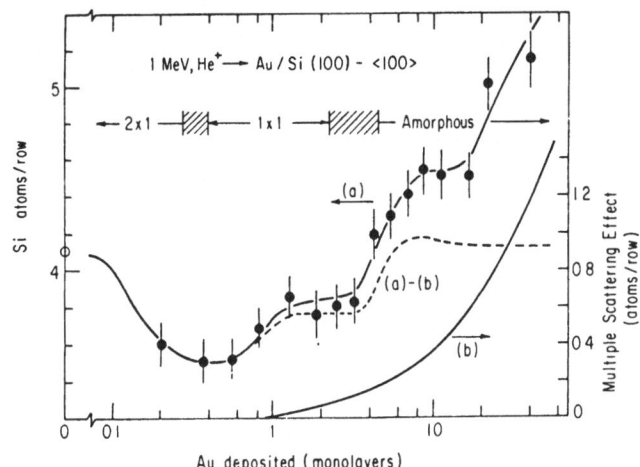

Fig.18.16. Silicon{100} SP intensity as a function of the average Au film thickness for 1 MeV He+ ions incident in a normal direction. The multiple scattering contribution (b) was calculated as described in [18.9]. For this direction 1 atom/row corresponds to 4 monolayers = $4 \times 6.8 \times 10^{14}$ Si/cm² [18.47]

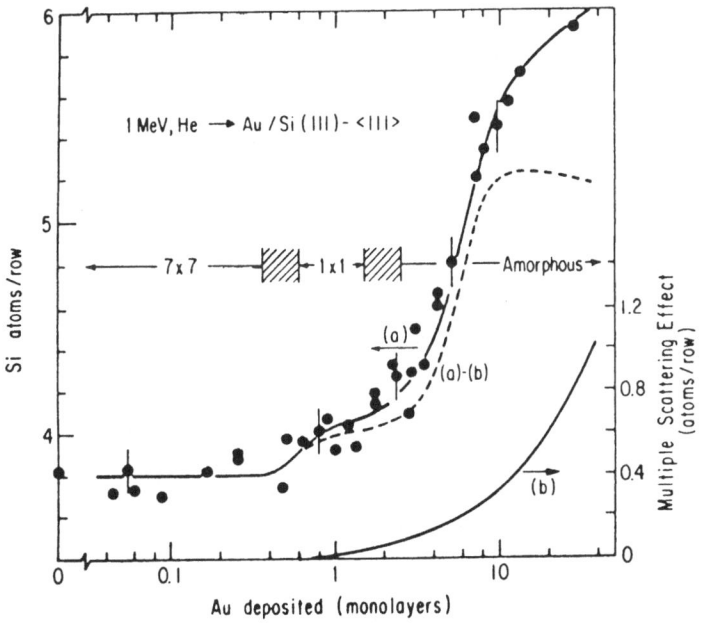

Fig.18.17. Si{111} SP intensity as a function of the average Au film thickness for 1 MeV He+ ions incident in a normal direction. Different symbols correspond to separate experimental determinations. For this direction 1 atom/row corresponds to 3 monolayers = $3 \times 7.8 \times 10^{14}$ Si/cm² [18.47]

447

these studies to take account of multiple scattering effects that broaden the direction distribution of the incident beam as it passes through the Au overlayer. This correction is based on theoretical calculations [18.9] and assumes that the Au overlayer is uniform.. The possible nonvalidity of this assumption will affect the quantitative estimates at Au coverages >5 monolayers but will not change the general conclusions. Measurements of the Au impurity peak in the scattered particle energy spectrum for gold deposition on samples at 550°C (well above the Au/Si eutectic at 350°C) shows that one monolayer on Si{111} surfaces and ~2 monolayers on Si{001} surfaces remain strongly bound to the surface while excess gold diffuses rapidly into the bulk [18.47]. This may be related to the increase in the SP at 1 monolayer coverage. The sudden increase at ~4 monolayer coverage is of particular interest since this indicates a spontaneous formation of an amorphous gold-silicide film as indicated by high-resolution Auger Electron Spectroscopy (AES) [18.48,49] and by loss of the electron diffraction pattern. The formation of the film appears to saturate at about 10 monolayers of gold and from the silicon SP change one can conclude that the nominal composition of the silicide film is Au_5Si. The apparent reaction threshold at ~4 monolayers coverage has been investigated by multiple metal evaporations [18.50] and may indicate the presence of an electronic screening mechanism [18.51] that may inhibit reaction of some metals with semiconductors at low coverage. A similar apparent (but weaker) threshold has been noted for interaction of gold with a GaAs{110} surface [18.22].

The position of gold atoms on the crystal during the initial stages of deposition can be investigated by using the transmission channeling geometry described in the introduction. The results of such measurements are shown in Fig.18.18 [18.52]. At coverage of less than one monolayer, the gold atoms appear to be confined in position (or motion) only along the {110} plane perpendicular to the surface. It is across this plane that Si-Si dimerization bonding is believed to take place that results in the (2×1) electron diffraction pattern observed for the clean surface. At about one monolayer coverage, shadowing of the Au is observed along the Si [100] surface normal direction, indicating confinement of Au atoms along these atomic rows. For coverage greater than four monolayers, all shadowing of Au by rows or planes in the Si lattice disappears. This is consistent with incorporation of gold into an amorphous silicide film. Further measurements are needed and are being carried out to establish more precisely the gold positions on the Si{001} surface.

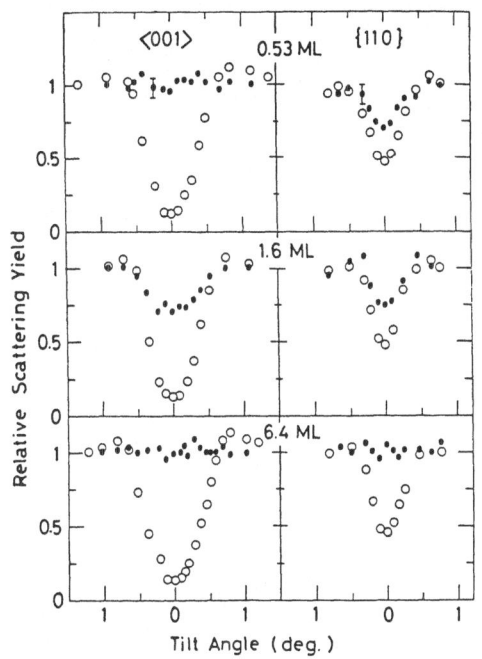

Fig.18.18. Relative scattering yield from the Au overlayer (closed circles) and the Si substrate (open circles) for 1.0 MeV He+ ions transmitted through a thin (~1 µ) Si{001} crystal for incidence parallel to the [001] atomic axis normal to the surface and a (110) plane at 5° to the surface normal. Measurements are shown for three different values of the average Au monolayer thickness in monolayers [18.51]

Similar transmission measurements have shown the registration of paladium atoms with the Si{111} substrate surface at submonolayer coverages [18.16]. Such registration is completely masked for normal backscattering HEIS measurements and for MEIS, LEED or UPS measurements by large local distortions around each Pd atom.

18.4.3 Epitaxy

One of the most exciting new applications of HEIS is in studies of the structural changes that take place during the initial stages of epitaxial crystal growth. The advantages of ion scattering in such studies was established by *Culbertson* et al. in studies of the Au/Ag{111} system [18.53] (lattice constant mismatch of 0.2%) in which it was shown that an almost ideal pseudomorphic Au crystalline layer is grown at room temperature to thicknesses exceeding 50 Å. Figure 18.19 shows the results of that study. Both the substrate SP change and the overlayer self-shadowing (the minimum yield χ_{min} is defined for this measurement as the intensity of the overlayer spectral peak for incidence in a particular low index direction divided by the intensity in a nonchanneling incident direction) show the integrity of the epitaxial Au film. The Au atoms act as a simple extension and occupy the ideal fcc lattice sites of the substrate.

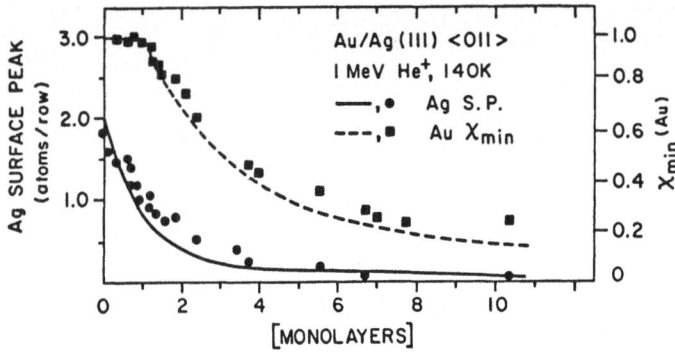

__Fig.18.19.__ Ag{111} SP intensity (closed circles) and Au minimum yield, χ_{min}, (closed squares) for 1 MeV He⁺ ions incident in a [110] direction as a function of the average Au overlayer thickness at 140 K. The dashed and solid curves show the result of computer simulation for Au χ_{min} and Ag SP respectively. Bulk-like Ag fcc structure was assumed for both the Ag substrate and Au overlayer [18.53]

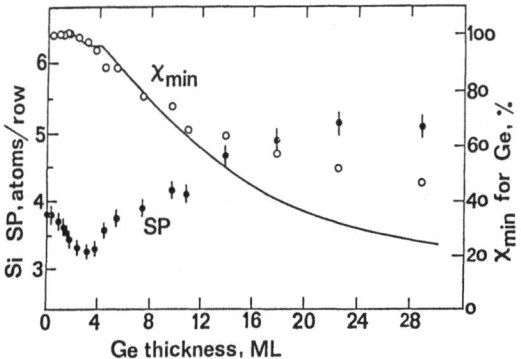

__Fig.18.20.__ Si{111} SP intensity (closed circles) and Geχ_{min} (open circles) for 1 MeV He⁺ ions incident in a normal direction as a function of the average Ger overlayer thickness deposited at 350°C. The experimental points were collected from three independent runs. The solid curve shows the result of a computer simulation for χ_{min} in which a bulk-like structure and dynamics ($\mu_1 = 0.88$ Å) were assumed for the Ge film [18.55]

This work was extended in a study of the epitaxial growth of germanium on Si{111} at 350°C [18.54,55]. At temperatures below 300°C the germanium overlayer was amorphous or polycrystalline and at temperatures above 400°C there was evidence for germanium diffusion into the silicon substrate. Measurements of the silicon SP and germanium minimum yield χ_{min} are shown in Fig.18.20. These measurements show that initially, the Ge atoms occupy the ideal diamond lattice sites of the Si lattice and that the coverage is uniform. Above a critical coverage of ~4 monolayers, the germanium atoms move out of registry with the substrate and considerable strain is introduced into the substrate lattice. This is seen from the sudden increase of the Si SP to a value higher than the initial clean crystal value. This is the first direct observation of such a transition from uniform pseudomorphic epitaxial crystal growth to a state with misfit dislocations and strain. This transition is in qualitative agreement with a calculation [18.56] of 5 monolayers for the critical thickness expected from elastic energy theory and known elastic constants. After growth of ~20 monolayers of germanium, the interface appears to be stable with ~4 monolayers of strain on the silicon substrate side and with considerable strain and extended defects in the germanium epitaxial overlayer. Very similar behavior has been observed for epitaxial growth of gold on Pd{111} where the lattice mismatch is 4.8% [18.56].

At higher deposition temperature where there is some intermixing of germanium with silicon, the interface is less strained and good epitaxial growth is possible [18.57,58]. Recent studies [18.59] of epitaxial growth of Ge on Si{111} at 520°C have shown the formation of a stable (5 × 5) LEED pattern. Systematic HEIS studies of the Ge_xSi_{1-x}{111} alloy epitaxial system are in progress [18.60].

18.5 Perspective

It is clear that HEIS techniques are useful in studying, at the atomic level, surface, adsorbate and interface structures. The number of groups actively involved in such studies is still small (approximately 6 worldwide) but increasing; many aspects of the application and limitations of the technique remain to be explored. The inherent ambiguity between static and dynamic displacement effects has received little attention and that principally by one group (at Chalk River Nuclear Laboratories in Ontario, Canada). This clearly must be addressed more carefully. The problems and possibilities presented by steps on the surface are only now beginning to receive attention and these topics along with more studies of both heterogeneous and homogeneous interface structures will be important for a future review. However, the most pressing need and greatest potential is for carefully conceived combinations of the HEIS technique for surface-structure determination with other traditional or nontraditional approaches to measure the structural, electronic, optical and chemical properties of surfaces and interfaces. The rapid acceptance of HEIS as a viable tool for surface and interface studies by the surface-science community is an encouraging indication that such a combination will be actively pursued.

References

18.1 E. Bogh: In *Channeling*, ed. by D.V. Morgan (Wiley, London 1973)
18.2 L.C. Feldman: Nucl. Inst. and Methods **191**, 211 (1981)
18.3 L.C. Feldman, J.M. Poate: Ann. Rev. Mater. Sci. **12**, 149 (1982)
18.4 L.C. Feldman: In *Chemistry and Physics of Solid Surfaces*, Vol.III, ed. by R.C. Vanselow (CRC, Boca Raton, Fla. 1981); Crit. Rev. in Solid State and Mater. Sci. **10**, 143 (1981)
18.5 L.C. Feldman, R.L. Kauffman, P.J. Silverman, R.A. Zuhr, J.H. Barrett: Phys. Rev. Lett. **39**, 38 (1977)
18.6 J.H. Barrett: Phys. Rev. B3, 1527 (1971)
18.7 I. Stensgaard, L.C. Feldman, P.J. Silverman: Surf. Sci. **77**, 513 (1978)
18.8 J.H. Barrett, D.P. Jackson: Nuclear Instrum. and Methods **170**, 115 (1980)
18.9 K. Kinoshita, T. Narusawa, W.M. Gibson: Surf. Sci. **110**, 115 (1980)
18.10 L.C. Feldman: Appl. of Surf. Sci. **13**, 211 (1982)
18.11 D.P. Jackson, T.E. Jackman, J.A. Davies, W.N. Unertl, P.R. Norton: Surf. Sci. **126**, 226 (1983)
18.12 D.S. Gemmell: Rev. Mod. Phys. **46**, 129 (1974)
18.13 J. Lindhard: K. Dan. Videnskab. Selsk. Mat. Fys. Medd. **34**, No.14 (1965)
18.14 N.W. Cheung, J.W. Meyer: Phys. Rev. Lett. **46**, 671 (1981)
18.15 L.C. Feldman, P.J. Silverman, J.S. Williams, T.E. Jackman, I. Stensgaard: Phys. Rev. Lett. **41**, 20 (1978)
18.16 R. Haight, T. Itoh, T. Narusawa, W.M. Gibson: J. Vac. Sci. Technol. **20**, 689 (1982)
18.17 N. Chueng: Rev. Sci. Instrum. **51**, 1212 (1980)
18.18 L.C. Feldman, P.J. Silverman, I. Stensgaard: Nucl. Instrum. and Methods **168**, 589 (1980);
       I. Stensgaard, L.C. Feldman, P.J. Silverman: Surf. Sci. **110**, 1 (1981)
18.19 R.J. Culbertson, L.C. Feldman, P.J. Silverman: Phys. Rev. Lett. **45**, 2043 (1980)
18.20 P.A. Bennett, L.C. Feldman, Y. Kuk, E.G. McRae, J.E. Rowe: Phys. Rev. (to be published)
18.21 H. Gossmann, W.M. Gibson, T. Itoh: J. Vac. Sci. Technol. A1, 1059 (1983)
18.22 H. Gossmann, W.M. Gibson: Surf. Sci. (to be published)
18.23 C.B. Duke, S.L. Richardson, A. Paton, A. Kahn: Surf. Sci. **127**, L135 (1983)
18.24 R.J. Meyer, C.B. Duke, A. Paton, A. Kahn, E. So, J.L. Yeh, P. Mark: Phys. Rev. B19, 5149 (1979)
18.25 S.Y. Tong, A.R. Lubinsky, B.J. Mrstik, M.A. van Hove: Phys. Rev. B17, 3303 (1978)
18.26 I. Stensgaard, L.C. Feldman, P.J. Silverman: Phys. Rev. Lett. **42**, 247 (1979)
18.27 L.C. Feldman, P.J. Silverman, I. Stensgaard: Surf. Sci. **87**, 410 (1979)
18.28 E. Bogh, I. Stensgaard: Phys. Lett. **65**, 357 (1978)
18.29 J.A. Davies, D.P. Jackson, N. Matsunami, P.R. Norton, J.U. Andersen: Surf. Sci. **78**, 274 (1978)
18.30 P.R. Norton, J.A. Davies, D.P. Jackson, N. Matsunami: Surf. Sci. **85**, 169 (1979)
18.31 P.R. Norton, J.A. Davies, D.K. Creber, C.W. Sitter, T.E. Jackman: Surf. Sci. **126**, 226 (1983)
18.32 T.E. Jackman, J.A. Davies, D.P. Jackson, W.N. Unertl, P.R. Norton: Surf. Sci. **120**, 398 (1982)
18.33 D.P. Jackson, T.E. Jackman, J.A. Davies, W.N. Unertl, P.R. Norton: Surf. Sci. **126**, 226 (1983)

18.34 B.R. Appleton, D.M. Zehner, T.S. Noggle, J.W. Miller, O.E. Schow, III, L.H. Jenkins, J.H. Barrett: In *Ion Beam Surface Layer Analysis*, Vol.2, ed. by O. Meyer, G. Linker, F. Kappeler (Plenum, New York 1976) p.607
18.35 R.J. Culbertson, L.C. Feldman, P.J. Silverman, H. Boehm: Phys. Rev. Lett. **47**, 657 (1981)
18.36 T. Narusawa, W.M. Gibson, E. Tornquist: Phys. Rev. Lett. **47**, 417 (1981) (1981); Surf. Sci. **114**, 331 (1982)
18.37 P.M. Marcus, J.E. Demuth, D.W. Jepsen: Phys. Rev. B**15**, 1460 (1975)
18.38 R.E. Allen, F.W. de Wette: Phys. Rev. **188**, 1320 (1969)
18.39 R.M. Tromp, R.J. Smeenk, F.W. Saris: Phys. Rev. Lett. **46**, 939 (1981)
18.40 T. Narusawa, K. Kinoshita, W.M. Gibson, L.C. Feldman (to be published)
18.41 J.A. Davies, D.P. Jackson, P.R. Norton, D.E. Posner, W.N. Unertl: Solid State Commun. **34**, 41 (1980)
18.42 J.A. Davies, P.R. Norton: Nucl. Instrum. and Methods **168**, 611 (1980)
18.43 L.C. Feldman, I. Stensgaard, P.J. Silverman, T.E. Jackman: In *Proceedings of the International Conference on the Physics of SiO$_2$ and its Interfaces*, ed. by S. Pantelides (Pergamon, New York 1978) Chap.VII
18.44 T.E. Jackman, J.R. MacDonald, L.C. Feldman, P.J. Silverman, I. Stensgaard: Surf. Sci. **100**, 35 (1980)
18.45 N.W. Cheung, L.C. Feldman, P.J. Silverman, I. Stensgaard: Appl. Phys. Lett. **35**, 859 (1979)
18.46 R. Haight, L.C. Feldman: J. Appl. Phys. **53**, 4884 (1982)
18.47 T. Narusawa, K. Kinoshita, W.M. Gibson, A. Hiraki: J. Vac. Sci. Technol. **18**, 872 (1981)
18.48 K. Oura, T. Hanawa: Surf. Sci. **82**, 204 (1979)
18.49 K. Okuno, T. Itoh, M. Iwami, A. Hiraki: Solid State Commun. **34**, 493 (1980)
18.50 T. Itoh, W.M. Gibson: To be published
18.51 A. Hiraki: J. Electrochem. Soc. **127**, 2662 (1980)
18.52 T. Narusawa, T. Itoh, R. Haight, W.M. Gibson: To be published
18.53 R.J. Culbertson, L.C. Feldman, P.J. Silverman: Phys. Rev. Lett. **45**, 133 (1980)
18.54 T. Narusawa, W.M. Gibson: Phys. Rev. Lett. **47**, 1459 (1981)
18.55 T. Narusawa, W.M. Gibson: J. Vac. Sci. Technol. **20**, 709 (1982)
18.56 Y. Kuk, L.C. Feldman, P.J. Silverman: Phys. Rev. Lett. (to be published)
18.57 L. Alexandrov, R.N. Lorygin, O.P. Pchelyakov, S.I. Stenin: J. Cryst. Growth **24/25** (1974)
18.58 A.G. Cullis, G.R. Booker: J. Cryst. Growth **9**, 132 (1971)
18.59 H.J. Gossmann, J.C. Bean, L.C. Feldman, W.M. Gibson: To be published
18.60 H.J. Gossmann: To be published

19. Surface Phonon Dispersion

H. Ibach and T. S. Rahman

With 12 Figures

19.1 Introduction

The concept of surface phonons has emerged from the lattice dynamics of the semi-infinite periodic solid. Therefore, it addresses a highly idealized situation which, at best, can be approached though never realized experimentally. Substantial and important experimental surface research is, however, directed towards less idealized systems. The study of reactions of gas-phase molecules with surfaces may serve as an example. Vibration spectroscopy has become the tool for investigating the chemical nature of surface species, to determine binding sites, and to study the chemical bonds within the adsorbed species as well as to the surface. The vibrating adsorbed species is thus an isolated entity with no coupling to neighboring molecules. The substrate essentially appears as a medium to support the adsorbates without participating in the vibrational motion. As successful as this concept was, and as it remains in many cases, it is bound to break down when lateral interactions between adsorbed species gain importance, e.g., in dense overlayers, or when adsorbate-substrate vibrations with frequencies comparable to the eigenmodes of the substrate are considered. If the adsorbed layer is disordered, the vibration spectrum will consist of intrinsically broad bands, such as one finds for a solid in an amorphous state. Obviously, this situation is not easily amenable to quantitative experimental and theoretical study. Fortunately, however, nature has provided us with an abundance of ordered overlayers of adsorbed species on single-crystal surfaces. It is there where the concept of surface phonons is brought to bear, and where it is going to provide us with new insights into the lateral coupling between surface species and into the nature of the surface chemical bond in densely packed overlayers. The results will also eventually relate to the geometrical structure and possibly also to the transitions between different surface phases.

As already mentioned, a surface phonon is a localized excitation of a semi-infinite periodic solid, with or without an ordered adsorbate overlayer, such that its amplitude has wave-like characteristics in the two directions parallel to the surface and an exponential decay in the direction perpendicular to the surface and into the bulk crystal. The wave-like nature of the solutions of the lattice dynamical equations is a natural consequence of translational symmetry of the crystalline solid. Thus bulk phonons are wave-like solutions in all three directions because of the translational symmetry in each direction. At the surface, however, translational symmetry is limited to the two directions parallel to the surface, so consequently surface phonons emerge as excitations localized at the surface of the crystalline solid. The vibrational amplitude $U(l,t)$ of an atom at position l, which describes a surface phonon, can be written as

$$U(l,t) = U(z) \exp[i(Q_{\parallel} \cdot l_{\parallel} - \omega(Q_{\parallel})t] , \qquad (19.1)$$

where $Q_{\parallel}$ is the wave vector parallel to the surface and $l = (l_{\parallel}, z)$. Here z is perpendicular to the surface and the crystal lies in the upper half-space. The amplitude $U(z)$ has the form $e^{-\kappa z} U_S$, where U_S is the amplitude of the vibration at the surface and the decay constant κ is such that $\mathrm{Re}\kappa > 0$. The frequency ω of the surface phonon in general depends on the wave vector $Q_{\parallel}$ and the relation $\omega(Q_{\parallel})$ is called the dispersion. For a dilute layer on noninteracting adsorbed molecules, the dispersion vanishes and the system can be equally well treated as an ensemble of noninteracting species as is continually done in surface chemistry.

Like bulk phonons, surface phonons can be acoustic, which means that the frequency varies linearly with $Q_{\parallel}$ at small $Q_{\parallel}$ (long wavelength), or they can be optic, where the frequency remains finite as $Q_{\parallel}$ approaches zero. According to this definition, adsorbate modes are always optic while the substrate can support acoustic as well as optical modes. The typical and most important example of an acoustic surface mode is the Rayleigh wave which will be discussed below.

Just as neutron scattering has proven to be a powerful technique for measuring bulk phonon dispersion, there are now two different methods available for determining the dispersion of surface phonons. Both techniques are based on the inelastic scattering of particles: electrons in one case and neutral atoms in the other. In either case, the usual energy conservation can be expressed as

$$E^S = E^I \pm \hbar\omega , \qquad (19.2)$$

where E^I and E^S are the particle energies before and after the scattering event and $\hbar\omega$ is the quantum of energy associated with the vibration. The plus

sign refers to a scattering process where a vibration quantum has been anni-
hilated, and the minus sign applies when a vibration quantum has been created.
As a result of the two-dimensional translational symmetry, the wave-vector
components parallel to the surface are conserved. Thus

$$k_\parallel^S = k_\parallel^I \pm Q_\parallel \quad , \tag{19.3}$$

where the particle scatters with a surface phonon of wave vector $Q_\parallel$. The wave-
vector component perpendicular to the surface is not conserved as the translational symmetry perpendicular to the surface is broken. Particles which
scatter from surfaces can, therefore, interact with bulk phonons of all perpendicular components $Q_\perp$, that is, with an entire frequency band for any
fixed $Q_\parallel$. In general, the frequencies of the surface phonons are outside
these bands (provided they have the same symmetry). We shall also see that
the dominating contribution to the amplitude of the surface atoms is from
surface phonons.

Until recently, no experimental technique was available actually to measure
the dispersion of surface phonons, despite an impressive number of vibrational
studies on well-defined and ordered surface systems. The main focus of those
studies, however, was directed towards adsorbate vibration with the vibrations
of neighboring adsorbates moving in phase, i.e., with $Q_\parallel = 0$. The reason for
this restriction is obvious for experimental techniques which use light, such
as infrared reflection-absorption spectroscopy and Raman spectroscopy, as the
photon wave vector is too small on the $Q_\parallel$ scale of phonon dispersion. Electron
energy loss spectroscopy, although used rather widely in surface studies, was
also not applied to measure phonon dispersion, for a different reason. In
electron energy loss spectroscopy a large differential cross section is provided by inelastic scattering from the dipole moment associated with the adsorbate vibrations. This dipole scattering is restricted to small $Q_\parallel$ similar
to optical techniques. To extend electron energy loss spectroscopy to large
$Q_\parallel$, scattering from the localized core potential must be employed. For this
type of scattering the term "impact scattering" is frequently used. The differential cross section for inelastic events mediated via impact scattering
is significantly smaller than for dipole scattering. Improvement of the experimental equipment was, therefore, required before the issue of dispersion
could be addressed.

This statement also holds for the other technique that enables measuring
surface phonon dispersion which is inelastic scattering of atoms, generally
of He atoms. It is only recently that highly monochromatic and intense beams

have become available [19.1-3] although the method was suggested and theoretically discussed as early as 1969 by *Cabrera* et al. [19.4].

In the following section we give a short review of the theory of surface lattice dynamics and describe the currently used methods to calculate surface phonon dispersion. In the third section we briefly describe the two experimental methods and compare their advantages and disadvantages. We shall see that depending on the issue addressed, one method is to be preferred over the other. We shall also comment on several factors which must be taken into account to perform experiments most rationally and successfully with a minimum amount of time. The final section presents experimental examples of dispersion curves, with their theoretical description and qualitative interpretation in terms of chemical bonding at surfaces.

19.2 Theoretical Calculation of the Phonon Dispersion Curves

In the past decade several theoretical techniques [19.5-10] have been used to calculate the dispersion of surface and bulk phonons in various types of crystals. The task has not been easy because a large number of atoms is involved, and one has very little knowledge about the exact nature of the interaction potential. Most of the methods have utilized a form of slab calculation where, because of computational limits, one has to cut off the slab after a certain number of layers (40 or so). The aim has been to achieve a convergence as fast as possible. One of the first attempts in this direction was the self-consistent slab calculations of *Allen* et al. [19.5], who used a Leonard-Jones type of interaction potential between the atoms and obtained dispersion curves for a wide variety of crystals with and without adsorbate overlayers. The displacement vectors, the polarization, and the amplitudes of the surface modes, as well as the effects of relaxing the atoms in the surface layers, were extensively studied by them.

Model calculations [19.8] in which the atoms are coupled together by mechanical springs, to different degrees of sophistication, have also been carried out. The criterion for the choice of the bulk force constants is agreement of the calculated bulk phonon dispersion curves with available data. The continued fraction method [19.11] using a real space Green's function technique [19.9] is yet another way of proceeding. In earlier calculations [19.12] it was found that the Fourier-transformed Green's function technique circumvented the problem of the large numbers of layers of atoms that enter any of the above-mentioned techniques. Instead one ends up with a Green's function hierarchy which can be solved by invoking an exponential ansatz.

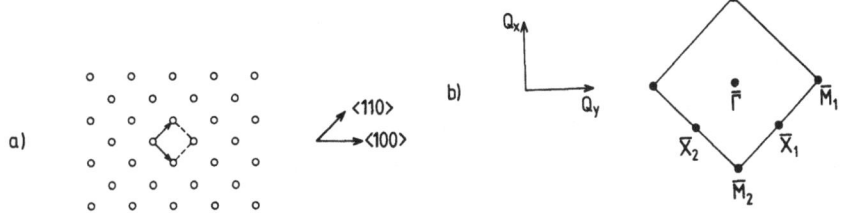

<u>Fig.19.1.</u> a) Top view of the {100} surface of an fcc crystal showing the primitive translation vectors and the unit cell with one atom. b) 2-dimensional Brillouin zone of the {100} surface

Here, we shall illustrate how this particular method can be used to calculate the bulk and surface phonon dispersion curves for a specific system.

We consider here the {100} surface of an fcc crystal and calculate the dispersion along the <100> direction (Fig.19.1) of the two-dimensional Brillouin zone. For the clean {100} surface there is only one atom per unit cell and the problem is considerably simpler than that with several atoms per unit cell, as would be the case for ionic crystals or for ordered overlayers, for example.

The equilibrium position of the atom is specified by $R_0(l_{\parallel} l_z)$, where $l_{\parallel}$ is the position of the atom in the plane parallel to the surface and l_z labels the layer in which the atom sits. The lattice-dynamical Hamiltonian in the harmonic approximation has the form

$$H = \frac{1}{2M} \sum_{l_{\parallel} l_z} P^2(l_{\parallel} l_z) + \frac{1}{2} \sum_{l_{\parallel} l_z} \sum_{l'_{\parallel} l'_z} \sum_{\alpha \beta} \Phi_{\alpha\beta}(l_{\parallel} l_z; l'_{\parallel} l'_z) u_{\alpha}(l_{\parallel} l_z) u_{\beta}(l'_{\parallel} l'_z) \quad ,$$

(19.4)

where $P(l_{\parallel} l_z)$ is the momentum of the atom at a given $l_{\parallel}$ and l_z, M is the mass of the atom, $u_{\alpha}(l_{\parallel}, l_z)$ is the αth Cartesian component of the displacement of the atom from equilibrium, and $\Phi_{\alpha\beta}$ is the two-body interaction potential which can be written more explicitly as

$$\Phi_{\alpha\beta}(ij) = \delta_{ij} \sum_{j'} \phi''_{ij'} \hat{n}_{\alpha}(ij')\hat{n}_{\beta}(ij') - \phi''_{ij} \hat{n}_{\alpha}(ij)\hat{n}_{\beta}(ij) \quad . \qquad (19.5)$$

Here i denotes the atom at $l_{\parallel}$ and in layer l_z, $\hat{n}(ij)$ is the unit vector from atom i to atom j, and ϕ''_{ij} is the second derivative of the pair potential between atoms i and j. In our model calculation we consider only nearest-neighbor interactions and central force, thus the sum in (19.5) extends over all nearest neighbors.

As a result of translational symmetry in the two directions parallel to the surface, we can write the atomic displacements in terms of the eigensolutions

$e_\alpha^{(S)}(\mathbf{Q}_\| ; 1_z)$ so that

$$u_\alpha(1_\| 1_z) = \frac{e_\alpha^{(S)}(\mathbf{Q}_\| ; 1_z)}{M^{\frac{1}{2}}} \exp[i\mathbf{Q}_\| \cdot \mathbf{R}_0(1_\| 1_z)] , \qquad (19.6)$$

where the wave vector $\mathbf{Q}_\|$ lies inside the two-dimensional Brillouin zone shown in Fig.19.1. The equation of motion for the eigenvectors for the mode S is

$$\omega_S^2(\mathbf{Q}_\|)e_\alpha^{(S)}(\mathbf{Q}_\| ; 1_z) - \sum_{1_z'} \sum_\beta d_{\alpha\beta}(\mathbf{Q}_\| ; 1_z 1_z')e_\beta^{(S)}(\mathbf{Q}_\| ; 1_z') = 0 , \qquad (19.7)$$

where the dynamical matrix has been defined in terms of the pair potential as follows:

$$d_{\alpha\beta}(\mathbf{Q}_\| ; 1_z 1_z') = \sum_{1_\|'} \Phi_{\alpha\beta}(1_\| 1_z ; 1_\|' 1_z') \exp[-i\mathbf{Q}_\| \cdot (\mathbf{R}_0(1_\| 1_z) - \mathbf{R}_0(1_\|' 1_z'))] . \quad (19.8)$$

To calculate the dispersion curves for the surface and bulk phonons, we need not go any further as the eigenvalues of (19.7) will give us the required dispersion. It is the sum over the number of layers that contribute that converts (19.7) into the problem of inverting a large matrix. This equation is the basis of the slab method where 20 layers (or so) are included and the matrix thus formed inverted on current computers.

The analytic solution of (19.7) is obtained under the assumption of only nearest-neighbor interaction, central forces, and along the high symmetry direction from the $\bar\Gamma$ to the $\bar M$ point of the Brillouin zone (Fig.19.1). Under these conditions the dynamical matrices are nonzero only for $1_z' = 1_z$ and $1_z' = 1_z \pm 1$. For the atoms in the bulk of the crystal the matrices $\overset{\leftrightarrow}{d}$ are

$$\overset{\leftrightarrow}{d}(Q_x ; 1_z 1_z) = \begin{pmatrix} 4k-2k\cos aQ_x & 0 & 0 \\ & 4k-2k\cos aQ_x & 0 \\ 0 & 0 & 4k \end{pmatrix} \qquad (19.9)$$

$$\overset{\leftrightarrow}{d}(Q_x ; 1_z , 1_z - 1) = \begin{pmatrix} -k\cos aQ_x & 0 & -ik\sin aQ_x \\ 0 & -k & 0 \\ -ik\sin aQ_x & 0 & -k(1+\cos aQ_x) \end{pmatrix} \qquad (19.10)$$

and

$$d(Q_x ; 1_z , 1_z + 1) = d^*(Q_x ; 1_z , 1_z - 1) . \qquad (19.11)$$

Here $a = a_0/\sqrt{2}$ (a_0 being the nearest-neighbor distance) and the effective force constant is given by $k = \phi''/M$.

In the expression for the dynamical matrices for the atoms in the crystal surface layer we have allowed the possibility of intralayer (k_{11}) and interlayer (k_{12}) coupling between the atoms to be different from the bulk values. The rationale for this is that one expects this force constant to be different from the bulk because of the reduced coordination number of surface atoms. This is important when we analyze experimental results in Sect. 19.4. Thus, with $k_{12} = \phi''_{12}/M$ and $k_{11} = \phi''_{11}/M$, we have

$$\vec{d}(Q_x;11) = \begin{pmatrix} 2k_{11}(1-\cos aQ_x)+k_{12} & 0 & 0 \\ 0 & 2k_{11}(1-\cos aQ_x)+k_{12} & 0 \\ 0 & 0 & 2k_{12} \end{pmatrix} \quad (19.12)$$

and

$$\vec{d}(Q_x;12) = \begin{pmatrix} -k_{12}\cos aQ_x & 0 & ik_{12}\sin aQ_x \\ 0 & k_{12} & 0 \\ -ik_{12}\sin aQ_x & 0 & -k_{12}(1+\cos aQ_x) \end{pmatrix} . \quad (19.13)$$

We now use (19.7) and the above expression for the dynamical matrices to write the displacements of the atoms in the surface and bulk layers. The task is facilitated by the fact that we are considering modes with a $Q_\parallel$ vector along a <100> direction. Vibrational modes then are either even or odd with respect to the symmetry plane spanned by this direction and the surface normal. The polarization of the odd modes must be normal to this symmetry plane, which here is the y direction. Thus the equations for the y component of the displacement decouple from those for the x and z components. For the motion of the surface layer atoms in the y direction, we have

$$[2k_{11}(1 - \cos aQ_x) + k_{12} - \omega^2]e_y(Q_x;1) - k_{12}e_y(Q_x;2) = 0 \quad , \quad (19.14)$$

for the atoms in the second layer

$$[3k - 2k\cos aQ_x + k_{12} - \omega^2]e_y(Q_x;2) - k_{12}e_y(Q_x;1) - ke_y(Q_x;3) = 0 \quad , \quad (19.15)$$

and for atoms in the bulk ($1_z > 2$)

$$[4k - 2k\cos aQ_x - \omega^2]e_y(Q_x;1_z) - ke_y(Q_x;1_z - 1) - ke_y(Q_x;1_z + 1) = 0 \quad . \quad (19.16)$$

To obtain the frequency spectrum of bulk phonon with this displacement pattern we seek a solution of the form

$$e_y(Q_x;1_z) = \xi_y(Q_x) \exp(iaQ_z 1_z) \quad . \quad (19.17)$$

From (19.16) we then get

$$\omega^2 = 4k - 2k\cos aQ_x - 2k\cos aQ_z \quad . \tag{19.18}$$

For a finite crystal with $Q_z = \pm \Pi/a$, the above equation gives us the bulk phonon band. At the M point $(Q_x = \frac{\Pi}{a})$ this band extends over the frequency range $\sqrt{4k}$ to $\sqrt{8k}$ and, as we shall see, is degenerate with the x motion.

The frequency of the surface phonon is similarly obtained by seeking solutions in the form of a decaying exponential, for $1_z > 2$

$$e_y(Q_x;1_z) = \xi_y(Q_x) \exp[-\alpha(1_z - 2)] \quad . \tag{19.19}$$

On substituting this into (19.16), we get the following expression for the decay constant α:

$$\cosh\alpha = 2 - \cos aQ_x - \frac{\omega^2}{2k} \quad . \tag{19.20}$$

To obtain the frequency of this surface mode we solve (19.14,15), substituting from (19.19), to give another equation for α. In general this leads to a messy equation, but for $k_{12} = k$ we obtain a simpler form. Thus

$$e^{-\alpha} = \frac{k}{3k - 2k \cos aQ_x - \omega^2} \quad . \tag{19.21}$$

The frequency of the surface mode obtained from (19.20,21) is

$$\omega_S^2 = 2k(1 - \cos aQ_x) \quad . \tag{19.22}$$

At $\bar{M}$ this mode is thus submerged in the bulk band of the same symmetry. Note that by making k_{12} smaller than k, this mode will emerge out of the bulk band.

Let us now consider the set of coupled equations for the displacements of the atoms along the x axis, which is here parallel to the wave vector $\mathbf{Q}_{\parallel}$, and those along the z axis, which is perpendicular to the crystal surface. These will provide the modes of even symmetry. For the surface-layer atoms

$$[2k_{11}(1 - \cos aQ_x) + k_{12} - \omega^2]e_x(Q_x;1) - k_{12} \cos aQ_x \, e_x(Q_x;2)$$

$$+ ik_{12} \sin aQ_x \, e_x(Q_x;2) = 0 \tag{19.23}$$

and

$$[2k_{12} - \omega^2]e_z(Q_x;1) - k_{12}(1 + \cos aQ_x) \, e_x(Q_x;2) + ik_{12} \sin aQ_x \, e_x(Q_x;2) = 0 \quad . \tag{19.24}$$

For the atoms in the second layer

$$[3k + k_{12} - 2k\cos aQ_x - \omega^2]e_x(Q_x;2) - k_{12} \cos aQ_x \, e_x(Q_x;1)$$

$$-k\cos aQ_x\; e_x(Q_x;3) - ik_{12}\;\sin aQ_x\; e_z(Q_x;1)$$

$$+iks\sin aQ_x\; ez(Q_x;3) = 0 \qquad\qquad (19.25)$$

and

$$[2k + 2k_{12} - \omega^2]e_z(Q_x;2) - k_{12}(1 + \cos aQ_x)\; e_z(Q_x;1)$$

$$-k(1 + \cos aQ_x)\; e_z(Q_x;3) - ik_{12}\;\sin aQ_x\; e_x(Q_x;1)$$

$$+iks\sin aQ_x\; e_x(Q_x;3) = 0 \quad . \qquad\qquad (19.26)$$

The equations of motion for $e_z(Q_x;1_z)$ and $e_x(Q_x;1_z)$, where 1_z refers to the layers in the bulk crystal, are obtained from (19.25,26) with the necessary changes, including putting $k_{12} = k$. The frequencies of the bulk and the surface phonons, with the displacement patterns considered in the above equation, (along any point along the $\bar{\Gamma}$ - $\bar{M}$ direction in the two-dimensional Brillouin zone) are obtained similarly as for $e_y(Q_x;1_z)$. The equations are straightforward but cumbersome. Here, the case for the high symmetry point $\bar{M}$, where all displacement patterns are decoupled, will be illustrated. This decoupling is a direct consequence of the higher symmetry of the $\bar{M}$ point which belongs to the C_{4v} group. Since the surface unit cell contains one atom lying on the fourfold axis, all modes belong to either the A or E representation. The A modes are polarized along the z axis while the E modes have arbitrary polarization parallel to the surface plane.

From (19.23,24) we get

$$[4k_{11} + k_{12} - \omega^2]e_x(Q_x;1) + k_{12}e_x(Q_x;2) = 0 \qquad\qquad (19.27)$$

and

$$[2k_{12} - \omega^2]e_z(Q_x;1) = 0 \quad . \qquad\qquad (19.28)$$

The latter is particularly interesting since it indicates that the perpendicular displacements of the surface layer atoms are localized in the surface layer and the frequency of this surface mode is $\sqrt{2k_{12}}$. This mode, which is also called the Rayleigh wave or the S_1 mode [19.5] is thus a direct measure of the force constant between the first- and second-layer atoms. The displacement pattern is depicted in Fig.19.2. At $\bar{M}$ the equations for the second-layer atoms are in turn, from (19.25,26)

$$[5k + k_{12} - \omega^2]e_x(Q_x;2) + k_{12}e_x(Q_x;1) + k_{12}e_x(Q_x;3) = 0 \qquad\qquad (19.29)$$

and

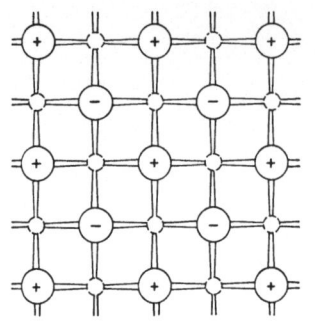

Fig.19.2. Displacement pattern of the z-polarized surface mode (S_1 mode) at $\bar{M}$. One sees that the forces on the second layer atoms cancel. The mode, therefore, is entirely localized to the first layer. This holds for any rotationally invariant nearest-neighbor force field. When second-nearest neighbors are included, the first layer couples to the third, the fifth, etc. [19.6]

○ first Layer

◌ second Layer

$$[2k + sk_{12} - \omega^2]e_z(Q_x;2) = 0 \quad . \tag{19.30}$$

Once again, the mode is decoupled from all other layers and is now localized in the second layer with displacement polarized perpendicular to the surface with a frequency of $\sqrt{2k + 2k_{12}}$. This is similar to the S_3 mode discussed by *Allen* et al. [19.5]. Note that at $\bar{M}$ this mode will be distinct from the bulk band only when k_{12} is different from k. It is trivial to see that the bulk phonons with polarization direction perpendicular to the surface have a distinct frequency of $\sqrt{4k}$ at the $\bar{M}$ point, independent of the perpendicular component of $\mathbf{Q}$. It is this complete absence of dispersion with $Q_\perp$ which allows one to construct a localized mode in each layer of the crystal and also at the surface. This is of course a feature of the nearest-neighbor model.

So far we have not said much about the modes with polarization parallel to $\mathbf{Q}_\parallel$, which belong to the E representation at $\bar{M}$. From (19.29) we see that in the bulk ($l_z > 2$) the displacement pattern satisfies

$$(6k - \omega^2)e_x(Q_x;l_z) + ke_x(Q_x;l_z - 1) + ke_x(Q_x;l_z + 1) = 0 \quad . \tag{19.31}$$

It is easy to see from the discussion for the mode $e_y(Q_x;l_z)$ that the bulk phonon band in this case also extends from $\sqrt{4k}$ to $\sqrt{8k}$ at the $\bar{M}$ point. In fact, the surface modes with polarization along the x axis are also degenerate with the corresponding y polarized modes, since at the $\bar{M}$ points all displacements parallel to the surface belong to the same E representation.

In summary, at the $\bar{M}$ point of the Brillouin zone bulk and surface phonons exist of both A- and E-type symmetry. The E-type bulk band extends in frequency from $\sqrt{4k}$ to $\sqrt{8k}$. When the interlayer coupling constant k_{12} is smaller than k, an E-type surface phonon lies below the bottom of the bulk band. In addition, regardless of whether k_{12} is different from or the same as k, there is an A-type

surface phonon (polarized parallel to z) with frequency $\sqrt{2k_{12}}$, localized in
the first layer. This is the Rayleigh wave. When k_{12} is different from k, we
also have a surface phonon of the same polarization with frequency $\sqrt{2k_{12}+2k}$,
which is localized in the second layer. The bulk band of this polarization
has a frequency of $\sqrt{4k}$ independent of $Q_\perp$. We note that the discussion of the
symmetry properties of the modes at $\bar{M}$ is also pertinent to the {100} surface
of a bcc crystal. The necessary modifications to calculate the frequencies
are easily made.

So far we have discussed the qualitative features of the eigenvectors
$e_\alpha(Q_{||};1_z)$ which satisfy the homogeneous equation (19.7). In particular, we
have been able to identify the frequencies of the various surface and bulk
modes at the $\bar{M}$ point of the Brillouin zone. To calculate the displacements
of the atoms and to make contact with experimental data (Sect.19.4) we find
that the Fourier-transformed Green's function [19.12] defined from these eigenvectors gives much more tractable results, even in cases when there are
several atoms per unit cell. As shown in previous work [19.12], the advantage of the Green's function technique is that the calculations are analytic,
except for solving a finite set of coupled equations.

Although this technique has been discussed in detail earlier [19.12], for
completeness some aspects of the calculation are reviewed here. The Fourier-
transformed Green's functions are defined as

$$U_{\alpha\beta}(1_z,1_z';Q_{||}\omega) = \sum_S \frac{e_\alpha^{(S)}(Q_{||};1_z)e_\beta^{(S)}(Q_{||};1_z')^*}{\omega^2 - \omega_S^2(Q_{||})} \quad , \tag{19.32}$$

where the sum is over all modes associated with the wave vector $Q_{||}$ and
$\omega_S(Q_{||})$ is the frequency of a particular mode. It is easy to see that these
functions satisfy an equation identical to (19.7) with an inhomogeneous term.
Thus

$$\omega^2 U_{\alpha\beta}(1_z,1_z';Q_{||}\omega) - \sum_{1_z''}\sum_\alpha d_{\alpha\gamma}(Q_{||};1_z,1_z'')U_{\alpha\beta}(1_z'',1_z';Q_{||}\omega) = \delta_{\alpha\beta}\delta_{1_z 1_z'} \quad . \tag{19.33}$$

Since it has been already described how the dynamical matrices are obtained,
the equations for the Green's functions, for the displacement patterns discussed earlier, can be written and, as before, the set of equations can be
solved by invoking an exponential ansatz. The details of such a calculation
will not be given here, as they are straightforward.

To make contact with the electron energy loss data there is one important
quantity that needs to be defined. This is the spectral density function
$\rho_{\alpha\beta}(1_z,1_z';Q_{||}\omega)$ defined as follows

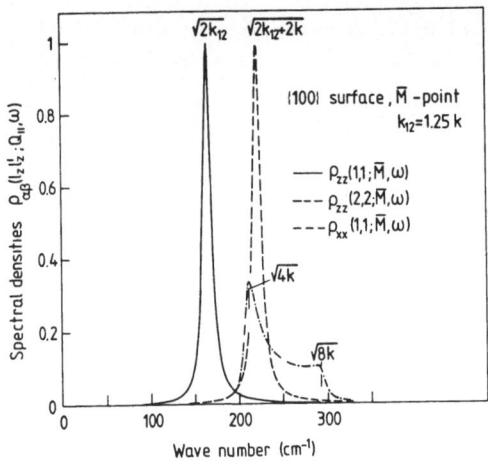

Fig.19.3. Spectral densities for
the M̄ point of a {100} surface
calculated in a nearest-neighbor
model. The A-type surface mode
localized in the first layer ap-
pears at $\omega = \sqrt{2k_{12}}$. We have assumed
k_{12} to be 1.25 k and k was chosen
to represent nickel. The A-type
surface mode of the second layer
appears at $\omega = \sqrt{2k_{12} + 2k}$, while
the bulk phonon frequency is $\sqrt{4k}$.
The bulk phonons appear as broad
features in the spectral density
of the third layer, and deeper
layers of course. The E-type bulk
bands extend from $\omega = \sqrt{4k}$ to
$\omega = \sqrt{8k}$. Some artificial broadening
[finite value of ε in (19.34)]
has been introduced to make the
δ functions appear as Lorentzians

$$\rho_{\alpha\beta}(1_z,1'_z;\mathbf{Q}_{\|},\omega) \equiv \sum_S e_\alpha^{(S)}(\mathbf{Q}_{\|};1_z)e_\beta^{(S)}(\mathbf{Q}_{\|};1'_z)^* \delta[\omega - \omega_S(\mathbf{Q}_{\|})]$$

$$= i\,\frac{\omega}{\pi}\,[U_{\alpha\beta}(1_z,1'_z;\mathbf{Q}_{\|},\omega + i\varepsilon) - U_{\alpha\beta}(1_z,1'_z;\mathbf{Q}_{\|},\omega - i\varepsilon)] \;. \quad (19.34)$$

The diagonal elements of the spectral density function are thus the square of
the amplitude of the displacements of the atoms. It has been shown [19.12,13]
that the scattering efficiency per unit frequency for the electrons is proportional to the spectral density. Figure 19.3 shows a plot of the spectral
densities for the surface phonons at the M̄ point of the Brillouin zone, discussed earlier. The features in the spectral density plot thus simulate the
electron energy loss spectra. The theoretical phonon dispersion curves are
obtained from the peak positions of the spectral densities calculated at a
number of points of the two-dimensional Brillouin zone.

19.3 Experimental Methods to Study Surface Phonon Dispersion

19.3.1 Electron Energy Loss Spectroscopy

The principle of electron energy loss spectroscopy, as well as many aspects
of the spectrometer design, have been described in a recently published mono-
graph by *Ibach* and *Mills* [19.14]. Here we focus our attention on the considerations specific to the measurement of surface phonon dispersion.

As already mentioned, one employs impact scattering to measure phonon dispersion. The inelastic intensity of such scattering events is distributed
(although not evenly) over the $Q_{\|}$ space. In any given spectrum, a small frac-

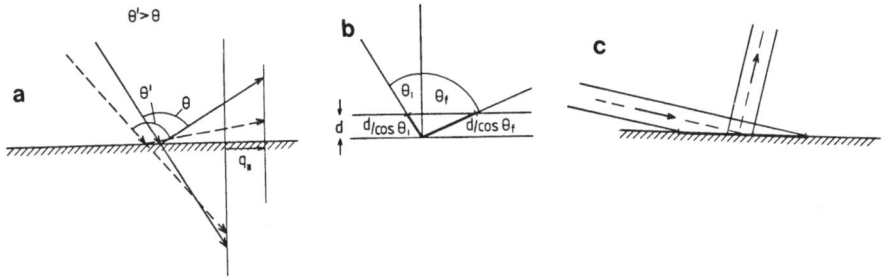

Fig.19.4a-c. Geometrical aspects of phonon spectroscopy with electrons.
a) To avoid beams with grazing polar angle the scattering angle θ should be
small. However, most spectrometers do not allow for θ < 90° since one also
wishes to have the lens elements close to the sample. b) Path length inside
the material as a function of θ_i and θ_f. c) Cross section of incident and
scattered beams proportional to cos θ_I and cosθ_S, respectively

tion of that $Q_{\parallel}$ space is sampled; the size of this space depends on the acceptance angle of the spectrometer. For reasons related to electron optics
[19.14], the acceptance angle of high-resolution spectrometers is rather
small, much smaller than that necessary for satisfactory resolution in $Q_{\parallel}$ space.
The design goal is therefore to have the acceptance angle as large as possible.
Larger acceptance angles are achieved by bringing the focal planes of the
lenses close to the sample. There are, however, limitations as to how far
one can go with this. One limit is set by the increasing image aberrations.
A second limit is set by the necessity to have a field free region around the
sample in order to have well-defined scattering angles. If the field penetrates
the last lens element into the space around the sample the actual scattering
angle deviates from the angle, which is read off the mechanical positioning
of the spectrometer. Finally, bringing the lens elements too close to the
sample reduces the minimum scattering angle θ. This has also an adverse effect
on the intensity of energy losses. As illustrated in Fig.19.4a, a larger minimum scattering angle θ requires the initial or final beam to be closer to
grazing incidence to achieve the same momentum exchange with the surface.
When the scattering is from deeper layers, the amount of material that has to
be penetrated by the incoming and outgoing beam is proportional to $\cos^{-1}\theta_i$
+ $\cos^{-1}\theta_f$ (Fig.19.4b). These beams will, therefore, experience an exponential
damping of the order of

$$I \sim \exp[-d(\cos^{-1}\theta_i + \cos^{-1}\theta_f)/\Lambda] \tag{19.35}$$

with Λ being the mean free path of the electron. With a mean free path of
about 5 Å for the electron energies relevant here and for a layer spacing of
1.8 Å, one begins to lose even second-layer information if either θ_i or θ_f

exceeds ~70°. A similar consideration applies also to the first layer: electrons scattered from there still have a nonvanishing probability to excite electronic "surface losses" [19.14] and may thus disappear from the scene of phonon-scattering events. The probability of surface losses is also proportional to $\cos^{-1}\theta_i + \cos^{-1}\theta_f$. The minimization of electronic surface losses, therefore, calls for a small minimum scattering angle θ as well.

In earlier work on vibrational losses of adsorbed molecules, where spectra at or near $Q_\| = 0$ were of primary interest and dipole scattering was employed, typical electron energies were between 2 eV and 10 eV. In the work described here, larger electron energies must be used to achieve sufficiently high $Q_\|$ (19.2,3). It is also useful to check the calibration of the angles θ_i and θ_s by observing at least one diffracted beam. Where this can be achieved, electron energies are above ~50 eV. Electron spectrometers suitable for measuring phonon dispersion, therefore, must allow for impact energies above 50 eV. On the other hand, we found that at about 300 eV impact energy multiphonon scattering events begin to dominate the spectrum [19.15,16]. Since such multiphonon events contribute a broad background to the spectrum with adverse effects on the signal-to-noise ratio, there is little reason in going up to electron energies higher than ~300 eV. In summary, one finds that an electron spectrometer optimized for work on surface phonon dispersion should feature a design slightly different from spectrometers employed for applications in surface chemistry and that one has to search for a compromise between acceptance angle, scattering angle, bending effects on trajectories near the sample, and impact energy.

One of the advantages of inelastic scattering of electrons is that it is sensitive also to vibrations parallel to the surface. Although observation of parallel modes is not favored by the typical scattering kinematics, one finds experimentally that for certain impact energies and scattering angles the cross section for parallel modes can be even larger than the cross section for perpendicular motions. This is in accordance with the scattering theory of *Tong* et al. [19.17] which accounts fully for the multiple-scattering nature of inelastic diffraction. A detailed comparison between theory and experiment regarding the intensity of phonon losses will be performed in the near future. A remark, however, about the experimental aspects of such a comparison seems to be appropriate at this stage. High-resolution electron spectrometers tend to be relatively unspecified in their acceptance angle. In fact, the acceptance angle is not even easy to determine. Absolute cross sections therefore may be off easily by a factor of two. Also one needs to consider that the theory assumes incident and scattered plane waves, while experimentally one works with beams and images of finite width. Typically electron spectrometers are built symmetrically, which means that the width w of the incident beam is equal to the beam width accepted by the analyzer. If the polar angle of the scattered electrons θ_S is smaller than the polar angle of the incident beam θ_I, the viewed area is smaller than the illuminated area (Fig.19.4c). The ratio of the number of scattered electrons per second and per interval $d\mathbf{k}_{||}$ to the number of incident electrons per second then is

$$\frac{dP}{d\mathbf{k}_{||}} = \frac{k^I}{k^S} |M(\mathbf{k}^I, \mathbf{k}^S)|^2 \frac{A}{(2\pi)^2} \quad , \tag{19.36}$$

where $\mathbf{k}^I, \mathbf{k}^S$ are the wave vectors of incident and scattered electrons, M is the appropriate matrix element and A is the surface area. The element $d\mathbf{k}_{||}$ may be converted into the solid angle through

$$d\mathbf{k}_{||} = (k^S)^2 \cos\theta_S \, d\Omega \quad . \tag{19.37}$$

Therefore one has

$$\frac{dP}{d\Omega} = \frac{k^I}{k^S} |M|^2 \frac{A}{(2\pi)^2} (k^S)^2 \cos\theta_S \quad \text{for} \quad \theta_I > \theta_S \quad . \tag{19.38}$$

If $\theta_S > \theta_I$ the viewed area is larger than the illuminated area. The number of scattered electrons then is not proportional to w, but to the width of the beam produced by the scattering from the illuminated area, which is $w \cdot \cos\theta_S / \cos\theta_I$. The scattering probability thus contains an extra factor $\cos\theta_S / \cos\theta_I$ and becomes

$$\frac{dP}{d\Omega} = \frac{k^I}{k^S} |M|^2 \frac{A}{(2\pi)^2} (k^S)^2 \frac{\cos^2\theta_S}{\cos\theta_I} \quad \text{for} \quad \theta_S > \theta_I \quad . \tag{19.39}$$

19.3.2 He Scattering

A beam of He atoms is formed by an adiabatic expansion from a high pressure (1-100 bar) cell through an orifice of 5-30 μm diameter. The supersonic expansion narrows the longitudinal velocity distribution down to roughly 1% of the width of the Maxwell distribution inside the pressure cell. The beam velocity is

$$v = \sqrt{5kT/m} \quad , \tag{19.40}$$

with T the temperature in the pressure cell and m the particle mass. Since only the longitudinal and not the transverse velocity distribution is narrowed by the expansion, the beam must be collimated to less than 1°. The formation of a monochromatic He beam, therefore, is even more inefficient than the production of a monochromatic electron beam. Unlike electrons, He atoms do not disappear from the vacuum unless pumped. Many stages of differential pumping are required between the expansion cell and the sample, and also between the sample and the mass spectrometer. In modern He-beam equipment [19.3], the equivalent background He pressure at the detector is $\sim 10^{-13}$ Torr, eighteen orders of magnitude lower than in the expansion cell. Unfortunately, the detection of He atoms is also rather inefficient since the atoms need to be ionized before detection and the ionization efficiency is only of the order of 10^{-6}. Elastic scattering events and the various inelastic events are separated by the time-of-flight technique, preferably using a cross-correlation chopper [19.18] which helps to minimize signal losses. Because of these obstacles, it is only recently that the technology to perform successful inelastic He scattering experiments became available.

At this point it seems useful to mention an important difference between an electron energy spectrum and a time-of-flight spectrum in He scattering, which has to do with the different masses of the particles involved. By combining energy and momentum conservation (19.2,3) one has the wave vector parallel to the surface

$$Q_{\parallel} = \frac{\sqrt{2m}}{\hbar} \left(\sqrt{E} \sin\theta_I - \sqrt{E \pm \hbar\omega} \sin\theta_S \right) , \tag{19.41}$$

where the minus and plus signs refer to energy loss and gain, respectively. Phonon dispersion relations are conventionally plotted with $Q_{\parallel}$ and ω as the x and y axis, respectively. In this plane, (19.41) is a parabola when plotted for fixed θ_I, θ_S and beam energy E. For electrons the mass is small and beam energies $E \gg \hbar\omega$ have to be used to reach $Q_{\parallel}$ values on a scale comparable to the size of the Brillouin zone. The parabola, in this case, degenerates to a vertical line. In other words, a loss spectrum, taken at fixed θ_I, θ_S and E, is a spectrum at (nearly) constant $Q_{\parallel}$. This does not hold for He scattering.

Figure 19.5 schematically illustrates (19.41) for a He beam of E = 20 meV and $\theta_I = \theta_S = 45°$ and $\theta_I = 34°$, $\theta_S = 56°$. The phonon dispersion curve corresponds to the so-called S_4 surface mode on a Ni{100} surface along the <110> direction. Positive and negative ω branches represent losses and gains, respectively. For (19.41) to intersect with the dispersion curve at all, the latter has to be extended through more than the first Brillouin zone. According to Fig.19.5, one finds a surface phonon loss at $\hbar\omega \sim 13.3$ meV, a phonon gain at $\hbar\omega \sim -17$ meV, and a quasielastic peak ($\omega = 0$) when $\theta_I = \theta_S = 45°$. Unlike in elec-

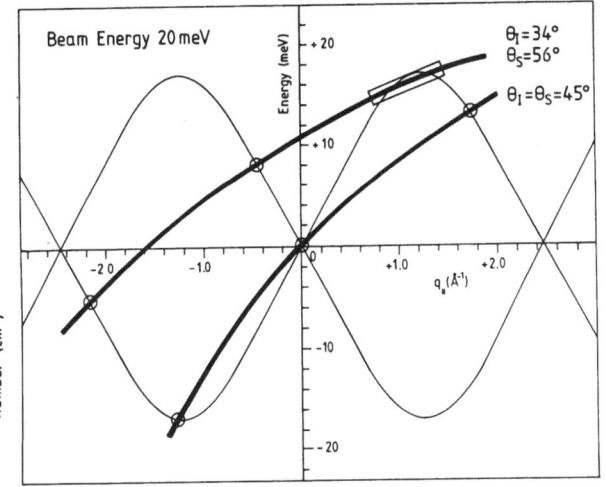

Fig.19.5. Kinematics of surface phonon spectroscopy with the He atoms of 20 meV energy. The kinematic curve (19.1) intersects with the phonon dispersion more than once and at different $Q_{||}$. In electron energy loss spectroscopy the kinematic equation is practically a vertical line which makes electron spectroscopy a "constant $Q_{||}$" spectroscopy

tron spectroscopy, where $Q_{||}$ is constant in a spectrum, loss and gain are not symmetric about the elastic events since they belong to different $Q_{||}$.

An interesting case is also $\theta_I = 34°$, $\theta_S = 56°$. There the kinematic equation intersects the dispersion curve at grazing incidence near $Q_{||} = 1.2$ $Å^{-1}$. Since the dispersion curve has a finite width due to finite lifetime of phonons and also the kinematic curve has a finite width due to the finite energy resolution of the He-beam experiment, one finds a rather broad peak extending from 14-18 meV in this case. The peak will also be particularly intense, since the intersection of the kinematic and phonon dispersion extends over a large Q space. The effect has been named "kinematic focusing" [19.19,20]. This and other subtle considerations regarding the time-of-flight technique have been described in detail recently by *Brusdeylins* et al. [19.3].

In reviewing the experimental data on surface phonons obtained by He scattering, one notices that with one or two exceptions mainly acoustical surface modes have been observed and only on those which carry displacements perpendicular to the surface. This is not accidental but relates to the nature of the inelastic scattering process. Helium atoms are scattered by the charge density of the substrate at an appreciable distance above the surface (4 to 8 a.u. [19.21]). The phonon-induced corrugation in the charge density at this distance practically vanishes for optical substrate modes and for modes polarized parallel to the surface plane. The effect has frequently been cited as the "Armand" effect. With recent inelastic scattering data on Ag{111} by *Harten* et al. [19.22] more quantitative information on this effect has become available. In Fig.19.6 we have plotted the inelastic inten-

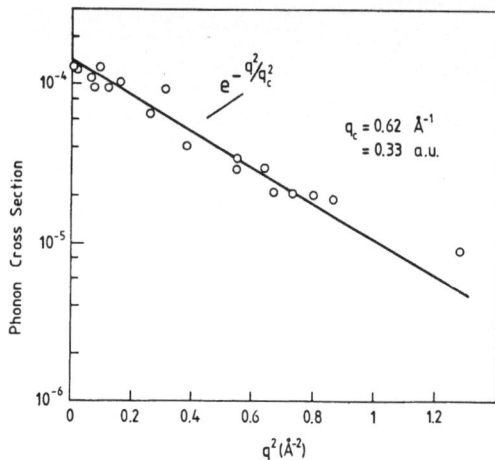

Fig.19.6. Intensities of inelastic He scattering from Rayleigh phonons on Ag{111} along the <112> direction. The experimental data of Harten et al. [19.22] have been corrected for the Bose factor $\bar{n}(\omega)$ and $\bar{n}(\omega)+1$ for phonon gain and loss events, respectively. The falloff of the intensity is in accordance with a theoretical model of Bortolani et al. [19.23,24]. The decrease in the cross section with Q results from the interaction of the He atoms with more than one surface atom

sity of phonon gain and phonon losses of the Rayleigh phonon on Ag{111} obtained for the <112> direction [19.22]. The data have been corrected for the Bose factor $\bar{n}(\omega) = [\exp(\hbar\omega/kT) - 1]^{-1}$ and $\bar{n}(\omega)+1$ for energy losses and gains, respectively. These data plotted on a log scale of phonon cross section versus Q^2, with **Q** the phonon wave vector, yield a straight line (Fig.19.6), indicating that the intensities fall off according to a Gaussian [$\sim\exp(-Q^2/Q_c^2)$]. This behavior was predicted by *Bortolani* et al. [19.23,24]. The calculated value for the critical wave vector Q_c for silver (0.737 Å^{-1}) agrees reasonably well with the experimental data (Fig.19.6). The large decrease in the intensities for phonon scattering requires long data accumulation times ($\sim$ 6h) when one approaches the zone boundary.

Up to now, vibrations within adsorbed species have not been found with helium scattering. As far as adsorbate modes of high frequency are concerned, the reason relates to the slow velocity of the helium beam. It is easy to appreciate that the vibration frequency of a mode ν must be small compared to the inverse of the scattering time of the particle, otherwise the forces exerted on the particle through the phonons cancel. Thus one has the condition

$$\nu_{phonon} \ll \nu_{beam}/\Lambda \quad ,$$

where ν is the beam velocity and Λ a typical interaction length ($\Lambda \sim 10^{-8}$ cm). A 20 meV He beam has a velocity $\nu \sim 1.0 \cdot 10^5$ cm/s which sets an upper limit of 10^{13} s^{-1} to the phonon frequency. The argument can be stated also more quantitatively. It has been shown by *Levi* and *Suhl* [19.25,26] that the matrix element for one-phonon scattering contains a frequency-dependent prefactor which is the Fourier transform of the phonon-induced forces on the He atoms along the trajectory. To explain the frequency dependence imposed by this

472

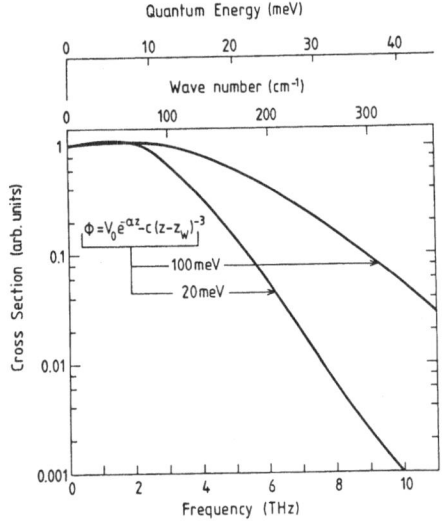

Fig.19.7. The "slow collision effect" in He scattering. The cross section decreases rapidly for higher phonon frequencies when frequency is larger than the inverse of the collision time

relation, one may study the simple case of scattering from a uniform jellium type surface, for which realistic potentials and potential parameters are available [19.16]. The result is plotted in Fig.19.7 for two different beam energies. The potential parameters are taken from *Harris* and *Liebsch* [19.21] and correspond to a copper surface. One sees from Fig.19.7 that for a 20 meV beam the phonon cross section has dropped by an order of magnitude already at a phonon energy of 21.4 meV (= 173 cm^{-1} = 5.2 THz) and decreases nearly exponentially for higher phonon energies. One might consider circumventing this problem by employing higher beam energies. Unfortunately, however, the multiphonon scattering of low-frequency phonons then causes a large background. It seems, therefore, that it will be increasingly more difficult to see vibration frequencies with He scattering the higher the frequency is.

19.3.3 Comparison of He Scattering and Electron Scattering

In the preceding two paragraphs, several distinct differences between the two techniques available for measuring surface phonon dispersion have already been pointed out. We continue the discussion by comparing data as obtained by the two techniques directly. First, however, another comment on the phonon cross section may be helpful. The one-phonon cross section contains the prefactors $\omega^{-1}(\bar{n}(\omega)+1)$ and $\omega^{-1}\bar{n}(\omega)$ for energy losses and gains, respectively (with $\bar{n}(\omega) = [\exp(\hbar\omega/kT) - 1]^{-1}$, the phonon occupation number). This introduces a significant frequency dependence of the cross section, which makes modes of higher frequency more difficult to detect. For helium scattering, this frequency de-

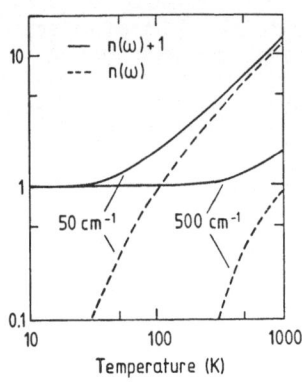

Fig.19.8. Cross section as a function of temperature, from the statistical factors $\bar{n}(\omega)$ = $[\exp(\hbar\omega/kT) - 1]^{-1}$ and $\bar{n}(\omega) + 1$ for phonon gain and loss events. The plot suggests that room temperature is preferred for scattering from low-frequency phonons whereas low temperatures are preferred for high-frequency phonons, to avoid a multiple-loss background

pendence appears on top of other adverse factors, as discussed before. This is particularly true for adsorbate layers where the motion is essentially confined to one layer.

For Rayleigh phonons the frequency dependence is not $\sim \omega^{-1}[n(\omega) + 1]$ and $\omega^{-1}\bar{n}(\omega)$, but roughly $\sim[\bar{n}(\omega) + 1]$ and $\sim \bar{n}(\omega)$, respectively, because at higher frequencies the Rayleigh modes become more localized to the surface. The characteristic decay length is (roughly) inversely proportional to the wave vector and thus approximately $\sim \omega^{-1}$. Since the eigenvectors of the atomic displacements are normalized, i.e.,

$$\sum_{l_z, \kappa, \alpha} |e_\alpha(Q, l_z\kappa)|^2 = 1 \quad,$$

one can see that the square of the amplitude of the surface layer must scale inversely with the number of layers participating in the motion and thus must be proportional to ω.

These considerations also lead to conclusions about the most suitable temperature at which experiments should be performed. In Fig.19.8 we have simply plotted $\bar{n}(\omega)$ and $\bar{n}(\omega) + 1$ for a 50 cm^{-1} and a 500 cm^{-1} loss, respectively, as a function of temperature. To observe a 50 cm^{-1} loss for a Rayleigh wave, for example, one should keep the sample temperature at 300 K or one may even stay above it, if possible. In electron energy loss spectroscopy where loss and gain of the same phonon appear in the spectrum, one has the additional advantage of being able to measure the difference between gain and loss peaks which enhance the effective resolution of the experiment by a factor of two (Fig.19.9). On the other hand, if one is interested in observing the higher frequencies of chemisorbed species, nothing is gained by using a higher temperature. On the contrary, it is preferable to work with lower temperature, since lower temperatures reduce the multiphonon background caused by low-fre-

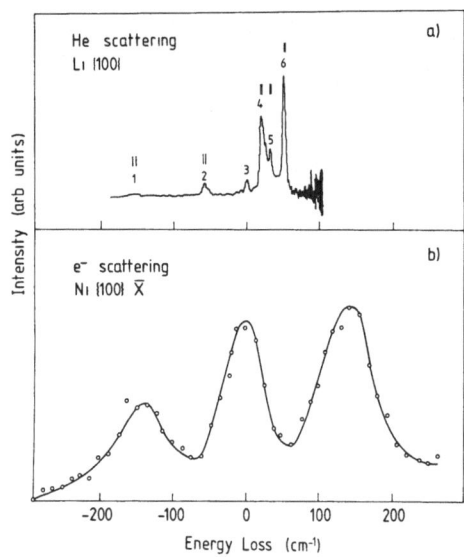

Intensity (arb units)

e⁻ scattering
Ni {100} X̄

b)

-200   -100   0   100   200

Energy Loss (cm⁻¹)

<u>Fig.19.9.</u> a) Surface phonon spectrum from LiF {100} obtained with He atoms after conversion from the time-of-flight scale to the energy scale [19.3]. Peaks in the spectrum represent phonons at different $Q_{\parallel}$. b) Surface phonon spectrum from Ni{100} obtained with electrons. The two peaks are due to gain and loss event of the same phonon at constant $Q_{\parallel}$

quency vibrations. In summary, as a rule of thumb, acoustical surface modes should be investigated at room temperature, while optical substrate and adsorbate modes should be investigated on a cooled sample.

We now turn to a direct comparison of experimental results as obtained from helium and electron scattering. In Fig.19.9 two spectra are displayed. The spectrum in Fig.19.9a was obtained with He scattering from LiF [19.3]. The time-of-flight spectrum is converted into an energy spectrum. The energy resolution depends on the time of flight, the kinematic factor discussed in the previous section, and the monochromaticity of the beam. The estimated energy resolution, according to *Brusdeylins* et al. [19.3], is plotted under each peak. One notices that the resolution varies between 4 and 20 cm⁻¹ (0.5 - 2.5 meV). As described in Fig.19.5, the various peaks in Fig.19.9a are phonon loss and gain events for phonons with different $Q_{\parallel}$. The peak labelled 2 is a spurious effect caused by a background of nonmonochromatic He atoms with which the He beam is polluted. These He atoms contribute a peak to the time-of-flight spectrum when their energy matches the condition for diffraction under the given angles of incidence and observation [19.27]. The scanning time for the spectrum was roughly 15 s per channel.

Figure 19.9b displays an energy loss spectrum for the clean Ni{100} surface. The two peaks here are phonon loss and gain at the same $Q_{\parallel}$ (1.26 Å⁻¹, X̄ point). The scan time was 5 s per channel. Obviously, the resolution of ~60 cm⁻¹ (7.5 meV) is significantly worse than in the helium spectrum. This lower re-

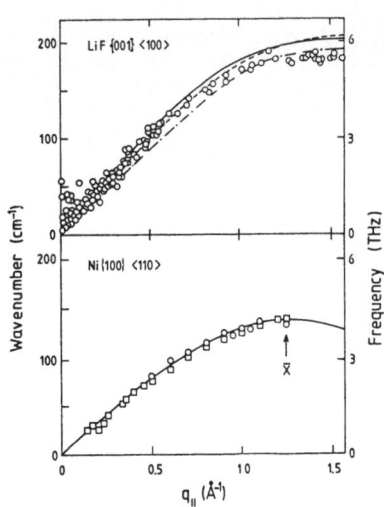

Fig.19.10. Comparison of dispersion data obtained by He scattering and electron scattering

solution is only partly compensated by the advantage of having only phonons of one $Q_{||}$ in the spectrum.

The two methods may also be compared with respect to the accuracy with which they determine the dispersion relation for surface phonons. In Fig. 19.10a,b we display data obtained by He scattering on the LiF {100} surface [19.3] and by electron scattering on the Ni{100} surface [19.15]. In both cases, the surface phonon is polarized within the sagittal plane, mainly perpendicular to the surface. Each data point on a LiF-dispersion curve represents the measured center position of a loss or gain peak in a time-of-flight spectrum with the $Q_{||}$ reduced to the first Brillouin zone. Data points for Ni{100} were obtained by measuring the difference between the phonon gain and phonon loss peak in electron energy loss spectra, such as Fig.19.9b. Considering the energy resolution of the He experiments, the scatter in the data shown in Fig.19.10a is unexpectedly high, particularly for the low Q range. The reason may be a slight misalignment of the sample orientation. Since in helium scattering data from several Brillouin zones are collected simultaneously, as described earlier, the technique is more sensitive to such misalignments. With the additional sources of error in He scattering, which are not present in electron scattering, the intrinsically better energy resolution of He scattering has not lead to a significantly better accuracy in determining the dispersion curve, at least so far.

19.4 Comparison of Some Experimental Results with Lattice Dynamical Calculations

19.4.1 Surface Phonons of Clean Surfaces

In Sect.19.2 we studied a simple lattice dynamical model which provided us with some qualitative insight into the surface phonon spectrum of the {100} surface of an fcc crystal. We now relate these considerations to experimental studies on the clean Ni{100} surface. Nickel is a particularly good example since the bulk phonon spectrum is reasonably well described by the nearest-neighbor central force field used in the model. The experimental data shown in Fig.19.11 have been obtained using electron energy loss spectroscopy [19.15,28]. For both directions the scattering plane was aligned with the sagittal plane. As shown earlier [19.14,17], only modes of even symmetry with respect to the sagittal plane can be observed when this plane is a symmetry plane of the crystal, as it is here. For the <110> direction, a second acoustic surface phonon of odd symmetry (S_1 mode) exists [19.5,14] which is, however, not observed. The results are compared with the theory using two sets of parameters. Firstly, the force constant between the atoms in the first and second layers k_{12} was assumed to be equal to the bulk constant k, which was matched to the maximum phonon frequency of nickel ($\sqrt{8k} = 295$ cm^{-1}). As can be seen from Fig.19.11, the model describes well the frequencies of the Rayleigh waves in the acoustic limit. As $Q_{\parallel}$ approaches the Brillouin zone ($\bar{X}$ or $\bar{M}$) the Rayleigh wave becomes more and more localized to the surface and, therefore, sensitive to the bonds between atoms near the surface. We have seen in Sect.19.2 that at $\bar{M}$ the Rayleigh wave is entirely a vibration of the first layer with the second at rest (Fig.19.2). At $\bar{X}$ the second layer participates in the motion, the amplitude, however, decays rapidly in deeper layers. As shown in Fig.19.11, both at $\bar{X}$ and at $\bar{M}$ the data points fall above the theoretical curve if one does not allow for an increased force constant between the first and second nickel layer. An optimum fit to both directions is obtained with $k_{12} = 1.2\ k$. This corresponds to an increased bonding between the atoms of the first and the second layer. The result is intuitively appealing, as one would expect an increase in the bond strength of the backbonds of the nickel surface atoms, as a consequence of the reduced coordination. The increased bonding should also lead to some contraction of the lattice spacing between the first and second nickel layers. Indeed a 3.5% contraction has been observed recently using ion scattering [19.29]. With the force constant k_{12} different from the bulk, a second layer mode at $\sqrt{2k_{12} + 2k}$ should also appear at $\bar{M}$ near 215 cm^{-1}. So far, this mode has not been clearly

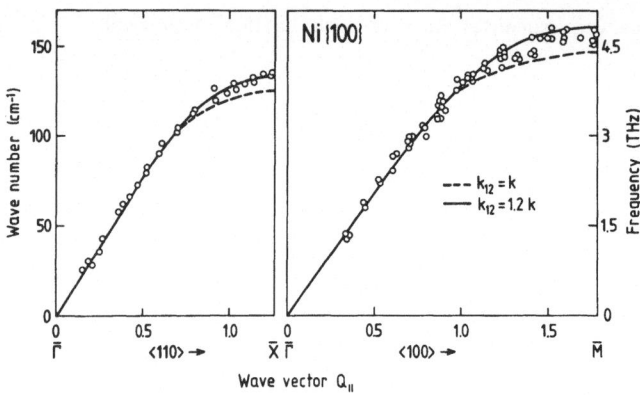

<u>Fig.19.11.</u> Dispersion curves of the Rayleigh phonons of a Ni{100} surface along the <110> and <100> directions. The dashed line is the result of the lattice dynamical model, see text, with the surface force constants unchanged. The full line is obtained if one allows the coupling constant between the first and second layer to be 20% larger than in the bulk

resolved. Here, we may recall that the resolution in electron energy loss spectroscopy (when applied to phonon dispersion measurement) is only 50 - 60 cm^{-1}. It is, therefore, not possible to separate the first and second layer modes unless one finds scattering conditions where the second layer mode appears at par if not stronger than the feature from the first layer. If the electron undertook only single elastic scattering events, such a condition would never occur, since, as remarked earlier, electrons scattered from the second layer experience a larger attenuation. Due to multiple elastic scattering, however, one might find appropriate energies and angles which bring out second-layer vibrational features. So far, only a broad peak centered between the frequency of the first and second layer mode has been observed for a few specific scattering conditions. From these data the frequency of the second layer mode could not determined.

The contraction of the interlayer spacing and the increased coupling constant is probably particularly large for the relatively open {100} crystal face, while the effect is expected to be smaller for the {111} surfaces. Current investigations of the {111} surfaces of silver [19.22] and gold [19.30] using the He-beam technique will explore this same issue further. Rayleigh surface phonons have also been studied on the alkali halides LiF, NaF, and KCl, and a detailed comparison between experiment and theory has been [19.3]. With these materials, however, a proper lattice dynamical description is substantially more complicated due to Coulomb and dipole contributions to the force field, and involves a number of parameters. At the surface these parameters may change, but it will take more than one surface mode in one direction to decipher what the changes are. Nickel provides a fortunate example as the lattice dynamics of the bulk and the surface is simple.

19.4.2 Dispersion of Adsorbate Modes

It appears that currently a number of adsorbate systems are being investigated, and by the time this publication is in print our information base will have greatly expanded, to which both electron energy loss spectroscopy and He scattering will contribute. The focus will be on higher frequency vibrations by the first and on the lower frequencies by the latter for reasons discussed in Sect.19.3. As far as the published literature is concerned, the dispersion of adsorbate vibrations has been measured only for two systems. In the first study [19.31] the dispersion of the CO-stretching frequency of CO molecules adsorbed on Cu{100} was measured and analyzed in terms of dipolar coupling between the molecules. Since dipole scattering of electrons was used in this study, only $Q_{\parallel}$ vectors near the $\bar{\Gamma}$ point were explored. The second study of adsorbate vibrations employed impact scattering of electrons in the energy range between 5-200 eV, and the entire dispersion curve at least in one crystallographic direction was measured. The study deals with the p(2×2) and c(2×2) ordered overlayers of oxygen on the Ni{100} surface [19.16,32]. Here we concentrate on the c(2×2) overlayer for which the dispersion is more interesting and a lattice dynamical analysis is also available. The experimental results are shown in Fig.19.12 together with the theoretical dispersion curves. The scattering plane was aligned along the sagittal plane, as in previous cases. The dispersion branches are, therefore, of even symmetry. Oxygen on Ni{100} is assumed to reside in the fourfold hollow site which has C_{4v} symmetry. The symmetry analysis of the electron energy loss spectra is consistent with this site. For example at $Q_{\parallel} = 0$ only a single oxygen vibration is seen at 320 cm^{-1} on a well-ordered c(2×2) structure, which is, therefore, assigned to the vertical A-type motion of oxygen. The parallel branch appears only at finite $Q_{\parallel}$. It has a frequency of about 450 cm^{-1} and is nearly dispersionless, while the branch arising from the vertical motion of oxygen near $\bar{\Gamma}$ exhibits an appreciable dispersion. The lattice dynamical calculations [19.33] based on central forces and nearest-neighbor interactions use the Fourier-transformed Green's function technique as discussed in Sect.9.2, to give good agreement with the observed dispersion of these modes. We refer the reader to [19.33] for details of the calculations and summarize only the essential features here. These calculations are quasianalytic, since only a finite number (twelve in this case) of coupled equations need be solved to ob-

479

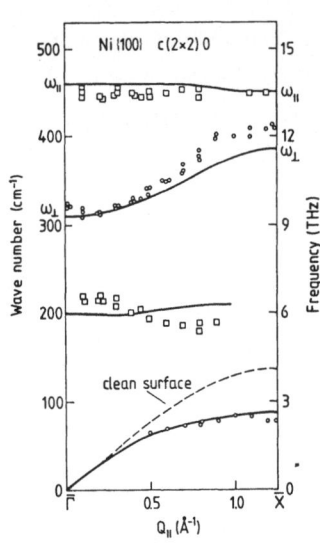

Fig.19.12. Dispersion curves of c(2 × 2) oxygen overlayer on Ni{100} [19.32]. The two upper branches are oxygen modes of even symmetry with respect to the sagittal ({110}) plane spanned by $\mathbf{Q}_{\parallel}$ and the surface normal. At $\bar{\Gamma}$ and $\bar{X}$ the modes are polarized strictly parallel and perpendicular to the surface, while along the $\bar{\Gamma}$-$\bar{X}$ direction the parallel and vertical motions couple. The acoustic branch is the Rayleigh surface phonon. Its frequency near $\bar{X}$ is substantially reduced, indicating a relaxation of the bonds between the nickel atoms in the first and second layers. The mode near 200 cm^{-1} is a surface resonance induced by the adsorbate layer. The full lines are the results of the lattice dynamical model as described in the text

tain the displacements of the adsorbate and substrate surface layer atoms. The force constant for the bulk nickel atoms are obtained from a one-parameter fit to the observed bulk phonon dispersion curve. Since the available ab initio calculations of the oxygen-nickel force constant [19.34] do not agree well with the data at the $\bar{\Gamma}$ point when oxygen atoms sit at 0.9 Å above the nickel surface [19.12], we have assigned the value of this force constant such that the calculated frequency for the vibration of the oxygen atom perpendicular to the surface near the $\bar{\Gamma}$ point ($\mathbf{Q}_{\parallel} = 0$) of the two-dimensional Brillouin zone is 310 cm^{-1}. The vibration frequency of the oxygen atoms parallel to the surface is then reproduced at 450 cm^{-1}, in accord with the experimental data. In fact, with the inclusion of a modest amount of oxygen-oxygen lateral interaction (a tenth of the oxygen-nickel interaction) the dispersion curves for both parallel and perpendicular vibration of oxygen atoms are in reasonable agreement with the experimental data, as seen in Fig.19.12. The value of ϕ_{01}'' (the second derivative of oxygen-nickel pair potential) that enters the calculation is 9.5×10^4 dynes/cm. This leads to a force constant whose value lies between that in bulk NiO and the NiO molecule. Although this seems like a reasonable choice, it is still not clear why this force constant should change by almost a factor of two from its value of $\phi_{10}'' = 16.29 \times 10^4$ dynes/cm for the p(2 × 2) overlayer of oxygen on Ni{100}.

In addition to the dispersion (or the lack) of the perpendicular and parallel modes of oxygen, two other substrate-associated features appear in the dispersion curves. One of them is the Rayleigh wave or the S_4 mode whose dispersion for the clean Ni{100} surface has been discussed previously in this section. The astonishing point about its dispersion, in the presence of an adsorbate like oxygen, is that at the $\bar{X}$ point (the zone boundary) its frequency drops to about 80 cm^{-1} from a clean surface value of 135 cm^{-1}. Since at the zone boundary this mode is most sensitive to the coupling between the first and second layer nickels atoms, this drastic reduction in the frequency of this mode can be accounted for, in our calculations, by assuming a reduction in the force constant between the first and second layer nickel atoms by 70% of its value in the bulk.

Finally, there is the feature around 200 cm^{-1}, with a small amount of dispersion, which disappears at about quarter of the way to the zone boundary ($\bar{X}$ point). This is a surface resonance mode of the adsorbate-substrate combined motion, as encountered before [19.12]. Near the zone center the motions of oxygen and nickel atoms associated with this mode are parallel to the surface. If the force constant between the first and second layer of nickel atoms is taken to be the same as its value in the bulk crystal, this mode extends, in the calculations, all the way to the zone boundary. The fact that the surface resonance mode disappears around the same region in the theoretical calculations as in the data when the interlayer force constant is reduced from the bulk value strengthens our proposal of the outward relaxation of the nickel surface layer in the presence of c(2×2) oxygen overlayer. Thus, for the clean Ni{100} surface, agreement between theory and experiment for the dispersion of the S_4 mode suggests a stiffening of the interlayer force constant by 20%, leading to an inward relaxation of the surface atoms, while with a c(2×2) oxygen overlayer on the same surface a softening of this force constant, implying an outward relaxation, is proposed. These conclusions are in qualitative agreement with recent ion scattering data [19.29].

Acknowledgements. We thank Professor V. Celli for many helpful conversations. One of us (TSR) would like to acknowledge partial support from U.S. Department of Energy under contract No. DEAT 0379-ER-10432 and also the hospitality of members of the Physics Department at the University of California, Irvine, during her stay there.

References

19.1 G.J.M. Horne, D.R. Miller: Phys. Rev. Lett. **41**, 511 (1978)
19.2 B. Feuerbacher, M.A. Adriaens, H. Thuis: Surf. Sci. **94**, L17 (1980)
19.3 G. Brusdeylins, R.B. Doak, J.P. Toennies: Phys. Rev. Lett. **46**, 1138 (1980); Phys. Rev. B27, 3662 (1983)
19.4 N. Cabrera, V. Celli, R. Manson: Phys. Rev. Lett. **22**, 346 (1969)
19.5 R.E. Allen, G.P. Alldredge, F.W. de Wette: Phys. Rev. B4, 1661 (1971)
19.6 S.W. Müsser, K.H. Rieder: Phys. Rev. B2, 3034 (1970)
19.7 G. Benedek: Phys. Status Solidi B58, 661 (1973)

19.8 J.E. Black, D.A. Campbell, R.F. Wallis: Surf. Sci. **105**, 629 (1981)
19.9 J.E. Black, B. Laks, D.L. Mills: Phys. Rev. B22, 1818 (1980)
19.10 V. Bortolani, F. Nizzoli, G. Santoro: Phys. Rev. Lett. **41**, 39 (1978)
19.11 F. Cyrot-Lackmann: Surf. Sci. **15**, 539 (1969);
      R. Haydock, V. Heine, M.J. Kelly: J. Phys. C5, 2845 (1972)
19.12 T.S. Rahman, J.E. Black, D.L. Mills: Phys. Rev. B25, 883 (1982); Phys.
      Rev. B27, 4059 (1983)
19.13 V. Roundy, D.L. Mills: Phys. Rev. B5, 1347 (1972)
19.14 H. Ibach, D.L. Mills: *Electron Energy Loss Spectroscopy and Surface
      Vibrations* (Academic, New York 1983)
19.15 S. Lehwald, H. Ibach, J. Szeftel, T.S. Rahman, D.L. Mills: Phys. Rev.
      Lett. **50**, 518 (1983)
19.16 H. Ibach: To be published
19.17 S.Y. Tong, C.H. Li, D.L. Mills: Phys. Rev. Lett. **44**, 407 (1980);
      Phys. Rev. B24, 806 (1981);
      C.H. Li, S.Y. Tong, D.L. Mills: Phys. Rev. B21, 3057 (1980)
19.18 G. Comsa, R. David, B.J. Schumacher: Rev. Sci. Instrum. **52**, 789 (1981)
19.19 G. Benedek: Phys. Rev. Lett. **35**, 234 (1975)
19.20 G. Benedek, G. Brusdeylins, J.P. Toennies, R.B. Doak: Phys. Rev. B27,
      2488 (1983)
19.21 See eg.g. J. Harris, A. Liebsch: J. Phys. C15, 2215 (1982);
      Phys. Rev. Lett. **49**, 341 (1982)
19.22 R.B. Doak, U. Harten, J.P. Toennies: Phys. Rev. Lett. **51**, 578 (1983).
      We gratefully acknowledge that Dr. Harten has made this data available
      to us prior to publication
19.23 V. Bortolani, A. Franchini, F. Nizzoli, G. Santoro, G. Benedek, V.
      Celli: Surf. Sci. **128**, 249 (1983)
19.24 V. Bortolani, A. Franchini, N. Garcia, F. Nizzoli, G. Santoro: To be
      published
19.25 A.C. Levi, H. Suhl: Surf. Sci. **88**, 221 (1979)
19.26 A.C. Levi: Nuovo Cim. **54**, 357 (1979)
19.27 W. Allison, R.F. Willis, M. Cardillo: Phys. Rev. B23, 6824 (1981)
19.28 M. Rocca, S. Lehwald, H. Ibach: To be published
19.29 F. Van der Veen: Private communication and to be published
19.30 M. Cates, D.R. Miller: J. Electron Spectrosc. Relak. Phenom. **30**, 157
      (1983)
19.31 S. Anderson, B.N.J. Persson: Phys. Rev. Lett. **45**, 1421 (1980)
19.32 J.M. Szeftel, S. Lehwald, H. Ibach, T.S. Rahman, J.E. Black, D.L. Mills:
      Phys. Rev. Lett. **51**, 268 (1983)
19.33 T.S. Rahman, D.L. Mills, J.E. Black, J. Szeftel, S. Lehwald, H. Ibach:
      To be published
19.34 T. Upton, W. Goddard: Phys. Rev. Lett. **46**, 1635 (1981)

20. Intrinsic and Extrinsic Surface Electronic States of Semiconductors *

J. D. Dow, R. E. Allen, and O. F. Sankey

With 10 Figures

20.1 Introduction

In the bulk of a tetrahedral semiconductor, a single substitutional s-p bonded impurity or vacancy will ordinarily produce four "deep" levels with energies near the fundamental band gap: one s-like (A_1) and three p-like (T_2) [20.1]. These deep levels may lie within the fundamental band gap, in which case they are conventional deep levels, or they may lie within either the conduction or the valence band as "deep resonances." A sheet of N vacancies produces 4N such deep levels —namely, the intrinsic surface-state energy bands, which may or may not overlap the fundamental gap (to a first approximation, insertion of a sheet of vacancies is equivalent to creating a surface).

   Intrinsic surface states share underlying physical principles with deep impurities because they too result from localized perturbations of a semiconductor [20.2], and so their energies can be predicted by extending to surfaces ideas developed by *Hjalmarson* et al. [20.1] for the deep impurity problem. This has been done by several authors [20.3-12], most notably by Allen and co-workers [20.13-17].

   Extrinsic and native-defect surface states are also governed by similar physical principles, and are especially interesting in the light of the Schottky barrier problem: *Bardeen* showed that modest densities of surface states on a semiconductor can "pin" the Fermi level [20.18], forming a Schottky barrier. The bulk Fermi energies of the semiconductor, the metal, and the semiconductor surface must align (Fig.20.1). If the semiconductor is heavily doped n-type, the surface Fermi energy is approximately the energy of the lowest empty surface state. The bands bend to accommodate this alignment of Fermi levels, forming the Schottky barrier. Thus the Schottky barrier height is given by the energy of the lowest naturally empty surface state, relative to the conduction band edge. In 1979 *Spicer* et al. proposed that the Bardeen

*We are grateful to the U.S. Army Research Office and the U.S. Office of Naval Research for supporting this work (Contract Nos. DAAG29-83-K-0122, N00014-84-K-0352, and N00014-82-K-0447).

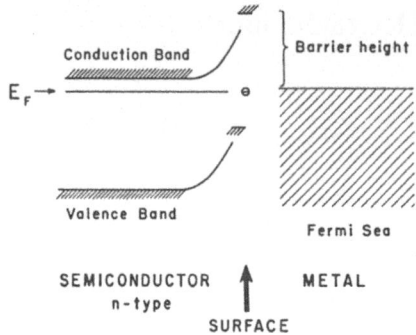

Fig.20.1. Fermi-level pinning.
Band edges of the bulk semiconductor and the semiconductor surface,
and the Fermi energy of the metal,
the surface of the semiconductor,
and the semiconductor are all shown
as functions of position. The
lowest energy surface-defect level
that is not fully occupied (before
charge is allowed to flow) is denoted by an open circle. This level
and the Fermi levels of the n-type
semiconductor and the metal align

surface states responsible for pinning the Fermi energy are due to native
defects [20.18-22].

Surface core excitons are similar to surface defect states, as can be seen
by using the optical alchemy approximation [20.23] or the $Z+1$ rule [20.24].
Consider core excitation of a Ga atom at the surface of GaAs; the radius of
the core hole is sufficiently small so that the hole can be assumed to have
zero radius (i.e., the hole is equivalent to an extra proton in the nucleus).
Thus the core-excited electron feels the potential of an atom whose atomic
charge Z is greater than that of Ga by unity, namely Ge. The Ga core exciton
spectrum is then approximately the same as the spectrum of a Ge impurity on
a Ga site. Hence the core exciton states in semiconductors can be either
"shallow" (Wannier-Mott excitons) or "deep" (Hjalmarson-Frenkel excitons),
as is the case for impurity states. The deep Hjalmarson-Frenkel excitons are
similar to the surface deep levels associated with impurities.

In this paper we show that the physics of deep impurity levels, intrinsic
surface states, surface impurity states, Schottky barriers, and Hjalmarson-
Frenkel core excitons are all similar.

20.2 Deep Defect Levels at the Surface: Schottky Barriers and Fermi-Level Pinning

The basic physics of most Schottky barriers can be explained in terms of the
Fermi-level pinning idea of *Bardeen* [20.18]: The Fermi energies of the metal, the bulk semiconductor, and the semiconductor surface all align in
electronic equilibrium. For a degenerately doped n-type semiconductor at low
temperature with a distribution of electronic states at the surface, the
Fermi level of the neutral surface is approximately equal to the energy of
the lowest states that are not fully occupied by electrons. Electrons diffuse, causing band bending near the semiconductor surface, until the surface

Fermi energy aligns with the Fermi levels of the bulk semiconductor and the metal. This results in the formation of a potential barrier between the semiconductor and the metal, the Schottky barrier (Fig.20.1). For an n-type semiconductor, the Schottky barrier height is essentially the energy separation between the surface state that represents the Fermi level and the conduction band edge. For a p-type semiconductor, the barrier height is the energy of the highest occupied electronic state of the neutral surface, relative to the valence band maximum. Thus the problem of determining Schottky barrier heights is reduced to obtaining the energy levels of the surface states responsible for the Fermi-level pinning.

In his original article, Bardeen focused his attention on intrinsic semiconductor surface states as the most likely candidates for Fermi-level pinning. But he also pointed out that deep levels in the gap associated with impurities or native surface defects could be responsible for the phenomenon.

Following Bardeen's work, a major advance occurred as a result of the experiments of *Mead* and *Spitzer* [20.25] who determined the Schottky barrier heights of many semiconductors, both n-type and p-type. Most of those old data have been confirmed by modern measurements taken under much more favorable experimental conditions. However, after this work, the Schottky barrier problem was widely regarded as understood [20.26] in terms of concepts quite different from Fermi-level pinning.

Recently Spicer and co-workers have revived the Fermi-level pinning model and have argued that pinning is accomplished by *native defects* at or near the surface. Their picture is that during the deposition of the metal, native defects are created at or near the semiconductor/metal interface, and that these semiconductor surface defects produce deep levels in the band gap that are responsible for Fermi-level pinning.

Spicer's viewpoint has been contested by *Brillson* and co-workers [20.27], who have emphasized the importance of chemical reactivity on barrier height. The Brillson viewpoint gains support from the observation of well-defined chemical trends in the variation of barrier height with the heat of reaction of the metal/semiconductor interface, as shown for n-InP by *Williams* et al. [20.28-30] (Fig.20.2). We believe, however, that the Spicer and Brillson viewpoints can be reconciled (see below).

Daw et al. [20.31] proposed that free surface vacancies account for some of the observed Schottky barrier heights in III-V semiconductors. *Allen* and co-workers have argued that antisite defects [20.32-36] "sheltered" [20.37] at the surface pin the Fermi energy for most Schottky barriers between III-V semiconductors and nonreactive metals, but that vacancies become the dominant

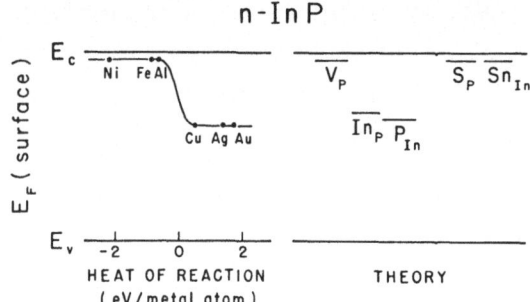

n-InP

E_c — Ni FeAl — Cu Ag Au

$\overline{V_P}$ $\overline{S_P}$ $\overline{Sn_{In}}$

$\overline{In_P}\,\overline{P_{In}}$

E_v -2   0   2

HEAT OF REACTION   THEORY
(eV/metal atom)

Fig.20.2. Surface Fermi energy of n-type InP versus heat of reaction of InP with the metals Ni, Fe, Al, Cu, Ag, and Au, extracted from data in [20.28], assuming Fermi-level pinning. The theoretical Fermi-level pinning defect levels for the surface P vacancy (V_P), the native antisite defects (In_P and P_{In}), and the extrinsic impurities S on a P site (S_P) and Sn on a surface In site (Sn_{In}) are given at the right. The n-InP data can be interpreted as follows: nonreactive metals produce only antisite defects as the dominant defects; reactive metals and treatment of the surface with oxygen and Cl produce P vacancies. Treatments with Sn and S produce surface Sn_{In} and S_P as dominant defects, respectively

pinning defect when the metal is reactive [20.36]. Thus the Brillson reactivity picture can be unified with Spicer's Fermi-level pinning picture: the chemical reaction merely changes the dominant pinning defect. The experimental results of *Mead* and *Spitzer* [20.25], *Wieder* [20.38-40], *Williams* [20.28-30], *Mönch* [20.41-45], their co-workers, and many others support this general viewpoint.

Moreover, the connection between the Schottky barriers formed at Si interfaces with transition metal silicides and the barriers between III-V semiconductors and metals appears to be provided by the recent work of *Sankey* et al. [20.46]: Fermi-level pinning can account for the silicide data as well. Thus a single unifying picture of Schottky barrier heights in III-V and homopolar semiconductors appears to be emerging. And although this Fermi-level pinning picture is no doubt oversimplified, it does provide a simple explanation of the first-order physics determining Schottky barrier heights, and how the physics changes when the dominant defect switches as a result of chemical reactivity.

It appears unlikely, however, that the Fermi-level pinning mechanism of Schottky barrier formation is universal. Layered semiconductors interfaced with non-reactive metals appear *not* to exhibit Fermi-level pinning, but rather seem to obey the original Schottky model [20.30]. This is probably because their surfaces are relatively impervious to defects and do not have defect levels in the band gap.

The Fermi-level pinning mechanism of Schottky barrier formation has the 'most advocates for the III-V semiconductor class, such as GaAs and InP. However, even for these materials there are other proposed mechanisms for Schottky barrier formation, most notably those of *Freeouf* [20.47] and *Ludeke* [20.48].

Studies of Si, especially Si/transition-metal silicide interfaces, have focused on the role of the silicide in Schottky barrier formation [20.49], in contrast to the studies of III-V semiconductors. Prior to the recent work of *Sankey* et al. [20.46], it was widely believed that Fermi-level pinning was *not* responsible for the Schottky barrier at these silicide interfaces.

Thus the present state of the field is that Fermi-level pinning has its advocates for some semiconductors, but is not generally accepted as a universal mechanism of Schottky barrier formation, especially at Si/transition-metal silicide interfaces.

A central point of this paper is the Fermi-level pinning *can* explain an enormously wide range of phenomena relevant to Schottky barrier formation in III-V semiconductors and in Si — which no other existing model can do. In fact, the authors believe that Fermi-level pinning by native defects is responsible for the Schottky barrier formation in III-V semiconductors and in Si.

Our approach to the problem is simple: we calculate deep levels of defects at surfaces and interfaces, and we use these calculations to interpret existing data in terms of the Fermi-level pinning model. To illustrate our approach, we first consider the Si/transition-metal silicide interface and Fermi-level pinning by dangling bonds, as suggested by *Sankey* et al. [20.46].

20.2.1 Si/Transition-Metal Silicide Schottky Barriers

A successful theory of Si/transition-metal silicide Schottky barrier heights must answer the following questions. (a) How are the Schottky barrier heights at Si/transition-metal silicide interfaces related to those at interfaces of III-V semiconductors with metals and oxides? (b) Why is it that Schottky barrier heights of Si with different transition metals do not differ by ~ 1 eV, although changes of silicide electronic structure on this scale are known to occur [20.50]? (c) What is the explanation of the weak chemical trends that occur on a ~ 0.1 eV scale [20.50]? (d) Why are the Schottky barrier heights of silicides with completely different stoichiometries, such as Ni_2Si, $NiSi$, and $NiSi_2$, equal to within ~ 0.03 eV? (e) Why are the Schottky barrier heights only weakly dependent on the silicide crystal structure? (f) Why is it that barriers form with less than a monolayer of silicide coverage? (g) Why do the Schottky barrier heights for n- and p-Si very nearly add up to the band gap of Si? (h) What role do the d electrons of the transition metal play in Schottky barrier formation?

The answers to these questions are simple and straightforward, if one proposes (as *Sankey* et al. [20.46] have done) that the Si/transition-metal silicide Schottky barriers are a result of Fermi-level pinning by Si dangling bonds at the Si/transition metal silicide interface. (a) The Fermi-level pinning idea unifies the Si/transition-metal silicide Schottky barriers with those found for III-V semiconductors. (b) The independence of the Schottky barrier heights of the transition-metal silicide comes from the fact that the causative agent, the Si dangling bond, is associated with the Si, and not with the silicide or transition metal. (c) The weak chemical trends in barrier heights occur because the different transition-metal silicides repel the Si dangling bond wave function somewhat differently, causing it to lie more or less in the Si. (d,e) The Schottky barrier heights vary very little with silicide stoichiometry and silicide crystal structure because the Si dangling-bond level is "deep-level pinned" in the sense of *Hjalmarson* et al. [20.1]: a large change in defect potential produces only a small change in the deep level responsible for Fermi-level pinning. The transition metal atoms act as inert encapsulants with the electronic properties of vacancies, because their energy levels are out of resonance with the Si. (f) Submonolayer barrier formation occurs because the Si dangling-bond defect responsible for the Fermi-level pinning is a localized defect that forms before a full interface is formed. (g) The Schottky barrier heights for n-Si and p-Si add up to the band gap because (in a one-electron approximation) the pinning level associated with the neutral Si dangling bond at the interface is occupied by one electron, and so can accept either an electron or a hole: it provides the surface Fermi level for both electrons and holes —both the lowest partially empty state and the highest partially filled state. (h) The d electrons of the transition metal atoms play no essential role in the transition-metal silicide Schottky barrier formation, except to determine the occupancy of the Si dangling bond deep level; they are out of resonance with the Si at the interface.

The physics of the Si dangling-bond, Fermi-level pinning mechanism is contained in the very simple model presented by *Sankey* et al. [20.46]: to a good approximation, a Si dangling bond at a Si/transition-metal silicide interface is the same as a vacancy in bulk Si with three of its four neighbors replaced by transition-metal atoms. To illustrate this physical concept, consider first a vacancy in bulk Si. This defect produces four deep levels near the band gap: a nondegenerate A_1 or s-like level deep in the valence band (a "deep resonance") and a threefold degenerate T_2 level in the band gap. The Si dangling bond defect at a Si/transition-metal silicide interface differs for the bulk Si vacancy in two ways: (a) some of the nearest neighbors of the inter-

488

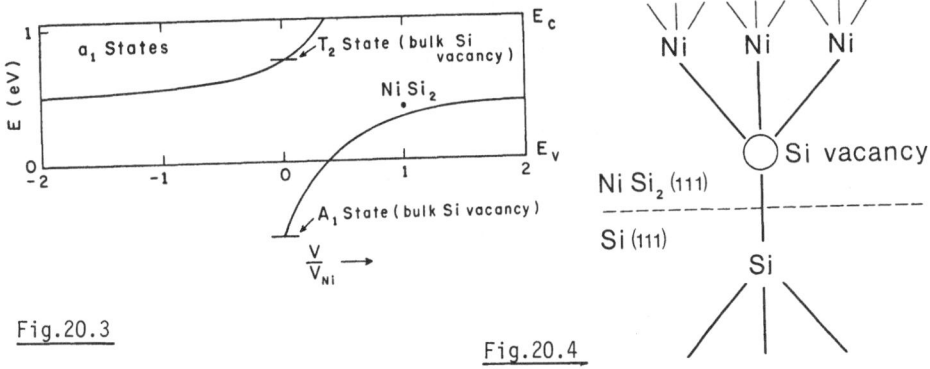

<u>Fig.20.3</u>

Fig.20.4

<u>Fig.20.3.</u> The totally symmetric (a_1) levels for a bulk Si vacancy, surrounded by one Si atom and three X atoms, as a function of the defect potential V, normalized to the Ni defect potential [20.46]. For V = 0, the X atoms are Si; for V = V_{Ni}, the X atoms are Ni. The parent levels of the isolated Si vacancy are shown for V = 0. The experimental Fermi-level pinning position for $NiSi_2$, extracted from [20.51] are denoted by a dot labeled $NiSi_2$

<u>Fig.20.4.</u> One type of interfacial vacancy "sheltering" a Si dangling bond [20.46]. The geometry is that determined for the $NiSi_2/Si\{111\}$ interface in [20.52]

facial vacancy are transition-metal atoms rather than Si atoms; and (b) more distant neighbors are also different atoms at different positions —but the experimental fact that Schottky barriers form at submonolayer coverages suggests that these differences in remote atoms are unimportant. Thus we can imagine constructing the Fermi-level pinning defect by slowly changing some of the Si atoms adjacent to a bulk Si vacancy into transition-metal atoms (Fig.20.3).

To be specific we consider a $Si/NiSi_2$ interface, with a missing Si-bridge atom. Thus (Fig.20.4) the Si bond dangles into the vacancy left by the removal of the Si bridge atom; this vacancy is surrounded by one Si atom and three Ni atoms.

How are the Ni atoms different from Si? First, their s and p orbital energies lie well above those of Si. Second, they each have an additional d orbital, with an energy that lies well below the Si s and p orbital energies (which is not particularly relevant here). The very positive Ni s and p energies act as a repulsive potential barrier to electrons, repelling the Si dangling bond electron from their vicinity in the silicide and forcing it to reside almost exclusively in the Si.

The effect of this positive potential barrier due to the Ni-Si difference, as it is turned on slowly in our imagination, is to drive the levels of the bulk vacancy *upward* in energy. In fact, for Ni, the potential is sufficiently positive to drive the T_2 bulk-Si vacancy level out of the gap into the conduction band. At the same time, the A_1 deep resonance of the Si bulk vacancy is also driven upward. For sufficiently large and positive potential, it pops into the fundamental band gap.

The A_1-derived level cannot be driven all the way through the gap by the potential though, because an (approximate) level-crossing theorem prevents this. A simple way to see that there is an upper bound within the gap for the perturbed A_1 level is to consider a paired defect of a vacancy V_{Si} with a neighboring atom X. If the atom X is Si, then the defect levels are the A_1 (s-like) valence band resonance and T_2 (p-like) band-gap deep level of the bulk Si vacancy. However, A_1 and T_2 are not good symmetries of the (V_{Si},X) pair; the A_1 level becomes σ bonded and the T_2 level produces one σ-bonded and two π-bonded orbitals, with the σ bond oriented along the (V_{Si},X) axis, and with the π bonds perpendicular to it. Thus the unperturbed (X = Si) σ levels of the (V_{Si},X) pair are the A_1 and T_2 bulk Si vacancy levels. The interlacing or no-crossing theorem [20.53] states that a perturbation cannot move a level further than the distance to the nearest unperturbed level. (It applies only approximately here.) Hence no matter how electropositive X is, the (V_{Si},X) level derived from the Si vacancy A_1 level cannot lie above the Si vacancy T_2 level. These considerations for general (V_{Si},X) pairs hold for the specific case of (V_{Si},Ni) pairs, and carry over to the dangling bond defect at the Si/transition-metal silicide interface, which is a vacancy surrounded by three Ni atoms and one Si. Thus the dangling-bond A_1 deep level is "deep-level pinned" (as distinct from Fermi-level pinned) in the sense of *Hjalmarson* et al. [20.1], and is insensitive to even major changes in the nearby transition-metal atoms. To a good approximation, the nearby-transition-metal atoms have the same effect as vacancies (which can be simulated [20.54] by letting the orbital energies of the transition-metal atoms approach $+\infty$, thereby decoupling the atoms from the semiconductor).

Thus the work of *Sankey* et al. [20.46] not only provides an explanation of the Si/transition-metal silicide Schottky barriers, it explains why calculations for defects at a free surface can often provide a very good description of the physics of Schottky barriers: the defects at interfaces are "sheltered" [20.37] or encapsulated by vacancies or by metal atoms that have orbital energies out of resonance with the semiconductor atoms; because of the deep-level pinning, the free-surface defects (which can be thought of as encapsulated by vacancies) have almost the same energies as the actual interfacial defects.

20.2.2 III-V Schottky Barriers

The Fermi-level pinning story for Si/transition-metal silicides holds for Schottky barriers formed on III-V semiconductors as well. Here we summarize the main predictions of the theory.

The basic approach of the theory was to calculate the energy levels in the band gap of thirty s- and p-bonded substitutional point defects at the relaxed [20.55,64] {110} surfaces of III-V semiconductors. With these results in hand, Allen et al. examined Schottky barrier data in the context of Fermi-level pinning and eliminated from consideration all defects that produced levels considerably farther than ~0.5 eV (the theoretical uncertainty) from the observed pinning levels. Interstitial defects were not considered; they have less of a tendency [20.56] to exhibit the deep-level pinning that may be associated with the experimental fact that different metals produce similar Schottky barrier heights. Moreover, extended defects were not considered initially, because it is known that paired-defect spectra are intimately related and similar to isolated-defect spectra [20.57]. (A more complete theory of Fermi-level pinning by paired defects, especially in GaSb where vacancy-antisite pairs are important, is in preparation.) For clean semiconductors, the native substitutional defects potentially responsible for the commonly observed Fermi-level pinning are vacancies and antisite defects (anions on cation sites or cations on anion sites).

In GaAs, the defects proposed by *Allen* et al. [20.32] as responsible for Fermi-level pinning and Schottky barrier formation are the antisite defects. The cation-on-the-As-site defect accounts for trends with alloy composition of the Schottky barrier heights of n-type $In_{1-x}Ga_xAs$ and $Ga_{1-y}Al_yAs$ alloys (Fig.20.5). The Fermi-level pinning of p-InAs [20.59], which shows quite different alloy dependences [20.60], is also explained.

This picture of Fermi-level pinning has been confirmed recently by *Mönch* and associates, who annealed Schottky barriers and showed that the Fermi-level pinning abruptly changed at the same temperature that the bulk (and presumably also the surface) antisite defect is known to anneal [20.61].

An even more interesting material is InP, because its Schottky barrier appears to depend on the heat of reaction of the interface [20.28-30]. This can be readily explained [20.36], however, in terms of switching of the dominant Fermi-level pinning defect from an antisite defect for nonreactive metals to a vacancy for reactive metals (Fig.20.2).

Moreover, surface treatments are known to alter the Schottky barrier height of n-InP in a manner that can be easily understood in terms of the theory [20.36]: surface treatments with Sn or S produce shallow donor levels asso-

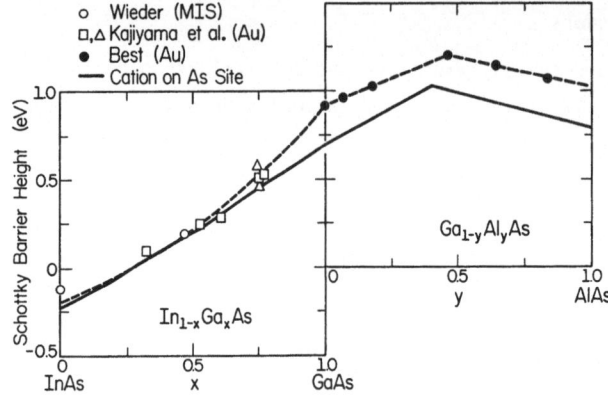

Fig.20.5. Predicted dependence of Schottky barrier height on alloy compositions x and y of $In_{1-x}Ga_xAs$ and $Ga_{1-y}Al_yAs$ alloys, compared with data, [20.58]

ciated with Sn_{In} or S_P at the surface, and these levels pin the surface Fermi energy for contacts between n-InP and the nonreactive noble metals. Likewise O and Cl treatments lead to reactions with P that leave P vacancies, so that the surface Fermi-level of treated n-InP interfaced with non-reactive metals lies near the conduction band edge — as though the metals were reactive.

Thus the Fermi-level pinning idea appears to provide a simple and unifying understanding of a wide variety of Schottky barrier data in common semiconductors.

20.3 Intrinsic Surface States

The calculations of surface defect levels for the Schottky barrier problem can be checked by simultaneously calculating surface state energies and comparing them with the considerable body of available data. The theory underlying surface state calculations is basically the same as that for bulk point defects or surface defects. It is quite simple, and requires only (a) the well-established empirical tight-binding Hamiltonian of the semiconductor [20.62] (the matrix elements of the Hamiltonian exhibit manifest chemical trends from one semiconductor to another), and (b) knowledge of the positions of the atoms at the surface. Thus a reliable treatment of the surface states of a semiconductor requires an adequate model of the geometrical structure of the surface. At present, no semiconductor surface structures are beyond controversy [20.63], but two seem to be rather well accepted; the {110} surface structure of III-V and II-VI semiconductors with zinc blende crystal structure [20.55,64-66], and the {10$\bar{1}$0} surface structure of II-VI semiconductors with wurtzite structure [20.65]. In particular, {110} zinc blende surfaces are characterized by an outward, almost-rigid-rotation relaxation of the anion (e.g., As in GaAs), with the bond between surface anion and surface cation rotating through about $27°$ (III-V's) or $33°$ (II-VI's), and with small bond length changes and subsurface relaxations.

20.3.1 {110} Surfaces of III-V and II-VI Zinc Blende Semiconductors

During the past five years, a number of groups have reported experimental and theoretical studies of intrinsic surface states at {110} zinc blende surfaces [20.3-17,67-76]. In Fig.20.6, we show the most recent calculation for the dispersion curves $E(\bar{k})$ at the GaAs {110} surface [20.14], together with the measured surface state energies of *Williams* et al. [20.68] and of *Huijser* et al. [20.69]. The calculation employs the ten-band sp^3s^* empirical tight-binding model of *Vogl* et al. [20.62]. The agreement between theory and experiment is excellent. For example, along the symmetry lines $\bar{X}'\bar{M}$ and $\bar{M}\bar{X}$ (i.e., the boundary of the surface Brillouin zone), the uppermost branch of observed states appears to be explained by A_5, the next branch by the overlapping resonances A_4 and A'_2, and the three lower branches by A'_1, A_3, and C_2. Here "A" and "C" refer to states localized primarily on anion and cation sites, respectively. A detailed comparison with previous theoretical studies of the GaAs {110} surface is given in [20.14]. The primary additional features are (i) the states $A_1 - A_5$ and $C_1 - C_4$ (in the notation of [20.7]) were located as bound states or resonances at all planar wave vectors $\bar{k}$ along the symmetry lines of the surface Brillouin zone, and (ii) two "new" resonances A'_1 and A'_2 were found. (The branch A'_1 was reported in [20.5,77], but not in the other theoretical studies. The branch A'_2 had not been previously reported.) The discovery of this additional resonant structure is apparently due to the "effective Hamiltonian" technique [20.14], which facilitates calculations of bound states and resonances.

In Fig.20.7, the theoretical dispersion curves of *Beres* et al. [20.14] are shown for the {110} surface of ZnSe, together with the measured surface state energies reported by *Ebina* et al. [20.11]. Again, the agreement between theory and experiment is quite satisfactory, being a few tenths of one eV near the band gap, and larger for more distant states. Some apparent discrepancies [20.11] between experiment and previous theory were found to be resolved by a more complete treatment of the resonances, using the approach described above.

Surface state dispersion relations have also been calculated for GaP, GaSb, InP, InAs, InSb, AlP, AlAs, AlSb, and ZnTe [20.14-17]. In none of the direct-gap materials were intrinsic surface states found within the band gap.

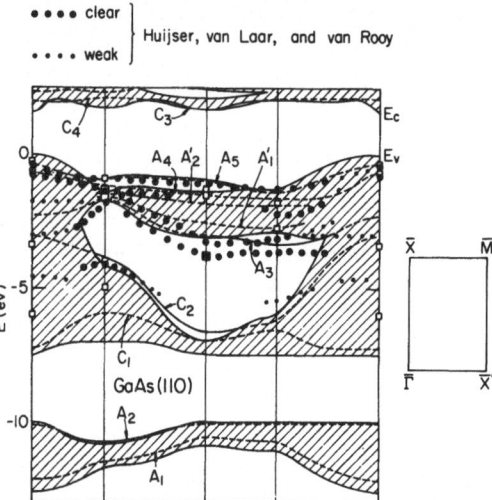

Fig.20.6. Predicted surface state dispersion curves $E(\bar{k})$ for surface bound states (solid lines) and surface resonances (dashed lines) at the relaxed {110} surface of GaAs [20.14]. The energy is plotted as a function of the planar wave vector $\bar{k}$ along the symmetry lines of the surface Brillouin zone, shown on the right. The labeling is the same as in [20.7], with A_1, A_2, C_1, and C_2 mainly s-like, and A_3, A_4, A_5, C_3, and C_4 mainly p-like. A_5 and C_3 are the "dangling-bond" states. A_3, A_1', and A_2' are largely associated with in-plane p orbitals in the first and second layers. The character of each state varies with the planar wave vector $\bar{k}$, and represents an admixture of all orbitals. The widths of the resonance are typically 0.5 to 1.0 eV, but in some cases are smaller than 0.1 eV or as large as 5.0 eV. The dots follow the continuous dispersion curves inferred by Huijser et al. [20.69] for the "clear" and "weak" experimental features. The open squares represent the states observed by Williams et al. [20.68]. The data reported in [20.6,67] are consistent with those shown here

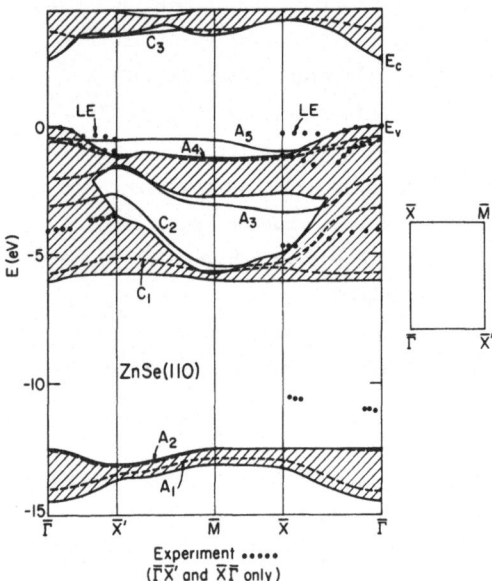

Fig.20.7. Predicted energies of surface bound states (*solid lines*) and surface resonances (*dashed*) for the {110} surface of ZnSe, as functions of the planar wave vector $\bar{k} = (k_1, k_2)$ [20.15]. The surface Brillouin zone is shown on the right; $\bar{\Gamma}$ is the origin, $\bar{k} = (0,0)$. The bulk bands are shaded. E_v and E_c are the valence and conduction band edges. The experimental features identified with bound and resonant surface states in [20.11], along the two symmetry lines $\bar{\Gamma}X'$ and $X\bar{\Gamma}$, are indicated by the dotted lines

However, GaP was found to have a band of unoccupied surface states that overlaps the fundamental band gap and extends below the bulk conduction band edge. This is in accord with the experimental facts: of these semiconductors only GaP has surface states in the gap [20.18,69-73]. Of the remaining indirect-gap materials, the theory indicates that intrinsic surface states may be observable near the top of the band gap in the indirect-gap Al-V compounds [20.16], although the theory is not sufficiently accurate to predict unequivocally that the states will lie within the gap.

20.3.2 Si{100} (2×1) Intrinsic Surface States

After many years of intensive study by numerous groups, there is still controversy over the geometrical structures of the most thoroughly studied semiconductor surfaces: Si{100} (2×1) and Si{111} (2×1). For example, four groups have recently given arguments for antiferromagnetic ordering of Si {111} surfaces [20.78], whereas *Pandey* has proposed replacing the conventional buckling model [20.79-81] of Si{111} (2×1) by a {110}-like chain model [20.82].

For Si{100}, arguments have recently been presented [20.83-85] against the (2×1) asymmetric dimer model of *Chadi* [20.86]. (In the asymmetric dimer model, adjacent rows of surface atoms dimerize, forming a pattern of paired atomic rows on the surface.) The most telling of these arguments involves the apparent disagreement between angle-resolved photoemission measurements of the surface-state dispersion curves [20.90,91] and theoretical calculations of these dispersion curves with conventional models of the electronic structure as applied to the asymmetric dimer geometry [20.86,87].

Very recently, two new calculations have been performed independently with improved models of the electronic structure [20.88,89]. The same con-

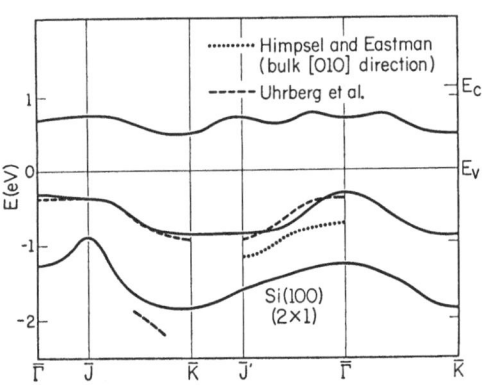

Fig.20.8. Dispersion curves for surface states and surface resonances at the {100} (2×1) surface of Si [20.88]. Energy E is shown as a function of the planar wave vector $\bar{k}$ around the symmetry lines of the surface Brillouin zone. Solid lines represent results of the present calculations; dashed lines are measurements from [20.90] and the dotted line from [20.91], which were taken from $\bar{\Gamma}$ to $\bar{J}'$, along the {010} direction, rather than along the symmetry line $\bar{\Gamma}$ to $\bar{J}'$. E_V and E_C are the Si valence and conduction band edges

clusion was reached in both of these studies: the electronic structure calculated for the asymmetric dimer model is in agreement with the measurements.
This is illustrated in Fig.20.8 [20.48], where both the theoretical bandwidth
of 0.65 eV and the detailed variation with the planar wave vector $\bar{k}$ agree
excellent with the experimental dispersion curves. In addition, there is
quite satisfactory agreement between the theoretical surface band gaps and
the 0.6 eV gap measured by *Mönch* et al. [20.92].

20.4 Surface Core Exciton States

The same calculations that predict native-defect surface deep levels for
the Schottky barrier problem also yield surface core exciton energies, because the optical alchemy or $Z + 1$ rule states that the Hjalmarson-Frenkel
core exciton energies are the energies of "impurities" that are immediately
to the right in the periodic table of the core-excited atom [20.23,24]. Thus
core-excited Ga produces a "Ge defect" and core-excited In yields "Sn."

   In Figs.20.9,10, the theoretical exciton energies for the {110} surfaces
of the Ga-V and In-V compounds are compared with experiment [20.58]. Note
that the experimental and theoretical exciton levels for InAs and InSb lie
above the conduction band edge, as resonances rather than as bound states.
In the present theory this result has a simple physical interpretation: like
a deep impurity state, the Hjalmarson-Frenkel exciton energy is determined
primarily by the high density-of-state regions of the bulk band structure.
There is only a small density of states near the low-lying direct conduction
band minimum (corresponding to the Γ point of the Brillouin zone), but a
large density of states near the higher, indirect X minima. Thus the conduction band minimum near Γ has relatively little influence on the position of
the exciton.

   The surface Hjalmarson-Frenkel core excitons have also been calculated
for the {110} surface of ZnSe and ZnTe [20.94] and agree well with the
measurements [20.95]. We conclude that the present theoretical framework
does a good job of explaining the basic physics of the "deep" Hjalmarson-
Frenkel core excitons, whether bound states or resonances.

20.5 Unified Picture

Thus, one interlocking theoretical framework successfully predicts the correct physics of (a) surface deep defect levels and Schottky barrier heights,
(b) intrinsic surface states, and (c) Hjalmarson-Frenkel core exciton states.

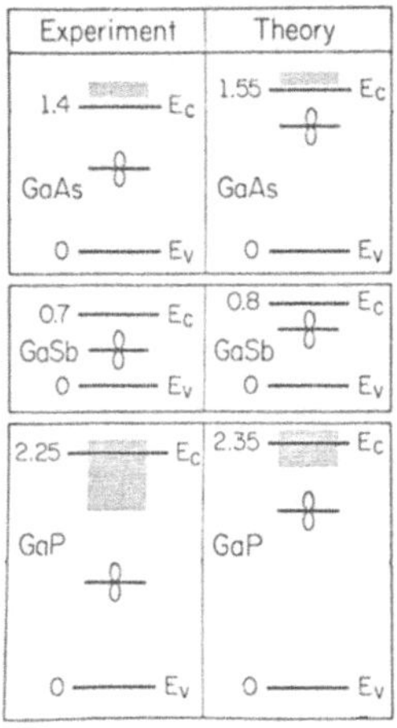

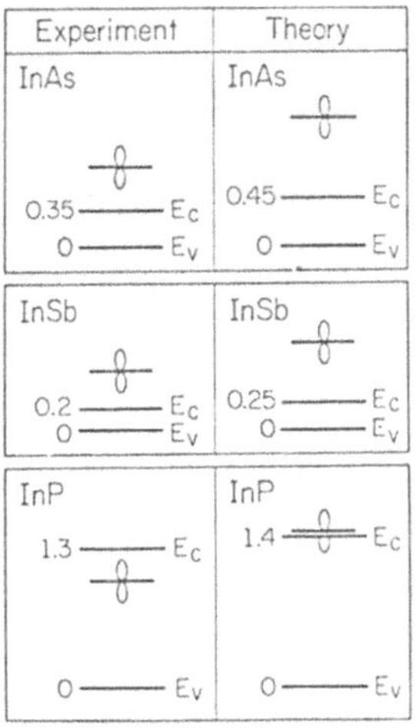

Fig.20.10. Predicted and observed In 4d core surface Frenkel excitons for InAs, InSb, and InP [20.32]

Fig.20.9. Predicted and observed Ga 3d core surface Frenkel excitons (double lobes) for GaAs, GaSb, and GaP [20.58]. The lower unoccupied surface states [20.13,14] are represented by closely spaced horizontal lines. E_v and E_c are, respectively, the top of the valence band and the bottom of the conduction band. The experimental results here and in Fig.20.7 are those of *Eastman* and co-workers [20.93]

References

20.1 H.P. Hjalmarson, P. Vogl, D.J. Wolford, J.D. Dow: Phys. Rev. Lett. **44**, 810 (1980)

20.2 H.P. Hjalmarson, R.E. Allen, H. Büttner, J.D. Dow: J. Vac. Sci. Technol. **17**, 993 (1980);
R.E. Allen, J.D. Dow: J. Vac. Sci. Technol. **19**, 383 (1981)

20.3 C. Calandra, F. Manghi, C.M. Bertoni: J. Phys. C**10**, 1911 (1977)

20.4 J.D. Joannopoulos, M.L. Cohen: Phys. Rev. B**10**, 5075 (1974);
E.J. Mele, J.D. Joannopoulos: Phys. Rev. B**17**, 1816 (1978)

20.5 D.J. Chadi: J. Vac. Sci. Technol. **15**, 631, 1244 (1978); Phys. Rev. B**18**, 1800 (1978)

20.6 J.A. Knapp, D.E. Eastman, K.C. Pandey, F. Patella: J. Vac. Sci. Technol. **15**, 1252 (1978)

20.7 J.R. Chelikowsky, M.L. Cohen: Phys. Rev. B**20**, 4150 (1979)

20.8 A. Mazur, J. Pollmann, M. Schmeits: Solid State Commun. **42**, 37 (1982)

20.9 A. McKinley, G.P. Srivastava, R.H. Williams: J. Phys. C**13**, 1581 (1980);
R.H. Williams et al.: To be published

20.10 M. Schmeits, A. Mazur, J. Pollmann: Solid State Commun. **40**, 1081 (1981)

20.11 A. Ebina, T. Unno, Y. Suda, H. Koinuma, T. Takahashi: J. Vac. Sci. Technol. **19**, 301 (1981)

20.12 F. Manghi, E. Molinari, C.M. Bertoni, C. Calandra: J. Phys. C**15**, 1099 (1982)

20.13 R.E. Allen, H.P. Hjalmarson, J.D. Dow: Surf. Sci. **110**, L625 (1981)

20.14 R.P. Beres, R.E. Allen, J.D. Dow: Solid State Commun. **45**, 13 (1983)

20.15 R.P. Beres, R.E. Allen, J.D. Dow: Phys. Rev. B**26**, 769 (1982)

20.16 R.P. Beres, R.E. Allen, J.P. Buisson, M.A. Bowen, G.F. Blackwell, H.P. Hjalmarson, J.D. Dow: J. Vac. Sci. Technol. **21**, 548 (1982)

20.17 R.P. Beres, R.E. Allen, J.D. Dow: Phys. Rev. B26, 5702 (1982)
20.18 J. Bardeen: Phys. Rev. 71, 717 (1947)
20.19 W.E. Spicer, P.W. Chye, P.R. Skeath, C.Y. Su, I. Lindau: J. Vac. Sci. Technol. 16, 1422 (1979), and references therein
20.20 W.E. Spicer, I. Lindau, P.R. Skeath, C.Y. Su: J. Vac. Sci. Technol. 17, 1019 (1980), and references therein
20.21 W.E. Spicer, I. Lindau, P.R. Skeath, C.Y. Su, P.W. Chye: Phys. Rev. Lett. 44, 520 (1980)
20.22 P. Skeath, C.Y. Su, I. Hino, I. Lindau, W.E. Spicer: Appl. Phys. Lett. 39, 349 (1981)
20.23 J.D. Dow, D.R. Franceschetti, P.C. Gibbons, S.E. Schnatterly: J. Phys. F5, L211 (1975), and references therein
20.24 H.P. Hjalmarson, H. Büttner, J.D. Dow: Phys. Rev. B24, 6010 (1981)
20.25 C.A. Mead, W.G. Spitzer: Phys. Rev. A134, 713 (1964)
20.26 S.M. Sze: *Physics of Semiconductor Devices* (Wiley, New York 1981)
20.27 L.J. Brillson: Phys. Rev. Lett. 40, 260 (1978), and references therein
20.28 R.H. Williams, V. Montgomery, R.R. Varma: J. Phys. C11, L735 (1978)
20.29 R.H. Williams, M.H. Patterson: Appl. Phys. Lett. 40, 484 (1982)
20.30 R.H. Williams: To be published
20.31 M.S. Daw, D.L. Smith: Phys. Rev. B20, 5150 (1979); J. Vac. Sci. Technol. 17, 1028 (1980); Appl. Phys. Lett. 36, 690 (1980); Solid State Commun. 37, 205 (1981);
20.32 R.E. Allen, J.D. Dow: Phys. Rev. B25, 1423 (1982)
20.33 R.E. Allen, J.D. Dow: J. Vac. Sci. Technol. 19, 383 (1981)
20.34 R.E. Allen, J.D. Dow: Appl. Surf. Sci. 11/12, 362 (1982)
20.35 R.E. Allen, J.D. Dow, H.P. Hjalmarson: Solid State Commun. 41, 419 (1982)
20.36 J.D. Dow, R.E. Allen: J. Vac. Sci. Technol. 20, 659 (1982)
20.37 R.E. Allen, R.P. Beres, J.D. Dow: J. Vac. Sci. Technol. B1, 401 (1983)
20.38 H.H. Wieder: Inst. Phys. Conf. Ser. 50, 234 (1980)
20.39 H.H. Wieder: Appl. Phys. Lett. 38, 170 (1981)
20.40 H.H. Wieder: To be published
20.41 W. Mönch, H.J. Clemens: J. Vac. Sci. Technol. 16, 1238 (1979)
20.42 H. Gant, W. Mönch: Appl. Surf. Sci. 11/12, 332 (1982)
20.43 W. Mönch, H. Gant: Phys. Rev. Lett. 48, 512 (1982)
20.44 W. Mönch, R.S. Bauer, H. Gant, R. Murschall: J. Vac. Sci. Technol. 21, 498 (1982);
      J. Assmann, W. Mönch: Surf. Sci. 99, 34 (1980)
20.45 W. Mönch: To be published
20.46 O.F. Sankey, R.E. Allen, J.D. Dow: Solid State Commun. 49, 1 (1983)
20.47 J.L. Freeouf, J.M. Woodall: Appl. Phys. Lett. 39, 727 (1981);
      J.L. Freeouf: Solid State Commun. 30, 1059 (1980);
      J.M. Woodall, G.D. Petit, T.N. Jackson, C. Lanza, K.L. Kavanagh, J.W. Mayer: Phys. Rev. Lett. 51, 1783 (1983)
20.48 R. Ludecke, L. Esaki: Phys. Rev. Lett. 33, 653 (1974);
      R. Ludecke, A. Kona: Phys. Rev. Lett. 34, 817 (1975); Phys. Rev. Lett. 39, 1042 (1977); and to be published
20.49 P.S. Ho, G.W. Rubloff: Thin Solid Films 89, 433 (1981)
20.50 J.M. Andrews, J.C. Phillips: Phys. Rev. Lett. 35, 56 (1975)
20.51 G. Ottavianai, K.N. Tu, J.W. Meyer: Phys. Rev. B24, 3354 (1981)
20.52 D. Cherns, G.R. Anstis, J.L. Hutchinson, J.C.H. Spence: Philos. Mag. A46, 849 (1982)
20.53 A.A. Maradudin, E.W. Montroll, G.H. Weiss: Solid State Phys. Suppl. 3, 132 (1963)
20.54 M. Lannoo, P. Lenglart: J. Phys. C30, 2409 (1969)
20.55 S.T. Tong, A.R. Lubinsky, B.J. Mrstik, M.A. van Hove: Phys. Rev. B17, 3303 (1978)
20.56 O.F. Sankey, J.D. Dow: Phys. Rev. B27, 7641 (1983)

20.57 O.F. Sankey, H.P. Hjalmarson, J.D. Dow, D.J. Wolford, B.G. Streetman: Phys. Rev. Lett. 45, 1656 (1980); O.F. Sankey, J.D. Dow: Appl. Phys. Lett. 38, 685 (1981); J. Appl. Phys. 52, 5139 (1981); Phys. Rev. B26, 3243 (1982)
20.58 R.E. Allen, J.D. Dow: Phys. Rev. B24, 911 (1981)
20.59 H. Wieder: To be published
20.60 H.H. Wieder: Private communication
20.61 W. Mönch: To be published
20.62 P. Vogl, H.P. Hjalmarson, J.D. Dow: J. Phys. Chem. Solids 44, 365 (1983)
20.63 W.M. Gibson: To be published; Gibson recently proposed that the relaxation of the {110} surface of GaAs is considerably less than once thought [20.55], based on his He-atom scatterinq data
20.64 A. Kahn, E. So, P. Mark, C.B. Duke: J. Vac. Sci. Technol. 15, 580 (1978)
20.65 C.B. Duke, R.J. Meyer, P. Mark: J. Vac. Sci. Technol. 17, 971 (1980)
20.66 C.B. Duke, A. Paton, W.K. Ford, A. Kahn, J. Carelli: Phys. Rev. B24, 562 (1981); C.B. Duke, A. Paton, W.K. Ford, A. Kahn, G. Scott: Phys. Rev. B24, 3310 (1981)
20.67 J.A. Knapp, G.J. Lapeyre: J. Vac. Sci. Technol. 13, 757 (1976); Nuovo Cimento 39B, 693 (1977)
20.68 G.P. Williams, R.J. Smith, G.J. Lapeyre: J. Vac. Sci. Technol. 15, 1249 (1978)
20.69 A. Huijser, J. van Laar, T.L. Rooy: Phys. Lett. 65A, 337 (1978)
20.70 J. van Laar, J.J. Scheer: Surf. Sci. 8, 342 (1967); J. van Laar, A. Huijser: J. Vac. Sci. Technol. 13, 769 (1976)
20.71 W. Gudat, D.E. Eastman: J. Vac. Sci. Technol. 13, 831 (1976)
20.72 A. Huijser, J. van Laar: Surf. Sci. 52, 202 (1975); A. Huijser, J. van Laar, T.L. van Rooy: Surf. Sci. 62, 472 (1977)
20.73 G.M. Guichar, C.A. Sebenne, C.D. Thualt: J. Vac. Sci. Technol. 16, 1212 (1979); D. Norman, I.T. McGovern, C. Norris: Phys. Lett. 63A, 384 (1977)
20.74 P. Chiaradia, G. Chiarotti, F. Ciccacci, R. Memeo, S. Nannarone, P. Sassaroli, S. Selci: Surf. Sci. 99, 76 (1980)
20.75 V. Dose, H.-J. Gassmann, D. Straub: Phys. Rev. Lett. 47, 608 (1981)
20.76 Further work is cited in the papers above and in the review by J. Pollmann: Festkörperprobleme 20, 117 (1980)
20.77 A. Zunger: Phys. Rev. B22, 959 (1980)
20.78 See the proceedings of the 9th International Conference on the Physics and Chemistry of Semiconductor Interfaces: J. Vac. Sci. Technol. 21, (1982)
20.79 D. Haneman: Phys. Rev. 121, 1093 (1961)
20.80 R. Feder, W. Mönch, P.P. Aver: J. Phys. C12, 2179 (1979)
20.81 J.P. Buisson, J.D. Dow, R.E. Allen: Surf. Sci. 120, L477 (1982), and references therein
20.82 K.C. Pandey: Phys. Rev. Lett. 47, 1913 (1981)
20.83 D.E. Eastman: J. Vac. Sci. Technol. 17, 492 (1980)
20.84 D.J. Chadi: Appl. Optics 19, 3971 (1980)
20.85 A. Redondo, W.A. Goddard: J. Vac. Sci. Technol. 21, 344 (1982)
20.86 D.J. Chadi: Phys. Rev. Lett. 43, 43 (1979)
20.87 J. Ihm, M.L. Cohen, D.J. Chadi: Phys. Rev. B21, 4952 (1980)
20.88 M.A. Bowen, J.D. Dow, R.E. Allen: Phys. Rev. B26, 7083 (1983). As discussed in this paper, the theoretical surface state energies of Fig. 20.8 have been lowered by 0.5 eV
20.89 A. Mazur, J. Pollmann, M. Schmeits: Phys. Rev. B26, 7086 (1982)
20.90 R.I.G. Uhrberg, G.V. Hansson, J.M. Nicolle, S.A. Flodstrom: Phys. Rev. B24, 4684 (1981)

20.91 J. Himpsel, D.E. Eastman: J. Vac. Sci. Technol. **16**, 1297 (1979)
20.92 W. Mönch, P. Koke, S. Krueger: J. Vac. Sci. Technol. **19**, 313 (1981)
20.93 D.E. Eastman, J.L. Freeouf: Phys. Rev. Lett. **33**, 1601 (1974); **34**, 1624 (1974);
W. Gudat, D.E. Eastman: J. Vac. Sci. Technol. **13**, 831 (1976);
D.E. Eastman, T.-C. Chiang, P. Heimann, F.J. Himpsel: Phys. Rev. Lett. **45**, 656 (1980)
20.94 R.P. Beres, R.E. Allen, J.D. Dow: Phys. Rev. B**26**, 769 (1982)
20.95 A. Ebina, T. Unno, Y. Suda, H. Koinuma, T. Takahashi: J. Vac. Sci. Technol. **19**, 301 (1981)

21. Work Function and Band Bending at Semiconductor Surfaces

W. Mönch

With 16 Figures

21.1 Introduction

Electrons are bound to solids, and thus energy is required to remove them from the solid. Usually, the difference between the potential energy "just outside" and the electrochemical potential or the Fermi energy inside the solid is defined as the work function:

$$\Phi = E_{vac} - E_F \quad .$$

(21.1)

As a surface property, the work function determines the functioning of such devices as thermal, field-effect and photocathodes.

A quantum-mechanical theory of the work function should take into account the lattice positions of all the positive ions and the n-1 remaining electrons and also include properly all many-body effects. Solutions of this problem are based on simplifying assumptions and approximations. Simple metals have been modelled by a gas of nearly free electrons embedded in an uniform background of positive charge abruptly terminated at the surface. In a somewhat more realistic approach this model has been improved by representing the effect of the metal ions by a pseudopotential. The calculated values of the work function deviate from the experimental data by only 5% to 10% [21.1]. However, such simple theories cannot be applied to transition metals because of the highly localised d electrons.

With semiconductors the difficulties in calculating the work function become even more severe since the valence electrons are concentrated in the bonds and are thus far from being uniformly distributed. Furthermore, semiconductor surfaces are reconstructed [21.2-4]: the surface planes are not only relaxed inward — as observed with metals, too — but the surface atoms are not arranged as in a bulk plane, and as a result the surface unit mesh is usually larger than the mesh of a corresponding bulk plane. The most prominent examples are the Si{111} 2×1 and 7×7 structures whose surface unit meshes measure twice and 49 times, respectively, the size of a unit mesh in a bulk {111} plane. The geometry of the metastable Si{111} 2×1 cleavage

structure is debated again; none of the many proposals regarding the stable
7×7 structure is generally accepted at present [21.2,3,5]. However, the
atomic arrangements in the Si{001} 2×1 surface and in the cleaved{110} surfaces of most III-V compounds are well established by now. We shall return
to this topic in Sects.21.3,4 where we shall discuss the orientational and
chemical trends of the work function observed with clean semiconductor
surfaces.

For many semiconductors the band structure of the electronic surface
states has been theoretically calculated taking into account the real arrangements of the surface atoms [21.6]. Good agreement has been achieved with the
energy dispersion curves of the surface states as determined by angle-resolved ultraviolet photoemission spectroscopy (ARUPS). However, most of these
theoretical approaches have not been concerned with the absolute energies
of the surface states with respect to the vacuum level but only with their
position relative to the edges of the bulk bands. Therefore, no systematic
calculations of the work function are available which consider the real geometry of semiconductor surfaces. Only occasionally has the work function
been evaluated in such studies. However, simple LCAO approaches provide insight into chemical trends although no exact values of the work function
are obtained [21.7]. This will be discussed in Sect.21.3.

Semiconductor surfaces — and this has been known for a long time — exhibit
surface states. When their energy levels are located within the bulk energy
gap, i.e., between the valence-band maximum and the bottom of the conduction band, their occupation varies as a function of the position of the band
edges, and by this of the surface states, relative to the Fermi level at the
surface. To maintain charge neutrality at the surface, an excess charge in
surface states has to be balanced by a space charge below the semiconductor
surface:

$$Q_{ss} + Q_{sc} = 0 \quad . \tag{21.2}$$

This is illustrated in Fig.21.1 which shows a simple energy diagram at the
surface of an n-type semiconductor containing acceptor type surface states.
Such surface states are neutral or charged negatively if their energy levels
are above or below the Fermi level. A negative charge in the surface states
needs a compensating positive space charge which is provided by positively
charged donors below the surface, i.e., the space-charge layer is depleted
of electrons. The associated band bending is calculated by solving Poisson's
equation. For a detailed and complete discussion the reader is referred to
[21.8,9].

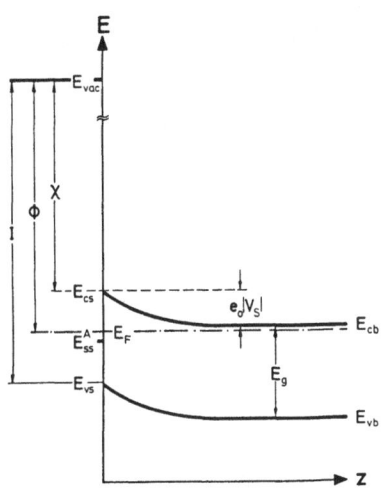

Fig.21.1. Energy-level diagram at the surface of an n-type semiconductor

Consequently, the work function of a semiconductor may be written as

$$\Phi = E_{vac} - E_F = I + e_0|V_s| - E_g + W_n \qquad \text{for} \quad \text{n-type} \quad \text{and} \qquad (21.3a)$$

$$\Phi = I - e_0|V_s| - W_p \qquad \text{for} \quad \text{p-type} \qquad (21.3b)$$

semiconductors. The distance $W_n = E_{cb} - E_F$ or $W_p = E_F - E_{vb}$ of the nearest band edge in the bulk to the Fermi level may be calculated from the doping. The surface band–bending, $e_0|V_s| = |E_{vs} - E_{vb}|$, is determined by the energy distribution of the surface states and also the bulk doping level. The ionisation energy $I = E_{vac} - E_{vs}$ contains the contribution of the surface dipole. Variations of the surface band–bending, observed with different reconstructions or caused by adsorbates, for example, immediately indicate changes in the spectrum of the surface states. An altered ionisation energy, on the other hand, reveals new contributions to the surface dipole. Examples of both effects are discussed in Sect.21.6.

Changes of the work function can be measured very precisely, locally resolved and without any damage to the surface by the Kelvin method [21.10]. The ionisation energy and the band bending, on the other hand, may be determined from photoemission spectra. Data obtained by both experimental techniques will be discussed here. For experimental details the reader is referred to the references cited and to the review [21.11].

21.2 Temperature Dependence of the Work Function

With semiconductors the surface terms, ionisation energy and surface band bending, as well as the bulk terms, width of the band gap and position of the band edges relative to the Fermi level, may contribute to variations of the work function with temperature. As an example, we discuss the temperature dependence of the work function measured with clean, cleaved GaAs{110} surfaces. Some of the experimental results are displayed in Figs.21.2,3.

Let us first analyse the data given in Fig.21.2 for a p-type sample [21.12]. The temperature variation of the bulk contribution W_p to the work function can be easily calculated from semiconductor statistics. In case all acceptors are ionised in a non-degenerately doped p-type sample, the hole concentration p_b is independent of temperature, and we obtain

$$W_p = E_F - E_{vb} = -k_B T \ln \frac{N_v}{p_b} \quad , \tag{21.4}$$

where N_v is the effective density of states in the valence band. The dashed line in Fig.21.2 gives this temperature dependence of $E_F - E_{vb}$ for the sample used. Up to 350°C the work function obviously decreases as the distance between the Fermi level and the valence-band maximum increases. Following (21.3b) this behaviour suggests that the ionisation energy remains unchanged and the bands are flat up to the surface in this temperature range. Above 350°C the work function approaches a constant value, or, in other words, the Fermi level becomes pinned within the band gap, and band bending develops. On GaAs{110} surfaces cleaved from p-type samples, surface states are thus present that are neutral up to 350°C and become noticeably charged above that temperature. These surface states have to be acceptor type, and a rough estimate places them by 0.3 eV + 4 $k_B T$ = 0.5 eV above the top of the valence band.

Such acceptor-type surface states should cause band bending at surfaces cleaved from n-type samples, and as Fig.21.3 shows this has indeed been observed [21.13]. Work function topographs measured by scanning surfaces with Kelvin probes have demonstrated that on n-type samples the work function may vary by up to 0.8 eV correlated with the local quality of the cleave, while on p-type crystals it is constant to within 50 meV although the cleaves are not perfect either [21.12,14,15]. The work functions measured on the most flat areas of n- and p-type samples differ by the band-gap energy minus the particular distances of the Fermi level to the conduction- and the valence-band edges, respectively. This proves the bands to be flat up to the surface on such regions, not only on p- but also on n-type samples, and thus indi-

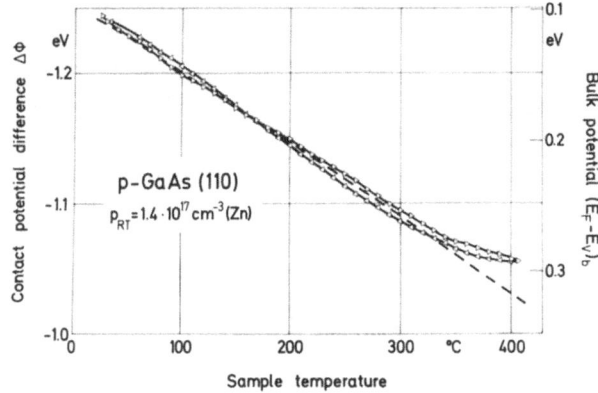

Fig.21.2. Temperature
dependence of the work
function of a {110}
surface cleaved from
p-type GaAs. Data
points ▷ and ◁ have
been measured with
increasing and de-
creasing temperature,
respectively. The
dashed line gives the
distance of the Fermi
level to the valence-
band maximum in the
bulk as a function of
temperature [21.12]

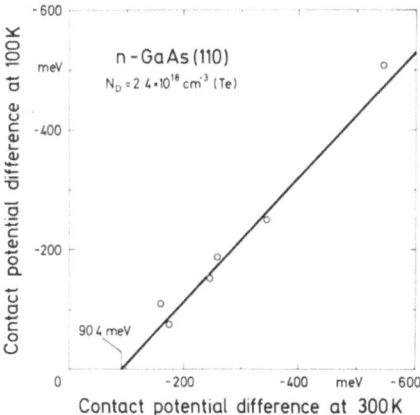

Fig.21.3. Correlation of contact potential differences measured at 100
and 300 K with six differently stepped {110} surfaces cleaved from an
n-type GaAs bar [21.13]

cates that the surface states originate in cleavage imperfections [21.12,14,
16,17]. Theoretical calculations [21.18,19] also revealed that the bulk
energy gap is free of intrinsic surface states since the rotational relaxation of the surface atoms on GaAs{110} surfaces drives the dangling-bond
surface states to below and above the edges of the valence and the conduction band, respectively. Dispersion curves of occupied surface states and
the energetic distribution of the empty surface states have been determined
experimentally by using angle-resolved photoemission with ultraviolet light
[21.20-22] and isochromat spectroscopy [21.23], respectively. The experimental results confirm the theoretical predictions [21.18,24], which were based
on the arrangements of the surface atoms evaluated from low-energy electron
diffraction [21.25].

Figure 21.3 contains the contact potential differences (CPD) measured
between a tungsten Kelvin probe and 6 GaAs{110} surfaces held at room tem-

perature and at 100 K. The surfaces were cleaved from the same n-type sample. A least-square fit to the data points gives

$$CPD(100\ K) = (1.08 \pm 0.06)CPD(300\ K) - (90.4 \pm 11.8)\ [meV]\quad.$$

This not only establishes a linear relation between the work functions at 100 and 300 K but also a constant difference

$$\Delta_T\Phi = CPD(100\ K) - CPD(300\ K) = -(90.4 \pm 11.8)\ meV\quad.$$

According to (21.3a) this difference may result from temperature variations of the ionisation energy, the surface band bending, the bulk band gap, as well as the bulk position of the Fermi level within the band gap:

$$\Delta_T\Phi = \Delta I + e_0\Delta|V_s| - \Delta E_g + \Delta W_n\quad. \tag{21.5}$$

The GaAs crystal used was degenerately doped n-type, and the change of the position of the Fermi level with respect to the bottom of the conduction band in the bulk was evaluated as $\Delta W_n = 0.7$ meV.

The band gap of GaAs varies with temperature as

$$E_g = 1.522 - \frac{5.8 \times 10^{-4} \cdot T^2}{T + 300}\ [eV]$$

between 0 and 973 K [21.26]. This gives a change in the width of the band gap between 100 and 300 K of $\Delta E_g = 72.5$ meV.

The temperature variation of the surface band bending is analysed in the following section. Finally, we shall find out whether or not the ionisation energy is temperature independent at GaAs{110} surfaces as anticipated from evaluation of the temperature dependence of the work function measured with p-type samples.

For simplicifcation, let us assume discrete surface states of acceptor type at an energy $W_{ss}^A = E_{cs} - E_{ss}^A$ below the bottom of the conduction band. Their density shall be N_{ss}^A. Their occupancy is determined by the Fermi-Dirac distribution function f_0, and the charge density in the surface states amounts to

$$Q_{ss}^A = - e_0 N_{ss}^A \cdot f_0(E_{ss}^A - E_F/k_BT)\quad. \tag{21.6}$$

On an n-type semiconductor, this negative charge in surface states is balanced by a positive charge in a depletion layer beneath the surface. Solving Poisson's equation we obtain the space-charge density

$$Q_{sc}^n = \{2\varepsilon\varepsilon_0 N_D[e_0|V_s| - k_BT - k_BT\ exp(-e_0|V_s|/k_BT)]\}^{\frac{1}{2}}\quad, \tag{21.7}$$

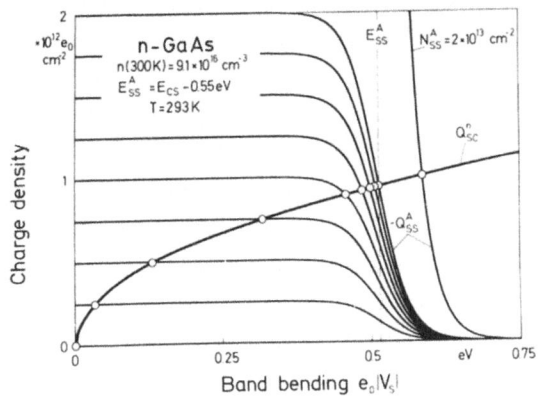

Fig.21.4. Space-charge density in a depletion layer on n-GaAs and occupancy of acceptor-type surface states as a function of surface band-bending

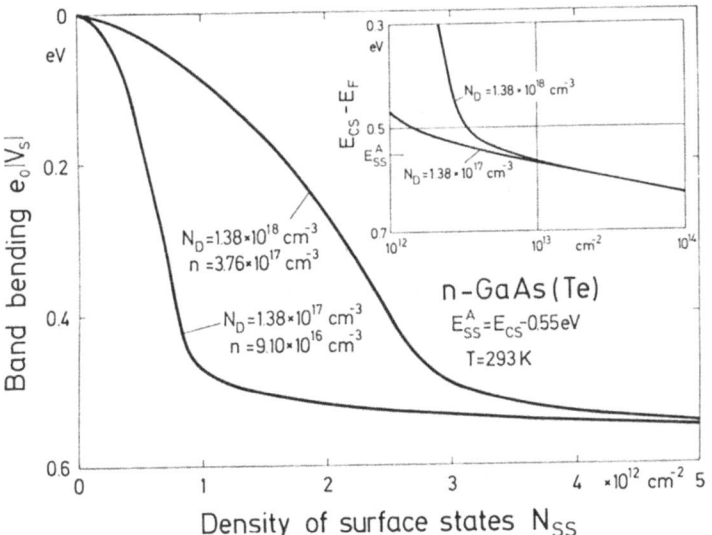

Fig.21.5. Surface band-bending in depletion layers on two differently doped n-GaAs samples as a function of the density of acceptor-type surface states

where N_D is the donor density in the bulk. For $e_0|V_s| > 3\ k_BT$ the exponential term in (21.7) contributes less than a percent and can be neglected.

The surface band-bending now adjusts so as to achieve charge neutrality $Q_{ss}^A + Q_{sc}^n = 0$ at the surface, Fig.21.4. At the intersections of the space-charge curve and the set of curves for different densities of surface states, the condition of charge neutrality is fulfilled. The resulting surface band-bendings are plotted as a function of the density of surface states in Fig. 21.5. Starting from flat bands, $e_0|V_s| = 0$, the surface band-bending first steeply increases with increasing density of surface states, and all surface states are occupied. Initially the surface band-bending varies as

$$e_0|V_s| = \frac{(e_0 N_{ss})^2}{2\varepsilon\varepsilon_0 N_D} + k_B T \quad . \tag{21.8}$$

With increasing surface-state density and thus surface band-bending, the surface states approach the Fermi level, and then the surface band-bending changes much less with further increasing surface-state density. The surface band-bending *seems* to saturate, and the Fermi level is *seemingly* pinned near the energetic position of the surface states. In this region, i.e., for $e_0|V_s| > W_{ss}^A - W_n$, the surface band-bending is given by

$$e_0|V_s| = W_{ss}^A - W_n + k_B T \ln\{e_0 N_{ss}[2\varepsilon\varepsilon_0 N_D(W_{ss}^A - W_n - k_B T)]^{-\frac{1}{2}}\} \quad . \tag{21.9}$$

From (21.8,9) it follows that the temperature dependence of the surface band-bending is not very pronounced. To good approximation we obtain

$$e_0 \Delta|V_s| = n\, k_B \Delta T \quad , \tag{21.10}$$

where $n = 1$ for $e_0|V_s| < W_{ss}^A - W_n$ and $n > 1$ for $e_0|V_s| > W_{ss}^A - W_n$ depending on the ratio of the density of surface states and the doping level.

The above analysis assumes non-degenerately doped semiconductors. However, the crystal used in the work function studies, whose results are shown in Fig.21.3, was doped degenerately. Therefore, (21.7) for the space charge density has to be modified [21.27]. From such a refined analysis one then obtains $e_0 \Delta|V_s| = 23$ meV for the samples under consideration; i.e., (21.7) gives an underestimate since $e_0|V_s| < W_{ss}^A - W_n$.

By now we are finally able to evaluate the change of the ionisation energy from the work functions measured with cleaved GaAs samples held at room temperature and at 100 K:

$$\Delta I = + (4.4 \pm 11.8)\text{ meV} \quad ,$$

and between 100 and 300 K the temperature coefficient amounts to

$$dI/dT = + (2.2 \pm 5.9) \times 10^{-5}\text{ eV K}^{-1} \quad .$$

The ionisation energy of GaAs{110} surfaces is thus almost temperature independent while the electron affinity $\chi = E_{vac} - E_{cs}$ then varies with temperature to almost the same extent as the band gap. For Si{111} - 7 × 7 surfaces the same behaviour was reported earlier [21.28].

On the other hand, theoretical calculations of the temperature dependence of the band structure of group IV and III-V compound semiconductors predict that the valence band goes up with increasing temperature and gives the most

508

Table 21.1. Temperature coefficients of transitions in GaP and GaAs

Transition	Temperature coefficient [meV·K^{-1}]	
	GaP	GaAs
Ga(3d) → X_6^c	-0.24 ± 0.05[a]	-0.25 ± 0.03[b]
Ga(3d) → L_6^c		-0.31 ± 0.03[b]
E_g	-0.34[c]	-0.36[c]
Ga(3d) → Γ_8^v	$+0.1 \pm 0.05$	$+0.13$ to $+0.19$
$I = E_{vac} - E_{vs}$		$+0.02 \pm 0.06$[d]

[a][21.31]; [b][21.32]; [c][21.26]; [d][21.13]

important contribution to the experimental value of the temperature coefficient of the band gap and the temperature coefficient of the conduction band is less important [21.29,30]. However, more recent studies of optical transitions from core levels to conduction-band states in GaP and GaAs strongly question this conclusion, but are, however, in very good agreement with the results of the surface studies.

Aspnes et al. [21.31] and *Skibowski* et al. [21.32] measured the temperature dependence of the reflectance with GaP and GaAs, respectively, between 20 and 22 eV. In these spectral regions transitions from the Ga(3d) core levels to the bottom of the conduction bands at L_6^c and X_6^c are observed. Their data are listed in Table 21.1.

The temperature variations of the band gaps, i.e., of the energy separations $\Gamma_8^v - X_6^c$ in GaP and $\Gamma_8^v - \Gamma_6^c$ in GaAs, are well known [21.26]. Their temperature coefficients between 100 and 300 K are listed in Table 21.1, too. For GaP the temperature coefficient for the Ga(3d) - Γ_8^v separation can be directly evaluated while it can be only estimated for GaAs since the optical transitions between valence and conduction band are direct. However, the values for GaAs and GaP are almost identical.

These temperature coefficients may arise from shifts in both the initial and the final states. The 3d levels are affected primarily by the changes in the electrostatic interaction between bond and core charges, and by the redistribution of the bond charge with increasing temperature. *Aspnes* et al. showed both effects to be quite small. They concluded that the temperature effects of the 3d levels are negligible compared to those of the sp^3 valence-conduction bands.

With increasing temperature the top of the valence band in GaAs thus moves to higher energies by approximately 0.05 to 0.1 meV per Kelvin. Considering

the limits of experimental error, this is in close agreement with the finding that the ionisation energy of GaAs{110} surfaces is almost temperature
independent.

21.3 Chemical Trends of the Ionisation Energy of Cleaved Surfaces

Chemical trends of the ionisation energy of semiconductors may be obtained
from LCAO-type calculations of the electronic structure [21.7]. In the LCAO
description the electronic band structure is derived from atomic term values
and interatomic matrix elements. For compound semiconductors crystallising
in the zinc-blende lattice, the energy of the valence-band maximum is given
by [21.7]

$$E_v = \frac{E_p^c - E_p^a}{2} - \sqrt{V_2^2 + V_3^2} \ . \tag{21.11}$$

Here E_p^c and E_p^a are the atomic term values of the p orbitals of the cation
and the anion in the CA compound, respectively; $V_3 = \frac{1}{2}(E_p^c - E_p^a)$ is the polar energy; and $V_2 = 2.16(\hbar^2/2m_0 \cdot 1/d^2)$ is the interatomic matrix element
or the covalent energy.

The atomic term values are measured from the vacuum level, and, therefore, (21.11) also gives the energetic position of the valence-band maximum
below the vacuum level, i.e., the ionisation energy. In Fig.21.6 the experimentally determined ionisation energies are plotted versus the associated
LCAO energy of the valence-band maxima calculated from (21.11) for 12 III-V
and II-VI compound semiconductors. The experiments were performed with {110}
surfaces prepared in situ by cleavage in ultrahigh vacuum. The data were determined as the threshold energies of the total yield in photoemission spectroscopy [21.33-36].

A least square fit to the data points in Fig.21.6 gives

$$I = 0.92 \ E_v - 3 \ [eV] \ .$$

The almost linear relationship with slope 1 indicates that the LCAO description indeed leads to the correct chemical trends. The offset by 3 eV is not
surprising since image forces are not included in the expression for the valence-band maximum. They can easily result in a reduction by 1.8 eV or even
more, depending on the approximations used [21.7].

All the data for the III-V and II-VI semiconductors were measured with
cleaved {110} surfaces. For GaP, GaAs, InP, InSb, ZnS, ZnSe, and ZnTe the
arrangements of the atoms in these surfaces have been determined and found
to deviate from the truncated bulk structure [21.4,25,37-41]. These surfaces
are reconstructed, and the anions are rotated outward and the cations inward

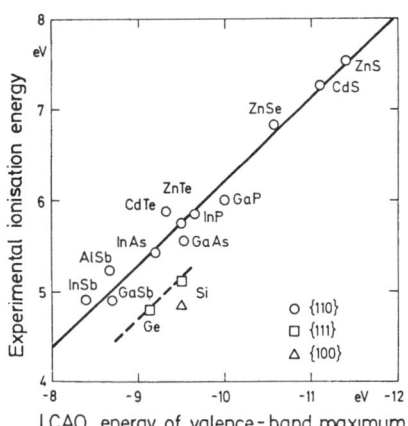

Fig.21.6. Experimental ionisation energies of cleaved surfaces plotted versus the LCAO energies of the valence-band maxima calculated from (21.11). After [21.6], but also considering some more recent experimental data

so that the bond lengths remain almost unaltered. The other semiconductors are believed to behave in the same way. This surface reconstruction does not contribute to the ionisation energy although the dangling bonds are doubly occupied on the anions, which are relaxed outward, while the cations exhibit empty dangling bonds since the surface atoms are neutral.

The ionisation energies of the cleaved {111} surfaces of the group IV elements silicon and germanium are also displayed in Fig.21.6 [21.42]. For the cleaved silicon surface we used an ionisation energy of 5.15 eV rather than 5.35 eV [21.43], because the work function and the position of the Fermi level in the band gap were determined as $\Phi = 4.83$ eV [21.42,43] and $E_F - E_{vs}$ = 0.32 eV [21.44-47], respectively, by several groups. The line connecting the Ge and Si data points is shifted by -0.6 eV compared with the line fitting the data points for the {110} surfaces of the compound semiconductors. The Si and Ge{111} surfaces exhibit a 2×1 reconstruction [21.48]. The atomic rearrangement on these surfaces was long considered to be adequately described by the buckling model [21.49,50]: adjacent rows of surface atoms are raised and depressed, respectively, with regard to a common plane which as a whole is relaxed inward [21.51,52]. This model was originally based on a simple picture of the rehybridisation of the sp^3 orbitals at these surfaces [21.49]. However, the buckling model of the Si{111} 2×1 reconstruction was questioned recently, and a π-bonded chain model has been proposed [21.53], containing 5- and 7-member rings in the top layers, and describable by two {110} layers terminating the {111} oriented crystal. Although the π-bonded chain model predicts dispersion of the dangling-bond surface states that is observed experimentally, this model is not compatible with the experimental data of low-energy electron diffraction from Si{111} 2×1 surfaces [21.54]. At present, this issue has not been resolved adequately.

511

The work functions of metals are known to depend on the crystallographic orientation of the surface. *Smoluchowski* [21.55] has given some simple arguments for the physical understanding of the trends observed. Two effects which are not independent contribute to the surface dipole. Since they oppositely influence the work function they shall be discussed separately, although they are actually mutually dependent.

The potential well at the surface is finite, and thus the electrons tail into vacuum. This leaves the atomic cores at the surface charged positively. Thus this surface "spreading" of the electrons into vacuum creates a dipole layer that increases the work function of a metal. On the other hand, the electrons may also redistribute laterally. In a simple picture, every atom in a metal "fills" a Wigner-Seitz cell of the particular lattice, rhombic dodecahedra for fcc and truncated octahedra for bcc lattices. The surfaces are thus rough. The electrons may therefore flow from "hills" into "valleys" formed by the polyhedra at the surface and thus leave the positive charge in their protruding parts partly uncompensated. This "smoothing" effect now decreases the surface dipole layer, and it should be more pronounced the rougher the surface is. Among the low-indexed faces of an fcc crystal, the {111} planes are the most densely packed, then come the {100} and finally the {110} surfaces, while the sequence is {110}, {100}, {112}, and {111} for bcc lattices. Indeed, the work functions of metals were found to decrease in that order.

Such simple model considerations cannot be directly applied to semiconductors since their surfaces are reconstructed. Relaxation and charge transfers normal to the surface are generally associated with such rearrangements of the surface atoms compared to an equivalent plane in the bulk. This effect then contributes to the surface dipole and thus the ionisation energy. Let us first consider the Si{001} 2×1 and then the Si{111} 2×1 and 7×7 surface structures

21.4.1 Si{001} 2×1

Already in 1959 *Schlier* and *Farnsworth* [21.56] studied low-energy electron diffraction with clean silicon {001} surfaces, and they found them to exhibit a 2×1 reconstruction which they attributed to a dimerisation between atoms of adjacent rows. In a simple picture, the formation of such dimers saturates one of the two broken bonds per surface atom in a {001} plane and leaves each surface atom with only one dangling bond. Theoretical calculations [21.57] have shown that such an arrangement of the surface atoms exhibits two overlapping, i.e., metallic, bands of surface states. A study of angle-resolved photoemission [21.58], however, detected a band of occupied surface states below the top of the valence band; the other, empty, band of surface states has been found by surface-photovoltage spectroscopy [21.59,60]. Such a semiconductor-like arrangement of the electronic surface states is achieved by an additional buckling of the dimers. This model is explained in Fig.21.7. This secondary Jahn-Teller effect lifts the degeneracy between the two dangling bonds at the symmetric dimers and causes the atom that relaxes inward to donate electronic charge to the atom moving outward [21.57,61]. The buckling and the associated charge transfer normal to the surface contributes to the surface dipole and increases the work function and also the ionisation energy. This has been established experimentally [21.59,60] and shall be considered in the following. By the way, the model of asymmetric dimers as the building blocks of the 2×1 reconstruction on silicon as well as germanium {001} surfaces is by now well established theoretically [21.57,61-63] and experimentally by quite a number of different experimental techniques [21.58-60,64-66].

On Si{001} surfaces the 2×1 reconstruction and thus the asymmetric dimers are removed by the adsorption of hydrogen [21.67]. First, both silicon atoms of the dimers are saturated with one hydrogen atom each resulting in a Si{001} 2×1:H monohydride structure; the dimers are now symmetric since both Si atoms are equivalent. Then the dimers are broken up, too, and the Si{001} 1×1:H structure forms in which the interplanar distance between the topmost two layers is decreased by a few percent compared to the bulk spacing [21.68-70]. In this dihydride structure two hydrogen atoms are bonded to each silicon surface atom. Finally, the surface becomes corroded [21.59,60, 71,72].

During the transition from the 2×1 to the monohydride structure, i.e. the transition from asymmetric to symmetric dimers, the work function decreases

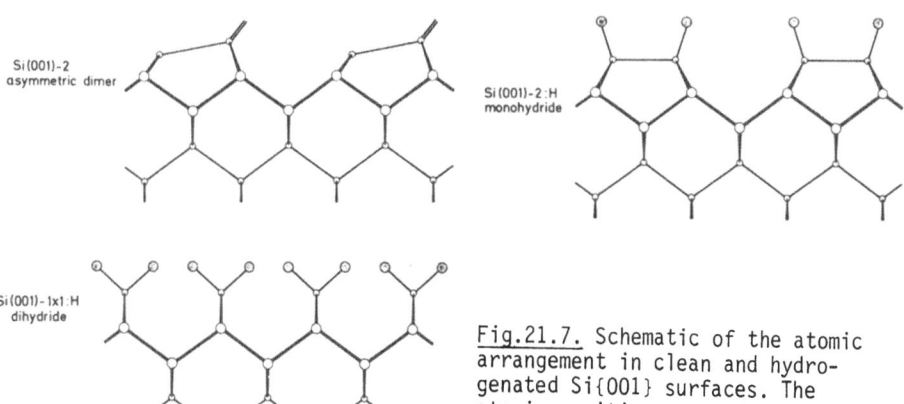

Si(001)-2
asymmetric dimer

Si(001)-2:H
monohydride

Si(001)-1x1:H
dihydride

Fig.21.7. Schematic of the atomic arrangement in clean and hydrogenated Si{001} surfaces. The atomic positions were from [21.57, 61,72]

513

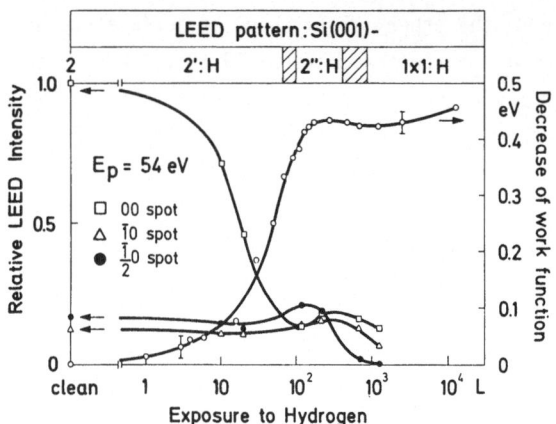

Fig.21.8. Change of work function and of LEED spot intensities during hydrogenation of Si{001} surfaces [21.60]

by (0.39 ± 0.05) eV as shown in Fig.21.8 and it remains almost unaltered by the subsequent formation of the 1×1:H structure [21.60]. This decrease comprises changes of the surface dipole as well as of the surface band-bending

$$\Delta\Phi = \Phi(1 \times 1:H) - \Phi(2 \times 1) = \Delta I + e_0 \Delta |V_s| \quad . \tag{21.12}$$

The change of surface band bending has been determined as $e_0 \Delta |V_s|$ $= (0.05 \pm 0.02)$ eV from measurements of the saturation of surface photovoltage as a function of the light intensity [21.60]. The remaining decrease of the ionisation energy $\Delta I = -(0.34 \pm 0.07)$ eV again may contain two contributions: one originates from the removal of the 2×1 reconstruction while the other may be caused by the possible introduction of dipoles associated with the newly formed Si-H bonds:

$$\Delta I = -\delta I_{recon} + \delta I_{Si-H} \quad . \tag{21.13}$$

In molecules silicon-hydrogen bonds are partly ionic with the H atom being negatively charged. Ab initio Hartree-Fock-LCAO calculations for hydrogen bound to silicon clusters indicated that the same should be true for hydrogen adsorbed on silicon surfaces [21.73]. However, the sign of the Si-H dipole immediately shows that the decrease of the ionisation energy by the chemisorption of H at Si{001} 2×1 surfaces originates in the removal of the surface dipole associated with this initial reconstruction and is not related to the Si-H bonds formed. This makes the lower bound for the contribution of the asymmetric dimers to the ionisation energy of clean Si{001} 2×1 surfaces $\pm(0.34 \pm 0.07)$eV.

Chemisorption-induced contributions to the ionisation energy may be evaluated in a simple but successful approach by considering each chemisorption bond as an additional surface dipole with a normal moment $\mu_\perp$. The potential

514

drop across a layer of N parallel dipoles per cm^2 is given by [21.74]

$$\delta I_{dip} = \frac{e_0}{\varepsilon \varepsilon_0} \cdot \frac{\mu_\perp}{1 + \kappa \alpha N^{3/2}} \cdot N \quad . \tag{21.14}$$

The denominator in the second factor on the right hand side reduces the normal component $\mu_\perp$ of the dipole moment due to the mutual interactions of the dipoles, where $\kappa \approx 9$ is a constant depending on the arrangement of the dipoles and α is their polarisability. In a point-charge model, the dipole moment of a chemical bond is given by the product of the partial charge on the bonding atoms and the bond length.

While the partly ionic character of Si-H bonds seems to be established, its magnitude can be only estimated. Following Pauling's concept of the relationship between electronegativity and the partial ionic character of chemical bonds, the hydrogen atom should be charged by $-0.02\ e_0$ in a Si-H bond [21.75]. In this model the bond length has to be taken as the sum of the covalent radii, i.e., $d_{Si-H} = 1.48$ Å. Assuming the bond angle to be a tetrahedral one and $\varepsilon = 1$, (21.14) then gives

$$\delta I_{Si-H} = 0.35 \text{ eV} \quad .$$

This value, however, seems to be an overestimate. Hydrogen adsorption on GaAs{110} surfaces was found to contribute to the surface dipole by only 50 to 100 meV [21.76], although the electronegativity differences between H and Si, Ga, and As, respectively, are almost the same.

The asymmetry of the dimers is removed during the formation of the 2×1:H monohydride structure. The subsequent formation of the 1×1:H dihydride structure does not further alter the work function. However, from this observation it cannot be concluded that the Si-H bonds do not exhibit a dipole moment and thus do not contribute to the work function, for the following reasons. The monohydride and the dihydride phases are not only distinguished by a doubling of the density of hydrogen atoms, but also by a change in the orientation of the Si-H bonds [21.72]. According to (21.14), both structures give almost the same contribution to the work function, namely 0.31 and 0.35 eV, respectively.

21.4.2 Si{111} 2×1 and 7×7

The changes of many surface properties caused by the adsorption of atomic hydrogen have also been studied with cleaved and with annealed silicon {111} surfaces. The structures of these two surfaces are characterised by the 2×1 and 7×7 LEED pattern [21.48]. Although both structures have been investigated for more than 20 years [21.2,3] even the atomic arrangement of the {111} 2 × 1 structure is still debated as has been mentioned above. Numerous models have been proposed for the 7 × 7 structure but none of them accounts for all the experimental results observed with surfaces exhibiting this structure.

On cleaved silicon surfaces hydrogen adsorption removes the 2 × 1 reconstruction and gives a 1 × 1:H LEED pattern [21.77]. With the 7 × 7 structure the effect of the exposure to atomic hydrogen is to quench most of the 7th-order spots, and only a ring of 6 of these extra spots around the normal spots remains clearly visible [21.78,79]. This is sketched in the upper panel of Fig.21.9. The lower diagram in this figure shows the decrease of the work function as a function of exposure to hydrogen. The change of band bending was evaluated from the intensity dependence of the surface photovoltage. The experimental data are plotted in Fig.21.10. Surface photovoltage is the change of the work function of a semiconductor due to illumination. Absorption of light with a photon energy larger than the band-gap energy produces electron-hole pairs. If a space-charge layer is present at the surface its electric field separates the photo-created electrons and holes, so decreasing the band bending. At very high intensities the bands finally are flat and the total change of the work function, i.e., the final value of the surface photovoltage, measures the surface band bending in the dark. For a more detailed discussion see [21.8,81]. Figure 21.10 shows that at low temperature the saturation of the surface photovoltage and thus flattening of the bands is easily reached with the light source used while it is approached with the samples held at room temperatures. The experimental data for clean and for hydrogenated Si {111} and Si{001} surfaces are listed in Table 21.2. The last column contains the difference between the ionisation energy measured with clean surfaces exhibiting the reconstructions given in the first column and its change effected by the hydrogenation

$$I_0^* = I_{clean} - \Delta I \quad . \tag{21.15}$$

Table 21.2. Ionisation energies, work function, and band bending of reconstructed and hydrogenated Si{001} and Si{111} surfaces. All values are in eV

Initial surface structure	Clean surface			Effects of hydrogenation			I_0^*
	Φ	$E_F - E_{vs}$	I	$\Delta\Phi$	$e_0\Delta|V_s|$	ΔI	
Si{001} 2 × 1	4.85[a]	0.3[b]	5.15	-0.39[g]	+0.05[g]	-0.34	4.81
Si{111} 7 × 7	4.6[c]	0.78[d]	5.38	-0.58[h]	+0.15[h]	-0.43	4.95
Si{111} 2 × 1	4.83[e]	0.32[f]	5.15	-0.4[h]	+0.23[i]	-0.18	4.97

[a][21.58,82], [b][21.58-60], [c][21.43,82,83], [d][21.47], [e][21.43,44], [f][21.44-47], [g][21.59,60], [h][21.80], [i][21.84]

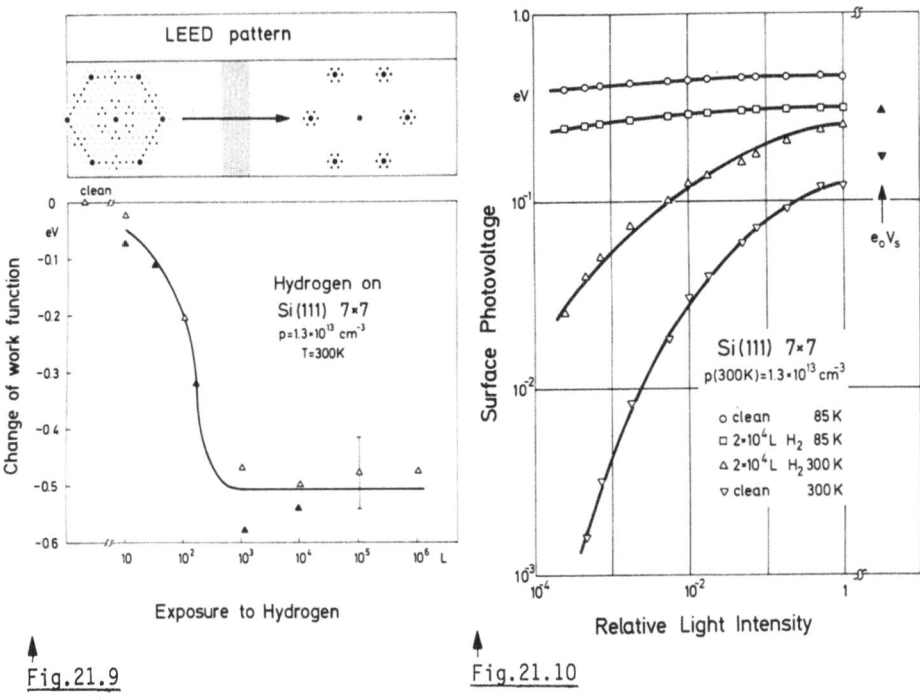

Fig.21.9

Fig.21.10

Fig.21.9. Change of work function and of LEED pattern during hydrogenation of Si{111} 7 × 7 surfaces [21.80]

Fig.21.10. Surface photovoltage as a function of light intensity observed with clean and hydrogenated Si{111} 7 × 7 surfaces at 85 K and at room temperature [21.80]

It is noteworthy that I_0^* is the same for both the Si{111} 2 × 1 and the Si{111} 7 × 7 structures and that I_0^* is larger on the {111} that on the {001} surface. The latter finding may be understood from Smoluchowski's spreading-and-smoothing picture discussed above: for a solid crystallising in the diamond structure, {100} planes are less densely packed than {111} planes and are thus to be expected to exhibit a lower ionisation energy, in analogy to the findings with fcc metals. In the evaluation of the I_0^* values, Si-H dipoles and their contribution to the ionisation energy have not been considered. Therefore, the "true" ionisation energies of Si{001} 1 × 1 and Si{111} 1 × 1 are expected to be larger by 0.1 to 0.3 eV than the values I_0^* given in Table 21.2.

21.5 Compositional Variations of the Ionisation Energy

21.5.1 GaAs{001}

With compound semiconductors a discussion of the work function and the ionisation energy is complicated for surfaces other than cleaved. Ideally, {111}, {$\overline{1}\overline{1}\overline{1}$}, and {001} surfaces of III-V compound semiconductors are expected to contain either anions or cations only. However, with GaAs{001} surfaces, for example, various surface reconstructions have been observed and each of them exhibits gallium and arsenic atoms in a particular ratio which ranges from a complete arsenic terminating layer to only approximately 20% of arsenic in the topmost layer [21.85,86]. A critical review of the coverage scale is found in [21.87]. The surface compositions can be controlled by varying the arsenic-to-gallium flux ratio during growth of the GaAs film by the molecular-beam epitaxy technique and during the subsequent cooling to room temperature [21.88]. Figure 21.11 displays the work function measured with differently reconstructed GaAs{001} surfaces [21.89] as a function of the surface composition [21.85]. The dopings of the films are not specified since the same work functions were reported for p- and n-type samples. This means that the Fermi level is pinned at these surfaces by surface states. They are most probably of extrinsic nature since no surface states have been detected in the bulk energy gap by using angular-resolved photoemission with one of the surfaces, namely GaAs{001} 2 × 4 which is equivalent to the c(2 × 8) structure [21.90]. In two other photoemission studies the separation between the Fermi

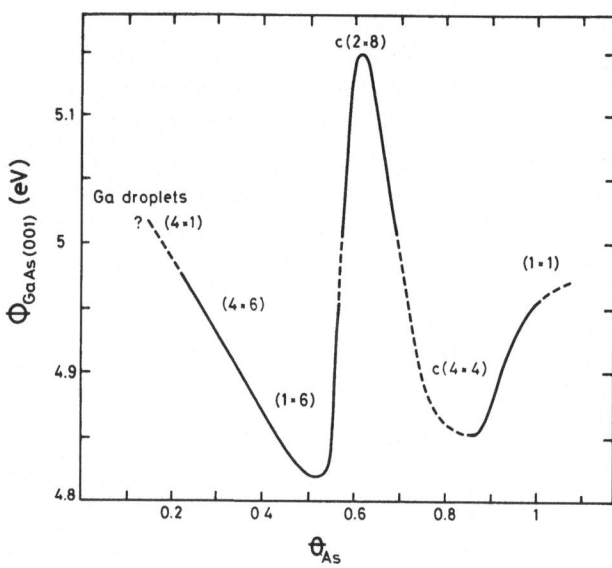

Fig.21.11. Work function of GaAs{001} surfaces exhibiting different surface structures as a function of arsenic surface coverage [21.89]. The coverage scale was adopted [21.85]

Table 21.3. Compositional dependence of electronic surface parameters of GaAs{001} surfaces

Reconstruction	θ_{As}	Φ in eV	$E_F - E_{vs}$ in eV	I in eV
4 × 6	≈ 0.2[a]	4.95[b]	0.73[c]	5.68
c(2 × 8)	1[a]	5.15[b]	0.4[d]	5.55
		-	-	5.40[f]
c(4 × 4)	> 1[e]	4.85[b]	0.4[d]	5.25
			0.73[c]	5.58

[a][21.85], [b][21.89], [c][21.92], [d][21.91], [e][21.89,93], [f][21.94] (corrected, see text).

level and the valence-band maximum has been determined recently [21.91,92] for three of the GaAs{001} surface reconstructions, Table 21.3. Unfortunately, the two values of $(E_F - E_{vs})$ given for the c(4 × 4) structure differ by 0.3 eV. This may be caused by different amounts of As chemisorbed [21.93]; however, the coverage was not specified in the publications. Nevertheless, the data compiled in Table 21.3 seem to suggest that the ionisation energies of these three structures with widely differing arsenic content vary only little with composition.

21.5.2 Cylindrical GaAs Sample

Quite recently, the orientation dependence of the ionisation energy was studied with a cylindrical GaAs single crystal by using photoemission with ultraviolet light [21.94]. With such a sample a number of low-indexed surfaces and vicinal faces can be prepared and investigated under the same conditions. The experimental results are displayed in Fig.21.12. The clean surface of this cylindrical sample was prepared by three different techniques: ion bombardment, ion bombardment followed by annealing in ultrahigh vacuum, and a film grown by molecular-beam epitaxy. After ion bombardment almost no variations of the ionisation energy are observed around the sample. This does not seem to be surprising since the surfaces are highly disordered and Ga-rich due to a preferential sputtering of As [21.95,96]. The low and constant ionisation energy may be explained if the excess Ga is described as adsorbed atoms which are charged partly positive. This seems to be plausible since Ga is more electropositive than As.

A subsequent annealing treatment generally increases the ionisation energy except for the {111} orientation. Finally, the MBE prepared surface shows

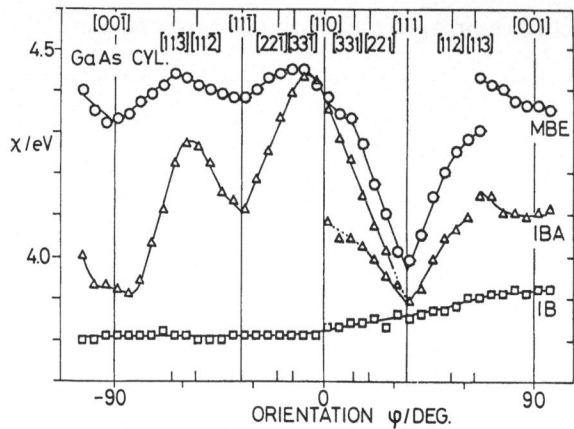

Fig.21.12. Variation of the electron affinity $\chi = E_{vac} - E_{cs}$ around a cylindrical GaAs sample with the surface prepared by ion bombardment (IB), ion bombardment followed by annealing (IBA) or molecular beam epitaxy (MBE) [21.94]. The electron affinities are too large by approximately 0.23 eV. For a detailed discussion, see text

only little variation around the cylindrical sample but again the {111} oriented part exhibits an ionisation energy which is lower than the average by 0.4 eV. These orientational trends of the results can be explained by considering the surface composition.

The conditions during MBE growth were chosen such that the As-terminated {001} c(2×8), {00$\bar{1}$} c(2×8), and {11$\bar{1}$} 2×2 surfaces were prepared. The {111} surface can only be prepared Ga-terminated and exhibiting a 2×2 reconstruction. The {110} surface always shows its 1×1 LEED pattern and equal densities of arsenic and gallium atoms. The evidently lower ionisation energy of the Ga-terminated {111} 2×2 surface compared with the As-terminated {11$\bar{1}$} 2×2 surface can again be understood from Pauling's concept of the ionisity of covalent bonds [21.74]. Gallium is electropositive compared to As, and from the difference of their electronegativities a 4% ionic character results for Ga-As bonds. This "chemical" surface dipole is oriented to decrease the ionisation energy on a Ga-terminated surface and to increase it on an As-terminated surface. These are indeed the experimental findings. As to be expected from Smoluchowski's speading-and-smoothing concept and in agreement with the results on the silicon surfaces discussed above, the ionisation energy is lower on the {00$\bar{1}$} than on the {11$\bar{1}$} surface, which are both As terminated. However, this trend is much less pronounced, and obviously close to the detection limit of UPS, than the chemical effect just discussed.

This recent study has demonstrated and elucidated the importance of chemical effects and their influence on the properties of even clean compound-semiconductor surfaces. However, the absolute values of the electron affinity were evaluated incorrectly. They are too large by approximately 0.23 eV. A comparison with data published earlier yields:

1) With cleaved GaAs{110} surfaces the ionisation energy has been determined
as 5.47 eV by *Gobeli* and *Allen* [21.97] and 5.4 eV by *Sebenne* and co-workers
[21.36] from photoelectric yield spectra and as 5.56 eV by *van Laar* et al.
from measurements of photoelectric yield [21.33] as well as of the work function of p- and n-type samples. The electron affinity at room temperature
is $\chi = E_{vac} - E_{cs} = 4.13$ eV. Ranke's value of 4.4 eV is larger by 0.27 eV.

2) In one of their earlier papers, *Ranke* and *Jacobi* [21.98] reported on photoemission studies with GaAs(111) and ($\bar{1}\bar{1}\bar{1}$) surfaces which they prepared by
ion bombardment and annealing. The ionisation energies of these surfaces were
determined as I(111) = 4.83 eV and I($\bar{1}\bar{1}\bar{1}$) = 5.27 eV, which values give the electron affinities at room temperature as χ(111) = 3.4 eV and χ($\bar{1}\bar{1}\bar{1}$) = 3.74 eV.
Figure 21.12 contains, as alread mentioned, one set of data measured with the
cylindrical sample after an IBA treatment. From the data plotted one reads
the electron affinities χ(111) = 3.9 eV and χ($\bar{1}\bar{1}\bar{1}$) = 4.1 eV, which are larger
by 0.5 and 0.35 eV, respectively, than the previously communicated ones.

In his recent paper, *Ranke* [21.94] evaluated the electron affinity from
the width W of the energy distribution of the photoemitted electrons

$$\chi = \hbar\omega - W - E_g \ . \tag{21.16}$$

By mistake, he took the band-gap energy E_g of GaAs at room temperature as
1.35 eV instead of the correct value of 1.43 eV [21.26]. Already the use of
the true band-gap energy increases the electron affinities by 0.1 eV. Furthermore, the resolution of the electron-energy analyser used, i.e., 0.16 eV, was
subtracted from the width W_{exp} evaluated from the recorded energy distribution curves of the photoemitted electrons. This procedure may add another
error to the determination of the electron affinities, resulting in a lowering of the values.

21.6 Adsorption-Induced Changes of the Work Function

Adsorption of foreign atoms or molecules alters the chemical composition of
a surface, and thus changes of the electronic surface properties are to be
expected. We will discuss two typical examples, Cs and Ge adsorbed on clean,
cleaved GaAs{110} surfaces. Experimental results are displayed in Figs.21.13,
14. As a function of coverage the work function behaves quite differently in
both cases. Cesium adsorption decreases the work function by 3.5 to 4 eV on
n- as well as on p-type substrates [21.99,100], while germanium adsorption
increases the work function on n-type but decreases it on p-type samples
[21.101]. These results will be analysed below.

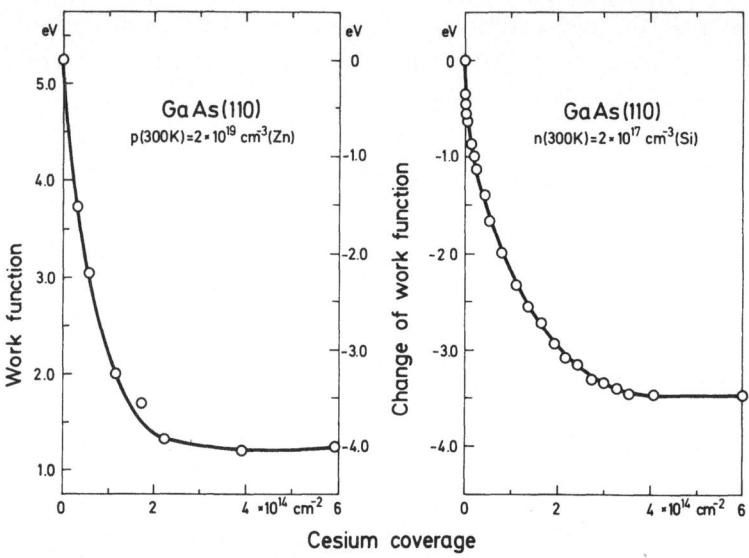

Fig.21.13. Work function as a function of Cs coverage measured with surfaces cleaved from p-type [21.99] and n-type [21.100] GaAs samples

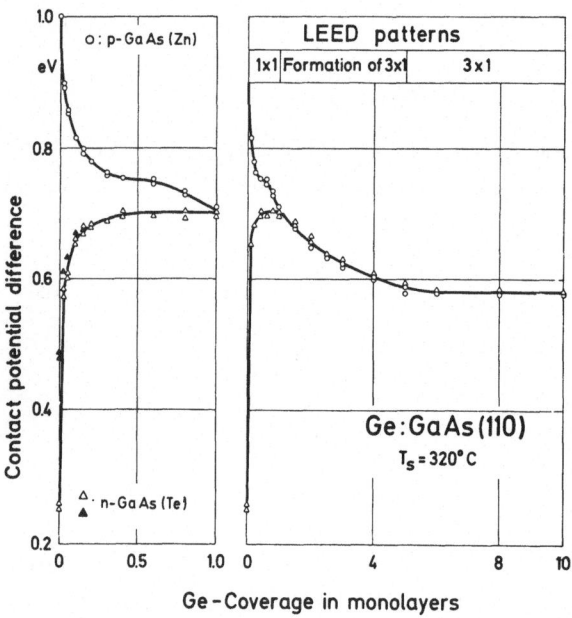

Fig.21.14. Contact potential difference and LEED patterns as a function of Ge coverage on {110} surfaces cleaved from p- and n-type GaAs [21.101]

21.6.1 Cesium on GaAs {110} Surfaces

The lowering of the work function by adsorbed cesium has been studied with metals as well as with semiconductors for a long time. Such investigations have been at least partly motivated by the technological interest in thermal as well as photocathodes exhibiting low work functions. Let us first consider the Cs-on-W system which was carefully studied by *Taylor* and *Langmuir* 50 years ago [21.102]. They found the work function of the tungsten substrate to decrease by up to 3 eV with increasing cesium coverage. This can be explained by the formation of a dipole layer at the W surface with the adsorbed Cs charged positively. The donation of an electron from the adsorbed cesium to the tungsten substrate is plausible since the ionisation energy of cesium atoms amounts to only 3.9 eV and the work function of tungsten measures at least 4.5 eV, depending on the orientation of the particular surface. Thus, the valence electron of the cesium atom may make a transition to the Fermi level of the metal. The closely spaced adsorbate interacts with the substrate, and this causes the original valence level of the adsorbate to broaden and shift. A portion of the perturbed energy level lies below the Fermi energy, and hence the level will be partially occupied by electrons shared by the adsorbate and the metal. Thus, the effective charge on the adsorbate is reduced to approximately 85% of the pure ionic charge [21.103].

Cesiation of the GaAs surfaces reduces the work function by 4 eV with p-type samples and by 3.5 eV with n-type samples. Since the decrease of the work function almost equals that observed with cesium-covered tungsten surfaces, again the formation of a dipole layer is to be concluded. This dipole layer should be the same on p- and n-type samples, and thus the difference in the reduction of the work function on both types of conductivity has to account for changes in surface band-bending:

$$\Delta\Phi_p - \Delta\Phi_n = e_0\Delta V_{sp} - e_0\Delta V_{sn} \quad . \tag{21.17}$$

By using ultraviolet photoemission spectroscopy, *Spicer* et al. [21.104] have determined the surface position of the Fermi level in the band gap with surfaces cleaved from n- as well as p-type GaAs samples as a function of Cs coverage θ. The coverages are measured in monolayers, and one monolayer is the total number of surface sites in a {110} plane of GaAs, i.e., $\theta = 1$ corresponds to $\sigma_{110} = 8.85 \times 10^{14}$ cm^{-2}. Above $\theta \approx 0.1$ the Fermi level was found to be pinned at 0.5 eV above the valence-band maximum on p-type and at 0.66 eV below the bottom of the conduction band on n-type samples. Since at clean surfaces on p-type samples the bands are flat up to the surface, the n-type sample used for the measurements reported in Fig.21.14 already exhibited an initial band bending due to cleavage-induced acceptor-type surface states, as discussed above.

The adsorption of cesium introduces depletion layers on both types of samples. Obviously, the charge stored in these depletion layers cannot compensate for the positively charged cesium ions responsible for the large reduction of the work function. Thus, different mechanisms have to be discussed for the surface band-bending and the dipole layer. Let us first consider the dipole layer.

Following Pauling's concepts of the partially ionic character of covalent bonds, the electronegativity difference gives an "average" ionic character of 26% for Cs - GaAs bonds. Consequently, the bond length should equal the sum of the covalent radii. The polarisability is taken as $\alpha = 1.3 \times 10^{-23}$ cm^{-3}, which was obtained from a fit of (21.14) to data measured with the system Cs-on-Si{111} 2×1 [21.46] and which equals the cube of the covalent radius of cesium. As Fig.21.14 shows, the saturation coverage amounts to 0.7 monolayers of cesium. With these data (21.14) yields a dipole-layer contribution of 3.7 eV, in very good agreement with the experimental value of $\Delta I = \Delta \Phi_p - e_0 \Delta V_s$ = 3.6 eV observed with cesiated p-type GaAs{110}.

21.6.2 Germanium on GaAs{110} Surfaces

As shown in Fig.21.14, the deposition of Ge on clean, cleaved GaAs{110} surfaces decreases the work function of p-type samples but increases it on n-type ones. Above a coverage of one monolayer the work function is independent of the type of substrate doping. As a function of coverage the work function becomes constant on films thicker than 5 monolayers. Above this thickness a fully developed 3×1 LEED pattern is observed. Its formation is related to segregation of arsenic on the growing Ge films [21.105,106]. In the following we shall discuss only the changes of the work function up to coverages of half a monolayer since we are interested in the surface properties of the GaAs substrate and its changes, rather than the development of the surface properties of the growing Ge film which is most probably covered with a GeAs$_x$ compound, as indicated by soft X-ray photoemission studies on such films and cleaved GeAs surfaces [21.107].

In the submonolayer coverage range the work function changes in opposite directions on p- and n-type samples. This behaviour indicates that possible changes of the ionisation energy must be less drastic than for Cs on GaAs, discussed in the previous section. Therefore, in addition to the measurements of the contact potential difference with a Kelvin probe, ultraviolet photoemission spectroscopy was used to determine the surface band-bending as a function of Ge coverage [21.101,108]. The results are plotted in Fig.21.15. The two ordinate scales are chosen such that the contact potential measured

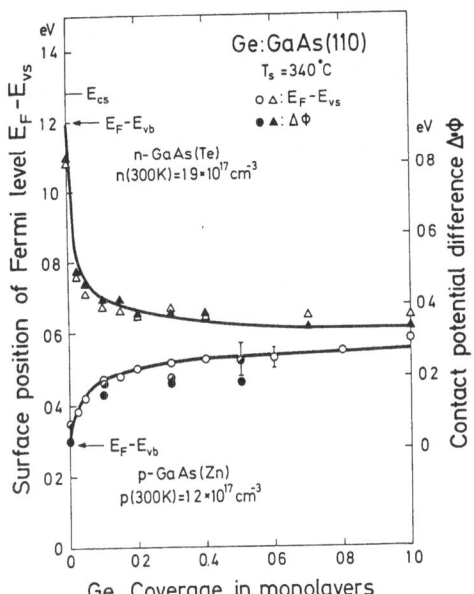

Fig.21.15. Comparison of the change in surface band bending and in work function as a function of Ge coverage as measured with {110} surfaces cleaved from p- and n-type GaAs [21.108]

with the clean p-type sample is adjusted to the position of the Fermi level in the bulk since the bands are flat up to the surface on such samples. The contact potential difference measured with the clean surface of the n-type sample differs by 0.1 eV from the position of the bulk Fermi level in this sample, indicating that this surface exhibits an initial band bending of 0.1 eV.

The data given in Fig.21.15 immediately show that within the limits of experimental error the changes of the position of the Fermi level in the band gap and of the work function are equal, i.e., no measurable changes of the ionisation energy as a function of coverage are to be detected. *Pauling's* concept [21.75] of the partially ionic character of covalent bonds again offers a plausible explanation for this observation. The electronegativity of Ge equals the mean of those values for Ga and As. Thus, on the average, no dipole layer is to be expected for the chemisorption of Ge on GaAs{110} surfaces which contain As and Ga in equal densities.

The data of Fig.21.15 also reveal that the adsorption of Ge on GaAs{110} surfaces causes depletion layers to form on both p- and n-type substrates. To fulfill the condition of charge neutrality at the surface, i.e. according to (21.2), negatively charged acceptor-type and positively charged donor-type surface states have to dominate on the n- and p-type samples, respectively. For an evaluation with respect to energetic position and density of these chemisorption-induced surface states, the much more precisely determined changes of the work function may be used and read as changes of the surface band-bending, since the ionisation energy remains unchanged. The limits of experimental error are ±20 meV for the Kelvin probe and ±100 meV for the UPS measurements.

For the analysis we have assumed discrete surface states. Since we are dealing with chemisorption-induced surface states, their density is assumed to increase in proportion to the coverage with Ge atoms

$$N_{ss}^{A,D} = q^{A,D}\sigma_{110}\theta \quad . \tag{21.18}$$

A least-square fit to the experimental data measured at $320^{\circ}C$ yields [21.108]

$$E_{ss}^{A} - E_{vs} = (0.72 \pm 0.05) \text{ eV} \quad , \quad q^{A} = 0.06 \pm 0.03 \quad ,$$

$$E_{ss}^{D} - E_{vs} = (0.42 \pm 0.05) \text{ eV} \quad , \quad q^{D} = 0.04 \pm 0.02 \quad .$$

It is noteworthy that the best fit was obtained by taking into account donor- and acceptor-type surface states on both p- and n-type samples [21.101]. The full lines drawn in Fig.21.15 were calculated with this set of parameters and doping of the samples. Measurements at room temperature gave the same generation factors but slightly shifted level positions [21.108]

$$E_{ss}^{A} - E_{vs} = (0.78 \pm 0.05) \text{ eV} \quad \text{and} \quad E_{ss}^{D} - E_{vs} = (0.51 \pm 0.05) \text{ eV} \quad .$$

The distance of 0.3 eV between both levels is independent of temperature. Both levels follow neither the bottom of the conduction band nor the top of the valence band as a function of temperature, and they thus show the behaviour of deep levels [21.109].

21.6.3 Chemisorption-Induced Defects in GaAs{110} Surfaces and Interfaces

The generation factors q of the chemisorption-induced states are quite small, and thus these states cannot be correlated with the bonds between the adsorbed Ge atoms and the GaAs substrate. Furthermore, the band bending which develops in the GaAs substrates during the submonolayer coverage with Ge was found to persist under thick Ge films, i.e. in a Ge:GaAs{110} heterojunction [21.107]. Similar observations were reported earlier by *Spicer* and co-workers [21.110, 111] for metal-GaAs contacts and the adsorption of oxygen on cleaved GaAs surfaces. The changes of the surface band-bending they observed on n- as well as p-type GaAs as a function of Al, Ga, In and oxygen coverage have been successfully described by using the above set of surface-state parameters

determined from the Ge-on-GaAs{110} experiments [21.112,113]. The most striking and technologically most important result is that already less than 0.2
of a monolayer of such different adsorbates as Cs, Al, Ga, In, Si, Ge, oxygen and hydrogen pin the Fermi level at the same positions near to midgap,
which differ by 0.27 eV on p- and n-type samples.

These findings indicate that the nature of the chemisorption-induced
states is independent of the particular adsorbate but related to an extrinsic property of the substrate. Already from less detailed experimental data,
Spicer et al. [21.114,115] concluded that the energy levels pinning the Fermi
levels are associated with native defects of the substrate that are created
during the early stages of adsorption. Similar conclusions were drawn by
Williams et al. [21.116] and by *Wieder* [21.117]. This defect model is most
directly supported by the result that the generation factor of the chemisorption-induced states amounts to only 0.05 and is thus far from unity
[21.101].

Earlier, *Bardeen* [21.118] proposed that intrinsic surface states of the
semiconductors might determine the barrier heights in metal-semiconductor
junctions since Schottky's electron affinity rule [21.119] could not correctly explain the experimental data. Spicer's picture now considers extrinsic
rather than intrinsic states and relates them to native defects of the semiconductor.

With regard to the nature of these defects, speculations based on theoretical models initially dominated [21.120-123]. The most widely discussed defects are vacancies and antisite defects. In particular, the V_{III} antisite
defects seem to be of characteristic importance in III-V compounds. The formation of chemisorption-induced vacancies in GaAs may be excluded. Studies
of radiation damage have shown vacancies to be stable up to only $220^{\circ}C$
[21.124], while chemisorption-induced states are observed at room temperature
and at $320^{\circ}C$ with the same density.

Increasing experimental evidence is gathering in favour of the involvement
of As_{Ga} antisite defects in the chemisorption-induced pinning of the Fermi
level at GaAs surfaces and interfaces. Such antisite defects have been identified by electron paramagnetic resonance (EPR) in as-grown [21.125] neutron-
irradiated [21.124], and plastically deformed GaAs [21.126]. The As_{Ga} defects
act as double donors with the charge states D^0, D^+, and D^{++} depending on the
position of the Fermi level with respect to their ionisation levels which
have been determined by photo-EPR [21.127] at 6 K

$$E(o/+) = E_c - 0.77 \text{ eV} = E_v + 0.75 \text{ eV} \quad \text{and}$$

$$E(+/++) = E_v + 0.52 \text{ eV} \quad .$$

Recent theoretical calculations found the levels of the As$_{Ga}$ defect to be
separated by 0.27 eV, in agreement with experiment, but to lie by 0.3 eV
closer to the bottom of the conduction band [21.128]. The positions of these
levels agree remarkably well with those determined for the chemisorption-
induced states at room temperature. However, As$_{Ga}$ antisite defects alone
cannot explain the Fermi-level pinning since they are double donors, while
the net charge states of the chemisorption-generated defects are C$^-$, C^0 and
C$^+$, depending on the position of the Fermi level. Most simply, these net
charge states can be achieved by adding a shallow acceptor or a double acceptor to the As$_{Ga}$ double donor. A more detailed discussion of this subject may
be found in [21.129].

An additional argument in favour of As$_{Ga}$ antisite defects being generated
by chemisorption on GaAs comes from recent annealing experiments. The EPR
signals from As$_{Ga}$ defects produced by neutron-irradiation [21.126] or plastic
deformation [21.127] are known to decay during isochronal anneals between
400 - 500°C. Experimental EPR results from As$_{Ga}$ defects [21.126] and photoluminescence data from the deep trap EL2 [21.130], which is thought to be an
As$_{Ga}$ antisite defect, are plotted versus the annealing temperature in the

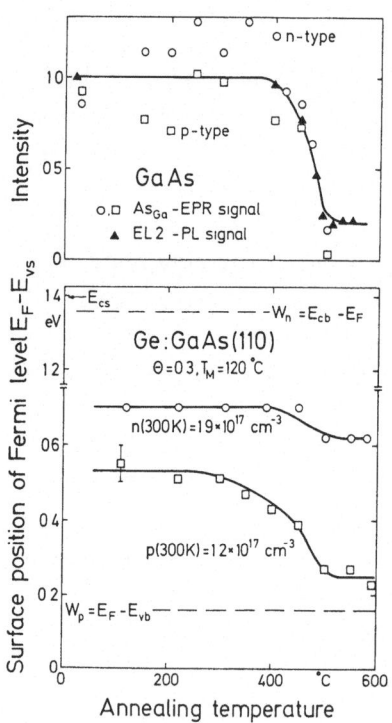

Fig.21.16. Annealing behaviour of the
EPR signal from As$_{Ga}$ antisite defects
[21.126] and the photoluminescence intensity from EL2 traps [21.130] and of
the surface band-bending observed on Ge-
covered GaAs{110} surfaces [21.108]

upper panel of Fig.21.16. The lower diagram in this figure contains the position of the Fermi level with respect to the band edges at a GaAs{110} surface covered with 0.3 monolayer of Ge as a function of the annealing temperature [21.108]. With increasing annealing temperature the EPR signal from the As_{Ga} antisite defects and thus the density of these imperfections evidently decays as the Fermi level at the surface of the Ge-covered p-type sample approaches the bulk position. This similar behaviour is easily understood since As_{Ga} antisite defects are double donors and donor-like surface states are necessary to balance the space charge in a depletion layer on p-type samples. The insensitivity of the large band bending on the Ge-covered n-type sample where acceptor states determine the formation of a depletion layer is in accordance with acceptors supplementing As_{Ga} double donors, as demanded above. These new results mean that the chemisorption-induced defects are not as simple-structured as proposed earlier.

While the As_{Ga} as well as the P_{Ga} antisite defects in GaAs and GaP have been positively identified [21.125-127,131], the same is not true for the Ga_V antisite defects. For the Ga_{As} defects two speculations have been proposed recently. *Schneider* [21.132] has tentatively assigned the well-known native hole traps A and B in GaAs to the Ga_{As} double acceptor. The associated energy levels are [21.133]

$$E(A) = E_v + 0.4 \text{ eV} \quad \text{and} \quad E(B) = E_v + 0.71 \text{ eV} .$$

Following Schneider's assignment, the B and A levels then control the three charging states A^{2-}, A^{1-}, and A^0 of the Ga_{As} double acceptor.

The assignment of the A and B trap levels to the Ga_{As} antisite defect is very appealing with respect to the annealing experiments with the Ge covered samples, the results of which are displayed in Fig.21.16. After annealing above 450°C the Fermi level has been found to be pinned at 0.62 eV above the top of the valence band on the n-type samples. As Fig.21.15 shows, the position of the Fermi level differs from the associated levels by approximately 100 meV. Thus, the acceptor-type levels determining the position of the Fermi level at the surface of the n-type sample should be at approximately 0.7 eV above the top of the valence band, and it might be identified with the above B level, which — in Schneider's proposal — is the upper one of the two levels of the double acceptor Ga_{As}.

The joint formation of As_{Ga} and of Ga_{As} antisite defects in GaAs substrates seems to be very likely since *van Vechten* [21.134] has estimated the same low enthalpy of formation of 0.7 eV for both types of antisite defects in GaAs.

21.7 Final Remarks

Clean semiconductor surfaces may be characterised by their ionisation energies and surface band-bendings. The chemical *trends* of the ionisation energies of the {110} surfaces of III-V as well as II-VI compounds are well described by a simple LCAO approach which has also been successfully used in predicting chemical trends of the valence-band discontinuities in semiconductor heterojunctions [21.135]. The compositional dependence of the ionisation energy is not understood to the same extent. For example GaAs{100} surfaces may be prepared so as to exhibit various reconstructions which differ in Ga to As ratios. As the data presented in Table 21.1 reveal, the average of the ionisation energies amounts to (5.54 ± 0.17)eV. However, the experimental data differ by 0.3 eV for one particular surface reconstruction and composition. Consequently, no precise conclusions can be drawn in this respect, and new data have to be waited for.

The surfaces of the elemental semiconductors do not cause any problems with regard to composition but the geometrical arrangements of the surface atoms are still debated in some of the most prominent cases. However, hydrogenation of differently oriented silicon surfaces has proven to be a very useful method for removing reconstructions or at least a large part of them. This experimental approach then yields an ionisation energy which is lower by 0.15 eV for the more rough {100} surfaces compared to the {111} surfaces. This difference is larger than the limits of experimental error of the individual values but close to it. Its sign, however, is predicted by the simple spreading-and-smoothing arguments that apply to metal surfaces.

In addition to the effect of hydrogen chemisorption on silicon surfaces I have discussed the change of ionisation energy and surface band-bending caused by the adsorption of cesium or germanium on cleaved gallium arsenide surfaces. In the latter cases, native defects are most probably created during the adsorption. This *adsorption-damage* is observed through the surface band-bending caused by the electronic states related to these imperfections, since clean, cleaved GaAs surfaces exhibit no intrinsic surface states within the bulk energy gap and thus no surface band-bending. At present, many experimental results are pointing to that As_{Ga} antisite defects being involved in the adsorption-induced imperfections. However, the other adsorption damage has not been identified. For semiconductors other than GaAs, the situation is even worse since only pinning positions of the Fermi level are known and thus speculations on the nature of the adsorption-induced defects prevail. Only for GaP do some experimental results again point to P_{Ga} antisite defects as being involved in the adsorption damage [21.108].

The experimental data compiled and reviewed in this chapter most clearly demonstrate that further systematic and detailed studies are needed for a more general understanding of such basic properties of semiconductor surfaces as the ionisation energy and defects induced by adsorption.

Acknowledgements. I should like to thank Dr. C. Sebenne (Paris) and Dr. H. Gant (now München) for sending a compilation of earlier results and for the assistance in collecting data, respectively. The paper was typed by A. Kohs, and the drawings were prepared by J. Krusenbaum. The numerical calculations for Figs.21.4,5 were done by Dr. H.J. Clemens.

References

21.1 N.D. Lang, W. Kohn: Phys. Rev. B3, 402 (1971)
21.2 W. Mönch: Surf. Sci. **86**, 672 (1979)
21.3 D.E. Eastman: J. Vac. Sci. Technol. **17**, 492 (1980)
21.4 C.B. Duke, R.J. Meyer, A. Paton, J.L. Yeh, T.C. Tsang, A. Kahn, P. Mark: J. Vac. Sci. Technol. **17**, 501 (1980)
21.5 E.G. McRae: Surf. Sci. **124**, 106 (1983)
21.6 J. Pollmann: Festkörperprobleme **20**, 117 (1980)
21.7 W.A. Harrison: *Electronic Structure and the Properties of Solids* (Freeman, San Francisco 1980)
21.8 D.R. Frankl: *Electrical Properties of Semiconductor Surfaces* (Pergamon, Oxford 1967)
21.9 A. Many, Y. Goldstein, N.B. Grover: *Semiconductor Surfaces* (North-Holland, Amsterdam 1965)
21.10 Sir W. Thomson: Nature **23**, 567 (1881)
21.11 M. Cardona, L. Ley: In *Photoemission in Solids I*, Topics in Appl. Phys., Vol.26, ed. by M. Cardona and L. Ley (Springer, Berlin, Heidelberg, New York 1978)
21.12 W. Mönch, H.J. Clemens: J. Vac. Sci. Technol. **16**, 1238 (1979)
21.13 W. Mönch, R. Enninghorst, H.J. Clemens: Surf. Sci. **102**, L54 (1981)
21.14 A. Huijser, J. van Laar: Surf. Sci. **52**, 202 (1975)
21.15 J.M. Palau, E. Testemale, L. Lassabatere: J. Vac. Sci. Technol. **19**, 192 (1981)
21.16 W. Gudat, D.E. Eastman, J.L. Freeouf: J. Vac. Sci. Technol. **13**, 250 (1976)
21.17 W.E. Spicer, I. Lindau, P.E. Gregory, C.M. Garner, P. Pianetta, P.W. Chye: J. Vac. Sci. Technol. **13**, 780 (1976)
21.18 D.J. Chadi: Phys. Rev. B**18**, 1800 (1978)
21.19 K.C. Pandey: J. Vac. Sci. Technol. **15**, 440 (1978)
21.20 J.A. Knapp, G.J. Lapeyre: J. Vac. Sci. Technol. **13**, 757 (1976)
21.21 G.P. Williams, R.J. Smith, G.J. Lapeyre: J. Vac. Sci. Technol. **15**, 1249 (1978)
21.22 A. Huijser, J. van Laar, T.L. van Rooy: Phys. Lett. **65A**, 337 (1978)
21.23 V. Dose, H.-J. Gossmann, D. Straub: Phys. Rev. Lett. **47**, 608 (1981); Surf. Sci. **117**, 387 (1982)
21.24 R.P. Beres, R.E. Allen, J.D. Dow: Solid State Commun. **45**, 13 (1983)
21.25 R.J. Meyer, C.B. Duke, A. Paton, A. Kahn, E. So, J.L. Yeh, P. Mark: Phys. Rev. B**19**, 5194 (1979)
21.26 M.B. Panish, H.C. Casey: J. Appl. Phys. **40**, 163 (1969)
21.27 R. Seiwatz, M. Greene: J. Appl. Phys. **29**, 1034 (1958)
21.28 R. Bachmann: Physik Kondens. Mater. **8**, 31 (1968)

21.29 D. Auvergne, J. Camassel, H. Mathieu, M. Cardona: Phys. Rev. B9, 5168 (1974)
21.30 J. Camassel, D. Auvergne: Phys. Rev. B12, 3258 (1975)
21.31 D.E. Aspnes, C.G. Olson, D.W. Lynch: Phys. Rev. Lett. 36, 1563 (1976)
21.32 M. Skibowski, G. Sprüssel, V. Saile: Appl. Opt. 19, 3978 (1980)
21.33 J. van Laar, A. Huijser, T.L. van Rooy: J. Vac. Sci. Technol. 14, 894 (1977)
21.34 R.K. Swank: Phys. Rev. 153, 844 (1967)
21.35 T.E. Fischer: Phys. Rev. 139, A1228 (1965)
21.36 G.M. Guichar, C.A. Sebenne, C.D. Thuault: In Proc. 3rd Intl. Conf. Solid Surf., ed. by R. Dobrozemsky, F. Rüdenauer, F.P. Vieböck, and A. Breth (Vienna 1977) p.623; Surf. Sci. 86, 789 (1979)
21.37 C.B. Duke, A.R. Lubinsky, M. Bonn, G. Cisneros, P. Mark: J. Vac. Sci. Technol. 14, 294 (1977)
21.38 R.J. Meyer, C.B. Duke, A. Paton, J.C. Tsang, J.L. Yeh, A.Kahn, P. Mark: Phys. Rev. B22, 6171 (1980)
21.39 C.B. Duke, R.J. Meyer, A. Paton, A. Kahn, J. Carelli, J.L. Yeh: J. Vac. Sci. Technol. 18, 866 (1981)
21.40 C.B. Duke A. Paton, W.K. Ford, A. Kahn, J. Carelli: Phys. Rev. B24, 562 (1981)
21.41 B.W. Lee, R.K. Ni, N. Masud, X.R. Wang, D.C. Wang, M. Rowe: J. Vac. Sci. Technol. 19, 294 (1981)
21.42 G.W. Gobeli, F.G. Allen: Phys. Rev. 137, A245 (1965)
21.43 C. Sebenne, D. Bolmont, G. Guichar, M. Balkanski: Phys. Rev. B12, 3280 (1975)
21.44 F.G. Allen, G.W. Gobeli: Phys. Rev. 127, 150 (1962)
21.45 M. Henzler: Phys. Stat. Sol. 19, 833 (1967)
21.46 W. Mönch: Phys. Stat. Sol. 40, 257 (1970)
21.47 J. Clabes, M. Henzler: Phys. Rev. B21, 625 (1980)
21.48 J.J. Lander, G.W. Gobeli, J. Morrison: J. Appl. Phys. 34, 2298 (1963)
21.49 D. Haneman: Phys. Rev. 121, 1093 (1961)
21.50 A. Taloni, D. Haneman: Surf. Sci. 10, 215 (1968)
21.51 W. Mönch, P.P. Auer: J. Vac. Sci. Technol. 15, 1230 (1978)
21.52 R. Feder, W. Mönch, P.P. Auer: J. Phys. C: Solid State Phys. 12, L179 (1979)
21.53 K.C. Pandey: Phys. Rev. Lett. 47, 1913 (1981); Physica 117B and 118B, 761 (1983)
21.54 R. Feder: Solid State Commun. 45, 51 (1983)
21.55 R. Smoluchowski: Phys. Rev. 60, 661 (1941)
21.56 R.E. Schlier, H.E. Farnsworth: J. Chem. Phys. 30, 1917 (1959)
21.57 D.J. Chadi: J. Vac. Sci. Technol. 16, 1290 (1979); and Phys. Rev. Lett. 43, 43 (1979)
21.58 F.J. Himpsel, D.E. Eastman: J. Vac. Sci. Technol. 16, 1297 (1979)
21.59 P. Koke, W. Mönch: Solid State Commun. 36, 1007 (1980)
21.60 W. Mönch, P. Koke, S. Krueger: J. Vac. Sci. Technol. 19, 313 (1981)
21.61 W.S. Verwoerd: Surf. Sci. 103, 404 (1981); 99, 581 (1980)
21.62 J. Pollmann, A. Mazur, M. Schmeits: Physica 117B and 118B, 771 (1983)
21.63 M.A. Bowen, J.D. Dow, R.E. Allen: Phys. Rev. B26, 7083 (1982)
21.64 R.M. Tromp, R.G. Smeenk, F.W. Saris: Phys. Rev. Lett. 46, 939 (1981)
21.65 W.S. Yang, F. Jona, P.M. Marcus: Solid State Commun. 43, 847 (1982)
21.66 M. Aono, Y. Hou, C. Oskima, Y. Ishizawa: Phys. Rev. Lett. 49, 567 (1982)
21.67 T. Sakurai, H.D. Hagstrum: Phys. Rev. B14, 1593 (1976)
21.68 S.J. White, D.P. Woodruff: Surf. Sci. 63, 254 (1977)
21.69 S.J. White, D.P. Woodruff, R.S. Zimmer: Surf. Sci. 74, 34 (1978)
21.70 R. Tromp, R.G. Smeenk, F. Saris: Surf. Sci. 104, 13 (1981)
21.71 H. Wagner, R. Butz, U. Backes, D. Bruchmann: Solid State Commun. 38, 1155 (1981)

21.72 W.S. Verwoerd: Surf. Sci. **108**, 153 (1981)
21.73 K. Hermann, P.S. Bagus: Phys. Rev. B**20**, 1603 (1979)
21.74 J. Topping: Proc. R. Soc. London A**114**, 67 (1927)
21.75 L. Pauling: *The Nature of the Chemical Bond* (Cornell University, Ihaca, New York 1940)
21.76 F. Bartels, H.J. Clemens, L. Surkamp, W. Mönch: J. Vac. Sci. Technol. B**1** (1983), in press
21.77 H. Ibach, J.E. Rowe: Surf. Sci. **43**, 481 (1974)
21.78 T. Sakurai, H.D. Hagstrum: Phys. Rev. B**12**, 5349 (1975)
21.79 E.G. McRae, C.W. Caldwell: Phys. Rev. Lett. **46**, 1632 (1981)
21.80 P. Koke, W. Mönch: Verh. DPG (VI) **17**, 941 (1982)
21.81 D.L. Lile: Surf. Sci. **34**, 337 (1973)
21.82 J.A. Dillon, H.E. Farnsworth: J. Appl. Phys. **29**, 1195 (1958)
21.83 F.G. Allen, G.W. Gobeli: J. Appl. Phys. **35**, 597 (1964)
21.84 G. Schulze, M. Henzler: Surf. Sci. **124**, 336 (1983)
21.85 P. Drathen, W. Ranke, K. Jacobi: Surf. Sci. **77**, L162 (1978)
21.86 R.Z. Bachrach, R.S. Bauer, P. Chiaradia, G.V. Hansson: J. Vac. Sci. Technol. **19**, 335 (1981)
21.87 W. Mönch: In *MBE and Heterostructures*, NATO Advanced Study Institute Series E, ed. by L.L. Chang (Nijhoff, The Hague 1983) in press
21.88 A.Y. Cho: In *MBE and Heterostructures*, NATO Advanced Study Institute Series E, ed. by L.L. Chang (Nijhoff, The Hague 1983) in press
21.89 J. Massies, P. Etienne, F. Dezaly, N.T. Linh: Surf. Sci. **99**, 121 (1980)
21.90 P.K. Larsen, J.D. van der Veen: J. Phys. C: Solid State Phys. **15**, L431 (1982)
21.91 J.F. van der Veen, L. Smit, P.K. Larsen, J.H. Neave: Physica **117**B and **118**B, 822 (1983)
21.92 S.P. Svensson, J. Kanski, T.G. Andersson, P.O. Nilsson: Surf. Sci. **124**, L31 (1983)
21.93 P.K. Larsen, J.H. Neave, B.A. Joyce: J. Phys. C: Solid State Phys. **14**, 167 (1981)
21.94 W. Ranke: Phys. Rev. B**27**, 7807 (1983)
21.95 G.D. Davis, E.E. Savage, M.G. Lagally: J. Electron Spectrosc. **23**, 25 (1981)
21.96 I.L. Singer, J.S. Murday, L.R. Cooper: Surf. Sci. **108**, 7 (1981)
21.97 G.W. Gobeli, F.G. Allen: Phys. Rev. **137**, A245 (1965)
21.98 W. Ranke, K. Jacobi: Solid State Commun. **13**, 705 (1973)
21.99 T. Madey, J. Yates, Jr.: J. Vac. Sci. Technol. **8**, 39 (1971)
21.100 H.J. Clemens, J. von Wienskowski, W. Mönch: Surf. Sci. **78**, 648 (1978)
21.101 H. Gant, W. Mönch: Appl. Surf. Sci. **11/12**, 332 (1982)
21.102 J.B. Taylor, I. Langmuir: Phys. Rev. **44**, 432 (1933)
21.103 J.W. Gadzuk: In *The Structure and Chemistry of Solid Surfaces*, ed. by G.A. Somorjai (Wiley, New York 1969) and references cited therein
21.104 W.E. Spicer, P.W. Chye, P.E. Gregory, T. Sukagawa, I.A. Babalola: J. Vac. Sci. Technol. **13**, 233 (1976)
21.105 W. Mönch, H. Gant: J. Vac. Sci. Technol. **17**, 1094 (1980)
21.106 R. Murschall, H. Gant, W. Mönch: Solid State Commun. **42**, 787 (1982)
21.107 W. Mönch, R.S. Bauer, H. Gant, R. Murschall: J. Vac. Sci. Technol. **21**, 498 (1982)
21.108 H. Gant: Thesis, Universität Duisburg, 1983
21.109 W. Jantsch, K. Wünstel, O. Kumagi, P. Vogl: Phys. Rev. B**25**, 5515 (1982)
21.110 P. Skeath, I. Lindau, P.W. Chye, C.Y. Su, W.E. Spicer: J. Vac. Sci. Technol. **16**, 1143 (1979); J. Electron Spectrosc. Relat. Phenom. **17**, 259 (1979)
21.111 W.E. Spicer, I. Lindau, P.W. Chye, C.Y. Su, C.M. Garner: Thin Solid Films **56**, 1 (1979)
21.112 W. Mönch, H. Gant: Phys. Rev. Lett. **48**, 512 (1982)

21.113 W. Mönch: Thin Solid Films **104**, 285 (1983)
21.114 W.E. Spicer, P.W. Chye, P.R. Skeath, I. Lindau: J. Vac. Sci. Technol. **16**, 1422 (1979)
21.115 W.E. Spicer, I. Lindau, P. Skeath, C.Y. Su: J. Vac. Sci. Technol. **17**, 1019 (1980)
21.116 R.H. Williams, R.R. Varma, V. Montgomery: J. Vac. Sci. Technol. **16**, 1418 (1979)
21.117 H. Wieder: J. Vac. Sci. Technol. **15**, 1498 (1978)
21.118 J. Bardeen: Phys. Rev. **71**, 717 (1947)
21.119 W. Schottky: Z. Phys. **118**, 539 (1942)
21.120 M.S. Daw, D.L. Smith: Phys. Rev. B**20**, 5150 (1979); Appl. Phys. Lett. **36**, 690 (1980); Solid State Commun. **37**, 205 (1981)
21.121 M. Nishida: Surf. Sci. **99**, L384 (1980); J. Phys. Soc. Japan **49**, Suppl. A, 1093 (1980)
21.122 R.E. Allen, J.D. Dow: J. Vac. Sci. Technol. **19**, 383 (1981); Phys. Rev. B**25**, 1423 (1982)
21.123 J.D. Dow, R.E. Allen: J. Vac. Sci. Technol. **20**, 659 (1977)
21.124 D.V. Lang: Inst. Phys. Conf. Ser. **31**, 70 (1977)
21.125 R.J. Wagner, J.J. Krebs, G.H. Stauss, A.M. White: Solid State Commun. **36**, 15 (1980)
21.126 R. Wörner, U. Kaufmann, J. Schneider: Appl. Phys. Lett. **40**, 141 (1982)
21.127 E.R. Weber, H. Ennen, U. Kaufmann, J. Windscheif, J. Schneider: J. Appl. Phys. **53**, 6140 (1982)
21.128 G.B. Bachelet, M. Schlüter, G.A. Baraff: Phys. Rev. B**27**, 2545 (1983)
21.129 W. Mönch: Proc. 2nd IUPAP-UNESCO Semiconductor Symp., Trieste, 1982, in Surf. Sci. **132**, 92 (1983)
21.130 T. Wosinski, A. Morawski, T. Figielski: Appl. Phys. A**30**, 233 (1983)
21.131 U. Kaufmann, J. Schneider, A. Räuber: Appl. Phys. Lett. **29**, 312 (1976)
21.132 J. Schneider: Proc. 2nd Conf. on Semi-Insulating III-V Materials, Evian, 1982
21.133 D.V. Lang, L.C. Kimerling: Inst. Phys. Conf. Ser. **23**, 581 (1975)
21.134 J.A. van Vechten: J. Electrochem. Soc. **122**, 423 (1975)
21.135 G. Margaritondo: Solid-State Electron. **26**, 499 (1983)

Subject Index

Contents of **Chemistry and Physics of Solid Surfaces IV**
(Springer Series in Chemical Physics, Vol. 20)

Applied Physics A
Solids and Surfaces

In Cooperation with the German Physical Society (DPG)

Applied Physics is a monthly journal for the rapid publication of experimental and theoretical investigations of applied research, issued in two parallel series. **Part A** with the subtitle "Solids and Surfaces" mainly covers the condensed state, including surface science and engineering. **Part B** with the subtitle "Photophysics and Laser Chemistry" mainly covers the gaseous state, including the application of laser radiation in chemistry.

Special Features:

- **Rapid publication (3–4 months)**
- **No page charges for concise reports**
- **50 complimentary offprints**

Fields and Editors:

Solid-State Physics

Semiconductor Physics: **H.J.Queisser**, MPI, Stuttgart
Amorphous Semiconductors: **M.H.Brodsky**, IBM, Yorktown Heights
Magnetism and Superconductivity: **M.B.Maple**, UCSD, La Jolla
Metals and Alloys, Solid-State Electron Microscopy: **S.Amelinckx**, Mol
Positron Annihilation: **P.Hautojärvi**, Espoo
Solid-State Ionics: **W.Weppner**, MPI, Stuttgart

Surface Sciences

Surface Analysis: **H.Ibach**, KFA Jülich
Surface Physics: **D.Mills**, UCI, Irvine
Chemisorption: **R.Gomer**, U. Chicago

Surface Engineering

Ion Implantation and Sputtering: **H.H.Andersen**, U.Copenhagen
Device Physics: **M.Kikuchi**, Sony, Yokohama
Laser Annealing and Processing: **R.Osgood**, Columbia U. NewYork
Integrated Optics, Fiber Optics, Acoustic Surface-Waves: **R.Ulrich**, TU Hamburg

Editor: **H.K.V.Lotsch**, Heidelberg

For further information write to:
Springer-Verlag, Journal Promotion Department,
P.O.Box 105280, D-6900 Heidelberg, FRG

Springer-Verlag
Berlin
Heidelberg
NewYork
Tokyo

Applications of the Monte Carlo Method

in Statistical Physics

Editor: **K. Binder**
1984. 90 figures. XIV, 311 pages. (Topics in Current Physics, Volume 36). ISBN 3-540-12764-X

Contents: *K. Binder, D. Stauffer:* A Simple Introduction to Monte Carlo Simulation and Some Specialized Topics. - *D. Levesque, J. J. Weis, J. P. Hansen:* Recent Developments in the Simulation of Classical Fluids. - *D. P. Landau:* Monte Carlo Studies of Critical and Multicritical Phenomena. - *K. E. Schmidt, M. H. Kalos:* Few- and Many-Fermion Problems. - *A. Baumgärtner:* Simulations of Polymer Models. - *K. W. Kehr, K. Binder:* Simulation of Diffusion in Lattice Gases and Related Kinetic Phenomena. - *Y. Saito, H. Müller-Krumbhaar:* Roughening and Melting in Two Dimensions. - *K. Binder, D. Stauffer:* Monte Carlo of "Random" Systems. - *C. Rebbi:* Monte Carlo Calculations in Lattice Gauge Theories. - Additional References with Titles. - Subject Index.

Chemistry and Physics of Solid Surfaces IV

Editors: **R. Vanselow, R. Howe**
1982. 247 figures. XIII, 496 pages. (Springer Series in Chemical Physics, Volume 20). ISBN 3-540-11397-5

Contents: Development of Photoemission as a Tool for Surface Science: 1900–1980. - Auger Spectroscopy as a Probe of Valence Bonds and Bands. - SIMS of Reactive Surfaces. - Chemisorption Investigated by Ellipsometry. - The Implications for Surface Science of Doppler-Shift Laser Fluorescence Spectroscopy. - Analytical Electron Microscopy in Surface Science. - He Diffraction as a Probe of Semiconductor Surface Structures. - Studies of Adsorption at Well-Ordered Electrode Surfaces Using Low-Engergy Electron Diffraction. - Low-Energy Electron Diffraction Studies of Physically Adsorbed Films. - Monte Carlo Simulations of Chemisorbed Overlayers. - Critical Phenomena of Chemisorbed Overlayers. - Structural Defects in Surfaces and Overlayers. - Some Theoretical Aspects of Metal Clusters, Surfaces, and Chemisorption. - The Inelastic Scattering of Low-Energy Electrons by Surface Excitations; Basic Mechanisms. - Electronic Aspects of Adsorption Rates. - Thermal Desorption. - Field Desorption and Photon-Induced Field Desorption. - Segregation and Ordering at Alloy Surfaces Studied by Low-Energy Ion Scattering. - The Effects of Internal Surface Chemistry on Metallurgical Properties. - Subject Index.

Theory of Chemisorption

Editor: **J. R. Smith**
With contributions by numerous experts
1980. 116 figures, 8 tables. XI, 240 pages. (Topics in Current Physics, Volume 19). ISBN 3-540-09891-7

Monte Carlo Methods

in Statistical Physics

Editor: **K. Binder**
With contributions by numerous experts
1979. 91 figures, 10 tables. XV, 376 pages. (Topics in Current Physics, Volume 7). ISBN 3-540-09018-5

Springer-Verlag
Berlin
Heidelberg
NewYork
Tokyo